Cytoplasmic Organization Systems

Cytoplasmic Organization Systems ~ Diana Malosinski

CYTOPLASMIC ORGANIZATION SYSTEMS

George M. Malacinski, Editor
Indiana University

McGraw-Hill Publishing Company
New York   St. Louis   San Francisco   Auckland   Bogotá
Caracas   Hamburg   Lisbon   London   Madrid   Mexico
Milan   Montreal   New Delhi   Oklahoma City
Paris   San Juan   São Paulo   Singapore
Sydney   Tokyo   Toronto

Library of Congress Cataloging-in-Publication Data

Cytoplasmic organization systems : a primer in developmental biology /
  George M. Malacinski, editor.
     p. cm. — (Primers in developmental biology)
    Bibliography: p.
    Includes index.
    ISBN 0-07-039749-X
    1. Cytoplasm.   2. Cellular control mechanisms.   3. Developmental
cytology.   I. Malacinski, George M.   II. Series.
  574.87′61—dc20   89-12366
                                           CIP

234567890 HDHD 99876543210

ISBN 0-07-039749-X

*The editors for this book were Jennifer Mitchell and Rita T. Margolies,
and the production supervisor was Dianne L. Walber. This book was
set in Century Schoolbook. It was composed by the McGraw-Hill
Publishing Company Professional and Reference Division composition
unit.*

*For more information about other McGraw-Hill materials,
call 1-800-2-MCGRAW in the United States. In other
countries, call your nearest McGraw-Hill office.*

This book is dedicated to the true embryologists of the world, present as well as past. It is they who labored so hard, often with refractory experimental systems and primitive techniques, to develop the concepts (e.g., nuclear-cytoplasmic interactions, morphogenetic determinants, cell-cell interactions, etc.) which are the foundation of modern developmental biology.

Primers in Developmental Biology

George M. Malacinski, Series Editor
Indiana University

Pattern Formation: A Primer in Developmental Biology, George M. Malacinski, Editor, and Susan V. Bryant, Consulting Editor

Molecular Genetics of Mammalian Cells: A Primer in Developmental Biology, George M. Malacinski, Editor, and Christian C. Simonsen and Michael Shepard, Consulting Editors

Developmental Genetics of Higher Organisms: A Primer in Developmental Biology, George M. Malacinski, Editor

Cytoplasmic Organization Systems: A Primer in Developmental Biology, George M. Malacinski, Editor

For more information about other McGraw-Hill materials, call 1-800-2-MCGRAW in the United States. In other countries, call your nearest McGraw-Hill office.

Contents

Contributing Authors

Stephanie H. Astrow — Department of Zoology, University of California, Berkeley, California 94720

Mary C. Beckerle — Department of Biology, University of Utah, Salt Lake City, Utah 84112

Marianne Bronner-Fraser — Developmental Biology Center, University of California, Irvine, California 92717

Shirley T. Bissen — Department of Zoology, University of California, Berkeley, California 94720

Jean-Claude Boucaut — Centre National de la Recherche Scientifique, Université Pierre et Marie Curie, Laboratoire de Biologie Expérimentale, Bâtiment C-30 7e étage 9 Quai Saint-Bernard, 75005 Paris, France

Ida Chow* — Jerry Lewis Neuromuscular Research Center, Department of Physiology, UCLA School of Medicine, Los Angeles, California 90024

Jonathan Covault — Department of Physiology and Neurobiology, University of Connecticut, Storrs, Connecticut 06268

Aaron W. Crawford — Department of Biology, University of Utah, Salt Lake City, Utah 84112

M. R. Dohmen — Department of Experimental Zoology, University of Utrecht, Padualaan 8, 3584 CH Utrecht, The Netherlands

Scott E. Fraser — Department of Physiology and Biophysics and Developmental Biology Center, University of California, Irvine, California 92717

B. C. Goodwin — Developmental Dynamics Research Group, Department of Biology, The Open University, Milton Keynes, MK7 6AA, United Kingdom

Gary W. Grimes — Department of Biology, Hofstra University, Hempstead, New York 11550

David P. Hill

Department of Biology and Institute for Molecular and Cellular Biology, Indiana University, Bloomington, Indiana 47405

Robert K. Ho

Department of Zoology, University of California, Berkeley, California 94720

Kurt E. Johnson

Department of Anatomy, The George Washington Medical Center, 2300 "I" Street NW, Washington, D.C. 20037

Martin H. Johnson

Department of Anatomy, Downing Street, Cambridge CB2 3DY, United Kingdom

Klaus Kalthoff

Center for Developmental Biology, Department of Zoology, The University of Texas at Austin, Austin, Texas 78712

Katherine Liu

Department of Zoology, University of California, Berkeley, California 94720

George M. Malacinski

Department of Biology, Indiana University, Bloomington, Indiana 47405

Norio Nakatsuji

Meiji Institute of Health Science, 540 Naruda, Odawara, 250 Japan

Anton W. Neff

Medical Sciences Program, Indiana University School of Medicine, Bloomington, Indiana 47405

Heather Perry-O'Keefe

Department of Biochemistry and Molecular Biology, Harvard University, 7 Divinity Avenue, Cambridge, Massachusetts 02138

Gary P. Radice

Department of Biology and Institute for Molecular and Cellular Biology, Indiana University, Bloomington, Indiana 47405

Michael Rebagliati

Center for Developmental Biology, Department of Zoology, The University of Texas at Austin, Austin, Texas 78712

Joshua R. Sanes

Department of Anatomy and Neurobiology, Washington University School of Medicine, St. Louis, Missouri 63110

Noriyuki Satoh

Department of Zoology, Faculty of Science, Kyoto University, Kyoto 606, Japan

Yan Shaoyi

The Institute of Developmental Biology, The Chinese Academy of Sciences, Beijing, China

J. E. Speksnijder

Marine Biological Laboratory, Woods Hole, Massachusetts 02543

Claudio D. Stern

Department of Human Anatomy, South Parks Road, Oxford OX1 3QX, United Kingdom

Susan Strome

Department of Biology and Institute for Molecular and Cellular Biology, Indiana University, Bloomington, Indiana 47405

Jitse M. van der Meer   Department of Biology, Redeemer College, Ancaster, Ontario, L9G 3N6 Canada

Masami Wakahara   Zoological Institute, Hokkaido University, Sapporo, 060 Japan

David A. Weisblat   Department of Zoology, University of California, Berkeley, California 94720

* Present address: Department of Biology, The American University, 4400 Massachusetts Ave., N.W., Washington, D.C. 20016.

Preface

PRIMERS IN DEVELOPMENTAL BIOLOGY was conceived as a series of volumes covering contemporary biologic research with the following goals:

1. To identify rapidly emerging and potentially important disciplines and subdisciplines of developmental biology.
2. To organize those disciplines for students and scholars who might be seeking both an introduction to current research problems and a broad overview of a discipline.
3. To provide insight into the thought processes of several of the important contributors to the discipline.

This fourth volume in the series—*Cytoplasmic Organization Systems*—was conceived with two specific aims: The first is to present a general explanation of the theme that cytoplasmic components, consisting of both informational molecules (e.g., mRNA, regulatory proteins, etc.) and morphological patterning elements (e.g., cytoskeletal structures, surface features, etc.), provide a major driving force in developing systems such as embryos. The second is to provide a description and analysis of key cytoplasmic organization systems, including several which function within individual cells and others which act at the cell surface.

The first aim was accomplished by soliciting a series of essay-style chapters (Section I, General Concepts). Most of those chapters include opinions as well as specific examples, in order to help the reader appreciate the role that many mechanisms which are not directly coupled to gene expression play in pattern formation. Section I, General Concepts, for example, attempts to generate an understanding of the mindset of contemporary developmental biologists who place their emphasis on signals, cues, and scaffolds which are separated in both time and space from gene expression.

The second aim was accomplished by inviting contributions which provide either general examples of cytoplasmic organization systems or detailed descriptions of model systems which are being exploited at both the cellular and molecular levels. This second aim—analysis of significant model systems—was accomplished by using the more traditional "review article" style. However, contributors were encouraged to express their opinions as well as simply collate data.

Once a tentative list of topics and contributors was assembled, it was circulated to colleagues for additional suggestions. The editor is appreciative of the enthusiasm with which colleagues, especially Brian Goodwin (The Open University, Milton Keynes, U.K.), made suggestions and helped finalize the thematic focus of the volume.

Chapters in this volume build upon foundations laid in two previous volumes in the series. Volume I (*Pattern Formation*), for example, explained the nature of morphogenetic programming in developing systems. Volume III (*Developmental Genetics*) offered explanations of the way in which genetic analyses can be employed to understand that pattern specification. This volume (IV) emphasizes the role played by specific cytoplasmic organization-information systems in specifying morphogenetic patterns.

Further, the following guidelines for contributors were formulated.

1. Chapters would be written at the level most meaningful to the graduate student, postdoctoral fellow, or new investigator in the field of developmental biology or embryology.
2. Authors would be urged to speculate and to provide their personal views or interpretations.
3. Literature references would be limited in scope to encourage authors to generalize.
4. Chapters would be cross-referenced, wherever possible, so that careful reading of the entire book would give the reader both a comprehensive and coherent view of contemporary studies in cytoplasmic organization systems. In several instances, draft manuscripts were circulated among contributors.

An "insider's view" of this area was generated by informal talks with several contributors. Those discussions are summarized in "Editor's Discussion with the Contributors." In addition, contributors' responses to two general questions about their subject are included at the end of each chapter. The net result is a book that should serve as an introductory textbook as well as a general reference manual for the field of contemporary developmental biology.

George M. Malacinski

Introduction

IN ORDER TO DEVELOP simple models for explaining how embryos work, the straightforward expedient of assuming that a horizontal description of the genome (i.e., nucleotide sequence) will lead to a vertical description of the mechanisms of tissue morphogenesis and cell differentiation is frequently adopted.

A more mature (realistic?) view, promoted in this volume, holds that developing systems depend as much on their prior experiences (i.e., recent history) as they do on the regulatory circuitry encoded in their genome. Development is viewed herein as an orderly and continuous series of building events. Each event emerges from the previous one, and in turn merges with the next one. Developing systems contain elaborate cytoplasmic organization systems which include blueprints (i.e., information systems—e.g., maternal mRNA) as well as physical scaffolds (i.e., cytoskeletons, extracellular matrices, etc.). This volume takes the position that those cytoplasmic organization systems drive each successive building event.

An extreme version of that point of view even maintains that gene expression is an *effect*, rather than a *cause*, of the orderly series of changes which characterize most developing and/or embryological systems. Mechanisms which cue cell-type-specific gene expression (i.e., cell differentiation) involve substances that synthesized, stored, and dispatched from the cytoplasm. In some instances those cytoplasmic blueprints and scaffolds are distantly separated, both in time and space, from their ultimate targets, specific genes. The germ plasm, for example, is stored during oogenesis in the polar or apical cytoplasm of many eggs and is not utilized until embryogenesis is well along. When finally deployed, it is believed to function in cells which have migrated long distances from the site of their origin (polar cytoplasm) to their ultimate destination (gonadal ridge).

Cytoplasmic rearrangements, no doubt driven by dynamic cytoskeletal systems, represent another example of organization systems. In several types of embyros, fertilization triggers a precise set of rearrangements of the egg cytoplasm. Those rearrangements in turn specify the fates of various regions of the embryo by virtue of the fact that specific maternally derived egg components are placed in individual blastomeres during subsequent cleavage.

More dramatic examples of cytoplasmic organization systems also exist. The morphology of the oral apparatus in many ciliated protozoa is largely derived from preexisting patterns. And the typical amphibian egg is preloaded with up to several thousand different template RNAs.

Cytoplasmic organization systems function in at least three different ways: (1) to fix the fate of particular cells (i.e., effect classical cell "determination"); (2) to program cell specialization pathways (i.e., initiate "cell differentiation"); or (3) to establish morphological patterns (i.e., generate arrangements, shapes, and sizes).

The key issues in the research area of *Cytoplasmic Organization Systems* are (1) appreciating the all-pervasive nature of those systems and (2) with understanding the molecular identity of signal or informational molecules which reside in the cell cytoplasm and elucidating their mechanism of action.

This volume approaches the first issue by presenting both an overview as well as a set of detailed examples of specific cytoplasmic organization systems. Section I presents a discussion of general features of cytoplasmic ordering systems.

Section II focuses on organization and information systems in embryos. A variety of embryos, including those of marine invertebrates, worms, insects, and mammals provide the context for discussing cytoplasmic control of embryonic pattern formation.

Section III considers the extracellular matrix as an information system. Gastrulation and neural crest systems provide focal points for discussion of the key features of cell-matrix interactions.

In Section IV the cell surface is discussed as both an organization and an information system. The importance of the cell surface to morphogenesis is reviewed as well as molecular details of several features of cell-cell and cell-substrate interactions.

In this volume, it is anticipated that readers will be provided with an understanding of the intellectual framework within which cytoplasmic organization systems are viewed, as well as knowledge of the experimental approaches and specific details of several important model systems.

George M. Malacinski

Editor's Discussion with Contributors

George M. Malacinski

On Definitions

What does the term "cytoplasmic organization system" mean?

In brief, a cytoplasmic organization system represents a set of components which usually reside in the cell cytoplasm and have an informational, signaling, cueing, or triggering role in pattern formation. That is, a cytoplasmic organization system (or "pattern specification mechanism" belonging to the cell cytoplasm) usually contributes either directly by affecting the genome or indirectly by guiding (i.e., informing) other cells to behave in a particular way.

The goal of contemporary research is to unravel the complex set of potential regulatory components of the cell cytoplasm and make sense out of how they work. No doubt several of those components function interactively, which will complicate matters somewhat.

On Reductionism

Will a thorough understanding of developmental systems emerge from studies of the molecular biology of gene expression?

No doubt some workers fully believe that a true reductionist approach, in which the focus is on smaller and smaller details of a developing system, such as the regulatory circuits involved in the expression of this or that gene, will be the most successful in understanding developing systems such as animal embryos. That attitude probably has its antecedents in the enormous progress made in understanding bacteriophage life cycles.

However, despite a gargantuan effort being made by literally thousands of researchers, it appears that (1) regulatory circuitry is extremely complex and not necessarily logical in a straightforward human sense (because evolution has been the major shaping process) and (2) cytoplasmic information systems, such as those which act at the cell surface, serve more of a regulatory role than previously believed.

The old cliche "the whole is greater than the sum of its parts" certainly applies here. There is just too much separation, in both time and space, between the genome and a morphological event, such as cell shape change or cell migration, for there to be direct gene control. As an extreme example, consider the growing neuron. Of course it has a nucleus. But major events occur way out in the growth cone, and they involve not only the neuron itself but also the substrate over which it is growing.

Although the results of development can certainly be traced back to the genome, so often there is no direct coupling between the expression of a specific gene and the morphological end product of a developmental pathway.

On Inferiority Complexes

Why do so many molecular biologists give the impression that they do not take cytoplasmic organization systems seriously?

In the past decade, the pendulum has swung away from emphasis on the cytoplasm as the site of control for pattern formation toward emphasis on genome studies (e.g., gene isolation, sequence analysis), probably because they are easier to carry out and generate such unambiguous results. The spectacular recent successes in isolating various genes has led to "gene probe" ego. Often the preparation of the first probe for this or that gene product has turned into a race. When the stakes are so high, the winner's ego is often inflated, which naturally colors attitudes about other areas of research. Since science is indeed a human endeavor, a sort of hierarchy or "pecking order" exists, not just among scientists but among subdisciplines as well. Gene expressionists are presently at the top.

Once we achieve a more complete understanding of the identity of the genes which are expressed at various times in developing systems, the pendulum will swing back. After all, genes don't regulate themselves; rather, they are regulated by components which originate in the cell cytoplasm. A simple nucleotide sequence does not provide clues as to how a gene is expressed; at best, it helps identify the protein product of the gene. But the regulatory system no doubt arises in the cytoplasm and is itself under either constraints or separate regulation. That is, much of the action and probably most of the information reside in the cell cytoplasm. In this regard, gene expression represents an *effect*, rather than a *cause*, of cell, tissue, and embryonic morphogenesis. In reality, therefore, there is no good reason for cell biologists who work on information-organization systems to have an inferiority complex.

On Stalled Projects

Why have some projects, e.g., elucidation of the molecular biology of germ plasm, stalled?

No genetics. Genetics can give clues (via mutant genes) for finding the "needle in a haystack" and also provide "cause and effect" (i.e., functional) tests.

There is no systematic way to approach the problem of the germ plasm. A genetic approach is lacking. A biochemical purification approach is not feasible because the only assay presently available is a very cumbersome bioassay in which fractions are microinjected at the vegetal pole (where germ plasm naturally resides) of eggs subjected to uv irradiation (which renders the eggs sterile) or into novel locations (e.g., equatorial cells) of normal eggs. The cytological approach has run its course. A "jump start" for the germ plasm project will probably not come from systematic probing. Most likely it will come from the *fortuitous* discovery that this or that heterologous cDNA probe or monoclonal antibody cross reacts with germ plasm, which will in turn provide a gateway to the isolation and/or functional test for germ plasm. However, although "luck" will probably play a role, discovery favors the prepared mind!

On Surprises

Progress toward accepting the existence of morphogen gradients has been both recent and rapid. Until recently, many developmental biologists only gave lip service to the concept; however, the evidence from Drosophila *developmental genetics now appears to be compelling. Would the famous cell biologist, E. B. Wilson, were he alive today, be surprised at recent discoveries?*

Perhaps the main surprise would be that the molecules which fit the definition of "morphogen gradient substances" are mundane, e.g., serine proteases or retinoic acid. They do not represent a new class of molecule. Frankly, many of us were quite surprised. One must, however, keep in mind that a molecule such as a serine protease may not itself be the actual "morphogen." It might serve to activate a "masked" morphogenetic determinant. So it is still possible that the "real" morphogen is a unique type of molecule.

As an aside on the subject of surprises, the excitement over growth factors can also be mentioned. It seems as if the embryology–developmental biology literature of the past few years reads "growth factors this," "growth factors that," "growth factors up, down, and around." The fact that embryos use growth factors, or growth factor–like proteins at all came as a surprise. Even more surprising, however, was their pervasiveness. It seems as if they are involved in the development of several embryonic stages and tissues.

Needless to say, the growth factor crowd would say "that's what we have been telling you for the past 20 years." Only recently, however, has the important role of growth factors been widely accepted.

Another area which I believe will generate some surprises soon is that of "receptors." Only recently have we begun to appreciate the widespread existence and major importance of specific receptors for molecules such as growth factors.

On Rules and Regulations

Considering the complexity of the genome (e.g., introns, exons, regulatory sequences, multigene families) and the elaborate cytoplasmic organization systems many cells possess, is there really any hope for developing a logical set of principles or rules which govern morphogenesis?

My intuition tells me that at best it will only be possible to formulate very general rules because all systems are products of evolution and must be viewed in that light. Evolution rarely involves the emergence of entirely new molecules. Rather, preexisting ones are modified slightly. The emerging "protein growth factor story," where literally dozens of growth factor–like proteins are turning up seemingly everywhere, is a *prime* illustration of the concept that a protein's structure and function can be altered along an evolutionary pathway.

Also, nothing ever suddenly "drops out" of a phylogenetic pathway. Instead, a structure, such as an organ or muscle, recedes slowly and is usually recognized by researchers as a "vestigial" structure. In the same way, "newness" is usually introduced at the termini of pathways, since a terminal structure or function is more easily modified than one which exists somewhere in the middle and is itself highly constrained by everything surrounding it.

To me this means that it is very unlikely—considering the random way in which the phylogenetic histories of individual developing systems have emerged—that meaningful generalizations or unifying principles can be devised. At best, generalizations such as the following will be possible: "yes, indeed, cell-cell interactions specify some morphogenetic events," *or* "growth factor–like molecules play a role in various early embryonic determination events," *or* "gradients specify pattern for certain tissues in some embryo types."

Beyond those "generalizations," however, each type of embryo will likely prove to be so idiosyncratic at the molecular level that unifying principles will not be possible. The elucidation of "regulatory circuits" will, perhaps, actually become a game of "trivial pursuit." A sort of molecular taxonomy. For each embryo type, a different "encyclopedia of mechanisms" will need to be written. That thought, of course, frightens many scientists. After having devoted an entire career to the study of one small phenomenon in a single esoteric organism, in the final analysis the findings may be so specialized that nobody else might be interested!

On Arrogance

Why is it that contemporary molecular biologists so arrogantly and tenaciously cling to the notion that knowledge of genes, which of course act through their protein products, will explain all we need to know about developmental biology?

So many of the spectacular recent successes, the "quantum jumps," so to speak, in contemporary developmental biology have come through gene isolation experiments, which are relatively easy to comprehend and make such good press, so perhaps we are overly impressed by them. However, some of the classical, long-standing questions in developmental biology, such as "What is the chemical nature of morphogenetic determinants (e.g., the germ plasm)?" remain untouched by gene isolation.

It should also be pointed out that the cytoplasmic environment, rather than the gene's nucleotide sequence, can be the final determinant in how an organism develops. The bacterial flagellum provides a paradigm. Understanding the nucleotide sequence of the gene which codes for the flagellar protein, flagellin, does not help us understand why some cells exhibit curly flagella while others exhibit wavy flagella.

On Remote Control Systems

Can large, immobile cytoplasmic molecules which are so distant from the nucleus, such as the extracellular matrix (ECM), serve as signals for regulating gene expression?

Yes, but only indirectly. Contact of ECM components with a cell's plasma membrane probably involves specific receptors on the cell's surface. That ECM-receptor interaction conceivably either (1) triggers the release of "second messengers" (soluble molecules) which act as regulators, eventually reaching the genome or (2) alters the physical properties of membrane lipids and thereby regulates the activity of ion channels or integral membrane enzymes.

Thus, the ECM can be regarded as being "instructive" and capable of influencing gene expression despite its remote location. Keep in mind, however, that substantial prior gene expression is required to construct the ECM and its cell-surface receptors. The ECM probably functions, therefore, as a tertiary or peripheral (no pun intended!) gene-expression signaling system.

On Combinations

Will the interactive usage patterns employed by developmental programs hinder progress in explaining cytoplasmic regulatory mechanisms?

Yes, no doubt. First, the genetics upon which we depend so heavily will not be as straightforward as one might hope. "Clean" mutants will be difficult to find. Many mutants are known to be pleiotrophic and many more, if carefully examined, would no doubt also turn out to be so. What that means is obvious. A single gene product is employed in different combinations (i.e., time and space), which will complicate the determination of straightforward "cause and effect" relationships.

Second, functional tests which are based on inactivation of target molecules (e.g., antisense RNA in situ inactivation of gene products) will give false positives. Developmental arrest might be misinterpreted as erasure of this or that *specific* function.

Hierarchies of signal molecules probably exist, so deletion of one signal molecule might not inhibit function. The extracellular matrix of many developing systems consists of several molecules (e.g., hyaluronic acid, proteoglycans, glycosaminoglycans). In vitro tests reveal that no single component is fully essential. Rather, the whole is greater than the sum of its parts.

On What's Out There?

Isn't it foolish to attempt to construct models for cytoplasmic organization systems without first having a thorough inventory of what's actually out there in the cytoplasm?

In a sense, yes. We know from two-dimensional protein gel electrophoresis that an egg, or for that matter even a typical somatic cell, contains several hundred different proteins. We know the true identity of only a relatively small proportion.

For the most part it is the major structural proteins whose identities are known. The proteins (and mRNAs?) which exist in relatively minor concentrations may, however, be critically important. For example, "capping proteins" may nucleate filament formation from actin subunits and thereby regulate an important feature of cytoplasmic organization. How many such "minor" proteins (or mRNAs) exist and what their functions are remain unknown.

We are, therefore, at a real disadvantage when it comes to building accurate models for cytoplasmic organization systems. Nevertheless, it is necessary to begin somewhere. Good models have predictive value, so today's models might lead to tomorrow's discovery of additional components of organization systems.

On Different Persuasions

Which specialist—the molecular biologist, geneticist, anatomist, etc.—is likely to make the most progress in understanding developing systems?

Some people believe that the problem of development can be reduced to a problem of "gene expression." Many chapters in Volume III of this series portrayed that view. Others are of a different persuasion. They feel that the problem of development can be reduced to a problem of cell behavior.

My opinion, however, is that development is too complex a process to be successfully comprehended by people who label themselves as being of this or that persuasion. Some features of development, e.g., differentiation of the exocrine pancreas or vitellogenic liver, clearly belong to the gene expressionists, but they cannot explain everything. For example, they are at a loss to explain how cells of the exocrine pancreas become arranged around ducts in the acinar configuration. Other features of embryonic systems, such as primordial germ cell or neural crest cell migration, belong to the cell biologist. Finally, still other features, such as cell-substrate interactions, belong to the biochemist.

Developmental biology is a thriving discipline, with room for many different approaches. A key—some say *the*—problem involves understanding how embryonic cells become determined (fixed) in their developmental fate. No doubt both "passive" mechanisms, such as inheritance of cytoplasmic components, and "active" mechanisms, such as gene expression, are involved. A multifaceted approach will be required to work out the details.

On Genetic Illusions

If the nucleotide sequence of an entire genome were known, could we explain how development proceeds?

Many molecular biologists think so. They believe that the control of development boils down to a problem of understanding how gene expression is regulated. Hence, sequence analyses will lead the way to understanding development. In my opinion, however, that is a grand illusion. The nucleotide sequence of the frog's genome will not provide an explanation of how a frog egg makes it through embryogenesis or how the embryo manages to undergo metamorphosis or how the adult grows to sexual maturity. Rather, it is elucidation of networks and combinatorial usage patterns which will be the most useful. My view is that accurate information on the phylogenetic history of the organism will be more informative than simple nucleotide sequences. From a phylogenetic history will come information about what a particular gene product means to development. That is, cause-effect relationships will become more clear when a particular phenomenon (e.g., lens development) is viewed in the context of its evolutionary antecedents. When the phylogenetic approach has been employed, it has often yielded surprises. For example, the vertebrate lens proteins (crystallines) comprises a set of enzymes (e.g., enolase, lactic dehydrogenase) which a nucleotide sequence analysis would have placed in roles of intermediary metabolism, not in vision.

Because of phylogenetic history, no single theory will be able to explain development. Hence, understanding the nucleotide sequence is unlikely to give rise to a coherent model which explains how an organism develops.

On Human Logic

Are developmental mechanisms intuitive, or do they defy human logic?

They appear to defy human logic. I firmly believe that anybody who claims that developmental mechanisms are intuitive is just good at rationalizing and does not want to admit that they have been fooled. For an example, 10 to 20 years ago a big search was begun for tissue-specific gene products because we believed that one cell or tissue was different from another because of the specific genes they expressed. Now, with the aid of ultrasensitive cDNAs and monoclonal antibodies we recognize that many cell types share the same gene products but in vastly different proportions. That is, interactive usage patterns are prevalent. That discovery conflicts with our original viewpoint.

Another counterintuitive discovery may be on the horizon. It concerns chaos theory, which is being refined with the assistance of supercomputers. Phenomena such as the interaction between two unlike molecules when they are mixed were previously viewed as random. Now, patterns of interaction are being revealed which previously escaped detection, and they appear to be neither random nor chaotic. Constraints generate micropatterns. In developing systems, such micropatterns may also exist, since evolution has no doubt woven its morphogenetic patterns around similar constraints.

On Complexity

Where do you expect to find the most complexity? At the level of gene expression, or at the level of organization of cytoplasmic components?

It is difficult at the moment to say. With metabolic pathways and gene expression regulator circuitry, the schemes which are emerging are highly complex and have arrows that go every which way. It's probably safe to say that the same will be demonstrated for cytoplasmic organization systems. Accumulating data is difficult because cytoplasmic organization systems are not nearly as amenable to dissection as are gene expression patterns. It is relatively easy to unravel a gene regulation system in one of the genetically amenable systems such as *Drosophila* or *C. elegans*. We can expect that integration-regulation systems which coordinate activities outside cells, e.g., in the extracellular matrix or at the cell surface, with activities within cells, and relationships between the nucleus and cytoplasm, also exist. It is just too difficult to dissect them with the present technology.

It is better to understand little than
to misunderstand a lot.

Anatole France

SECTION I

General Concepts

CHAPTER 1

Morphogenesis: Gene Action within the Context of Cytoplasmic Order

B. C. Goodwin

WHY ARE SO MANY of us so preoccupied with the notion that embryonic development results almost exclusively from gene expression?

Dobzhansky (4a) declared that "nothing in biology makes sense except in the light of evolution." Since the basic stuff of evolution from a neo-Darwinian perspective is hereditary variation in organisms, and since genes are the main vehicles of heredity, it is natural to focus on genes as the determinants of the organismic variety that underlies evolution. But this variety arises from hereditary differences in the developmental processes that generate organisms. Therefore, genes are the determinants of developmental processes. This is the first reason for the view that embryonic development results from gene activity.

The second reason comes from the belief that molecular composition is a primary determinant of embryonic state. Genes code for macromolecules, and gene interactions, via their products, result in specific spatial and temporal patterns of gene expression in the developing embryo. Therefore, embryonic development is to be understood in terms of these changing patterns of gene-determined molecular composition.

There is no doubt that classical and molecular genetic analyses have been enormously valuable in the study of developmental processes. These techniques, applied to the investigation of gene activity patterns in early *Drosophila* development, have revealed an unprecedented wealth of molecular detail which is being interpreted almost exclusively in terms of gene interactions. However, both arguments presented above for the view that embryonic development is to be understood primarily in terms of gene expression patterns are logically flawed.

First, the correlation of genetic differences with inherited developmental and morphological differences tells us that gene products can cause diversions from one developmental pathway to another. However, this fails to explain the organized processes that generate morphology, which requires a quite different type of understanding. This brings us to the limitation of the second argument: A knowledge of macromolecular composition in developing organisms does not explain morphogenesis, a process which takes place over characteristic spatial distances of micrometers to millimeters and involves long-range order that is not explained simply by a knowledge of the molecules involved. Rather, what is required is an understanding of structures such as the cytoskeleton and the extracellular matrix as components of morphogenetic fields, together with the modifying role of molecular differences, the properties of cell sheets, and the role of cell adhesion molecules in altering their morphogenetic behavior. Gastrulation, for example, is common to the members of several phyla of metazoa, and the basic mechanical principles involved are likely to be similar in all of them, differences being important, but secondary. Neither a genetic analysis of hereditary differences nor a knowledge of molecular composition will provide an explanation of this most basic of metazoan developmental events.

Another example is provided by the study of the morphogenetic processes underlying phyllotaxis patterns in plants. Here the focus is not on the differences between species but on the similarities: the different spirals observed all belong to patterns that obey the Fibonacci number sequence and involve the golden section of a circle in the sequential positioning of leaf primordia. Neither details of molecular composition nor a genetic analysis of differences provides an explanation of this universal plant patterning process.

Major problems in developmental biology center on an understanding of such regularities of morphogenetic process, which provide the themes on which evolutionary variations are founded. Making morphology intelligible, and hence discovering the generative principles that underlie development, depends on understanding these morphogenetic principles. Thus it could be argued that nothing in evolution makes sense except in the light of morphogenetic principles, which are not explained by patterns of gene expression.

Introduction: The Problem of Biologic Form

Over the biologic realm as a whole, morphology varies much less than the genes that influence it. This is exemplified by what is known as the C value paradox: closely related species with very similar morphologies can nevertheless have widely different amounts of haploid DNA. One of the best known examples of this comes from studies of the *Plethodontid* salamanders, an extremely successful, ecologically diversified group whose members have retained a basic urodele morphology (Fig. 1) that is very similar to that of their fossil ancestors of 60 million years ago (33). Despite this morphological stasis,

FIGURE 1. Two species of *Plethodontid* salamanders whose DNA C values differ by a factor of 2, (*A*) *Desmognathus welteri* and (*B*) *Gyrinophilus porphyriticus*.

the DNA C values of contemporary species vary greatly, reflecting divergent genealogical histories as measured by the constantly ticking molecular clock.

To give this paradox sharper focus, consider another example that illustrates the same point but identifies more precisely the level at which we need to look for a resolution. The ciliated protozoa have always provided a rich source of cautionary epigenetic tales, which is evident from Chap. 2. They remind us that morphogenesis is not the prerogative of multicellular organisms. A very interesting illustration of the paradox of a changing genome within an unchanging cytoplasmic morphology is provided by the 26 or so species of *Tetrahymena* that are distinguishable on the basis of isozyme patterns and breeding tests but morphologically are all essentially identical. In particular, the oral apparatus (OA) of these species has a distinctive form which is defined by the shape and disposition of an undulating membrane and three membranelles (Fig. 2). These are organized patterns of cilia in a depression on the cell surface that leads into the gullet. The cilia and their basal bodies are anchored in the specialized cortical cytoskeleton of the cell, which consists of an ordered array of microtubules and filamentous structures. One might have expected that the morphological constancy of the OA in the 26 species would be a result of constancy of the molecular composition of the constituent structural proteins. However, this has not been the case. A detailed analysis of the cytoskeletal proteins of the OAs and of the total proteins of different species (35) revealed that the molecular heterogeneity between species for the two categories of protein were similar. These differences were comparable to those observed between different *genera* of mice. Morphological similarity between species is thus perfectly compatible with considerable variability in the DNA which codes for the protein constituents of the morphology. It is therefore clear that there cannot be any simple and direct correspondence between the information coded in the DNA of a species and its form.

A basic implication of the genetic program concept is that genes, via their products, directly determine macroscopic form. What is meant by this is that a knowledge of the information in the genes is *sufficient* to determine the form produced by their products. We have just seen that different proteins can produce the same form so that there can be degeneracy of state between compo-

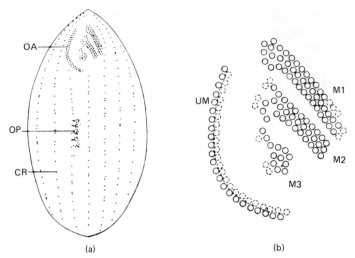

FIGURE 2. (a) The general form of a *Tetrahymena* cell, showing the oral apparatus (OA), the formation of an oral primordium (OP) prior to cell division, and the ciliary rows (CR). (b) The detailed structure of the oral apparatus, made up of cilia organized into an undulating membrane (UM) and three membranelles (M1, M2, and M3). (From Ref. 6, with permission.)

sition and structure. However, the opposite can also occur. Consider the case of flagella formation in the bacterium *Salmonella*. Since flagellin is the sole protein constituent of these structures, the proposition would be that the information in the DNA that determines the primary structure of this protein is sufficient to determine its assembled form. However, this is not the case, since flagellin from normal ("wavy") flagella can assemble into either wavy or curly forms, depending on what nucleation center or seed is present to direct the assembly process (27). Such a result is not in the least surprising since the same type of crystal polymorphism occurs in inorganic materials; one has only to think of diamond and graphite or rhombic and monoclinic sulfur. The simple conclusion is that atomic or molecular composition does not in general determine form in crystallization or self-assembly processes. Thus the molecular composition of organisms does not in general determine their form, even when this form can be reduced to pure self-assembly mechanisms. The state of any nucleation centers would also have to be specified.

The genetic program concept might then be extended to include inherited nucleation centers in the organism, but this leads to a terminological contradiction. Nucleation centers are usually metastable structures, capable of taking on more than one configuration. Thus strictly environmental effects could cause a change in their form which is not reversed when the environmental influence is eliminated. The organism would then continue to produce an altered macroscopic form even though no change occurred in its genome. This type of inherited change is observed in ciliated protozoa, where aspects of morphology (i.e., orientation of kineties and number of oral apparatuses) can be controlled by local nucleating processes and longer-range cytoplasmic order, rather than gene products (26, 33). Describing this as part of the genetic program involves a contradictory use of terms because we are now talking about inherited patterns of cytoplasmic forces which are not programmed in the genes.

Similar conclusions emerge from an analysis of morphogenesis in the multicellular organisms, except that we cannot yet describe the mechanistic details so clearly and so the argument is inevitably more abstract and subtle. Consider first the phenomenon of phenocopying. The effect of a great many morophological mutants in *Drosophila* can be copied by nonspecific stimuli (heat, cold, ether shock, acid exposure, etc.) applied to genetically wild-type embryos at particular stages of their development (10). A well-known example is the transformation of haltere to wing by either the bithorax mutation or ether shock to *Drosophila* embryos at 2½ to 3½ h after egg deposition (11). The latter treatment at 1½ to 2½ h was also found to result in unhatched larvae with cuticle patterns that resemble certain categories of gap and pair-rule mutant phenotypes (Fig. 3) (11). The analogue of this type of result in a computer would be to obtain the same output from either a specific program alteration (o.g., substitution of ✶ by −, or changing exponentiation into multiplication) or from a nonspecific perturbation (e.g., transient overheating or

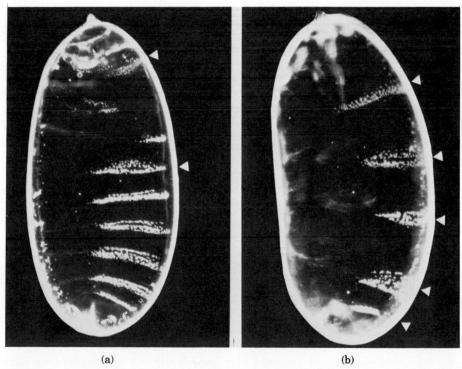

(a)   (b)

FIGURE 3. Segmentation defects in ether-treated *Drosophila* embryos showing (a) gap and (b) pair-rule type modifications. Arrowheads indicate complete segments that flank gaps in the pattern.

reduced power supply). This type of phenomenon obviously has a vanishingly small probability of occurring in computation because the output is a direct causal result of the structure of the program. Programming errors and transient errors are not causally equivalent. However, in *Drosophila*, mutant genes and transient environmental perturbations can exert the same effect on various developmental processes and so have causal equivalence: they perturb development so that one organized morphology occurs instead of another. Furthermore, the set of possible perturbed morphologies is very restricted, much smaller than the diversity of perturbations. Therefore, the developmental process is governed by principles of organization, the results of which are a constrained set of spatial forms which are not generated by genes in the same sense that a computer output is generated by a program. Genes act within an organized context whose dynamic order must be exactly defined (not simply explained away as a result of "complex interactions" or "pleiotropy") before we have a mature science of morphogenesis. An understanding of this dynamic spatiotemporal order is required to resolve the problem of biologic form.

Metabolic and Developmental Pathways

Let us approach this problem of biologic form via a familiar route: metabolic pathways. Metabolism and development have often been compared, notably by Goldschmidt and Waddington, two eminent explorers of the relations between development and evolution. I want to use the comparison to make a specific point about the relationships between general laws and their particular expression in organisms. What makes metabolic pathways *possible* is the difference in chemical potential between substrates and products. These thermodynamic relationships are measured under conditions of constant temperature and pressure, primarily by free energy differences. There is no way that a metabolic sequence can run up a free energy gradient. Like the proverbial river that provides us with such a rich source of metaphors, metabolism always runs downhill, with the products having less energy than the substrates. A metabolic sequence is a series of such downward steps from one or more (relatively) stable metabolite(s) to the next. The set of possible metabolic sequences is determined by these thermodynamic properties.

The rate at which different steps in a possible sequence occur is, however, dependent not on the free energy difference between substrate and product but on the activation energy involved in converting one metabolite into another. In the real world of process, rates are where the action is, and organisms control them by enzymes which reduce activation energies and by ligands that secondarily affect an enzyme's ability to influence these energies. Thus there is a clear distinction between the laws of thermodynamics that make metabolic pathways possible and gene products (enzymes) that alter rates of metabolic transformation. Consequently, gene products do not make metabolism possible; they stabilize particular expressions of the laws of thermodynamics in organisms by influencing specific transitions and by cross-linking rates in different pathways via control signals. The universal features of biochemistry are dependent on basic chemistry and thermodynamics; the particulars arise from gene product specificities.

Now let us apply this argument to developmental pathways. Since I am concerned in this chapter with morphogenesis, it is the shape-determining aspects of development that I want to consider. I shall argue that just as gene products do not make metabolism possible, that being a result of physical and chemical laws, gene products do not make morphogenesis possible, this also being a result of the laws of physics and chemistry. But what laws exactly? Clearly, if we knew this in detail, we would have a knowledge of morphogenesis as exact and rigorous as our knowledge of metabolism. But we already know enough to give a general answer. Since morphogenesis concerns making shapes, the laws on which it is based are those which describe how forms are generated in systems with particular types of space-time organization. Technically, these are the symmetry-breaking processes that result in the emergence of more complex structures from simpler (more symmetric) structures. In addition, it is necessary to have a general description of the basic building blocks out of which spatial forms are constituted, the analogues of metabolites which are the units of metabolic pathways. Finally, a complete description of morphogenesis requires some understanding of the energetic relations between these different building blocks, the spatial elements of morphogenetic sequences.

The basic components of a morphogenetic field theory, those concerned with breaking spatial symmetry, are well characterized. Spontaneous symmetry breaking or bifurcation is the process in which a spatially uniform state, subject to random perturbation, develops into a stable nonuniform pattern as a result of the balance of forces acting within the system, which make the initial spatially homogeneous state unstable. The first demonstration of how this could occur in a biochemical system was given by Turing (32) in his celebrated paper "The Chemical Basis of Morphogenesis." In it he showed how enzyme-catalyzed biochemical reactions, together with diffusion, can result in an instability of spatially uniform states, which spontaneously transform into spatial patterns described by stationary waves of chemical concentration. That remarkable result showed how the laws of physics and chemistry, operating within a biologic context, can generate patterns. The forces involved are those of chemical reaction together with diffusion. In relation to this theory, gene products (e.g., enzymes) do not themselves generate the patterns. In models of these processes, they determine parameter values in the equations describing the potentially bifurcating system and so can determine whether bifurcations occur; they also influence the wavelengths and amplitudes of any spatial patterns that arise.

This brings us to the second component of a theory of morphogenetic pathways: the spatial patterned elements out of which a morphogenetic sequence is constructed. Like all theories dealing with spatiotemporal organization, Turing's is a field theory. His equations describe the behavior of a reaction-diffusion field. There are many different types of fields in physics and chemistry, each characterized by different equations and describing spatiotemporal patterns with distinctive features of waveform and rate of pattern initiation or transformation. But all field theories have certain properties in common: the solutions of such equations in their linearized forms are known as harmonic functions. These are spatially periodic functions that differ in wavelength, and any pattern of the field is initially described by some set of harmonic functions. However, as the pattern develops, the nonlinear features of any particular field are expressed and distinctive waveshapes emerge. In general, this results in a discrete set of stable forms, which characterize the set of possible spatial patterns which a particular system can display. In the case of Turing's equations there are many solutions whose variables are concentrations of "morphogens", and a variety of examples have been described by Meinhardt (20) and Murray (21) in their analysis of biologic pattern formation based on Turing's theory. These are the elements out of which any morphogenetic sequence is constructed if it is based on this particular theory. However, it is important to remember that other, equally plausible theories are now available, a prime candidate being the mechanochemical field model (9, 21, 26, 28). Solutions are again harmonic functions describing spatial patterns whose variables in this case are calcium concentration and mechanical strain in the cytoskeleton.

Finally, what about the energetic relations between successive steps in a pattern-forming process? This is the least well characterized aspect of field theories, except in their simplest (linear) form. In fact it is here that the anology with metabolic pathways begins to break down if taken too literally. I shall now give a morphogenetic example that illustrates the whole set of general principles discussed so far.

Axis Formation and Cleavage Patterns

When laid, the spherical amphibian egg has a single axis, from animal to vegetal pole. It is not clear how this arises during oogenesis, but it appears to be a case of spontaneous symmetry breaking; the spherical oocyte generates an axis as a result of its own dynamics by a process that is at least qualitatively similar to that described by Turing. Fertilization of the egg by a sperm normally initiates the second symmetry-breaking event, which generates a bilaterally symmetric embryo, as revealed by the formation of the gray crescent on the side of the egg opposite to that of sperm entry. This gray crescent is the site where gastrulation will begin in the late blastula. That sperm entry is simply a nucleating event, a small stimulus that initiates a global response from a system in unstable equilibrium, is evident from the observation that other stimuli can override this and reset the axis of bilateral symmetry (see Chap. 7, this volume). An example of such a stimulus is gravity acting on rotated eggs, as described by Ancel and Vitemberger (1) and more recently in detailed studies by Scharf and Gerhart (29). This stimulus appears to act by causing a flow of the heavier vegetal material within the egg, initiating a different axis of bilateral symmetry. Depending on when this secondary stimulus is given, embryos can develop with either one (new) axis of bilateral symmetry or two axes, resulting in twinned embryos.

Another indicator of bilateral symmetry in the fertilized egg is the plane of first cleavage, which in many species normally divides the egg into the presumptive right and left halves of the future organism. This is the first of a series of cell divisions that often follow a well-defined geometric pattern (Fig. 4), subdividing the initial giant egg cell into a few thousand cells before the next stage of morphogenesis, gastrulation. This highly ordered cleavage sequence follows an orthogonal pattern of vertical cleavage planes followed by horizontal cleavage planes. It has been described by a series of harmonic functions. They comprise particular members of the set of possible solutions of a field equation used to describe the cleavage process (8). This sequence of cleavage planes is selected by the intrinsic asymmetries of the egg after fertilization: the animal-vegetal and presumptive dorsoventral axes. They function in the model as selection rules. In addition, the cleavage sequence follows a minimum energy pathway through the set of possible transitions. However, the energy function used has not been identified with any experimentally measured quantity. It is a surface-energy function that has been interpreted in relation to the work of cleavage. It has, however, not been rigorously derived from a consideration of the actual forces generated by the cystoskeleton and calcium ion gradients during this process. It is the intrinsic dynamics of the morphogenetic field that drives the whole process, from the initial symmetry breaking of the developing oocyte to the sequence of cleavages that result in the embryonic blastula. This is itself dependent on metabolic energy and its conversion into the mechanical work of morphogenesis; thus the process runs down a potential energy gradient. The "building blocks" of the sequence are field solutions which describe the particular ways in which the laws of physics and chemistry operate in the model. Hunding (12) has shown how Turing's type of reaction-diffusion equations, when solved for spherical geometry, can give rise to similar initial patterns of the cleavage process. Bjerknes (4) has presented a much more detailed account of the mechanical forces involved in cleavage.

There are many variations on the cleavage sequence described in Fig. 4. Different species show variations with respect to rates of cell division and the exact positions of the cleavage planes which result in the specific sizes and positions of the blastomeres. Since these differences are species specific, they

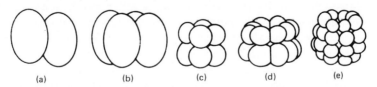

FIGURE 4. A typical pattern of holoblastic radial cleavage, showing the sequence from (a) 2, (b) 4, (c) 8, (d) 16, and (e) 32 cells.

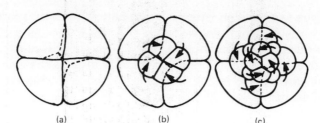

FIGURE 5. The spiral cleavage pattern in the snail *Limnea* seen from the animal pole, showing the sequence of (*a*) 4, (*b*) 8, and (*c*) 16 cells.

are a result of the influence of gene products on the cleavage process. For example, there are species in which the cleavage planes are skewed relative to one another so that the patterns generated have a spiral form (Fig. 5) rather than the vertical and horizontal order shown in Fig. 4. This is characteristic of snails, for example, which retain a spiral form in the adult and exhibit a lack of bilateral symmetry. On the other hand, there are species, such as certain annelids, which have spiral cleavage patterns but develop bilateral symmetry in the later embryo. All these inherited variations are a result of specific gene influences on the sequences of morphogenetic field solutions that particular species follow as they generate the adult form. Clearly a great diversity of sequences can be selected or stabilized from the set of possible solutions. However, if there are constraints on morphology, there must be some constraints on the possible sequences.

Constraints on Morphogenetic Patterning

One possible source of such constraint arises from the variational principle used by Goodwin and Trainor (8) to describe the set of possible solutions of the cleavage process. The principle implies that all the possible morphogenetic field solutions are states that minimize "energy" according to a particular criterion. However, within this set there are energy differences, and one of the selection rules specifies that at each cleavage the pattern selected has a lower energy than the other possible alternatives. This implies that organisms move through stable generic morphological states during their development; i.e., the forms generated are the ones that arise "naturally" by a principle of least action. Evolution should not then be viewed as a process in which natural selection drives organisms into highly improbable states of increasing adaptation. Rather, ontogeny should be considered to be restricted to morphogenetic patterns, such as the cleavage sequences previously described, which differ only slightly from one another in their energies. It is from that limited set of possibilities that evolutionary variation arises. In this view, morphogenetic sequences are never displaced significantly from these generic states. This is the morphological analogue of Kauffman and Levin's (14) argument that organisms can be described as strongly epistatic genetic networks that cannot be moved far from their generic states by natural selection. The implication of both propositions is that natural selection is a weak force in determining organismic states, which are limited by the intrinsic dynamic organization of the living condition. However, it is important to recognize the conjectural nature of both propositions. What they define is a research program that is an alternative to that pursued within the neo-Darwinist assumption that the external force of natural selection rather than the dynamics of development, is the major determinant of the organismic state. The alternative proposed in this essay belongs within the tradition defined by D'Arcy Thompson (31).

Harmonic Sequences of Gene Transcripts in *Drosophila*

In the example of cleavage described above, the ordered sequences of spatial forms are observed, but the spatial patterns of gene products that are involved in generating these forms are not. If they were visible, what would we expect to observe? In general, since gene products could influence field parameters, could themselves act as primary spatial patterning variables, or could modulate the response of other genes to morphogens, almost anything is possible. However, in instances where mutations are known to result in specific morphological modifications, one expectation is that there is a correlation between the spatial pattern of gene products and the pattern of disturbed morphology. The use of recombinant DNA techniques to prepare cDNA probes to specific gene transcripts in *Drosophila*, together with the use of antibodies against the protein products of genes, has made it possible to investigate these correlations directly, using in situ cytochemical methods. The observations have been dramatic. Let us consider the less expected results that provide extremely interesting insights into the basic dynamics of spatial patterning processes in development.

Eleven genes are currently known to be involved in the development of the dorsoventral (D-V) axis in *Drosophila* (2). One of these genes, called *Toll*, (see Chap. 14, this volume) was regarded as a likely candidate for the role of the primary determinant of the axis. Not only were there a series of mutant alleles whose phenotypes exhibited a loss of ventral characters (i.e., dorsalization), but a dominant allele ($Toll^D$) resulted in ventral characters appearing around the whole D-V axis (ventralization). The expectation was that the $Toll^+$ product would be found distributed in a gradient from ventral to dorsal in the early embryo. However, it turned out not to be so: $Toll^+$ transcripts are uniformly distributed around the axis, as are all the other transcripts of the *dorsal* series that have been studied.

This is consistent with the perspective of gene activity within a dynamically ordered cytoplasmic context. The reason is perfectly simple. The variables of the morphogenetic field need not be gene products. In a reaction-diffusion model they are small molecular weight morphogens (e.g., cAMP, serotonin, retinoic acid, peptides), while in a viscoelastic or mechanochemical model they are cytosolic free calcium and the mechanical state of the cytoskeleton ("strain"). Gene products (enzymes, repressors, calcium-binding proteins) affect the parameters of these fields and hence can influence the bifurcation conditions that make spatial patterns possible as well as the wavelength and amplitude of these patterns. But the variables are diffusible molecules or ions and mechanical strain. Taking the mechanochemical field model as an illustration, it is perfectly possible that a gradient in cytosolic free calcium develops from ventral to dorsal either in the egg or in the early embryo, and this is the primary determinant of the D-V axis. The gene products involved in consolidating this axis then act as amplifiers of the graded signal by means of a cascade mechanism that could be compared with the process of blood clotting, which is also triggered by calcium. The fact that one of the D-V genes, *snake* (see Chap. 14, this volume) codes for a protein with the characteristics of a serine protease adds plausibility to this conjecture. This example makes more specific the concept of gene products as stabilizers and/or amplifiers of morphogenetic field patterns. The particular harmonic field solution of the mechanochemical model that could act as the primary gradient is a half wave with a maximum value of cytosolic free calcium and strain along the ventral midline that decreases dorsally. This is likely to be rather labile and subject to particular types of disturbances during early development; i.e., there should be ways of phenocopying the effects of the dorsal genes by agents that disturb calcium distribution. The hypothesis that calcium, other ions, and the cytoskeleton may play such a role in the determination of the D-V axis comes from the work of A. Matheson in our laboratory, and results supporting it have been obtained (18).

Analysis of gene products influencing the anteroposterior (A-P) axis has revealed graded distributions of activity, as determined by injection of cytoplasm from specific regions of donor embryos into recipients and its capacity to induce particular A-P pattern elements (20). This accords more with expectations, though it does not explain how the gene products come to be spatially distributed in these graded patterns. Many of the genes involved act maternally, so they are pumped into the oocyte by the nurse cells at the anterior pole and get appropriately distributed during oogenesis. Detailed transcript and protein studies have been carried out on those maternal components (4*b*, 16). It is of considerable interest that *caudal* protein (see Chap. 14, this volume), which has both maternal and zygotic sources, develops into an A-P gradient during early stages despite uniform distribution of maternal transcripts even in unfertilized eggs (13). Hence the gradient generating mechanism in this case involves a cytoplasmic translation control mechanism. The zygotic gene products are then involved in subsequent changes of the *caudal* pattern, which become more complex as development proceeds.

Probably the most striking of the phenomena to emerge from the use of cDNA probes in situ is the sequence of spatial patterns of the pair-rule transcripts. The initial observations of Nüsslein-Volhard and Wieschaus (25) on the segmentation genes had shown the startling characteristics of pair-rule mutants, revealing an unexpected two-segment repeat unit prior to the final segmental pattern. The transcript patterns reveal that a number of genes in this class pass through a series of harmonics of decreasing wavelengths, as shown for *even-skipped* (*eve*) and *fushi-tarazu*(ftz) (see Chap. 14, this volume) in Fig. 6. The approximate sequence is 1 band, 2 bands, 4 bands, 7 bands, with some members of this group (e.g., *paired*) then going to 14 bands at cellular blastoderm and others going later (e.g., *ftz*). The 1, 2, and 4 band patterns appear to be nonfunctional. So we seem to be seeing dynamic transient patterns that obey a wave doubling bifurcation sequence prior to the functional 7 band pattern, which is itself also transient but is longer-lived than the others. The different members of the pair-rule set follow similar though not identical patterns, with characteristic phase shifts relative to one another. Such sequences are precisely what is to be expected from a morphogenetic field perspective. Since the elements of spatial patterning in such fields are harmonic functions, one expects to see some evidence of these during the development of a particular form, such as segmental periodicity. This does not, of course, prove that the patterns are actually field solutions. Meinhardt (20) has argued that the pair-rule transcripts achieve their spatial patterns as a result of local induction by gap gene interactions. However, it is not at all evident how such a model can account for the distinctive sequence of bands shown in Fig. 6.

What is implied by a morphogenetic field interpretation of the harmonic sequence of Fig. 6 is that there are global field variables that underlie the observed periodic spatial patterns. In general, it is to be expected that such variables should show periodic patterns over the whole of the embryonic axis. The pair-rule transcripts appear, however, to be confined to about two-thirds of the embryo, there being a larger anterior and smaller posterior region without transcripts. It was therefore of considerable interest when Edgar et al. (5) showed that cycloheximide treatment of embryos 1 h before the appearance of the normal striped pattern of *ftz* RNA resulted in the emergence of two additional bands in the anterior region. This suggests that a periodic variable controlling *ftz* transcription extends into this domain, but a protein localized toward the anterior pole normally represses *ftz* gene activity. Cycloheximide may stop production of the repressor, allowing the anterior periodicity to be revealed by the additional *ftz* bands. The normal pattern of gene expression at any stage of development is then a result of interactions between field variables and gene products, analyzable into superimposed global harmonics of different wavelengths together with local inductive and repressive effects of gene products. An analysis of the data on *Drosophila* in these terms, providing a consistent interpretation of the observed gene expression hierarchy and of the mutant data on deletions and mirror symmetries, is given in Goodwin and Kauffman (7) and Kauffman and Goodwin (13).

Morphogenetic Transformations

The final phenomenon to be considered from the morphogenetic field perspective developed in this essay is a particular category of morphological transformations. These are most clearly represented by a periodic series of elements in an organism, all of which have features in common, though each has a distinct character. Examples are legion, for this is one of the major morphogenetic

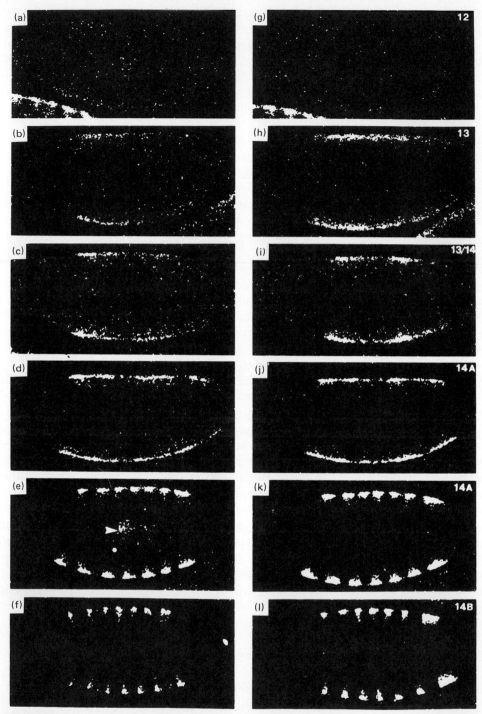

FIGURE 6. Transcript patterns at successive developmental stages of *Drosophila* embryos as revealed by [³H]-labeled cDNA probes for *even-skipped* (a–f) and *fushi tarazu* (g–l). (from Ref. 17, with permission.)

strategies that organisms use. Leaf series in plants, segments in arthropods, somites in vertebrates, and digits in tetrapod limbs are all familiar examples. According to a genetic program or a "naming" procedure in which there is a well-defined genetic interpretational code for positional states that lead to particular structures, one would anticipate that disturbances to the normal sequence would be confined to a reordering of well-defined elements. Examples include homeotic mutants in *Drosophila* and mirror-symmetric tetrapod limbs induced by either ectopic Zone of polarizing activity (ZPA) grafting or exposure to retinoic acid (15). In both cases the altered structures are usually identified as normal but positionally misplaced elements of the periodic sequence. An element of the series can even be made up of parts normally arising in different positions, such as the occurrence of the distal part of a first thoracic leg in place of distal antenna elements in the antennapaedia mutant of *Drosophila*. This gives a modular view of morphological transformation, since fixed rules of genetic interpretation operate on positional values (36).

If periodic elements arise from global periodicities, together with modifying influences that are also spatially continuous (as in superimposed harmonics of different variables with different wavelengths), then one expects a much more extensive set of transformational possibilities when normal morphogenesis is perturbed by either environmental stimuli or genetic mutation. A detailed examination of the denticle band pattern in mutant *Drosophila* embryos such as *Kruppel* or *bicaudal* reveals patterns that are not identifiable as members of the normal set, even though they are given identifying names, such as abdominal 6, because of their relative position within the overall set. "Phenocopies" of gap mutants have similarly abnormal denticle bands (see Fig. 3a). These altered forms do not belong to the set of "normal" structures, with well-defined "names" or genetic codes as predicted by a genetic program model of pattern formation. In such a model, no elements of mixed or intermediate character relative to normal structures are expected (37).

The same conclusion that elements of mixed character arise from perturbations comes from a study of the oral apparatus in *Tetrahymena*. Normal cells have three oral membranelles arranged as a series of oblique bands, each with a very regular arrangement of basal bodies (Fig. 2). Frankel et al. (6) identified mutants which have four or five membranelles. The additional ones run parallel to the original three so that the periodic series is extended. Detailed examination revealed that in such mutants the original membranelles were altered, with characteristics intermediate between the normal ones, while the additional membranelles also had some normal features. As with the *Drosophila* denticle bands, there are transformations that extend over several members of the periodic set, resulting in novel patterns of constituent elements: denticles in the case of the segmental series in *Drosophila*, basal bodies in the case of the oral apparatus in *Tetrahymena*. Thus it is common for members of a periodic set of elements, each with identifiable characteristics, to undergo transformations that extend over many members of the set, resulting in elements with patterns distinct from normal ones. Bateson (3) studied such periodic arrays in great detail, and he described many morphological variants with global disturbances of the type described above. These variants clearly do not exclude homeotic-type transformations, which are simply restricted members of the more general class of morphological transformations that allow for novel patterns of elements outside the normal set.

Conclusion: The Evolution of Generic Forms

These observations on mutant OA membranelle patterns in *Tetrahymena* take us back to the earlier discussion of the stability of the normal three-membranelle form, despite extensive variation in the structural proteins making up the oral apparatus. Evidently this pattern is mutable. Why, then, is it so faithfully conserved in the 26 *Tetrahymena* species? There are essentially two types of answers, which must ultimately be resolved into one. From the external perspective of selective forces, the pattern is maintained by stabilizing selection. From the internal perspective of morphogenetic fields, the stability is due to an energy minimum for this structure. These need to be combined into a unified account which recognizes morphogenesis as the generative process underlying the life cycles of species; these cycles must also be dynamically stable in a range of environments if species are to survive. The hypothesis of natural selection acting on genetic variants as the means of generating adapted phenotypes is an inadequate explanatory foundation for a dynamic analysis of species for two reasons. First, as has been argued in detail, a knowledge of gene activity cannot itself explain morphogenesis, so genetics is an insufficient foundation for an understanding of species characteristics. Secondly, stability principles of the type described in natural selection cannot explain organismic morphology any more than stability testing of a bridge can account for the principles used in its construction or its basic structure. Only morphogenetic principles can explain the *generative* origins of species. The study of these principles and the transformations they undergo is founded experimentally on the study of developmental processes. Thus, development provides the rigorous logical and empirical basis for understanding evolutionary processes, which otherwise are subject to the vagaries of historical narratives that constitute such a predominant aspect of current evolutionary theorizing.

From the perspective that sees morphogenetic principles as the generative origin of species (34a), the primary biologic problem is not, however, evolution itself. Rather, it is to understand the logical order revealed in biologic taxonomy, evolution being the time-dependent process in which this order becomes manifest. The origin of this order is postulated to be morphogenesis itself, whose hierarchical nature leaves its distinctive mark on organismic morphologies and their taxonomic relationships as a reflection of morphogenetic transformations. Such an approach to the problems of biologic form and transformation was described by Bateson (3) and D'Arcy Thompson (31), among others and is again emerging as a significant biologic research program. This focuses on the study of the generic properties of morphogenesis as the basis of robust, adaptable life cycles. Genes act within the context of this robust and orderly dynamic, contributing to the hereditary stability of reproductive cycles in particular habitats. Biology is then firmly grounded on its generative foundations.

General References

NEEDHAM, J.: 1968. *Order and Life*. M. I. T. Press, Cambridge.

GOODWIN, B. C., SIBATANI, A. and Webster, G. C., (eds).: 1989. *Dynamic Structures in Biology*. Edinburgh University Press, Edinburgh.

References

1. Ancel, P., and Vintemberger, P. 1948. Recherches sur le déterminisme de la symmétrie bilaterale dans l'oeuf des amphibiens. *Bull. Biol. Fr. Belg.* (Suppl.) **31**: 1–182.

2. Anderson, K. V., and Nüsslein-Volhard, C. 1984. Genetic analysis of dorsal-ventral embryonic pattern in *Drosophila*. In: *Pattern Formation: A Primer in Developmental Biology*. (G. M. Malacinski and Sue V. Bryant, eds.) pp. 269–289. Macmillan, New York.

3. Bateson, W. 1894. *Materials for the Study of Variation*. Cambridge Univ. Press.

4. Bjerknes, M. 1986. Physical theory of the orientation of astral mitotic spindles. *Science* **234**:1413–1416.

4a. Dobzhansky, T., 1973. Nothing in biology makes sense except in the light of evolution. *Am. Biol. Teacher*. March, 125–129.

4b. Driever, W., and Nüsslein-Volhard, C. 1988. A gradient of *bicoid* protein in *Drosophila* embryos. *Cell* **54**:83–93.

5. Edgar, B. E., Weir, M. P., Schubiger, G., and Kornberg, T. 1986. Repression and turnover pattern of *fushi tarazu* RNA in the early *Drosophila* embryo. *Cell* **47**:747–754.

6. Frankel, J., Nelson, E. M., Bakowska, J., and Jenkins, L. M. 1984. Mutational analysis of patterning in oral structures in *Tetrahymena*: II. A graded basis for the individuality of intracellular structural arrays. J. Embryo. Exp. March. **82**:67–95.

7. Goodwin, B. C., and Kauffman, S. A. 1988. Spatial harmonics and pattern specification in early *Drosophila* development: I. Bifurcation sequences and gene expression. *J. Theor. Biol.* (submitted).

8. Goodwin, B. C., and Trainor, L. E. H., 1980. A field description of the cleavage process in embryogenesis. *J. Theor. Biol.* **85**:757–770.

9. Goodwin, B. C., and Trainor, L. E. H. 1985. Tip and whorl morphogenesis in *Acetabularia* by calcium-regulated strain fields. *J. Theor. Biol.* **117**:79–106.

10. Goldschmidt, R. B. 1945. Additional data on phenocopies and gene action. *J. Exp. Zool.* **100**:193–201.

10a. Ho, M. W., Bolton, E., and Saunders, P. T. 1983. Bithorax phenocopy and pattern formation. *Exp. Cell Res.* **51**, 282–290.

11. Ho, M. W., Matheson, A., Saunders, P. T., Goodwin, B. C., and Smallcombe, A. 1987. Ether induced segmentation defect in *Drosophila inelanogasta*. *Roux's Arch. Dev. Biol* **196**:511–521.

12. Hunding, A. 1984. Bifurcations of nonlinear reaction-diffusion systems in oblate spheroids. *J. Math. Biol.* **19**:249–263.

13. Kauffman, S. A., and Goodwin, B. C. 1988. Spatial harmonics and pattern specification in early *Drosophila* development: II. The four color wheel model. *J. Theor. Biol.* (in press).

14. Kauffman, S. A., and Levin, S. 1987. Towards a general theory of adaptive walks on rugged landscapes. *J. Theor. Biol.* **128**:11–45.

15. Maden, M. 1982. Vitamin A and pattern formation in the regenerating limb. *Nature (Lond.)* **295**:672–675.

16. Macdonald, P. M. and Struhl, G. 1986. A molecular gradient in early *Drosophila* embryo and its role in specifying body pattern. *Nature (Lond.)* **324**:537–545.

17. Macdonald, P. M., Ingham, P., and Struhl, G. 1986. Isolation, structure, and expression of *even-skipped*: a second pair-rule gene of *Drosophila* containing a homeo box. *Cell* **47**:721–734.

18. Matheson, A. D., and Goodwin, B. C. 1988. Ionic perturbations of axial order in *Drosophila* embryos. (in preparation).

19. Meinhardt, H. 1982. *Models of Biological Pattern Formation.* Academic, London.

20. Meinhardt, H. 1986. Hierarchical inductions of cell states—a model for segmentation in *Drosophila*. *J. Cell Sci.* (Suppl.) **4**:357–381.

21. Murray, J. D. 1977. *Nonlinear Differential Equation Models in Biology.* Clarendon Press, Oxford.

22. Murray, J. D., and Oster, G. 1984. Generation of biological pattern and form. *IMA J. Math. Appl. Med. Biol.* **1**:51–75.

23. Ng, S. F., and Frankel, J. 1976. 180°-rotation of ciliary rows and its morphogenetic implications in *Tetrahymena pyriformis*. *Proc. Nat. Acad. Sci. U.S.A.* **74**: 1115–1119.

24. Nüsslein-Volhard, C., Frohnhöfer, H. G., and Lehmann, R. 1987. Determination of anteroposterior polarity in *Drosophila*. *Science* **238**:1675–1681.

25. Nüsslein-Volhard, C. and Wieschaus, E. 1980. Mutations affecting segment member and polarity in *Drosophila*. *Nature (Lond.)* **287**:795–801.

26. Odell, G., Oster, G. F., Burnside, B., and Alberch, P. 1981. The mechanical basis of morphogenesis. *Dev. Biol.* **85**:446–462.

27. Oosawa, F., Kasai, M., Hatano, S., and Asakura, S. 1966. In: *Principles of Biomolecular Organisation* (G. E. W. Wolstenholme and M. O'Connor, eds.), pp. 273–303. Little, Brown, Boston.

28. Oster, G. F., Murray, J. D., and Harris, A. 1983. Mechanical aspects of mesenchymal morphogenesis. *J. Embryol. Exp. Morphol.* **78**:83–125.

29. Scharf, S. R., and Gerhart, J. C. 1980. Determination of the dorso-ventral axis in eggs of *Xenopus laevis*: Complete rescue of UV-impaired eggs by oblique orientation before first cleavage. *Dev. Biol.* **79**:181–198.

30. Sonneborn, T. M. 1970. Gene action in development. *Proc. R. Soc. Lond. B. Biol. Sci.* **176**:347–366.

31. Thompson, D'Arcy W. 1916. *On Growth and Form.* Cambridge Univ. Press.

32. Turing, A. M. 1952. The chemical basis of morphogenesis. *Philos. Trans. R. Soc. Lond. B. Biol. Sci.* **237**:37–72.

33. Wake, D. B., Roth, G., and Wake, M. H. 1983. On the problem of stasis in organismal evolution. *J. Theor. Biol.* **101**:211–224.

34. Webster, G. C. 1984. The relations of natural forms. In: *Beyond Neo-Darwinism*. (M.-W. Ho and P. T. Saunders, eds.). pp. 193–218. Academic, London.

34a. Webster, G. C., and Goodwin, B. C. 1982. The origin of species: a structuralist approach. *J. Social Biol. Struct.* **5**:15–47.

35. Williams, N. E. 1984. An apparent disjunction between the evolution of form and substance in the genus *Tetrahymena*. *Evolution* **38** (1):25–33.

36. Wolpert, L. 1971. Positional information and pattern formation. *Curr. Top. Dev. Biol.* **6**:183–224.

37. Wolpert, L., and Stein, W. D. 1984. Positional information and pattern formation. In: *Pattern Formation: A Primer in Developmental Biology*. (G. M. Malacinski and S. V. Bryant, eds.). pp. 3–21. Macmillan, New York.

Questions for Discussion with the Editor

1. *Do you believe that "symmetry breaking" occurs only once, during reorganization of the egg cytoplasm following fertilization? Or is morphogenesis the product of a sequence of major to minor symmetry-breaking events?*

Every time an embryonic domain develops a new type of spatial order, a symmetry-breaking process is involved. Initially, these processes occur over the whole egg or embryo, such as the establishment of the A-P and D-V axes, resulting in bilaterally symmetrical forms. However, there is a hierarchical sequence of symmetry-breaking events that result in finer and finer spatial detail. The subdivisions of the *Drosophila* embryo into the spatial domains which are most readily identified morphologically by the effects of genes in the categories called maternal, gap, pair-rule, and segment polarity all involve processes in which new spatial domains are established, one inside the other. It is very interesting to note that on each of these spatial scales along the A-P axis, strong alleles in the different categories can result in embryos with regions of mirror symmetry that reflect (literally) the spatial domains of the patterning process. Thus bicaudal mutants can result in mirror-symmetrical embryos that reflect up to half the body length; gap mutants, up to one-quarter; pair rule, about one-eighth; and segment polarity, about one-sixteenth. An explanation for these observations is suggested in the papers by Goodwin and Kauffman (7) and Kauffman and Goodwin (13).

But there are other symmetry-breaking processes as well. The imaginal discs of *Drosophila* involve a new spatial order, with an axis that will become the proximodistal axis of the segmental appendages such as limbs, wings, eyes, and antennae. Similarly in vertebrates, the limb and eye fields arise as a result of further symmetry-breaking events within the context of the A-P and D-V axes. These domains retain the order of the original axes locally but add a new broken symmetry, resulting in mirror-symmetrical structures in three dimensions, such as right and left limbs. And within the limbs, more detailed spatial patterning arises from the same type of nested bifurcations as found along the main body axis, resulting in segments in insects and the sequence of bones from the humerus to the phalanges in tetrapods.

If symmetry-breaking events proceed too far, order begins to turn into disorder. This is the celebrated "march to chaos" through progressive bifurcations, resulting in phenomena such as turbulence in liquids. Developing organisms usually stop before reaching this state spatially, though there are instances in which apparently chaotic pigment patterns arise in restricted domains in certain species of lizard.

Finally, as a footnote, it is important to remember that the same progression that is found in space occurs also in time. The developing embryo undergoes temporal symmetry-breaking events, which result in periodicities in the time dimension— i.e., oscillations of different frequency and complexity. These can also become chaotic: deterministic chaos, as it is called. Thus embryonic development can be seen as an ordered sequence of symmetry-breaking events in space and in time, both of which can be described as properties of what are called excitable media. We thus end up with the wholly appropriate concept that life arises in excitable media with memory: gene action in the context of (dynamic) cytoplasmic order.

2. *Do electric currents fit into your scheme of things? Wouldn't they be candidates for establishing gradients or setting up harmonic waves?*

Electric currents resulting from free ion gradients are expected accompaniments of symmetry-breaking events that arise from dynamic interactions between ions, such as calcium, and the cytoskeleton; this is the basis of the mechanochemical models of Oster et al. (28) and Goodwin and Trainor (9). Furthermore, calcium gradients will result in amplified fluxes of other ions, such as H^+, K^+, Na^+, and Cl^-, as a consequence of calcium-regulated ion pumps and channels in membranes. But in considering the question of candidates for setting up gradients or harmonic waves, it is always necessary to bear in mind that the mechanism involved must be self-activating. Electric currents themselves do not have this property. However, there are many ways in which ions could be components of a spontaneous symmetry-breaking process, including the spatial movement of charged ion channels in membranes in response to the currents they facilitate. In general, any process that produces spatial order is likely to have an effect on ion distributions and so result in local electric currents. How important these are depends on the exact molecular mechanism involved.

Generalizing from specific mechanisms, past and present evidence points to a fundamental set of variables that are involved over and over again in the initiation of embryonic order. These include ions (particularly Ca^{2+}), small metabolites such as second messengers and inositol phosphates, and the cytoskeleton. These define an excitable electromechanochemical system in which all components are intimately coupled so that there are no primary variables, and, when in a responsive (excitable) state, the system can respond to any of a large number of perturbations. Hence the bewildering multiplicity of agents that can influence axis formation and induction in competent tissue. If one can speak of a fundamental spatiotemporal matrix that defines the characteristic responsiveness of the living state, then we seem to be getting much closer to an understanding of its composition and dynamics. This is the initial generator of cytoplasmic order, and electric currents accompany and contribute to its activity.

CHAPTER 2

Inheritance of Cortical Patterns in Ciliated Protozoa

Gary W. Grimes

Introduction

THE TITLE OF THIS chapter probably has different meanings to different people in the fields of genetics and developmental biology. To the molecular geneticist, the phrase most likely has little or no meaning, because the molecular geneticist automatically equates inheritance to "genetics" and thus to the study of DNA and gene expression in organisms.

On the other hand, the developmental biologist might automatically think about homeotic mutants, polar granules, etc., all of which are (presumably) based on the direct action of nucleic acids (wherever they may operate). The significant point to be made here is that to most biologists, genetics and inheritance are synonymous with the study of genes. It is therefore assumed that the study of *genes* will provide answers to *all* important questions regarding cellular and developmental biology.

To clarify the goal of this chapter and to emphasize the need for a broader definition of "genetics" and hence the necessity of a more integrated approach to major questions in developmental biology, we should recall the classic statement of Virchow (1855): "Omnis cellula e cellula" (all cells from cells). Modern biologists should be reminded that life is a continuum and that all cells are derived from existing cells. The fact that spontaneous generation does not occur emphasizes that, as developmental biologists, we must remember to look "beyond the gene."

Cells: An Overview

A typical course in cell biology begins with a detailed description of cellular morphology, including the nucleus, Golgi apparatus, endoplasmic reticulum, ribosomes, mitochondria, etc.; a summary of the relative functions of each; and a discussion of the bioenergetics of the cell. Also common is a very detailed description of the "central dogma," the process of gene function, i.e., the information flow from DNA to RNA which results in a specific polypeptide sequence. As a result, the student ends the course with a reasonable understanding of the basic mechanics of cellular function.

A glaring omission in such a course is, however, that the nature of the information that determines the spatial fate of polypeptide chains within the cell once they are released from the ribosome is not usually addressed. The cell obviously is not simply a randomly arrayed "bag of enzymes." Some information *must* exist within cells for determining the spatial fate (i.e., the organization) of gene products. That information is what specifies *intracellular* patterning of gene products.

This chapter deals with how the ciliated protozoa have been used by investigators first to demonstrate and second to analyze the presence of nonnucleic acid–based mechanisms of inheritance in the cell's cytoplasm. Those mechanisms play key roles by complementing the information contained in the genome.

The Ciliates: Their Utility for Studies of Cortical Inheritance

Why would one choose to study ciliates? This is a reasonable and provocative question that deserves a straightforward answer.

Ciliates offer a series of unique advantages for the study of intracellular patterning. They possess highly organized arrays of ciliary (and accessory) structures which serve as easily *visible markers of the precise localization of a specific series of gene products.* Additionally, they are easily cultured in the laboratory and have a relatively short generation time. Standard Mendelian genetics can be performed with relative ease, and the cells are large enough to be manipulated microsurgically. Thus, heritable differences between cells (both within and between clones) can be analyzed thoroughly. Because the clearest examples of cortical inheritance were analyzed most thoroughly in the ciliates *Paramecium* and *Tetrahymena* (2–5, 23, 25), those examples will be considered first, followed by a discussion of other types of cortical "mutants" which have been described in other ciliates.

The Classical Cortical "Mutants": The Doublet and Inverted Ciliary Rows of *Paramecium*

Analysis of these cortical mutants first requires a brief decription of the organization of a "normal" *Paramecium*. A typical *Paramecium* has a prolate spheroid shape, with an oral apparatus located midventrally on the surface. The oral apparatus is located in the middle of a series of longitudinally oriented rows of cilia, and each ciliary unit possesses its own characteristic polarity and asymmetry. Thus, the typical *Paramecium* (or *Tetrahymena*) has a definite polarity and asymmetry as well as defined dorsal-ventral axes. Therefore, an observer can identify all axes by looking at the entire cell or at any one of the ciliary units which are characteristically arrayed on the cell surface.

The first anomaly in this basic pattern to be analyzed thoroughly in a Mendelian approach was the doublet phenotype. These cells are in essence Siamese twins; they possess two complete sets of ciliary structures which share a common cytoplasm and a *common set of gene products*. Laborious genetic analysis demonstrated that the phenotype was stable through sexual and asexual reproduction. Therefore, the nuclear genes did not specify the doublet phenotype. Additionally, Sonneborn (25) eliminated the possibility that some soluble (or even particulate) cytoplasmic factor exchanged during conjugation was responsible for the stability of the doublet phenotype. The only rational explanation of the data was that the simple presence of the structures themselves was sufficient to perpetuate the existing ciliary pattern on the cell surface (2, 23, 25).

A more easily understood example of this type of phenomenon is represented by the inheritance patterns of modifications within individual ciliary rows (a linear series of repeating ciliary units forms a ciliary row) (4, 23). A clever experiment was performed in which one ciliary row (or many rows) was made to undergo a planar 180° rotation on the surface of the cell, thus "inverting" it, i.e., making the anterior-posterior axis of the row opposite to that of its neighboring ciliary rows (Fig. 1*a* to *d*). *If* the nuclear genotype determined the polarity and asymmetry of these units, this rotational "mutation" should revert during either sexual or asexual reproduction. However, such "inversions" did not revert but were perpetuated as a stable phenotypic alteration for years, through repeated rounds of sexual and asexual reproduction! Typically, new basal bodies arise *anterior* to the existing basal body of the ciliary unit, but as might be predicted, in inverted rows the new basal bodies always arose *posterior* to the existing basal body (1, 2, 23, 25; Fig. 1*a* to *d*).

Hence, both types of these phenotypic "mutants" appear to be dependent *solely* on the orientation and organization of the assembled gene products rather than on the characteristics of the gene products themselves. The word "mutant" has been bracketed by quotation marks until this point because the word *mutant* has implied a genetic change involving an alteration in a nucleic acid sequence. The phenotypes described above *do not* appear to be a function of a change in gene structure. For the remainder of this chapter the word will

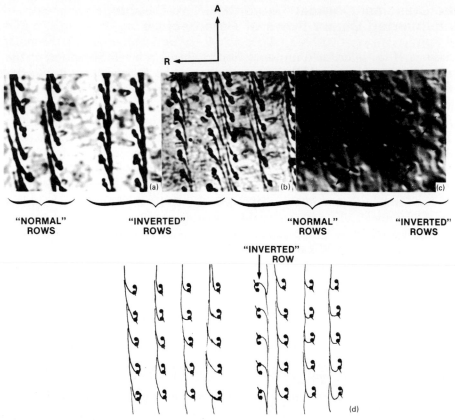

FIGURE 1. Cellular axes are designated by the perpendicular arrows (a = anterior; r = right). (*a*) and (*b*) Fernandez-Galiano silver impregnation of "inverted" ciliary rows in *Paramecium* illustrating the standard polarity of the rows and the difference between the right and left junctures between typical and "inverted" adjacent rows. (*c*) DIC optics of "inverted" rows in vivo. Arrows indicate the positions of basal bodies. (*d*) A highly schematic diagram of material presented in parts *a* to *c*. (All photographs courtesy of Dr. Karl Aufderheide.)

be used routinely without quotation marks in a simple sense to indicate an inherited change in a cell's phenotype.

Cortical Mutants and Developmental Patterns in the Oxytrichid Ciliates

Paramecium lacks developmental flexibility; it reproduces only by linear propagation of existing ciliary structures (both rows and oral apparatus) and is not easily amenable to microsurgical experimentation. Other organisms, therefore, that have different cortical mutants as well as a greater degree of developmental flexibility were chosen for further analysis. The oxytrichid ciliates have a highly complex life cycle and a great deal of developmental flexibility. The key aspects of their morphogenetic cycles are illustrated in Figs. 2 and 3. As is the case with all ciliates, they divide transversely, but compared to *Paramecium* and *Tetrahymena*, they utilize a very different strategy. A highly ordered sequence of intradevelopmental events leading to a faithful reproduction of the ciliary pattern occurs during each cell cycle (14), rather than a simple linear reproduction of repeating cortical units. Development in these cells is invariant (as it is for *Paramecium*), regardless of whether the cells undergo prefission morphogenesis or morphogenesis associated with other phases of their life cycle (12, 14–16, 18, 22). Figure 3 describes standard prefission morphogenesis in a singlet cell. The primordium for the new membranelles first develops near the posterior limit of the cell (Fig. 3a), and, during later addition of basal bodies, migrates anteriorly, where it is subsequently organized into membranelles (see Fig. 3b to d). As alignment of the membranelles proceeds anteriorly and posteriorly and from right to left, the

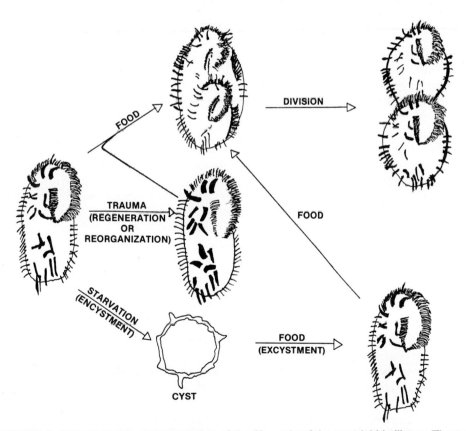

FIGURE 2. Diagrammatic representation of the life cycle of the oxytrichid ciliates. The typical morphostatic cell (i.e., a cell not undergoing morphogenesis) has three major options discussed in detail in the text: (1) standard prefission development (see Fig. 3 for a diagrammatic representation of those processes), (2) regeneration, and (3) reorganization.

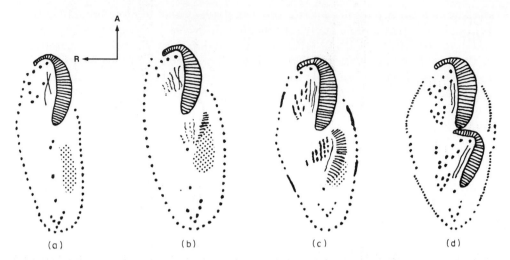

FIGURE 3. Standard prefission morphogenesis in a singlet cell. Cell axes are designated by arrows (A = anterior; R = right).

primordia for the ventral ciliature develop by disaggregation of some of the existing ciliature, and the remaining ventral ciliature is resorbed (Fig. 3*b*). Finally, the primordia for the marginal rows develop within existing rows, while differentiation of the other primordia continues (Fig. 3*c*). The result is a duplication in the cortical pattern for both the anterior and posterior fission products (Fig. 3*d*).

Other aspects of the life cycle prove to be critical for the analysis of cortical inheritance in oxytrichids. They possess two important developmental alternatives, *regeneration* (14) and *reorganization* (12). However, if the cell is subjected to various types of trauma (e.g., microsurgery), it undergoes a developmental sequence similar to that observed during prefission morphogenesis, but one which results in replacement of its missing structures. With sufficient food, the cell then reenters the growth and division cycle. However, when the food supply is exhausted, the organism encysts, loses *all* of its ciliature, and is completely morphostatic until favorable conditions are again present. Once this happens, a fully differentiated cell emerges from the cyst and reenters the growth and division cycle. Regeneration differs from reorganization in that the former is a morphogenetic sequence initiated by cell trauma as opposed to physiological stress (Fig. 2).

Another essential feature of the oxytrichids that makes them useful for cortical mutant analysis is their ability to encyst and excyst. During periods of starvation, these cells encyst by forming a surrounding protective coat. When the environmental conditions are again favorable, they excyst and resume a typical vegetative growth and division cycle (16). Since encystment involves the *complete* resorption of all visible ciliature (15), it can be determined whether the inheritance of the cortical mutants is dependent on the continued presence of visible ciliature or on changes in the genome.

Cortical Pattern Mutants in the Oxytrichids: The Typical Doublet Phenotype

The typical doublet phenotype was the first such mutant to be analyzed (18). Predictably, the phenotype was inherited as a stable phenotype, both sexually and asexually, in the same way it is inherited in *Paramecium*. The obvious question was asked: Is the phenotype inherited through the cyst stage of the life cycle, when there is a physical discontinuity of the ciliature? Surprisingly, the phenotype is indeed stable through the cyst stage: Doublets always excyst as doublets. Additionally, if the doublets are transversely bisected while encysting, thus creating two fragments (each of which possesses parts of both ventral components of the doublet), both fragments continue the process of encystment and *both* excyst as typical doublets with complete complements of their doublet ciliature! However, if the cells are cut longitudinally during encystment so that the two ventral surfaces of each doublet are separated, the resulting fragments emerge from the cyst stage as typical *single* cells (10). It appears, therefore, that the cell cortex contains some sort of information which, in the absence of visible ciliary structures, nevertheless specifies them. The information is divisible yet localized and is therefore separable (10).

One cortical mutation which is *not* inherited through the cyst stage but *is* inherited stably through both sexual and asexual reproduction has typical rows of cilia which appear as if they have been "transplanted" from the ventral surface to the middle of the dorsal surface. The typical dorsal ciliature is structurally different from that found on the ventral surface, and its developmental pattern is quite distinctive (13, 19). Inheritance of the dorsally located ventral structures (which maintain their distinctive morphology) is *not* a function of their position on the cell surface but is rather a function of their inherent organization. However, once the structures are lost (as they are during encystment), the information which specifies the presence of those structures is also lost (15). Apparently, whatever is inherited through the cyst stage can specify the number of *complete sets* of cilia produced but cannot specify information for each *individual* component of the set. Therefore, the contrast between phenotypes which are inherited through encystment and those which are not indicates a duality of information systems. One system, local patterning, is *dependent* on the continued presence of the visible ciliature, whereas the other, global patterning, is *not* (see below regarding patterning hierarchy; Refs. 6–9).

Axial Conflicts and Cortical Mutants

Significant data are obtained when axial conflicts in the arrangement of the ciliary pattern of these organisms is created experimentally. The types of axial conflicts described below involve microsurgical alterations only in the arrangement of the ciliary pattern, a procedure not generally considered to be "mutagenic" in nature, since such gross microsurgical alterations *do not* in themselves change nucleic acid sequences.

The two types of axial conflicts to be described result in mirror-imaged cells with different phenotypes. Recall, the typical cell possesses a highly defined left-right asymmetry. When such cells are cut longitudinally so that the *right* fragment folds completely upon itself, the old right margin becomes the new left margin of the cell, with the wound margin in the center of the cell (16). If the cell's phenotype were determined exclusively by the nuclear genome, the cell would regenerate as a typical single cell does and produce one typical cell. However, this is not the case! Rather, regeneration gives rise to a cell which retains its original right-left axis but regulates its longitudinal axis. The product of this sequence of events is a major cytogeometric alteration resulting in a *planar*, mirror-imaged cell. Apparently, the phenotype is not controlled by the genomic content of the cell alone but also by the pattern of existing structures. Obviously, the gene products synthesized in these cells are identical to those synthesized by typical single cells. Therefore, the alteration is in the pattern according to which those gene products are assembled rather than in the gene products themselves.

Unfortunately, the stability of this phenotype during asexual reproduction and through the cyst stage cannot be assessed because these cells possess nonfunctional mouths and cannot feed. Therefore, they eventually starve to death, in spite of successive attempts to repair themselves through physiologically induced reorganizations. Throughout these reorganizations, however, the mirror-image phenotype *is* maintained (17).

The other type of axial conflict is essentially the reciprocal of the first. In this case, the *left* side is typical and has a functional mouth but the *right* half is the mirror image of the typical left side (Fig. 4a to d). Because this type of cell possesses a functional mouth, it can feed and reproduce asexually, and it does so *true to type*! Thus, a clone of these cells can be obtained (18, 22).

An obvious question to ask regarding these cells would be: Is the mirror imagery inherited true to type through the cyst stage? Yes, it is. Although the analysis has not been completed in as much detail as has the analysis for typical doublets, cultures of these cells which have been through encystment and excystment contain freshly emerged mirror-imaged cells. Thus, the *overall pattern*, as well as the determinants for the number of sets of ciliature, appears to be inherited even in the absence of visible ciliature. This result not only confirms the conclusions derived from the study of typical doublets but extends them to the specification of the global pattern (12).

In addition to the study of the cyst stage of these reproductive mirror-imaged cells, a detailed analysis of their organization and development during asexual reproduction has provided valuable information regarding the patterning and assembly of gene products (12; Figs. 4–9).

The sequence of development for both halves of the mirror-imaged cell is the same as that for typical single cells (compare Fig. 3a to d with Fig. 4a to d). Predictably, the arrangement of the localized areas of developing ciliary primordia reflects the pattern of the typical cell. In other words, the primordia on the typical half develop sequentially from left to right, whereas the primordia on the right (symmetry-reversed half) develop from right to left. Furthermore, the organization within these primordia proceeds from anterior

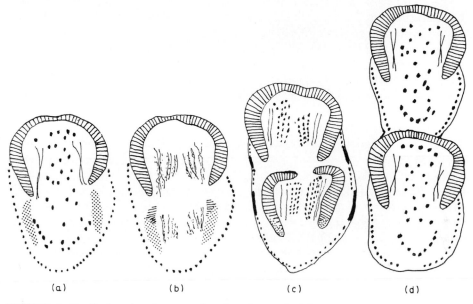

FIGURE 4. Prefission development in mirror-imaged cells.

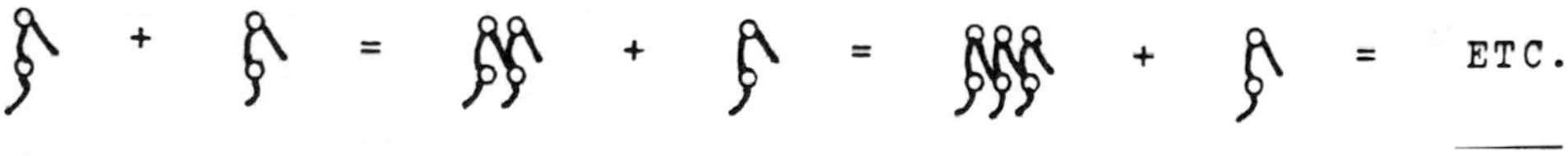

FIGURE 5. Formation of membranelles.

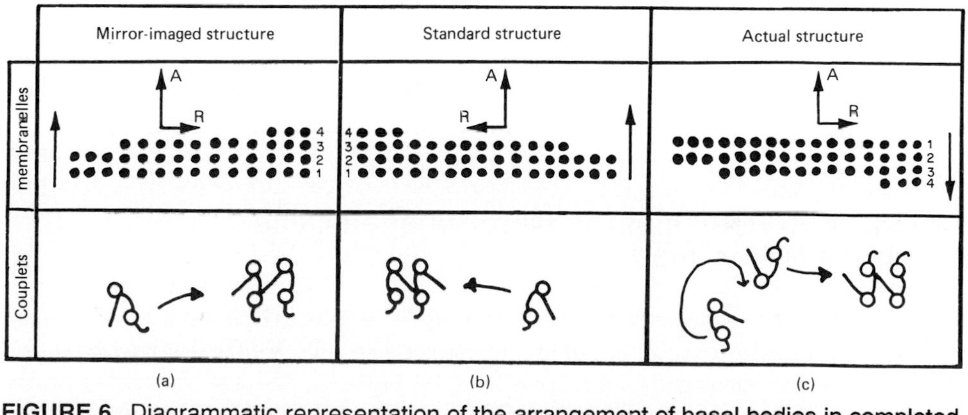

FIGURE 6. Diagrammatic representation of the arrangement of basal bodies in completed membranelles in typical cells and in mirror-imaged halves of cells.

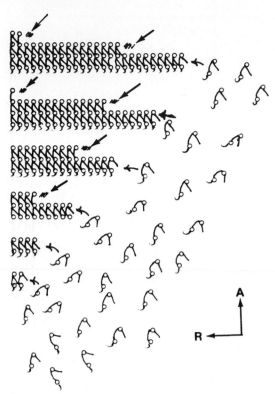

FIGURE 7. The standard assembly sequence of membranelles both in "typical" cells and in the "typical" half of a mirror-imaged cell. As indicated previously, new basal bodies are added for rows three and four in situ and progressively from anterior to posterior as well as right to left. Arrows indicate the sites of addition of couplets, which occurs simultaneously to form the first two rows. Cellular axes are defined by A and R.

to posterior on both halves, reflecting the common polarity of the two halves (Fig. 3). Thus, both halves of the cell undergo morphogenesis simultaneously, utilizing the same gene products but organizing those products in a manner that leads to the perpetuation of the mirror-imaged phenotype. Hence, pattern begets pattern, and the organization of the old structures is perpetuated. That phenomenon represents a perfect example of *"cortical inheritance."*

Are the Two Halves of the Mirror-Imaged Cells Complete Mirror Images?

Surprisingly, the answer is *no*. Even though the overall pattern is obviously mirror-imaged, the details of some structures are not. Ventral structures, although they are arranged in a mirror-imaged pattern, are not themselves mirror images of the structures in the typical single cell. Rather, the internal relations of the ciliature are the same on both halves of the bilaterally

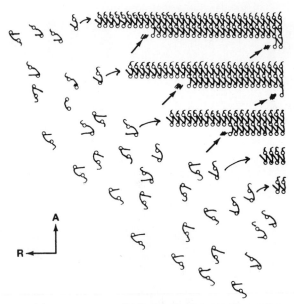

FIGURE 8. The assembly sequence observed in the membranelles of the symmetry-reversed half of the mirror-imaged cell. The only manner in which the couplets can conform to the reversed symmetry of the *overall pattern* of the ciliature is by assembling "upside-down." Thus, the addition of new basal bodies is constrained to the posterior-right margin of the newly assembled basal bodies which comprise rows three and four. Couplets therefore must undergo a 180° planar rotation *prior* to their initial assembly, as indicated by the curved arrows.

symmetrical cell. This observation is not surprising. One would predict that certain types of structures can be assembled in only one way, or at least in a strictly limited number of ways, especially when the complexity of the compound ciliary organelles is taken into account. This complexity results in some fascinating consequences in regard to the details of architecture in mirror-imaged cells.

Although the overall pattern of the midventral ciliature (cirri) of the symmetry-reversed half of these cells is arrayed as a mirror image of the typical (left) half of the cell, the detailed structure of each cluster of hexagonally packed basal bodies (cirrus) is not mirror-imaged. Cirri on both halves of the cell possess the *identical* internal relations of their ciliary components.

A significant difference exists between the cirri and the ciliature of the oral apparatus. As with the cirri, the overall pattern of the oral apparatus is mirror-imaged. On the typical half, the mouth is shaped as a *left-handed* comma, as expected. However, the ciliature in the oral apparatus of the symmetry-reversed half is not (in detail) organized as a mirror image. As illustrated in Figs. 5 to 8, each band of ciliature in the oral apparatus has its own definite polarity and asymmetry. Typically, each band contains four rows of cilia, the shortest of which is at the *anterior right* of the band. In the symmetry-reversed half of the cell, the shortest row of cilia is at the posterior

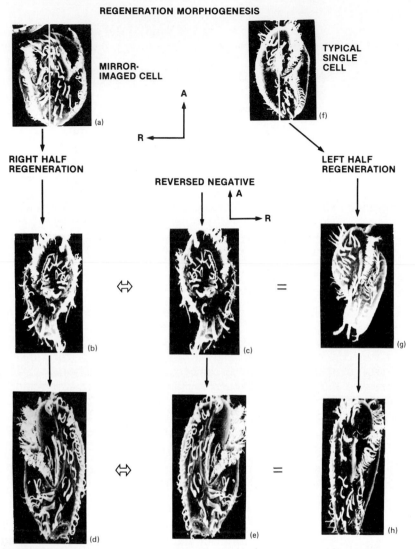

FIGURE 9. Scanning electron micrographs of a longitudinally bisected mirror-imaged cell and of a typical cell. Left column: (a) The morphostatic cell prior to microsurgery; (b) a late stage of development; (d) the mirror image of a "typical cell." Center column (c) and (e): Photographs which were merely reversed in the enlarger; photos in parts *b* and *c* and parts *d* and *e* are apparent mirror images of each other. Right column: (f) comparable stage in the development of *either* a typical left fragment, or the *left* fragment of a mirror-imaged longitudinal fragment. (g), (h) The left half of the cell regenerates as a typical cell, with all axes corresponding to those which define the "type species." When parts *d* and *h* are compared at the completion of regeneration, it can be seen that complete mirror-imagery of the overall pattern of the celiature is achieved.

left of each band of cilia, not at the *anterior left*, as would be expected if complete mirror imagery existed. It is as if each band of the *entire* oral apparatus underwent a 180° planar rotation, similar to the inverted rows in *Paramecium* (Figs. 6 to 8).

However, this planar rotation of the *entire* oral apparatus apparently does not occur, and this can be explained easily by a short discussion of the development of these bands and a consideration of the constraints placed upon the assembly of the ciliary components within each band.

When the sequence of assembly of a typical band of cilia in the mouth is considered in detail, the first step, after the assembly of the basal bodies themselves, involves the formation of couplets, pairs of basal bodies (Fig. 5; Refs. 7, 21). The first step in the formation of each band of the membranelles is the formation of couplets of basal bodies, each of which has its own polarity *and* asymmetry. Couplets in the typical half assemble from right to left, apparently because of their inherent asymmetry. The formation of each membranellar band thus begins with alignment of the couplets, with subsequent addition of basal bodies in situ. Thus, the first two rows of basal bodies of a membranelle are assembled as a unit, and the other basal bodies *must* conform to the geometric constraints placed on them. They first start to organize as bands from *right to left* and from *anterior to posterior* in the field of couplets (Fig. 3b and c; Fig. 6). The third and fourth rows of basal bodies are then added sequentially from *right to left*, resulting in each band having a characteristic polarity and asymmetry (Fig. 6). Note also that these rows are shorter than the rows formed by assembly of couplets.

In Fig. 6, the center column represents the standard arrangement as visualized in typical single cells. The right column represents the actual structure of membranelles in the mirror-imaged half of the cell. This structure is the consequence of the 180° rotation, which must take place for the couplets to conform to the cell's asymmetry.

In the mirror-imaged half of the cell, however, the *same* sequence of assembly events leads to a mouth which is *not* a true mirror image of the typical mouth. Although the initial step in the assembly of the bands is the formation of couplets, as in the typical membranellar bands, and the assembly of the membranelles is initiated at the anterior margin of the primordium and proceeds posteriorly, during the formation of the mouth the couplets must undergo a 180° planar rotation in order to match the overall geometry of the mirror-imaged half of the cell (Fig. 8). If this initial assembly of the bands begins at the *anterior left* portion of the couplet field and progresses to the *right* (rather than to the *left*) and *posteriorly* (as appears to be the case; see Fig. 7), then it follows that the third and fourth rows *must* have their basal bodies added to each new band on the *posterior right* side (Fig. 7). In that way, the basal bodies retain their typical relations to one another but ultimately produce an "upside-down" mouth (Figs. 7 and 8). The gullet, in both halves, forms at the *posterior* limit of each oral apparatus. Nevertheless, the cell has one functional mouth (on the typical half) and thus can feed and divide.

Therefore, a strict limit exists on the way in which the basal bodies of the mouth can be assembled, and in order for a cell to have a "true" mirror-imaged mouth, there would have to exist D- and L-forms of basal bodies and hence D- and L-forms of tubulins, etc. (see left column of Fig. 6). Obviously, the more straightforward explanation is that "global" mechanisms of patterning are the primary determinants of the overall pattern and that the basal bodies *must* fit together as best as they can to accommodate those global instructions. Thus, this one aspect of assembly appears to be "subservient" to the instructions for the *overall* arrangement of the ciliature. Hence, there appears to be a hierarchy in cytoplasmic information systems.

Are the Two Halves of the Mirror-Imaged Cell Distinct Entities?

In the same way that the separability of the two halves of the typical doublet cell was demonstrated, the independence of the two fields was investigated in mirror-imaged cells. Specific microsurgical alterations showed that the two fields were distinct entities (Fig. 9). When the cells were cut transversely so that two fragments were created, each possessing portions of the ventral surfaces of the two halves, each fragment underwent a regeneration sequence which resulted in two mirror-imaged cells. However, as with the typical doublets, when the two halves of the ventral surface were *completely* separated by longitudinal bisection, the resulting fragments underwent regenerative morphogenesis which resulted in two singlets. One fragment (the left half) in a late stage of development possessed the typical symmetry, whereas the other fragment (the right half) *retained its overall pattern reversal*, as shown in Fig. 9b and c).

Therefore, the two fields are distinct yet can be subdivided, the same conclusion that was reached from studies of "typical" doublet cells. The fate of these fragments is illuminating because it indicates that *no* change occurred in the genes of the organism; rather, inheritance of the phenotype is determined *exclusively* by cytogeometric alterations in the surface organization. The *left* fragment gives rise to a completely typical singlet cell, as if it had never been half of a Siamese twin (Fig. 9h). On the other hand, the *right* half (which retains its pattern reversal) typically undergoes a series of reorganization events but always retains the pattern reversal (Fig. 9c). Eventually, these cells starve and die, presumably because the ciliary bands of the oral apparatus are "upside-down," and the cilia beat the food away from the gullet rather than into it. The cilia beat in the opposite direction, consistent with their internal organization. However, a symmetry-reversed cell has been observed to undergo three divisions prior to death, thus proving itself capable of reproduction. The reversed pattern is therefore apparently quite stable and independent of the nuclear genome.

What Is the Basis of Cortical Inheritance?

The mechanisms of cortical inheritance are not understood at the molecular level. They can, however, be viewed in rather simple terms as the result of interactions (reciprocal?) among multiple levels of assembly mechanisms. As indicated previously, the organization of a cell is often considered to reflect the results of *self-assembly*. Carried to its extreme, that idea leads to the conclusion that if all the components of the cell were mixed together, a viable cell would result. Clearly, this is not the case. Therefore, some other mechanism(s) *must* be present in a cell which serves to specify the spatial fate and organization of gene products.

Cortical inheritance in the ciliated protozoa currently is visualized as the end product of reciprocal interactions among multiple "levels" of assembly mechanisms. These levels include self-assembly, aided assembly, directed assembly, and directed patterning. Each is considered individually below.

Self-Assembly: Its Limitations

Examples of *true* self-assembly, i.e., cases where reconstitution of isolated components results in a particular functional structure, are rare. Perhaps the most legitimate example of *true* self-assembly, sensu stricto, is the case of ribosomes. Ribosomal components (e.g., proteins and RNAs) can be purified individually and reconstituted in vitro to form a functional ribosome. The same is also true for other nucleic acid complexes such as snRNP particles.

However, even the most simple cases of self-assembly that are often cited (such as assembly of the tobacco mosaic virus) seem to require information regarding length determination, and the formation of rods versus sheets of assembled capsid proteins. Another example typically cited in support of self-assembly is the bacterial flagellum. The monomeric flagellin *can* self-assemble in a strict sense. However, mutant forms of the flagellar proteins dictate the way in which wild-type flagellin assembles. The wild-type bacterial flagellum has a characteristic sine wave periodicity. The mutant form (curly) has a sine wave periodicity approximately one-half that of the wild-type flagellum. Wild-type purified flagellin polymerizes in vitro to yield a typical flagellum. Likewise, "curly" flagellin, when polymerized in vitro yields curly flagella. If a small fragment of curly flagellum is added to wild-type flagellin, a *curly* flagellum results, showing that a conformational change in the wild-type flagellin was induced by the curly "seed." The "seed" thus dictates the conformation of the wild type during "self-assembly" (20).

What Is Aided Assembly and Where Does It Work?

Aided assembly is best exemplified, perhaps, by an analysis of the morphogenesis of the bacteriophage T4. Elegant genetic studies have demonstrated that certain proteins, even though not present in the mature virion, are *essential* for the successful assembly of the phage. Therefore, proteins other than those in the functional virus are necessary for successful virion formation. Self-assembly by itself *cannot* completely account for T4 assembly (even though certain steps in the sequential formation of the virus represent authentic cases of self-assembly). Instead, "aided assembly" best describes the phenomenon because extrinsic factors (such as scaffolding proteins) are required to produce a functional virion (27).

Many such examples of posttranslational modifications of proteins which demonstrate the phenomenon of aided assembly (e.g., tubulins, proinsulin, fibrin) are available. However, even when coupled with self-assembly, these two levels of assembly control are not sufficient to explain cortical inheritance in the ciliates. Another level, called directed assembly, must be invoked to explain these observations. Directed assembly works coordinately with both self-assembly and aided assembly to yield a highly organized and biologically functional system. As will be discussed below, this concept of directed assembly has broad biologic implications.

Directed Assembly: The Concept and Applicability to General Problems of Developmental Biology

The concept of directed assembly is really quite straightforward. Stated simply, directed assembly means that the *temporal sequence and spatial organization* of new structures at the time of their assembly are determined by the organization of the existing structure. This definition has extremely broad applicability at both the "molecular" and "supramolecular" levels of organization.

The cases of cortical inheritance discussed above can be easily understood within the framework of directed assembly. In the case of the "inverted" rows in *Paramecium*, the existing organization within the row dictates where the new structures are assembled. The existing row, therefore, apparently serves as a template which determines the organization of the new parts.

In principle, directed assembly does not differ from the tenets of the "central dogma," in which the sequence of nucleotides serves as a template for the assembly of the new nucleotide chain. Although this process is typically described simply as DNA synthesis, it is actually an example of directed assembly at the "molecular" level. Just as the sequence of nucleotide assembly is directed by the specific sequence of another nucleotide sequence (transcription), the sequence of amino acids in a polypeptide chain is determined by a specific sequence of nucleotides (translation).

Self-assembly is not sufficient to explain the vast majority of events which occur during gene expression, because each step in the process depends on the existing organization of the system (1–4, 6, 11, 12, 16, 17, 23–27; N. Williams, personal communication). Hence, the "molecular basis" of heredity should be understood in terms of directed assembly *because* the genes can determine *only* the sequence of an amino acid chain. More information *must* exist for the organization of those polypeptide chains.

Directed Patterning: Another Level of Cytoplasmic Information

The ciliated protozoa were used in experiments which have led to the concept of directed assembly, as discussed above in detail. However, when the *overall* pattern of the ciliature is considered (which is highly variable among species but not within species), logic dictates that a "higher" level of information exists which determines the final shape and arrangement of the ciliature. This higher level is *directed patterning*, and in the case of the ciliates it also appears to be a heritable information system and governs the overall organization of the cell. This makes it the highest level of cortical inheritance in these organisms. Data obtained from studies of cyst-stage and mirror-imaged cells support this interpretation. In the case of the cyst, the cells "remembered" what they were when they encysted—they "remembered" not only the number of sets of ciliature to form during excystment but also the *original polarity and asymmetry* of those sets.

Since the mouth on the symmetry-reversed half of the cell forms "upside-down," the "master plan" of patterning in the cells must be independent of the parts (in this case, the ciliature) and it must dictate the pattern in which the ciliary components assemble. However, the ciliary components are restricted in how they *can* assemble by their own inherent structural polarity and asymmetry. It is as if the ciliary components were pieces on a chessboard while the information system of directed patterning played the "game" of organization.

Are These Principles of Assembly Valid Only for the Ciliated Protozoa?

All cells encounter the same fundamental challenges. Data obtained from studies on ciliates often are ignored because ciliates are not considered to represent "ordinary" cells. What is an "ordinary" or "typical" cell? Such a cell probably does not exist. Many cellular and developmental biologists are now becoming active in studying this problem from a perspective other than that of strict genic control.

Many exciting results are being reported, especially from studies of membrane "trafficking." Particularly noteworthy is the following observation by Dr. George E. Palade (quoted with permission): "Gene products are not self assembled or assembled de novo in membranes. They are sorted and integrated into preexisting membranes. The patterns of sorting and integration are inherited from ancestral genomes—they are not created by current genomes." Similarly, a recent paper by Dr. Poyton (24) fully supports the concept of directed assembly as developed here. The membrane is fluid, yet it retains (i.e., self-perpetuates) certain microdomains that are essential for cellular function.

Obvious comparisons can also be made to the mosaic nature of the eggs of many metazoans. The mosaic character of the insect egg (e.g., polar granules) has been well documented. Mammalian eggs also establish polarity and asymmetry early in their development (see Chap. 15, this volume). The common problems faced by all cells (including eggs) most likely would have brought about the evolution of common mechanisms to cope with those challenges. The ciliated protozoa present to the investigator a series of advantageous characteristics which can be manipulated at will to investigate these questions. Future results promise to be exciting and should eventually lead to a full understanding of the mechanisms that determine the spatial fate of gene products. The ciliates presumably will continue to be instrumental in providing some of the "pieces of the puzzle."

Acknowledgments

The author gratefully acknowledges the editorial assistance of and advice during preparation of this manuscript from Drs. Karl Aufderheide, Noël de Terra, Joseph Frankel, Susan Duhon and George Malacinski. The research by the author was supported by grants from the National Science Foundation and the Hofstra College of Liberal Arts and Sciences.

General References

AUFDERHEIDE, K. J., FRANKEL, J., and WILLIAMS, N. E.: 1980. Formation and positioning of surface-related structures in protozoa. *Microbiol. Rev.* **44**:252–302.

GRIMES, G. W.: 1982. Pattern determination in hypotrich ciliates. *Amer. Zool.* **22**:35–46.

POYTON, R. O.: 1983. Memory and membranes: The expression of genetic and spatial memory during the assembly of organelle macrocompartments. *Mod. Cell Biol.* **2**:15–72.

SONNEBORN, T. M.: 1970. Gene action in development. *Proc. Roy. Soc. Lond. B Biol. Sci.* **176**:347–366.

References

1. Aufderheide, K. J. 1980. Mitochondrial associations with specific microtubular components of the cortex of *Tetrahymena themophila*: II. Response of the mitochondrial pattern to changes in the microtubule pattern. *J. Cell Sci.* **42**:247–260.

2. Aufderheide, K. J., Frankel, J., and Williams, N. E. 1980. Formation and positioning of surface-related structures in protozoa. *Microbiol. Rev.* **44**:252–302.

3. Beisson, J., and Sonneborn, T. M. 1965. Cytoplasmic inheritance of the organization of the cell cortex of *Paramecium aurelia*. *Proc. Natl. Acad. Sci. U.S.A.* **53**:275–282.

4. Dippell, R. V. 1968. The development of basal bodies in *Paramecium. Proc. Natl. Acad. Sci. U.S.A.* **61**:461–468.

5. Gall, J. G. 1986. *The Molecular Biology of Ciliated Protozoa.* Academic Press, New York.

6. Gavin, R. H. 1984. *In vitro* reassembly of basal body components. *J. Cell Sci.* **66**: 147–154.

7. Grimes, G. W. 1972. Cortical structure in nondividing and cortical morphogenesis in dividing *Oxytricha fallax. J. Protozool.* **19**:428–445.

8. Grimes, G. W. 1973. Morphological discontinuity of kinetosomes during the life cycle of *Oxytricha fallax. J. Cell Biol.* **57**:229–232.

9. Grimes, G. W. 1973. Differentiation during encystment and excystment in *Oxytricha fallax. J. Protozool.* **20**:92–104.

10. Grimes, G. W. 1973. An analysis of the determinative difference between singlets and doublets of *Oxytricha fallax. Genet. Res.* **21**:57–66.

11. Grimes, G. W. 1976. Laser microbeam induction of incomplete doublets of *Oxytricha fallax. Genet. Res.* **27**:213–226.

12. Grimes, G. W. 1982. Pattern determination in hypotrich ciliates. *Amer. Zool.* **22**: 35–46.

13. Grimes, G. W., and Adler, J. A. 1976. The structure and development of the dorsal bristle complex of *Oxytricha fallax* and *Stylonychia pustulata. J. Protozool.* **23**:135–143.

14. Grimes, G. W., and Adler, J. A. 1978. Regeneration of ciliary pattern in longitudinal fragments of the hypotrichous ciliate, *Stylonychia. J. Exp. Zool.* **204**:57–80.

15. Grimes, G. W., and Hammersmith, R. L. 1980. Analysis of the effects of encystment and excystment on incomplete doublets of *Oxytricha fallax. J. Embryol. exp. Morphol.* **59**:19–26.

16. Grimes, G. W., and L'Hernault, S. W. 1979. Cytogeometrical determination of ciliary pattern formation in the hypotrich ciliate *Stylonychia mytilus. Dev. Biol.* **70**: 372–395.

17. Grimes, G. W., Knaupp-Waldvogel, E., and Goldsmith-Spoegler, C. M. 1981. Cytogeometrical determination of ciliary pattern formation in the hypotrich ciliate *Stylonychia mytilus*: II. Stability and field regulation. *Dev. Biol.* **84**:477–480.

18. Grimes, G. W., McKenna, M. E., Goldsmith-Spoegler, C. M., and Knaupp, E. A. 1980. Patterning and assembly of ciliature are independent processes of hypotrich ciliates. *Science* **209**:281–283.

19. Hammersmith, R. L., and Grimes, G. W. 1981. Effects of cystment on cells of *Oxytricha fallax* possessing supernumerary dorsal bristle rows. *J. Embryol. exp. Morphol.* **63**:17–27.

20. Iino, T. 1974. Assembly of *Salmonella* flagellin *in vitro* and *in vivo. J. Supramol. Struct.* **2**:372–384.

21. Jerka-Dziadosz, M. 1981. Ultrastructural study on development of the hypotrich *Paraurostyla weissei*: III. Formation of the paroral membranelles and an essay on comparative morphogenesis. *Protistologica* **17**:83–97.

22. Jerka-Dziadosz, M. 1983. The origin of mirror-image symmetry doublet cells in the hypotrich *Paraurostyla weissei. Roux's Arch. Dev. Biol.* **192**:179–188.

23. Ng, S. F., and Frankel, J. 1977. 180° rotation of ciliary rows and its morphogenetic

implications in *Tetrahymena pyriformis*. *Proc. Natl. Acad. Sci. U.S.A.* **74**: 1115–1119.

24. Poyton, R. O. 1983. Memory and membranes: The expression of genetic and spatial memory during the assembly of organelle macrocompartments. *Mod. Cell Biol.* **2**: 15–72.

25. Sonneborn, T. M. 1970. Gene action in development. *Proc. Roy. Soc. Lond. B Biol. Sci.* **176**:347–366.

26. Williams, N. E., Honts, J. E., and Lu, C. 1989. Identification and localization of major cortical proteins in the ciliated protozaan, *Euplotes eurystomies*. *J. Cell. Sci.*, **92**:433–439.

27. Wood, W. B. 1980. Bacteriophage T4 morphogenesis as a model for assembly of subcellular structure. *Q. Rev. Biol.* **55**:353–367.

Questions for Discussion with the Editor

1. *Do you think that cytoplasmic-based information systems are more prominent in protozoans than metazoans?*

This question is quite broad, and the appropriate answer really depends upon how the word *prominent* is perceived. My opinion is that these information systems simply are more *conspicuous* in the ciliates because we are looking at highly organized supramolecular arrays which can be visualized with standard microscopic techniques. So my answer is yes, but I am qualifying that response by saying that the *importance* of cytoplasmic information systems is most likely equivalent in metazoa and protozoa. The real problem is that an adequate technology does not exist to allow an investigator to analyze intracellular localizations of an extremely small number of molecules or subtle changes in the cytoskeletal structures which must occur during morphogenesis (e.g., minor allosteric shifts). Cytoplasmic localizations are extremely important in metazoa, as exemplified by the localization of determinants in eggs, so I think that once we have the appropriate technology, we will find common mechanisms for most (if not all) of these phenomena.

2. *Is a concerted effort for developmental genetics studies in the cards? Why haven't hundreds of patterning mutants been discovered? Ciliates are easy to culture, aren't they?*

This is a multifaceted question, and I need to answer it in a reversed sequence. *Some* ciliates are indeed easy to grow. However, most require a lot of tender, loving care. Additionally, only a few ciliates have been "tamed," so that detailed Mendelian genetics is limited to those few species (e.g., *Paramecium* and *Tetrahymena*). The species that are tamed do not have the developmental flexibility, such as encystment, reorganization, and regeneration that the others possess (e.g., *Stylonychia* or *Stentor*). This leads to the first and second parts of the question. I do indeed see a major effort being put forth in the genetics of development in ciliates. Because only a few species can be manipulated this way, very few patterning mutants have been described; however, data from other labs on those species indicate that such studies are indeed fruitful. I think the major reason why "hundreds of patterning mutants" have not been described is that such mutants are lethal. In order to study the genetics of a patterning mutant, it *must* be viable, whether it is a protozoan or a metazoan. Patterning mutants in the ciliates typically lead to cells that alter their behavior and cannot feed efficiently, and therefore cannot divide. I think that there will be only a highly restricted number of genic mutants (as well as cytoplasmic mutants) which will be "discovered" for these reasons. I predict that those mutants which are discovered will be simple Mendelian mutants of minor components of the "cytoskeleton" (or cortex) of cells that are still capable of asexual reproduction. I also think that more mutants have not been identified because relatively few investigators are currently involved in this specific field of research.

CHAPTER **3**

Examples of Localized Cytoplasmic Factors in Animal Development

Heather Perry-O'Keefe

Introduction

ONE OF THE MOST fascinating aspects of developmental biology concerns the way in which a single cell, the egg, grows and develops into an intricate pattern of differentiated cell types. To help explain this phenomena, it has been hypothesized that eggs contain cytoplasmic information or determinants that determine the fates of cells in developing embryos. Because fate maps have shown that different regions of the egg give rise to different cell and tissue types, it follows that the cytoplasmic determinants must be localized to specific regions of the egg. In this way, for example, one part of the egg receives the cues to make muscle and another part the cues to make skin. Thus, as the egg divides, the different localized determinants are separated into different cells and thus act to direct the differentiation of those cells and their daughter cells. Variations on this general idea include the proposal that the localized information is present in the form of a gradient with different concentrations specifying different cell fates (see Chap. 14, this volume).

  This generalized view of how early development proceeds dates back to the turn of the century (52). At that time embryologists carefully described the development of a great variety of organisms, including many marine animals. One of the first observations was that the external environment of a developing embryo is not an important factor in determining what will develop from a fertilized egg. For instance, if fertilized eggs from a variety of marine organisms are all placed in a bowl of seawater, they will all develop according to their specific pattern and at their own speed (53). From such simple experiments it was concluded that the morphology of the organism is controlled by factors located within the egg itself and is not dependent on external cues such as light and temperature.

Many of the organisms whose early development was studied between 1888 and 1920 were invertebrates and protochordates, which tend to have few cells in the larval forms and are easily observed with simple microscopes. In addition, the cleavage pattern and relative positions of blastomeres is relatively constant from embryo to embryo. It soon became evident that certain blastomeres *always* gave rise to the same tissues, and this made it possible to establish cell lineages. That is, it is possible to establish the fates of the daughter cells of each blastomere and to know which cell and tissue types will form from each blastomere.

The Classic Example: Ascidian Embryogenesis

Work on ascidians, primarily that of E. G. Conklin, will be used as one example of the kind of detailed study that was undertaken to understand the factors that direct the development of blastomeres and determine cell lineages. Conklin's earliest work on the ascidian was a study of cell lineage and is considered by many to be a classic set of experiments (6). Of the ascidians he used (*Ciona intestinalis, Molgula manhattensis*, and *Styela partita*) *Styela partita* has the distinct advantage of having four different colored cytoplasms which become visible shortly after fertilization (Fig. 1). The most notable of these is the yellow crescent which becomes localized to the vegetal hemisphere on what will become the posterior side of the embryo. Once Conklin had established which tissue types developed from which blastomeres, he found that the cell lineages of different blastomeres could be correlated with the four localized colored cytoplasms (6). He determined that blastomeres containing yellow cytoplasm form muscle and mesenchyme, clear cytoplasm forms the ectoderm, dark gray cytoplasm forms endoderm, and light gray cytoplasm forms chordoneuroplasm. However, Conklin noted that while the yellow cytoplasm is ultimately partitioned to the muscle cells, it is not found exclusively in muscle or premuscle cells until the 32-cell stage (Fig. 1).

Although Conklin was unsuccessful at getting the individual cytoplasmic pieces to develop in isolation, he did show that at any developmental stage, when blastomeres are removed from the cleaving embryo an incomplete embryo-larvae develops. In this way he confirmed his cell lineage study, proving that when he ablated blastomeres which he had characterized as premuscle, an embryo formed which was missing its muscle cells.

One of the more puzzling aspects of ascidian development is that cytoplasm in an ascidian, or any organism, for the most part *looks* homogeneous. In an attempt to identify the nature, or source, of localized determinants in the egg, it is not surprising that one of the ideas tested was whether the colored pigments of the various cytoplasms had a role in determining cell fate. Conklin (7) centrifuged fertilized eggs to redistribute the colored cytoplasms into horizontal stripes, yet development occurred normally. He therefore con-

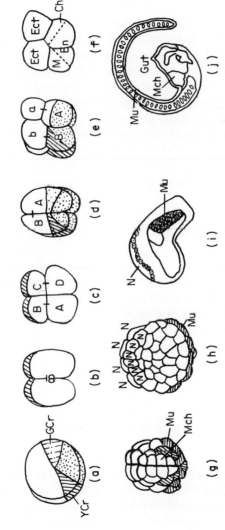

FIGURE 1. Cytoplasmic localization during early ascidian development (*a, d, e, f*) Right view. (*g*) Vegetal pole view. (*b, c, h*) Animal pole view. (*a–e*) First to third cleavage. Y Cr = yellow crescent; G Cr = gray crescent. (*f*) The presumptive cytoplasms are shown localized after the third division into blastomeres destined to become specific larval structures. Ch = chorda; Ect = ectoderm; En = endoderm; M = mesoderm. Note that yellow crescent cytoplasm is localized to the presumptive mesoderm. (*g, h*) 64-cell stage; presumptive mesoderm stippled; Mch = mesenchyme; Mu = muscle cells; N = neural tissue. (*i*) Mature larva. Gut = gut; Mch = mesenchyme; Mu = mucle cells. [Based on Conklin (6): redrawn from Grant (16).]

47

cluded that pigment, yolk, free cytoplasm, and mitochondria, all substances which are reoriented as a result of centrifugal force, do not act as the localized information or determinants. A similar type of experimental manipulation of egg cytoplasm was performed on many organisms, and in each case the results were similar; centrifuging eggs so that the pigment granules were stratified had no ill effect on the developing embryo (23; see Ref. 8 for review).

Other embryologists also studied cell lineage and gave special attention to the question of how localized determinants specify cell fate. For example, Zelny (55) established the early lineage of the nemertean worm *Cerebratulus* and showed the important role of the upper quartet of cells in the formation of the apical organ. Wilson's work on the molluscs *Dentalium* and *Patella* emphasized the fact that the localized determinants in eggs are of maternal origin (50, 51). These and other studies (reviewed in Refs. 8 and 52) led to several general conclusions about early development. First, cell lineage studies demonstrated that specific cell and tissue types are generated from specific regions of the egg and/or blastomeres of the cleaving embryo. Cleavage of the egg (cytokinesis) separates one organ-forming substance from another. Second, the developmentally important localized information is too small to be observed with conventional microscopes; i.e., the determinants are not the large inclusions or pigment granules that are reoriented by centrifugation. Finally, the localized cues are present in unfertilized eggs and are therefore of maternal origin.

While the issue of localized information was vigorously pursued around the turn of the century, by 1920 the topic had declined in popularity. Although work continued in this area, it was not until the early 1970s that it began to be investigated with renewed energy. More recent work on the topic will be reviewed by examining studies that use ascidians, ctenophores, insects, and amphibians as experimental organisms.

The role of localized determinants in the specification of muscle and other cell types in ascidians has been investigated using a variety of modern techniques. Histo-specific enzymes have been used to identify cell types and extend the cell lineage studies of Conklin (45, 46). Lineage relationships have also been investigated by injecting horseradish peroxidase (HRP) into different blastomeres (9, 30, 31). And monoclonal antibodies against both cell surface molecules and cytoplasmic components have been used to distinguish one cell type from another (27, 32).

Muscle cell lineage in ascidians has been extensively studied. Whittaker and colleagues have used the muscle-specific enzyme marker acetylcholinesterase (AChE) and various metabolic inhibitors to establish when, during early embryogenesis, cells are committed to form muscle. Their findings confirm earlier work that showed two posterior vegetal blastomeres at the 8-cell stage giving rise to most of the muscle in the swimming larva. Studies with inhibitors of RNA and protein synthesis suggest that new embryonic transcription is required to turn on the genes encoding AChE; i.e., the localized expression of this muscle-specific protein is not simply the consequence of translating a prelocalized maternal mRNA for AChE (46). The synthesis of

AChE mRNA by the embryonic genome has been directly demonstrated by testing for the mRNA (24, 35). The simplest conclusion from all these studies is that a cytoplasmic factor(s) is segregated to the two premuscle blastomeres by the 8-cell stage and that this factor commits those cells to form muscle.

Recently the exactness of the muscle cell lineage has been called into question by the work of Satoh and colleagues, who used HRP to follow the descendants of the individual blastomeres from the 8-cell embryo. Nishida and Satoh (31) demonstrated that in some cases a small number of muscle cells are derived from an additional four blastomeres at the 8-cell stage. This brings the number of premuscle blastomeres in the 8-cell embryo to six. Nishida (30) continued this investigation and refined the cell lineage studies even further. Using 64- and 110-cell stage embryos, he injected one blastomere per embryo and then followed the daughter cells which contained HRP to determine the tissue type they formed in the larvae. He concluded that the fate map of the ascidian egg and early cleavage stage embryo is much more complicated than previously realized. Some of these apparent discrepancies may be due to the fact that different species of ascidians have been used. Despite these complications, it can be stated that most, but not all, of the muscle cells in the swimming larvae are derived from the two posterior vegetal blastomeres at the 8-cell stage of development.

Whittaker et al. (49) and Meedel and Whittaker (25) showed that the two premuscle blastomeres of the 8-cell stage could be separated from the rest of the embryo and still develop AChE. Similarly, muscle-specific antigens (32) are expressed when the two premuscle cells develop in isolation. These studies show that the two posterior vegetal blastomeres of the 8-cell embryo can differentiate into muscle cells in the absence of cell contact with the rest of the embryo. Although this does not rule out the action of an inducer which acts before the 8-cell stage, it does suggest that the action of an inducer is not necessary.

Whittaker (47, 48) changed the cell lineage by applying pressure on the top of the embryo just prior to the 8-cell stage, causing the plane of cleavage to be altered (Fig. 2). In this way the cytoplasm that normally goes to the premuscle blastomeres was redistributed, causing an increase in the number of cells which display AChE activity in the larvae. The histo-specific enzyme assay was also used during cytoplasmic transfer experiments (10). Cytoplasm from the premuscle blastomeres was assayed for its ability to direct the synthesis of AChE in cells not normally destined to become muscle cells. Although the percentage of cells that survived in these experiments was small, a few cases did show additional AChE in "nonmuscle" cells.

In summary, these experiments support the view that a cytoplasmic determinant that commits cells to form muscle (or at least muscle-specific enzymes and antigens) is present in ascidian eggs. This determinant is present in the egg and is segregated during early cleavage divisions so that the highest concentration of the determinant is found in the two posterior vegetal cells at the 8-cell stage. The molecular nature of the factors comprising this muscle determinant is unknown.

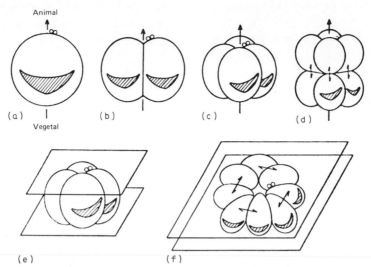

FIGURE 2. Normal distribution and redistribution of the yellow crescent cytoplasm of *Styela* embryos after compression during the third cell cleavage. (*a*) Fertilized egg; (*b*) 2-cell stage; (*c*) 4-cell stage; (*d*) 8-cell stage demonstrating the animal-vegetal axis; (*e*) orientation of the compression of 4-cell embryos; (*f*) the result of compression, a flat plate of 8 cells. [Taken from Whittaker (47).]

An Alternative Strategy: *Cerebratulus*

The second series of experiments on cytoplasmic localizations to consider make use of *Cerebratulus*, a marine nemertean worm. *Cerebratulus* differs from ascidians in that upon fertilization it does not undergo any cytoplasmic rearrangement. Whereas ascidians show a general rearrangement and localization of determinants prior to the first cell cleavage, the localization of determinants in *Cerebratulus* occurs more gradually and is dependent on cytokinesis. In an impressive set of experiments, Freeman (12) demonstrated that the determinants which direct the formation of the apical tuft and of the gut are gradually segregated through the first three cell cleavages. In the unfertilized egg, the gut determinants are uniformly distributed, whereas the determinants for the apical tuft are more concentrated in the vegetal hemisphere of the egg. Once fertilization occurs, the gut determinants move out of the animal hemisphere and concentrate in the vegetal half. The apical tuft determinants move to the animal hemisphere (Fig. 3). These two movements do not appear to be coupled. In addition, a distinction should be made between the gradual movement of these determinants and the rapid whole-scale cytoplasmic movements that many embryos undergo during the first few minutes of their development. The gut and apical tuft determinants become fully segregated from one another as cell division occurs, and division is complete by the time the embryo consists of eight distinct blastomeres. Thus, the timing with which the cytoplasmic determinants move in *Cerebratulus* is clearly dif-

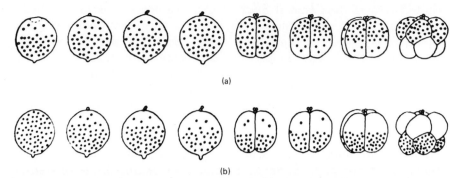

FIGURE 3. Diagrammatic representation of the distribution of factors specifying apical tuft and gut formation. (*a*) Dots represent factors that specify apical tuft formation. (*b*) Dots represent factors that specify gut formation. This shows (moving from left to right) the development of the fertilized egg through the 8-cell stage.

ferent from the timing in the ascidian. In *Cerebratulus* there is a slow redistribution which is complete relatively late. In ascidians, the muscle cell determinants are localized to a particular portion of the fertilized egg as a consequence of the cytoplasmic streaming that occurs before the first cell cleavage.

Freeman also demonstrated that the localization of the apical tuft and gut determinants is somehow triggered by the formation of the asters during mitosis (12). He reasoned that since the movements of the cell determinants occur after fertilization there must be something which occurs at the time of fertilization to induce this change. His studies showed that if the meiotic apparatus is surgically removed from the egg or if aster formation is inhibited, the localization of determinants does not occur. Conversely, if asters are induced precociously in unfertilized eggs, the gut and apical tuft determinants are localized prematurely. These studies demonstrate that in *Cerebratulus* the localization of the cytoplasmic determinants is associated with the action of some part of the mitotic machinery.

Brief Review of Insects

Among the insects, both the choronomid midge and the fruit fly will be considered, but only briefly since they are both extensively discussed in Chap. 14, this volume. Experiments by Yajima (53, 54) and Kaltoff (18) have shown that the development of the head and thorax requires cytoplasmic factors localized in the anterior portion of choronimid eggs. Ultraviolet (uv) irradiation destroys the determinant, resulting in an embryo with a duplicated mirror image of the abdomen; hence the name "double abdomen" (18, 19). It was reasoned that the relevant cytoplasmic determinants are nucleic acids both because of the spectrum of sensitivity to the uv irradiation and because the effects of uv irradiation are photoreversible (17). Moreover, Kandler-Singer and Kaltoff (20) showed that if *Smittia* eggs are submerged in a solution of RNase and are punctured at their anterior end, a significant percentage of the developing embryos have a "double abdomen." Submersion in RNase followed by puncturing of the embryo in the middle or posterior region was followed by normal development. Although the cytoplasmic components have not been individually identified, in the case of *Smittia* a strong case has been made for the involvement of ribonucleic acid in the determination of anterior structures of the embryo (11).

In the case of *Drosophila melanogaster*, transplantation studies and genetic analyses are leading to a rather complete picture of the mode of action and the number of gene products involved in establishing the anterior-posterior (A-P) and dorsal-ventral (D-V) axis. The A-P axis of the fly embryo is controlled by two organizing centers, one at the posterior pole and the other at the anterior pole (34). The existence of these organizing centers is demonstrated by experiments in which cytoplasm from one or another pole is removed or transplanted to an ectopic position. For example, removal of a small amount of cytoplasm (5 to 10 percent of egg volume) from the anterior end of a fly egg leads to a loss of more anterior structures that normally form from the region where the cytoplasm was removed. From such simple experiments it was inferred that the cytoplasm that was removed is required at a distance for the correct development and patterning of embryonic regions. Similarly, the transplantation of anterior cytoplasm to a posterior position can suppress normal posterior development and induce the formation of a second head and thoracic parts. Similar experiments have demonstrated the existence of a cognate organizing center at the posterior pole.

Systematic searches for maternal effect mutants that affect A-P patterning have been most informative (33). From these studies it has been concluded that fewer than 40 different genes are involved in setting up the A-P pattern. The gene *bicoid* (13) provides the best candidate for a gene product that forms the organizing center at the anterior end, while *oskar* (14) apparently has a similar role at the posterior end. It is important to note that neither of these genes acts singly in determining either the anterior or the posterior organizing center of the egg. Both *bicoid* and *oskar* require action from other genes in order to set up the appropriate gradients (14, 34; also see Chap. 14, this volume).

Another level of control which has been suggested for maternal genes is that of controlling the embryonic areas in which the zygotic genes act. In other words, some of the zygotic genes may be directly affected by the A-P gradients that are set up by the maternal products of genes such as *bicoid* and *oskar* (1; see Chap. 14, this volume for more detail).

The D-V axis is also determined by a set of maternal genes, and many of these genes have been thoroughly characterized by genetic analyses (3). Experiments by Anderson et al. (3, 4) have strongly suggested the the *toll* gene product is the principal determinant of the D-V axis; i.e., *toll* is the first step in the determinantion of the axis. As in the A-P axis, it is thought that the gradient set up by *toll* acts in a hierarchal fashion to control the expression of zygotically active genes involved in D-V patterning.

A strong case can be made that development of the insect systems on the molecular level is perhaps the best understood. In this respect, much work has been done in different species in an attempt to identify homologues of important fly genes. However, it remains to be seen how generally applicable the insect models will be to other systems.

Thus the experiments and genetic studies which have been conducted on these insect systems have given a clearer idea of the large number of genes which may function in setting up an axis polarity. In addition, there is at least a hypothesis for the order and hierarchy of these genes. This helps us recognize the complex nature of the way in which these genes may function and the extent to which they must be coordinated with each other.

Cytoplasmic Information in Amphibian Eggs

Cytoplasmic information systems in amphibian eggs have been analyzed for decades. Recently, Gerhart and colleagues (15) collated earlier data and added several observations of their own to develop a relatively coherent view of how the D-V axis in the *Xenopus* egg is specified. That view is summarized as follows. The unfertilized egg is radially symmetric, and the dorsal axis can form at any point on the equator, i.e., on any meridian. The exact meridian that will mark the dorsal side is usually determined by the site of sperm entry; that is, the point at which invagination of the dorsal lip of the blastopore first occurs is usually opposite the point of sperm entry. After sperm entry the vegetal cytoplasm undergoes a rotation relative to the cortex. It has been suggested that the mechanical force driving this rotation may involve the egg's newly formed aster. In any case, as a result of the rotation, vegetal cytoplasm on the future dorsal side comes in contact with an equatorial-animal region of the cortex. Somehow this rotation creates an active organizing center, or "Spemann organizer." A molecular or cellular explanation of this process is not yet available. It is possible that the organizer region is activated as a result of the translation of a stored maternal RNA. This is an attractive hypothesis because the Spemann organizer is formed long before transcriptional activation of the embryonic genome occurs. Alternatively, an inactive maternal protein(s) that is radially distributed at or near the equator could be activated by rotational movements. Further work is needed to clarify the molecular basis of dorsal axis specification.

The differences in developmental potential along the animal-vegetal (A-V) axis in frog eggs has also received a lot of attention. There are clear differences between the contents of the animal and vegetal hemispheres including, for example, yolk platelets, mitochondria, and pigmentation. There is also an important difference in the developmental potential of the animal and vegetal hemispheres. Chung and Malacinski (5) have shown that explants made from the top, or animal pole, will develop into ciliated epithelium, or ectoderm, whereas explants made from the vegetal region will form endoderm, or gut. Thus, the animal pole region has the potential to make ectoderm, and the vegetal pole region has the potential to make endoderm. These and other studies have suggested that the difference in developmental potential along the A-V axis is due to localized maternal factor(s).

In order to identify candidates for such localized molecules, Moen and Namenwirth (28) sectioned eggs and translated the RNA isolated from the sections. Their study suggested that 25 percent of the translatable RNA was unequally distributed. A more recent study by King and Barklis (22), however, found that 96 percent of the translatable mRNA was evenly distributed along the A-V axis. The apparent discrepancies between these studies are not easily resolved.

It is important to consider the composition of the population of maternal mRNA. A number of studies have used hybridization kinetics to determine the size of the mRNA pool in many different animals. When these values are compared, there is both a similarity of egg mRNA complexity and a relatively constant ratio of complex mRNA : ribosomal RNA in eggs from a variety of species (for review see Ref. 8). In the case of amphibians, estimates of maternal mRNA range from 40 to 90 ng per egg (40, 41). Since the maternal mRNA directs the early development of the embryo, one is interested in the number of unique, or rare, mRNA sequences that are available in the egg. This issue is complicated by the large portion of interspersed repeats (2) reported in this mRNA, 60 to 70 percent, which makes this RNA functionally untranslatable (8, 38). This is also true of mRNA isolated from sea urchin eggs. Some of the repeats isolated from sea urchin mRNAs have been sequenced and shown to encode stop codons in all reading frames (36). At this point it is unclear whether these repeats are in the 5' or 3' untranslated region or whether they are in the coding region and are spliced out to form a translatable mRNA. It is unclear whether 30 percent of the total mRNA, i.e., that mRNA which does *not* contain interspersed repeats, would be sufficient to carry the embryo through to the midblastula transition.

While it is informative to know where the general population of poly A^+ RNAs and proteins are localized, it is difficult to construct a detailed model based on whole populations of messages or proteins. More recently work has been carried out to identify *specific* poly A^+ RNAs that are localized in frog eggs. Rebagliati et al. (37) screened for cDNAs that represent mRNAs localized to either the animal pole or the vegetal pole of *Xenopus* eggs. Although the majority of RNAs did *not* show a differential distribution, four clones were identified, one vegetal specific (Vg1) and three animal specific (An1, An2, An3). These clones were all shown to maintain their pattern of localization in fertilized eggs and early cleavage stage embryos. The developmental pattern of expression for these RNAs showed that two are strictly maternal (Vg1 and An1) while the other two (An2 and An3) are reexpressed by the embryonic genome.

The way in which these four mRNAs are localized is being analyzed. Thus far, however, there is data only for Vg1. In situ hybridization data (26) show that the Vg1 RNA is first found distributed equally throughout stage I and II oocytes. As the oocytes enter stage III, localization ensues. There appears to be a gradual movement of the Vg1 message toward one pole of the oocyte, the vegetal pole. As oogenesis proceeds, this localization becomes more and more dramatic, so that by the time the oocyte has fully matured, the Vg1 message is found localized as a crescent hard against the vegetal pole of the oocyte. This localization is maintained throughout early embryogenesis (see Fig. 4).

The identity of those maternal mRNAs is also being investigated. Sequence data have been obtained for two of the cloned localized *Xenopus* RNAs.

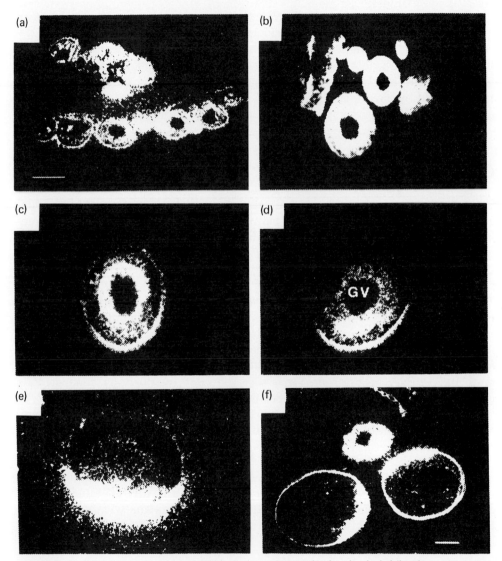

FIGURE 4. Translocation of Vg1 RNA during oogenesis. In situ hybridizations were performed with antisense Vg1 RNA probe. Albino oocytes at various stages are shown. The black hole present in the center of most of the sections is the nucleus, or germinal vesicle. (a) Stage I oocytes; (b) stage II oocytes; (c, d) stage III oocytes, (e) stage IV oocytes; (f) stage II and IV oocytes from a wild-type female. Note that the white ring around the pigmented oocytes (in part f) is due to the pigment granules in the cortex of the egg. [Taken from Melton (26).]

An2 has a strong homology with a mitochondrial ATPase (44), and it has been proposed that its localization could be used to differentially activate the maternal mitochondria in the animal hemisphere. If so, this would help explain the observed respiration gradient in the A-V axis. Based on DNA sequence homology, Vg1 is a member of a family of growth factors typified by transforming growth factor β (43).

Nieuwkoop (29) had shown earlier that the vegetal end of the amphibian egg has the capacity to induce mesoderm in overlying ectodermal tissue. Recent studies by a number of groups have shown that at least two known growth factors, fibroblast growth factor (21, 42) and transforming growth factor β2 (39), can induce mesoderm in isolated animal pole tissue. Thus the protein encoded by the vegetally localized Vg1 mRNA is a good candidate for a cytoplasmic determinant involved in mesodermal induction.

Concluding Remarks

This chapter has discussed selected examples which have generated the concept that localized cytoplasmic determinants play a key role in regulating embryonic gene expression and establishing body pattern. A general theme which seems to be emerging is that this developmental information is stored in an inactive form in the egg, probably as mRNAs. As mentioned above, insect systems are perhaps the best understood (also see Chap. 14, this volume). The details still need to be worked out for each of the other systems described in this chapter. One can, however, entertain the following simple possibilities as plausible explanations. After fertilization the localized RNA is translated into protein which in turn acts in the cells which contain the RNA, as, for example, a trans-acting factor that regulates gene expression. Alternatively, the maternal mRNA might code for a secreted protein, perhaps generating a graded signal which acts on neighboring cells by induction.

At this point in the history of this important area of developmental biology, it is a reasonable assumption that rapid progress in describing the molecular action of cytoplasmic determinants will come from many of the experimental systems outlined here, because of the excellent background information which exists from the work of several generations of embryologists. What is needed as a catalyst for progress is the further elucidation of other localized cytoplasmic determinants as well as means with which to determine the function of those molecules.

General References

WILSON, E. B.: 1925. *The Cell in Development and Heredity*. Macmillan, New York.

DAVIDSON, E. H.: 1986. *Gene Activity in Early Development*, 3d ed. Academic Press, New York.

References

1. Akam, M. 1987. The molecular basis for metameric pattern in the *Drosophila* embryo. *Development (Camb.)* **101**:1–22.

2. Anderson, D. M., Richter, J. D., Chamberlin, M. E., Price, D. H., Britten, R. J., Smith, L. D., and Davidson, E. H. 1982. Sequence organization of the poly(A) RNA synthesized and accumulated in lampbrush chromosome stage *Xenopus laevis* oocytes. *J. Mol. Biol.* **155**:281–309.

3. Anderson, K. V., Jurgens, G., and Nusslein-Volhard, C. 1985. Establishment of dorsal-ventral polarity in the *Drosophila* embryo: genetic studies on the role of the *Toll* gene product. *Cell* **42**:779–789.

4. Anderson, K. V., and Nusslein-Volhard, C. 1984. Information for the dorsal-ventral pattern of the *Drosophila* embryo is stored as maternal mRNA. *Nature (Lond.)* **311**:223–227.

5. Chung, H. M., and Malacinski, G. M. 1983. Reversal of developmental competence in inverted amphibian eggs. *J. Embryol. Exp. Morphol.* **73**:207–220.

6. Conklin, E. G. 1905. The organization and lineage of the ascidian egg. *J. Acad. Natl. Sci. (Phila.)* **13**:5–119.

7. Conklin, E. G. 1931. The development of centrifuged eggs of ascidians. *J. Exp. Zool.* **60**:1–77.

8. Davidson, E. H. 1986. *Gene Activity in Early Development*, 3rd ed. Academic Press, New York.

9. Deno, T., Nishida, H., and Satoh, N. 1984. Autonomous muscle cell differentiation in partial ascidian embryos according to the newly verified cell lineages. *Dev. Biol.* **104**:322–328.

10. Deno, T., and Satoh., N. 1984. Studies in the cytoplasmic determinant for muscle cell differentiation in ascidian embryos: an attempt at transplantation of the myoplasm. *Dev. Growth & Differ.* **26**:43–48.

11. Elbethieha, A. and Kaltoff, K. 1988. Anterior determinants in embryos of *Chironomus samoensis*: characterization by rescue bioassay. *Development (Camb.)* **104**:61–76.

12. Freeman, G. 1978. The role of asters in the localization of the factors that specify the apical tuft and the gut of the nemertine *Cerebratulus lacteus*. *J. Exp. Zool.* **206**:81–108.

13. Frohnhofer, H. G., and Nusslein-Volhard, C. 1986. Organization of anterior pattern in the *Drosophila* embryo by the maternal gene *bicoid*. *Nature (Lond.)* **324**:120–125.

14. Frohnhofer, H. G., and Nusslein-Volhard, C. 1987. Maternal genes required for the anterior localization of bicoid activity in the embryo of *Drosophila*. *Genes & Dev.* **1**:880–890.

15. Gerhart, J., and Keller, R. 1986. Region specific cell activities in amphibian gastrulation. *Annu. Rev. Cell Biol.* **2**:201–229.

16. Grant, P. 1978. *Biology of Developing Systems*. Holt, Rinehart and Winston, New York.

17. Kalthoff, K. 1973. The realm of the ultraviolet. *Photochem. Photobiol.* **18**:353–364.

18. Kalthoff, K. 1979. Analysis of a morphogenetic determinant in an insect embryo

(*Smittia Spec., Chironimidae, Diptera*). In: *Determinants of Spatial Organization.* (S. Subtelny and I. R. Konigsberg, eds.), pp. 97–126, Academic Press, New York.

19. Kalthoff, K., and Sander, K. 1968. Development of the malformation "double abdomen" in eggs of *Smittia parthenogenetica* (Dipt., Chironomidae) partially irradiated by UV. *Roux's Arch. Dev. Biol.* **161**:129–146.

20. Kandler-Singer, I., and Kaltoff, K. 1976. RNase sensitivity of an anterior morphogenetic determinant in an insect egg (*Smittia sp., Chironomidae, Dipters*). *Proc. Natl. Acad. Sci. U.S.A.* **73**:3739–3743.

21. Kimelman, D., and Kirschner, M. 1987. Synergistic induction of mesoderm by FGF and TGF-β and the identification of an mRNA coding for FGF in the early *Xenopus* embryo. *Cell* **51**:869–877.

22. King, M. L., and Barklis, E. 1985. Regional distribution of maternal messenger RNA in the amphibian oocyte. *Dev. Biol.* **112**:203–212.

23. Lillie, F. R. 1909. Polarity and bilaterality of the annelid egg. Experiments with centrifugal force. *Biol. Bull.* **16**:54–79.

24. Meedel, T. H., and Whittaker, J. R. 1983. Development of translationally active mRNA for larval muscle acetylcholinesterase during ascidian embryogenesis. *Proc. Natl. Acad. Sci. U.S.A.* **80**:4761–4765.

25. Meedel, T. H., and Whittaker, J. R. 1984. Lineage segregation and developmental autonomy in expression of functional muscle acetylcholinesterase mRNA in the ascidian embryo. *Dev. Biol.* **105**:479–487.

26. Melton, D. A. 1987. Translocation of a localized maternal mRNA to the vegetal pole of *Xenopus* oocytes. *Nature (Lond.)* **328**:80–82.

27. Mita-Miyazawa, I., Nishikata, T., and Satoh, N. 1987. Cell- and tissue-specific monoclonal antibodies in eggs and embryos of the ascidian *Halocynthia roretzi*. *Development (Camb.)* **99**:155–162.

28. Moen, T. L., and Namenwirth, M. 1977. The distribution of soluble proteins along the animal-vegetal axis of frog eggs. *Dev. Biol.* **58**:1–10.

29. Nieuwkoop, P. D. 1969. The formation of the mesoderm in urodelean amphibians. *Roux's Arch. Dev. Biol.* **162**:341–373.

30. Nishida, H. 1987. Cell lineage analysis in ascidian embryos by intracellular injection of a tracer enzyme: III. Up to the tissue restricted stage. *Dev. Biol.* **121**:526–541.

31. Nishida, H., and Satoh, N. 1983. Cell lineage analysis in ascidian embryos by intracellular injection of a tracer enzyme: I. Up to the eight-cell stage. *Dev. Biol.* **99**:382–394.

32. Nishikata, T., Mita-Myazawa, I., Deno, T., and Satoh, N. 1987. Muscle cell differentiation in ascidian embryos analyzed with a tissue-specific monoclonal antibody. *Development (Camb.)* **99**:163–171.

33. Nusslein-Volhard, C., and Wieschaus, E. 1980. Mutations affecting segment number and polarity in *Drosophila*. *Nature (Lond.)* **287**:795–801.

34. Nusslein-Volhard, C., Frohnhofer, H., and Lehmann, R. 1987. Determination of anteroposterior polarity in *Drosophila*. *Science* **238**:1675–1681.

35. Perry, H. E., and Melton, D. A. 1983. A rapid increase in acetylcholinesterase mRNA during ascidian embryogenesis as demonstrated by microinjection into *Xenopus laevis* oocytes. *Cell Differ.* **13**:233–238.

36. Posakony, J. W., Flytzanis, C. N., Britten, R. J., and Davidson, E. H. 1983. Interspersed sequence organization and developmental representation of cloned poly(A) RNAs from sea urchin eggs. *J. Mol. Biol.* **167**:361–390.

37. Rebagliati, M. R., Weeks, D. L., Harvey, R. P., and Melton, D. M. 1985. Identification and cloning of localized maternal RNAs from *Xenopus* eggs. *Cell* **42**:769–777.

38. Richter, J. D., and Smith, L. D. 1984. Interspersed poly(A) RNAs of amphibian oocytes are not translatable. *J. Mol. Biol.* **173**:227–241.

39. Rosa, F., Roberts, A., Danielpour, D., Dart, L., Sporn, M., and Dawid, I. 1987. Mesoderm induction of early *Xenopus* embryos: the role of TGF- β2-like factors. *Science* **239**:783–785.

40. Rosbash, M., and Ford, P. J. 1974. Polyadenylic acid-containing RNA in *Xenopus laevis* oocytes. *J. Mol. Biol.* **85**:87–101.

41. Sagata, N., Shiokawa, K., and Yamana, K. 1980. A study on the steady state population of poly(A)+ RNA during early development in *Xenopus laevis. Dev. Biol.* **77**:431–448.

42. Slack, J., Darlington, B., Heath, J., and Godsave, S. 1987. Mesoderm induction in early *Xenopus* embryos by heparin binding growth factors. *Nature (Lond.)* **326**: 197–200.

43. Weeks, D. L., and Melton, D. M. 1987. A maternal mRNA localized to the vegetal hemisphere in *Xenopus* eggs codes for a growth factor related to TGF-β. *Cell* **51**: 861–867.

44. Weeks, D. L., and Melton, D. A. 1987. A maternal mMRA localized to the animal pole of *Xenopus* eggs encodes a subunit of mitochondrial ATPase. *Proc. Natl. Acad. Sci. U.S.A.* **84**:2798–2802.

45. Whittaker, J. R. 1973. Segregation during ascidian embryogenesis of egg cytoplasmic information for tissue-specific enzyme development. *Proc. Natl. Acad. Sci. U.S.A.* **70**:2096–2100.

46. Whittaker, J. R. 1979. Cytoplasmic determinants of tissue differentiation in the ascidian egg. In: *Determinants of Spatial Organization.* (S. Subtelny and I. R. Konigsberg, eds.), pp. 29–52, Academic Press, New York.

47. Whittaker, J. R. 1980. Acetylcholinesterase development in extra cells caused by changing the distribution of myoplasm in ascidian embryos. *J. Embryol. Exp. Morphol.* **55**:343–354.

48. Whittaker, J. R. 1982. Muscle lineage cytoplasm can change the developmental expression in epidermal lineage cells of ascidian embryos. *Dev. Biol.* **93**:463–470.

49. Whittaker, J. R., Ortolani, G., and Farinella-Ferruzza, N. 1977. Autonomy of acetylcholinesterase differentiation in muscle lineage cells of ascidian embryos. *Dev. Biol.* **55**:196–200.

50. Wilson, E. B. 1904. Experimental studies on germinal localization. I. The germ-regions in the egg of Dentalium. *J. Exp. Zool.* **1**:1–71.

51. Wilson, E. B. 1904. Experimental studies on germinal localization. II. Experiments on the cleavage-mosaic in patella and dentalium. *J. Exp. Zool.* **1**:199–267.

52. Wilson, E. B. 1925. *The Cell in Development and Heredity.* Macmillan, New York.

53. Yajima, H. 1960. Studies on embryonic determination of the harlequin-fly, *Chironomus dorsalis* I. Effects of centrifugation and of its combination with constriction and puncturing. *J. Embryol. Exp. Morphol.* **8**:198–215.

54. Yajima, H. 1964. Studies on embryonic determination of the harlequin-fly, *Chironomus dorsalis*. II. Effects of partial irradiation of the egg by ultraviolet light. *Embryol. Exp. Morphol.* **12**:89–100.

55. Zeleny, C. 1904. Experiments on the localization of developmental factors in the nemertine egg. *J. Exp. Zool.* **1**:293–329.

Questions for Discussion with the Editor

1. *What are the advantages of cytoplasmic determinants being composed of mRNA versus proteins?*

Although it has not been proved that protein cytoplasmic determinants do not exist, there is compelling evidence that in at least some cases maternal mRNA is stored in the egg and may act as cytoplasmic determinants throughout early development. There are two important things to remember. First, it seems likely that at least in some cases the proteins that act to direct development may be secreted and act as diffusible substances. This is one possible mechanism for inductive events. Second, a cytoplasmic determinant that is an mRNA allows an extra level of control. For example, the mRNA might be masked or be under translational control.

2. *Given that the four different species outlined in this chapter vary so much in their development, is it reasonable to assume that there may be a general pattern to the way in which all organisms develop?*

The general premise in this kind of approach is that if genes are conserved through evolution, their functions may also be conserved. The difficulty lies in actually proving the function of any gene. However, the fact remains that from insects to mammals the segmented body plan is retained. Because development is so complex, it seems likely that some of the mechanisms for directing development may have been retained from one species to the next. At least on a broad scale one expects that there will be some similarity from one developmental program to the next.

CHAPTER **4**

The Nucleo-cytoplasmic Interactions as Revealed by Nuclear Transplantation in Fish

Yan Shaoyi

Introduction

A BASIC BIOLOGIC PROBLEM, which has been investigated and discussed by several generations of cytologists, geneticists, and embryologists, is the interaction between the nucleus and the cytoplasm. It is well known that in higher organisms a fertilized egg or zygote results from the combination of a sperm nucleus surrounded by a small amount of cytoplasm and an egg nucleus surrounded by a large amount of cytoplasm. These cytoplasmic components contribute to the zygote and directly or indirectly affect the development of the fertilized egg, the differentiation of embryonic cells, and the expression of the hereditary characteristic of the individuals.

Among the various nuclear and cytoplasmic components, the chromosomes in the nucleus, especially their structure and behavior, first attracted the attention of biologists. Half of the chromosomes in the zygote originate from the female parent and the other half from the male. During cleavage of a fertilized egg, the chromosomes undergo distinct division and are equally distributed to the daughter cells. This observation led Roux (25) to suggest that the chromosomes were the essential material for the determination of development, differentiation, and inheritance in organisms. Later, Weismann (36) proposed that there are various different determinants contained in the nucleus which enter into the daughter cells during cell division and regulate cell differentiation (see Chap. 11, this volume).

Since Weismann's time, many embryologists have examined his theory of determinants. The most impressive evidence bearing on the problem was first provided by Hans Spemann (28, 29). In one of Spemann's experiments, the fertilized egg of the urodele *Triturus* was tied prior to its first cleavage with a hair loop so that its animal and vegetal poles were bisected. The egg was thus divided into two parts, right and left, but not separated completely. The parts remained connected by a fine cytoplasmic bridge. Subsequently, only the half of the egg possessing the nucleus was able to divide; the half without the nucleus did not. When the nucleated half reached the 16-cell stage, however, the volume of the nucleus in each daughter cell became very small, approximately equal in size to one-sixteenth of the original zygote nucleus. The nucleus of one cell adjacent to the nonnucleated half was, therefore, able to pass across the cytoplasmic bridge into the nonnucleated half of the original egg cell, which was now able to divide. Subsequently the hair loop was closed until the two parts were completely separated from each other, and each part of the egg developed into a whole embryo. These results indicated that, although the nucleus in the nucleated half of the zygote cell was reduced to one-sixteenth of its original volume and the cytoplasm of the zygote cell was reduced by half, each still retained the potential for development. This was also distinctly different from what would be predicted by the theory of determinants proposed by Weismann. In order to study this problem further, Spemann (reviewed in Ref. 5) proposed another experiment. He suggested that if the nucleus of an embryonic cell could be isolated and transplanted into an egg devoid of a nucleus, then it would be possible to investigate whether cell nuclei from embryos of different developmental stages, including the nuclei of differentiated cells, could initiate normal development in the egg cytoplasm. At that time, however, Spemann was unable to perform his experiment because appropriate techniques were not yet available.

Successful nuclear transplantation was first accomplished by Commandon and de Fonbrune (7) in the amoeba. The nuclear transplanted amoeba was able to live and to divide, and it successfully produced daughter amoebas. Danielli (8, 9) performed nuclear transplantation experiments on two different kinds of amoebas and observed three new systems of inheritance: (1) nuclear inheritance, (2) cytoplasmic inheritance, and (3) blended inheritance from both the nucleus and the cytoplasm. Danielli concluded that both the nucleus and the cytoplasm were of equal importance in inheritance. Nevertheless, the simple structure, short life cycle, and limited differentiation of the unicellular amoeba suggested that the mode of inheritance of amoebas might be quite different from that of higher organisms. Later research would indicate that the relative importance of nuclear inheritance and cytoplasmic inheritance was not the same in all organisms. Therefore, this early research on a unicellular organism failed to clarify the functions of the nucleus and cytoplasm and their actions on the process of development and differentiation.

Fortunately, significant advances in understanding the problem followed successful nuclear transplantation in amphibian eggs. Briggs and King (2, 3) transferred the nuclei from cells of the animal hemisphere of a *Rana pipiens* blastula into enucleated eggs of the same species. Several of these eggs developed into intact embryos.

Briggs (4) and DiBerardino (13) extensively reviewed the results of nuclear transplantation experiments in several kinds of urodeles and anurans (*Ambystoma, Triturus, Bufo, Pleurodeles, Rana*, and *Xenopus*), and they drew the following conclusions:

1. In general, when the nuclei of cells of later developmental stages (gastrula and neurula) are transplanted into enucleated eggs, a smaller percentage of individuals develop normally as compared to transplantation of nuclei from cells of earlier developmental stages (e.g., blastula).

2. A few nuclei from late developmental stages, such as those from endoderm cells of hatching *Xenopus* larvae (reviewed in Ref. 13) maintain their developmental totipotency. That is, they retain their ability to direct the production of a complete embryo. Most nuclei from late developmental stages are, however, unable to initiate normal development in enucleated eggs.

3. Developmental totipotency of nuclei of some differentiating cells (such as neural cells and some endodermal cells) are severely restricted and are seldom restored even when they are serially transferred in enucleated eggs.

4. Restriction of the developmental potency of the nucleus is not corrected by parabiosis of nuclear transplanted embryos with normal ones, nor can it be restored when a haploid set of egg chromosomes is combined with the diploid set from the endodermal nucleus.

5. In most nuclear transplanted embryos, a change in the karyotype of the cells occurs to varying degrees. The later the developmental stage of nucleus, the more severe the abnormality of the karyotype and, thereby, the more severe the abnormality of the developing embryo.

The above conclusions indicate that the nuclei of amphibian embryonic cells gradually lose their developmental totipotency as cell differentiation progresses. However, the cytoplasm of mature recipient eggs, prior to cleavage of the egg, is able to restore the developmental totipotency of the nucleus to varying degrees. Thus, whether the nuclear transplanted egg can initiate its development anew appears to be related to the interaction of nucleus and cytoplasm in a cell.

In nuclear transplantation between members of the same species, it is difficult to distinguish which characteristics are due to nuclear contributions and which are due to cytoplasmic contributions. However, the analysis of hybrids obtained by transplanting the nucleus of one species into the enucleated egg of a different species with identifiably different characteristics can help separate the influence on development of the nucleus from that of the cytoplasm where these characteristics are concerned. For example, Kawamura and Nishioka (18, 19) transplanted nuclei from blastula cells between the

Japanese pond frogs, *Rana nigromaculata* and *Rana brevipoda*, and between *Rana japonica* and *Rana ornativentris*. The characteristics of the hybrid frogs were mostly similar to the species that donated the nucleus, but they also showed certain features which were similar to the species that contributed the cytoplasm. Some characteristics were intermediate in character. These experiments confirmed the importance of cytoplasmic factors on development and differentiation in amphibians.

Nevertheless, the interpretation of these experiments is restricted because nuclear transplantation in amphibians can be accomplished only between closely related species. Hence, it is necessary to look at nuclear transplantation in other systems to study further the interactions of the nucleus and cytoplasm between more distantly related species with greater differences in hereditary character. In addition, from the standpoint of the improvement of economically important genetic stocks, amphibians are obviously of little value. In contrast, the application of nuclear transplantation techniques to create clones of commercially important animals would be useful for agricultural purposes. Fish show great promise as an experimental system for nuclear transplantation (33) because they satisfy both requirements. This chapter describes the technologies developed for nuclear transplantation in fish and various experiments and their results.

Some Special Aspects of Nuclear Transplantation in Fish

Eggs of freshwater teleosts, mainly the crucian carp (*Carassius auratus*), common carp (*Cyprinus carpio*), goldfish (*Carassius auratus*), bitterling (*Rhodus sinensis*), grass carp (*Ctenopharyngodon idellus*), and blunt-snout bream (*Megalobrama amblycephala*) of the order Cypriniformes, and the tilapia (*Oreochromis niloticus*) of the order Perciformes, have been used as experimental material in our laboratory. All these fishes spawn, are fertilized, and develop in appropriate laboratory aquariums as well as under natural conditions. At present, advances in fish farming in China have led to the development of techniques, such as artificial insemination and the application of hormones to promote sexual maturity and spawning, which facilitate the use of these fishes for nuclear transplantation and other experiments. The properties of these eggs and the methods of fertilization and culture with respect to nuclear transplantation are described in the appendix.

The quality of the mature eggs used as enucleated recipients as well as of the fertilized eggs providing donor blastula cell nuclei is of great importance. Successful nuclear transplantation also depends on the technical skill of the researcher doing the operation. Finally, the proper combination of species used as nucleus donors and recipients should also be carefully considered, since species that are phylogenetically distant are more likely to exhibit incompatibility between the nucleus and cytoplasm.

Observations on the Nucleo-cytoplasmic Interaction as Revealed by Nuclear Transplantation in Fish

The nuclei of the blastula, gastrula, and neurula stages of the developing eggs of goldfish (*Carassius auratus*) were transferred into enucleated eggs using procedures similar to those used in amphibian nuclear transplantation with minor modifications (34). The proportion of normal larvae obtained from the eggs provided with the nucleus of blastula cells was 9.5 percent. Eggs into which the nuclei of gastrula- and neurula-stage cells were transplanted yielded a very low proportion of normal larva (0.1 to 0.5 percent: unpublished data). These results are comparable to those reported in amphibians. Therefore, rather than continuing to duplicate the amphibian experiments, we focused our attention on the following questions:

1. Can nucleo-cytoplasmic hybrid fish be obtained by nuclear transplantation between distantly related fishes? If so, are they fertile?
2. If such hybrid fish are obtained by nuclear transplantation, what does the pattern of inheritance of parental characteristics reveal about nucleus and cytoplasm compatibility?
3. When agriculturally important fishes are used for the experiments, do the nucleo-cytoplasmic hybrid fish display potentially useful features?

In our research, we used blastula-stage fish embryos as nucleus-donor cells in order to take advantage of the higher rate of successful development. The key results follow.

Nucleo-cytoplasmic Hybrids Yield Viable Adult Fish

Adult nucleo-cytoplasmic hybrid fish were obtained in two combinations of eggs and nuclei from fishes of different genera and in one combination of fishes from different subfamilies.

COMBINATIONS OF NUCLEUS AND CYTOPLASM FROM FISHES IN DIFFERENT GENERA

Blastula nuclei from the common carp, *Cyprinus carpio L.*, were transferred into enucleated eggs of the crucian carp, the wild-type *Carassius auratus L.*, which belongs to the same subfamily, Cyprininae. In this combination, 2 to 3.2 percent of the total number of nuclear transplanted eggs produced viable adult hybrid fish (35, 42) (Fig. 1). The number of chromosomes was the normal diploid number ($2n = 100$). With regard to morphology, the oral barbs and pharyngeal tooth type of these nucleo-cytoplasmic hybrid fish were identical to those of the nucleus-donor fish, the common carp. However, the number of vertebrae was the same as in the crucian carp, suggesting that this characteristic was specified or in some way affected by the recipient egg

FIGURE 1. Hybrid between fish of different genera, obtained from the combination of the nucleus of common carp (*Cyprinus carpio*) and the cytoplasm of crucian carp (*Carassius auratus*).

cytoplasm. The number of lateral line scales was intermediate between the common carp number and the crucian carp number. The morphology of their air bladders differed from expected morphology for both the common carp and the crucian carp (42). Biochemical analysis showed that the electrophoretic patterns of serum proteins and the zymograms of lactate dehydrogenase (LDH) and of malate dehydrogenase (MDH) from erythrocytes of the hybrids were basically similar to those of the common carp (i.e., the nucleus-donor type), but there were some differences (e.g., banding patterns and migration rates of bands) (unpublished data).

The nucleo-cytoplasmic hybrids developed to sexual maturity and produced four generations of offspring. Among those four generations, the main morphological characteristics are retained. The growth rate shown by these nucleo-cytoplasmic hybrid fish was around 20 percent higher than that of the common carp. The protein content of its muscles was 3.78 percent higher and the fat content was 5.8 percent lower than in the common carp (unpublished data).

The reciprocal nucleo-cytoplasmic transplantation combination, in which the nucleus from crucian carp (*Carrasius auratus L.*) was transferred into the enucleated egg of common carp (*Cyprinus carpio L.*), also produced adult hybrid fish (39) (Fig. 2). The adult hybrid fish derived from this combination comprised 0.9 percent of the total number of the nuclear transplants. They had no oral barbs and their pharyngeal tooth type, lateral line scale number, number of vertebrae, and body size were similar to those of the nucleus-donor fish (crucian carp). The hybrid fish developed to sexual maturity, but, unfortunately, they died because of careless cultivation conditions, and we were unable to analyze them further.

In both of these experiments, there was little doubt that the resulting fish represented true nucleo-cytoplasmic hybrids. The hybrid fish had the normal diploid chromosome number ($2n = 100$ for both crucian carp and common

FIGURE 2. Hybrid between fish of different genera, obtained from the combination of the nucleus of crucian carp (*Carassius auratus*) and the cytoplasm of common carp (*Cyprinus carpio*).

carp), indicating that the eggs had been successfully enucleated, and transplantation resulted in a single set of diploid chromosomes, which can be used to identify the origin of the nucleus of the nuclear transplanted hybrid. Furthermore, since the main morphological features of the hybrids were most similar to those of the nucleus-donor species, we are quite certain that the hybrid fish developed from a nuclear transplanted egg.

COMBINATIONS OF NUCLEUS AND CYTOPLASM FROM FISHES IN
DIFFERENT SUBFAMILIES

Nuclear transplantation experiments were also conducted using fish from different subfamilies. A nucleus derived from the grass carp (*Ctenopharyngodon idellus*), subfamily Leucinae, was transplanted into an enucleated egg from a blunt-snout bream (*Megalobrama amblycephala*), subfamily Abramidinae. The survival rate of the nucleo-cytoplasmic hybrid fish (Fig. 3) in this combination was 3.6 percent of the total number of transplanted eggs. The chromosome number was the same as that of grass carp and blunt-snout bream ($2n = 48$). The chromosome pattern of the hybrid fish was the same as that of the grass carp (the nucleus donor), as confirmed by the presence of its eleventh marker chromosome. Some morphological characteristics, such as the body length to head length ratio; body length to body height ratio; body length to body width ratio; presence of the dorsal fin spine and abdominal keel; and the number of anal fins, gillrakers, pharyngeal teeth, lateral line scales, and vertebrae were similar to those of the nucleus donor (the grass carp). The hybrid also reached sexual maturity by the fourth year of growth, as do normal grass carp. At present, one nucleo-cytoplasmic hybrid fish has developed to sexual maturity and has produced normal sperm (41). Biochemical analysis showed that the electrophoretic patterns of hemoglobin, isozymes of LDH and

FIGURE 3. Hybrid between fish of different subfamilies, obtained from the combination of the nucleus of *Ctenopharyngoden idellus* and the cytoplasm of *Megalobrama amblycephala*.

MDH, and of serum proteins were essentially similar to those of the grass carp as well. A few differences were, however, observed. The electrophoretic patterns of LDH isozymes isolated from the hybrid fish's gill and kidney revealed five bands; the grass carp has six bands. In addition, the serum immunoelectrophoretic patterns for hybrid fish showed new precipitate lines. More precipitate lines appeared on the gels of hybrid fish serum than had been observed on the gels of either grass carp serum or blunt-snout bream serum when rabbit antiserum directed against hybrid fish serum was tested with these three serums (40, 21).

Nucleo-cytoplasmic Combinations That Produce Embryos or Larvae Only

COMBINATIONS OF NUCLEUS AND CYTOPLASM FROM FISHES OF DIFFERENT SUBFAMILIES

The nuclei of blastula-stage embryos of the goldfish (domestic *Carassius auratus*), subfamily Cyprininae, were transplanted into enucleated eggs of the bitterling (*Rhodus sinensis*), subfamily Acheilognathinae. Both fish belong to the same family, Cyprinidae. A total of 59.6 percent of all transplanted eggs of this combination developed to the blastula stage, but only 0.4 percent of them reached the late embryo or larval stage. The morphology of the hybrid larvae was similar to that of bitterling larvae (small head, eyes, and brain; slender embryo body; two yolk outgrowths on either side of the head region; and single caudal fin). Therefore the expression of larval characteristics in nucleo-cytoplasmic hybrid fish in this combination were affected by the egg cytoplasm (34).

COMBINATIONS OF NUCLEUS AND CYTOPLASM FROM FISHES FROM DIFFERENT ORDERS

The nuclei of blastula-stage embryos of the tilapia (*Oreochromis nilotica*), order Perciformes, were transplanted into the enucleated eggs of the common carp (*Cyprinus carpio*), order Cypriniformes. From this combination, 50 percent of the transplanted eggs developed to the blastula stage, but only a few of these developed into larvae. Two larvae between 6 and 18½ days old died as a result of poor blood circulation. The rate of early development in these embryos was similar to that of carp but faster than that of tilapia. Therefore, the development of these hybrid embryos was also affected by the cytoplasm. Other characteristics, however, could not be identified distinctly (unpublished data).

Nuclear Transplantation Using Nuclei from In Vitro Cultured Cells

Chen et al. (6) reported the transplantation of nuclei from an in vitro subculture of blastula cells of the crucian carp (*Carassius auratus*). The nuclei of the subcultured cells of the 53rd to 59th subcultures (every generation of subcultured cells was maintained for 5 days) were transplanted into enucleated eggs from the same species. In this experiment, 83 percent of the nuclear transplanted eggs developed to the blastula stage, and 7.9 percent of them developed to the gastrula stage. When normal blastula cells obtained from the transplanted crucian carp eggs described above were used as donor cells and their nuclei transplanted again into new batches of enucleated crucian carp eggs, one nuclear transplanted adult fish with normal morphological features resembling the crucian carp was obtained. This fish died after 3 years. The chromosome number of that fish was heteroploid ($2n = 136$), and it was discovered, upon dissection, that its gonad was undeveloped.

In a second experiment, kidney cells of the crucian carp were cultured for a week, and nuclei taken from the cultured cells were transplanted into enucleated eggs of crucian carp. In this experiment, 41 percent of these transplanted eggs developed into normal and abnormal blastulas. The nuclei of the blastula cells of normal appearance from these transplanted eggs were serially transplanted into a second batch of enucleated crucian carp eggs. One fertile, female, nuclear transplanted, adult fish with normal morphological features of crucian carp was obtained. Its chromosome number was abnormal ($2n = 150$).

Successive Nuclear Transplantation in Goldfish

Wu et al. (38) transplanted blastula cell nuclei of double-caudal-fin goldfish into enucleated eggs of single-caudal-fin goldfish. The fish obtained from the first nuclear transplantation had double-caudal fins similar to the nucleus-donor fish, and the expression of the characteristic seemed to be controlled by the nucleus. The nuclei of blastula cells derived from the development of the egg of the first transplantation were retransplanted into a second batch of enucleated eggs of single-caudal-fin goldfish. Some single-caudal-fin fish were obtained. After each of four serial transplantations, more single-caudal-fin goldfish were obtained. The authors suggested that the cytoplasmic effects were strengthened through serial transplantations. Perhaps during serial nuclear transplantation the cumulative effects of the cytoplasm of several recipient eggs on the nucleus becomes strong enough to modify the original function of the donor nucleus.

Conclusions from Nucleo-cytoplasmic Transplantation Experiments in Fishes

The following conclusions can be drawn from these experiments:

1. Blastula nuclei of fish possess developmental totipotency. As development proceeds, the capability of a nucleus to promote the successful development of an enucleated egg is gradually restricted. These findings are similar to the results obtained from experiments on amphibians (reviewed in Ref. 13).

2. Hybrid fishes derived by nuclear transplantation between species of different genera or subfamilies were able to develop to sexual maturity. Certain hybrid individuals appeared identical in their expressed characteristics to the nucleus-donor species, other individuals had some of the same characteristics of the enucleated recipient species, while still other individuals had intermediate characteristics. A few hybrid fishes also showed new characteristics, which did not occur in either original species. The expressed characteristics varied among different combinations of species providing the nucleus and cytoplasm and also among different individual hybrid fish derived from the same combination. Certain nucleo-cytoplasmic hybrid fish, such as those derived from the combination of a common carp nucleus and crucian carp cytoplasm, may pass on some unique characteristics to their offspring. This unexpected result was observed in F2, F3, and F4 offspring (unpublished data), which all exhibited the same characteristics as the initial nucleo-cytoplasmic hybrid. This result is potentially important for improvement of fish resources. The mechanism of such inheritance needs further confirmation and investigation.

3. Some nuclear transplanted eggs, into which donor blastula nuclei from fish belonging to different subfamilies or to different orders had been transferred, could develop only to the embryo or larval stage. These hybrid embryos or larvae also revealed characteristics of the recipient species (which contributed cytoplasm only) to varying degrees.

4. Subcultured kidney cell nuclei were serially transplanted twice into the enucleated eggs of fish of the same species. Two nuclear transplanted adult fish were obtained, indicating developmental totipotency in these nuclei. These fish, however, did not have a normal number of chromosomes.

5. Serially transplanting nuclei of blastula cells into enucleated eggs may enhance the cytoplasmic effects on the nucleus. Such serial transplants may gradually strengthen the expression of the characteristics of the recipient species.

Discussion

To date most studies on nuclear transplantation have used amphibians. Although the results obtained have proved that the nuclei of embryonic amphibian cells possess developmental totipotency in varying degrees, so far no nuclei taken directly from an adult amphibian somatic cell have been proved to possess developmental totipotency (reviewed in Ref. 13). In fishes, results obtained from the experiments of transplantation of cell nuclei from early developmental stage embryos into enucleated recipient eggs duplicate the results of amphibian experiments in terms of developmental totipotency. Chen and his coworkers (6) also reported that the transplantation of the nuclei of short-term in vitro cultured kidney cells of crucian carp into enucleated eggs of the same species successfully produced crucian carp which developed to normal sexual maturity. Therefore the nuclei of somatic cells of adult fish may also need to be cultured in vitro for a short time to restore their capability to divide. Similar results have been observed in amphibians (reviewed in Ref. 13).

In mammals, nuclear transplantation studies have also been undertaken to determine the developmental totipotency of nuclei. In the first such experiment, Illmensee and Hoppe (17) reported that nuclei of inner cell masses of mouse embryos transplanted into the enucleated eggs of a different strain of mice resulted in successful hybrid development. McGrath and Solter (22) also reported that mouse embryonic cell nuclei can be introduced into enucleated mouse eggs by means of cell fusion, and the resulting embryo will develop. More recently, Willadsen (37) reported that complete development of enucleated sheep eggs can be obtained when donor nuclei from 8- or 16-cell embryos are used. These results show that the nuclei of undifferentiated cells in early mammalian embryos also have developmental totipotency. Other nuclear transplantation experiments conducted on mammals have failed to produce viable adults (23, 30). In a recent review, Solter (27) concluded that the ability of the enucleated mouse egg to reprogram the transplanted nucleus may be limited and is unlikely to result in complete development. In addition, the different methods used for nuclear transplantation in mammals may have produced the varying results.

Several aspects of nuclear transplantation in fish are worthy of note:

1. In order for the transplanted nuclei of kidney cells to promote normal development in an enucleated egg, they not only had to undergo short-term culture to restore their ability to divide, but they also had to undergo two successive transplantations. DiBerardino et al. (12) showed in *Rana pipiens* that successive nuclear transplantation could promote the developmental potency of transplanted nuclei in enucleated eggs. Wu et al. (38) showed in goldfish that successive nuclear transplantation could strengthen the expression of characteristics normally found only in the enucleated recipient species, which contributed cytoplasm only. These studies demonstrated the possibility of a cumulative effect of enucleated egg cytoplasm on the donor nucleus, which may act to reprogram the process of gene expression within the nucleus.

2. The one fish obtained by Chen et al. (6) from the transplantation of cultured kidney cells of crucian carp into enucleated eggs of the same species was fertile, even though it had a triploid chromosome set. Moreover, although the nuclei of blastula cells of crucian carp that underwent subcultures for as many as 53 to 59 generations developed heteroploid chromosome sets, the nuclei could nevertheless also be transplanted into an enucleated recipient egg and result in an adult fish of normal appearance. In contrast, in amphibians abnormalities of the chromosome set always cause abnormal development or death of nuclear transplanted eggs (10, 11). How can these results be explained? Many species of fishes exhibit heteroploid chromosome sets under natural conditions (20, 26). Arai (1) and Goodier et al. (15) reported that in salmonid hybrids, some heteroploid individuals were obtained as a result of chromosome fragmentation and loss or because the different haploid nuclei of the eggs and sperm contained chromosome sets that differed in number or type. It seems that, unlike in the case of higher vertebrates, heteroploidy may not preclude normal or nearly normal development in many fishes. The precise influence of heteroploid or triploid chromosome sets on the development of nuclear transplanted fish is not known. Current theories of gene expression and regulation in higher vertebrates do not explain such phenomena in fish. Solter (27) concluded that the presence of paternal and maternal genomes is required for complete development in mice, and suggested that more than the normal number of chromosomes in the genome of the zygote may have unexpected influences on development in some special cases. Obviously, further investigation of this most interesting subject at the cell biologic as well as at the molecular biologic level is necessary.

3. Although there is evidence of developmental totipotency of the nuclei of early embryonic cells in amphibians and fishes, interspecific nuclear transplantation among frogs generally does not result in the recipient eggs developing into adult hybrids (14). However, Kawamura and Nishioka reported (18, 19) that when they transplanted nuclei of blastula cells between different species in the genus *Rana*, normal diploid nucleo-cytoplasmic adult hybrid frogs were obtained. The characteristics of the nucleo-cytoplasmic hybrid frogs in these combinations were identical to those of the nucleus-donor species in most cases. However, some characteristics of several individuals resembled those of the recipient species or were intermediate between the two species. Although some of those hybrid frogs matured, they did not produce fertile sperm or eggs. These results indicate that, in at least one interspecific combination, the developmental totipotency of the nucleus of the amphibian blastula cell is not restricted. The reasons for the different results obtained by Gallein (14) and by Kawamura and Nishioka (18, 19) remain to be clarified.

4. In fish, adult nucleo-cytoplasmic hybrids with normal diploid chromosome sets were obtained through nuclear transplantation between fishes of different genera and even subfamilies. The donor nuclei were taken from blastula cells, and the resulting hybrid zygotes developed to sexual maturity and were able to produce viable germ cells. Some of these hybrid fish had characteristics identical to those of the species of the nucleus donor. Some were identical to those of the species contributing cytoplasm only, and several individuals had intermediate characteristics, or even exhibited some apparent novel characteristics not found in either parental species. Those experiments also demonstrated that there was no species specificity for the developmental totipotency of the nuclei of fish blastula cells because adult hybrid fishes were obtained from even distantly related combinations. These results from experiments with fish are similar to the results obtained in experiments with amphibians (18, 19), except for the stronger compatibility among distantly related species of fishes. Since it is possible to create adult nucleo-cytoplasmic hybrid fish whose parents are from distantly related species, it is easier to identify and separate the effects or influences of the nucleus or of the cytoplasm on the development of the hybrid by means of expressed characteristics.

It appears that the developmental pattern of early embryos from nuclear transplanted eggs is determined in part by functional materials, perhaps maternal mRNAs stored in egg cytoplasm (5). It has also been confirmed that egg cytoplasmic components are of several types; some act merely as sources of energy or as raw materials for growth, while others have a much more specific role, e.g., cytoplasmic polynucleotides (i.e., mitochondrial DNA and maternal RNA) and preformed structures and patterns (24). Tung et al. (32) showed that there are some materials in the cytoplasm of early developing goldfish eggs that move from the vegetal pole toward the animal pole and act in the determination of normal egg development. Regulation of gene expression in the development of nucleo-cytoplasmic hybrid eggs may be related to such cytoplasmic factors. However, the details of how cytoplasmic factors influence the characteristics of the nucleo-cytoplasmic hybrid adult, and whether cytoplasm-derived characteristics can be inherited, remain to be explained.

The quantity of genetic material contained in the chromosomes of higher organisms greatly exceeds the quantity of known genes actually expressed. For example, it is estimated that about 50,000 genes are expressed in humans, and yet the amount of DNA contained in human chromosomes could theoretically involve at least 2 million genes (5). Some of this genetic material is involved in the regulation of structural genes or represents multiple copies of certain genes, but the identity of the remaining unexpressed portion of the genome is unknown. On the other hand, we also know that the vertebrate immune system reveals a remarkable capacity for gene rearrangement, which increases the coding capacity of the genome manyfold (16). In addition, there is evidence from other eukaryotes that gene rearrangement is possible, and there are indications that rearrangement of structural genes with respect to controlling sequences may take place during normal cytodifferentiation in vertebrates (24). Therefore we may speculate that in the chromosomes of animals, including fish, in addition to the genes which specify the animal's own characteristics, there may be related genes conserved during the evolutionary process yet not presently expressed. When the nucleus of one kind of animal is placed in a heterospecific cytoplasm, the incompatibility and interaction between the foreign nucleus and the cytoplasm may reprogram or rearrange the action of the normally expressed genes to produce either normal or abnormal development in the enucleated egg. On the other hand, the stimulus of heterospecific cytoplasm may initiate the expression of these "silent genes" of the nucleus, which were not active before. Moreover, in the nuclear transplanted eggs, because of the differences between the nucleus and the cytoplasm and because of the heterogeneity of the individual enucleated recipient eggs, the differences in developmental potency and in the expression of their hereditary characteristics could be revealed in the nucleo-cytoplasmic hybrids of different species combinations or could appear among different individuals from the same combination. The existence and possible rearrangement of such "silent genes" has profound theoretical consequences for the process of evolution and raises the possibility of using nuclear transplantation as a method of probing the evolutionary history of a particular species. Certainly, this idea deserves further investigation.

It may be possible to obtain superior clones of fishes for agriculture from nucleo-cytoplasmic hybrids of distantly related species with remote affinities. This goal, however, faces the following problems:

1. The probability of obtaining a nucleo-cytoplasmic hybrid fish by taking the donor nucleus and enucleated recipient cell from the same blastula is small (mainly owing to technical difficulties).

2. Nuclear transplanted fish derived from nuclei from a single blastula will all be of the same sex, making a self-reproducing cloned fish line impossible.

3. It has yet to be shown that nucleo-cytoplasmic hybrids will breed true in every combination. The influence of the donor nucleus (i.e., the original regulative mechanism of its gene expression system) may be eventually recovered, since the genomic material will be replicated in each subsequent generation. The cytoplasm has no such mechanism, and its influence may be correspondingly diluted.

In summary, it is common knowledge that all cells of an early developing embryo contain identical genes, which must be affected from outside the nucleus during the course of embryonic development and cell differentiation. A great deal of evidence has accumulated that the cytoplasm which surrounds the nucleus affects the expression of genes in the nucleus. Wilson (reviewed in Ref. 5) has described the cell as a "reaction system," in which the nucleus and the cytoplasm each play their own role in the determination of the cell's characteristics. This view has been confirmed by nuclear transplantation experiments in fish. The great challenge of developmental biology is to understand how the nucleus and the cytoplasm interact.

APPENDIX
Some Special Features and Behavior of Fish Eggs Cultured In Vitro

Eggs of freshwater teleosts—mainly the crucian carp, goldfish (the domestic strain of the same species as the wild-type crucian carp), common carp, bitterling, grass carp, and blunt-snout bream of the order Cypriniformes, and the tilapia of the order Perciformes—have been used as experimental materials in our laboratory. Eggs of these fishes show the following properties which affect their culture and nuclear transplantation operation.

1. Activation of fish eggs differs from that of amphibian eggs. The eggs of these fishes, once ovulated and in contact with water, are spontaneously activated (31). Thereafter they cannot be fertilized. Successful artificial insemination can be obtained in two ways:

   a. The matured eggs are first squeezed into a clean and dry container to maintain them without activation; the sperm of the male fish are then squeezed out and gently mixed with the eggs. Finally, the now fertile eggs are rinsed with clean water.

   b. The sperm are first squeezed from the male fish into a container of clean water, which is shaken to disperse the sperm. Then the mature eggs are immediately squeezed from the female fish into the container with the sperm. The eggs will be activated almost synchronously with fertilization.

   Unfertilized eggs can be obtained simply by squeezing the mature eggs from female fish into clean water and waiting for spontaneous activation.

2. Eggs of most fishes differ from the eggs of amphibians in that the former generally lack pigment and are semitransparent (the yolk in several fishes is a light yellowish color). This lack of pigmentation makes it difficult to distinguish the animal and vegetal poles just before or after fertilization and prior to blastoderm formation. Moreover, unlike the eggs of amphibians, the eggs of most fishes do not necessarily rotate after activation, and the animal pole is not predictably oriented upward. The difficulty in distinguishing the animal and vegetal poles, along with the random orientation of the egg, increases the difficulty of the nuclear transplantation operation.

3. There is a period of about 1 to 3 h following final maturation of the oocytes (either naturally matured or artificially induced by hormone treatments) during which viable eggs may be produced. Within this period eggs can be successfully spawned at the convenience of the experimenter.

4. Several factors may affect egg quality and thereby affect the success of nuclear transplantation. Most fishes spawn during a well-defined spawning season. Females in their first reproductive season following sexual maturity usually produce eggs with low rates of fertility. The experience of our laboratory indicates that eggs obtained from females during the second reproductive season following sexual maturation are best for nuclear transplantation experiments. Other factors that may contribute to low fertility rates are incompletely mature or overripeeggs, which may result from artificial induction of sexual maturity by hormone treatments.

The quality of the eggs can be determined by observing their color and appearance through a microscope and by testing their elasticity using a soft hair loop. Sampling a batch of eggs to determine the fertility rate is helpful. For example, a small quantity of eggs can be artificially inseminated. If many of the eggs prove to be fertile, the remaining eggs from the same spawning fish can be used for nuclear transplantation.

5. Eggs derived from some kinds of fishes contain oil droplets. Such eggs will float under the water's surface rather than sink to the bottom of the container, causing practical difficulties during the nuclear transplantation operation. For this reason, we used only fish eggs without oil droplets. Those eggs may be classified into two kinds:

a. Sticky eggs. Eggs such as those from the carp (*Cyprinus carpio*) are sticky. The capsule of the egg sticks to the bottom surface of a container when the eggs are squeezed into it.

b. Nonsticky eggs. When the eggs are squeezed into the water, they remain suspended (e.g., the eggs of the grass carp, *Ctenopharyngodon idellus*) because the capsule is not sticky and because the two-layered capsule soon becomes swollen. Such eggs will sink to the bottom of the container immediately after their capsules are removed. Another kind of nonsticky egg is ellipsoidal in shape, such as eggs from bitterling and tilapia. It has only a single layer and a closely enveloping capsule, and it is rich in yolk. These eggs are inadequate for use as the enucleated recipient egg because they lose their elasticity and become flat in shape after the enveloping capsule is removed and because they frequently exhibit abnormal development. However, the blastula cells of embryos from species with such eggs may be used as nucleus donors.

Nuclear Transplantation Procedures

The diameter of the common carp (*Cyprinus carpio*) egg is generally around 1 mm. Nuclear transplantation is carried out with a stereomicroscope (X50 to X100). The optimum temperature for the operation is 18 to 20°C. This temperature is most suitable for the normal development of the various kinds of fish eggs used in our laboratory and is also close to the water temperatures typically found during the optimal spawning seasons of those fishes. Standard amphibian Holtfreter's solution is used as an operation medium.

The following procedures must be followed during nuclear transplantation:

1. During the operation the fish eggs are placed on the bottom of a petri dish, which is covered with a layer of 1.2 percent agar to prevent friction between the egg surface and the bottom of container and possible injury to the eggs after their capsules are manually removed with a pair of forceps.

2. After activation of the fish egg, the cytoplasm coalesces and then

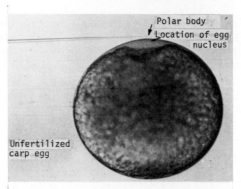

FIGURE 4. The unfertilized carp egg (*Cyprinus carpio*) showing the locations of polar body and nucleus.

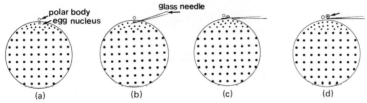

FIGURE 5. Enucleation of a fish egg. (*a*) Lateral view showing the locations of polar body and nucleus. (*b*) A glass needle is inserted into the egg cytoplasm at the location of the nucleus. (*c*) The nucleus is removed with the glass needle. (*d*) The nucleus is separated from the egg.

moves toward the animal pole, where it forms the blastoderm, which constitutes approximately one-sixth of the total volume of the whole egg. Formation of the blastoderm takes about 30 min. At this time, the fertilized egg is said to be in the 1-cell stage. After about 10 min, the blastoderm starts to cleave. Thereafter, egg cleavage occurs about once every 15 min. Nuclear transplantation must be carried out prior to completion of the formation of the blastoderm. It is best to complete the transplant operation within 10 to 15 min after activation of the egg.

3. Because the orientation of spawned eggs in water is random, it is necessary to assure the correct orientation of each egg by turning it over with a soft hair loop to determine the position of the egg nucleus (see Fig. 4). The nucleus is plucked out with a fine glass needle (Fig. 5). Attempts to destroy the nucleus in situ by irradiation with ultraviolet or laser light, as has been done in amphibians, have generally proved difficult in fish.

Acknowledgments

The author wishes to thank Professor Huang Gefang of the Institute of Developmental Biology, The Chinese Academy of Sciences, Beijing, China, for his help in preparing this manuscript and Dr. Thomas Hourigan, Department of Zoology, University of Hawaii, for his valuable comments and advice. Thanks are also given to Professor Yoshitaka Nagahama, who read this manuscript at the National Institute for Basic Biology, Okazaki, Japan, and to Susan Duhon (Indiana University Axolott Colony) for editorial assistance. This research project was supported by research grants from the Chinese Academy of Sciences, the Chinese Natural Science Foundation, the Chinese Science and Technology Committee, the Rockefeller Foundation Grants RF79008 and RF84031, and the United Nation for Population Activities Grant CPR/80/P08.

General References

BRIGGS, R.: 1979. Genetics of cell type determination. *Int. Rev. Cytol.* (Suppl.) **9**:107–125.

DIBERARDINO, M. A.: 1987. Genomic potential of differentiated cells analyzed by nuclear transplantation. *Am. Zool.* **27**:623–644.

References

1. Arai, K. 1984. Developmental genetics studies on salmonids: morphogenesis, isozyme phenotypes and chromosomes in hybrid embryos. *Mem. Fac. Fish. Hokkaido Univ.* **31**:1–94.

2. Briggs, R., and King, T. J. 1952. Transplantation of living nuclei from blastula cells into enucleated frogs' eggs. *Proc. Natl. Acad. Sci. U.S.A.* **38**:455–463.

3. Briggs, R., and King, T. J. 1953. Factors affecting the transplantability of nuclei of frog embryonic cells. *J. Exp. Zool.* **122**:485–505.

4. Briggs, R. 1979. Genetics of cell type determination. *Int. Rev. Cytol.* (Suppl.) **9**:107–125.

5. Browder, L. W. 1980. *Developmental Biology.* Saunders College, Philadelphia.

6. Chen, H., Yi, Y., Chen, M., and Yang, X. 1986. Studies on the developmental potentiality of cultured cell nuclei of fish. *Acta Hydrobiol.* **10**(1):1–7.

7. Commandon, J., and de Fonbrune, F. 1939. Greffe nucleair total, simple ou multiple, chey une amibe. *C. R. Soc. Biol.* (Paris) **130**:744–748.

8. Danielli, J. F. 1958. Cellular inheritance as studied by nuclear transfer in amoebae. In: *New Approaches in Cell Biology.* pp. 15–22, Academic Press, London.

9. Danielli, J. F. 1959. Some theoretical aspects of nucleo-plasmic relationships. *Exp. Cell Res.* (Suppl.) **6**:252–267.

10. DiBerardino, M. A., and King, T. J. 1967. Development and cellular differentiation of neural nuclear-transplants of known karyotype. *Dev. Biol.* **15**:102–128.

11. DiBerardino, M. A., and Hoffner, N. J. 1971. Development and chromosomal constitution of nuclear transplants derived from male germ cells. *J. Exp. Zool.* **175**:61–72.

12. DiBerardino, M. A., Hoffner, N. J., and Mckinnell, R. G. 1986. Feeding tadpoles cloned from *Rana* erythrocyte nuclei. *Proc. Natl. Acad. Sci. U.S.A.* **83**:8231–8234.

13. DiBerardino, M. A. 1987. Genomic potential of differentiated cells analyzed by nuclear transplantation. *Am. Zool.* **27**:623–644.

14. Gallein, C. L. 1979. Expression of nuclear and cytoplasmic factors in ontogenesis of amphibian nucleocytoplasmic hybrids. *Int. Rev. Cytoplasm.* In: Progress in Developmental Biology, Part A. pp. 35–38, Alan R. Liss, Inc. New York.

15. Goodier, J., Hai-Fei, M., and Fumio, Y. 1987. Chromosome fragmentation and loss in two salmonid hybrid. *Bull. Fac. Fish. Hokkaido Univ.* **38**(3):181–184.

16. Gough, N. 1981. The rearrangement of immunoglobulin genes. *Trends Biochem. Sci.* **6**(8):203.

17. Illmensee, K. and Hoppe, P. C. 1981. Nuclear transplantation in *Mus musculus*: Developmental potential of nuclei from preimplantation embryos. *Cell* **23**:9–18.

18. Kawamura, T. and Nishioka, M. J. 1963. Reciprocal diploid nucleo-cytoplasmic hybrids between two species of Japanese pond frogs and their offspring. *J. Sci. Hiroshima Univ. Ser. B Div.* 1 (*Zool.*) **21**:65–84.

19. Kawamura, T. and Nishioka, M. J. 1963. Nucleo-cytoplasmic hybrid frog between two species of Japanese brown frog and their offspring. *J. Sci. Hiroshima Univ. Ser. B Div.* 1 (*Zool.*) **21**:107–134.

20. Kosswig, C. 1973. The role of fish in research on genetics and evolution. In: *Genetics and Mutagenesis of Fish.* (J. H. Schroder, ed.), pp. 3–16, Springer-Verlag, New York.

21. Lu, T., Shih, Y., and Yan, S. 1984. Animal models for cloning. In: *Human Fertility, Health and Food. Impact of Molecular Biology and Biotechnology.* (D. Puett, ed.), pp. 241–250, UNFPA, New York.

22. McGrath, J. and Solter, D. 1983. Nuclear transplantation in the mouse embryo by microsurgery and cell fusion. *Science* **220**:1300.

23. McGrath, J. and Solter, D. 1984. Inability of mouse blastomere nuclei transferred to enucleated zygotes to support development *in vitro*. *Science* **226**:1317–1319.

24. Pritchard, D. J. 1986. *Foundations of Developmental Genetics*, p. 68, Taylor and Francis, Ltd., New York.

25. Roux, W. 1883. *Über die Bedeutung der Kerntheilungsfiguren.* Engelmann, Leipzig.

26. Schaltz, R. J. 1980. Role of polyploidy in the evolution of fishes. In: *Polyploidy: Biological Relevance.* (W. H. Lewis, ed.), pp. 313–340, Plenum, New York.

27. Solter, D. 1988. Nuclear transfer in mammalian embryos: Role of paternal, maternal and embryonic genome in development. In: *Developmental Genetics of Higher Organisms.* (G. M. Malacinski, ed.), pp. 441–457, Macmillan Publishing Company, New York.

28. Spemann, H. 1914. Über verzogerte Kernversorgung von Keimteilen. *Verh. Dtsch. Zool. Ges.* 00:216–221.

29. Spemann, H. 1943. *Embryonic Development and Induction.* Hafner Publishing Company, New York.

30. Surani, M. A. H., Barton, S. C., and Norris, M. L. 1986. Nuclear transplantation in the mouse: Heritable differences between parental genomes after activation of the embryonic genome. *Cell* **45**:127–136.

31. Tchou-Su. 1937. Activation spontaneous de l'oeuf du poisson rouge (*Carassius auratus* L.) au contact de l'eau douce. *C. R. Acad. Sci.* (Paris) **204**(22):1676–1677.

32. Tung, T. C., Chang, C. Y., and Tung, Y. F. Y. 1945. Experiments on the develop-

mental potencies of blastoderm and fragment of teleostean eggs separated latitudinally. *Proc. Zool. Soc. Lond.* **115**:175–188.

33. Tung, T. C., Wu, S. C., Tung, Y. F. Y., Yan, Y. S., Tu, M., and Lu, T. Y. 1963. Nuclear transplantation in fishes. *Scientia (Peking)* **14**(8):1244–1245.

34. Tung, T. C., Tung, Y. F. Y., Lu, T. Y., Tung, S. M., and Tu, M. 1973. Transplantation of nuclei between two subfamilies of teleosts (goldfish—domesticated *Carassius auratus*, and Chinese bitterling, *Rhodeus sinensis*). *Acta Zool. Sin.* **19**(3):201–212.

35. Tung, T. C. 1980. Nuclear transplantation in teleosts. I. Hybrid fish from the nucleus of carp and the cytoplasm of crucian. *Scientia (Peking)* **23**(4):517–523.

36. Weismann, A. 1892. *Das Keimplasma.* [Translated by Parker and Rounnfeldt, 1898. Scribner, New York.]

37. Willadsen, S. M. 1986. Nuclear transplantation in sheep embryos. *Nature (Lond.)* **320**:63–65.

38. Wu, S., Cai, N., and Xu, Q. 1980. Nuclear transplantation for several generations between different varieties of goldfish, *Carassius auratus*. *Acta Biol. Exp. Sin.* **13**(1):65–74.

39. Yan, S., Lu, D., Du, M., and Li, G. 1984. Nuclear transplantation in teleosts. II. Hybrid fish from the nucleus of crucian and the cytoplasm of carp. *Scientia (Peking)* *27*(8):729–732.

40. Yan, S., Wu, N., Yan, J., Xue, G., and Li, G. 1984. Some evidence of cytoplasmic influences on the gene expression of hybrid fish *CtMe* obtained from the combination of nucleus and cytoplasm from two subfamilies of fresh water teleosts, *Ctenopharygoden idellus* (Ct) and *Megalobrama amblycephala* (Me). *J. Embryo. Exp. Morphol.* **82**(Suppl.):105.

41. Yan, S., Lu, D., Du, M., Li, G., Han, Z., Yang, H., and Wu, Z. 1985. Nuclear transplantation in teleosts. Nuclear transplantation between different subfamilies—hybrid fish from the nucleus of grass carp (*Ctenopharyngodon idellus*) and the cytoplasm of blunt-snout bream (*Megalobrama amblycaephala*). *Chin. Biotech.* **1**(4):15–26.

42. Yan, S., Du, M., Wu, N., Yan, J., Jin, G., Qin, Y., and Zhang, X. 1986. Identification of intergeneric nucleo-cytoplasmic hybrid fish obtained from the combination of carp nucleus and crucian cytoplasm. In: *Progress in Developmental Biology*, Part A. pp. 35–38, Alan R. Liss, Inc. New York.

Questions for Discussion with the Editor

1. *What are the prospects for employing distinct nuclear markers for tracing the fate of the transplanted nucleus? Without such markers won't a measure of skepticism always surround claims about viable hybrids?*

As a matter of fact, the nuclear markers are necessary for identifying the true viable hybrids obtained from the combinations of nuclei and cytoplasm of different origins using the method of nuclear transplantation. There are many kinds of fish, for instance, which belong to different varieties, species, genera, or even more remotely related groups, but have the same number of chromosomes. In addition, it is also very difficult to distinguish viable hybrids on the basis of chromosome patterns because of the large number of chromosomes in many species and the lack of proper methods for such kinds of identifications.

For example, the common carp and crucian carp belong to different genera, but have the same number of chromosomes, i.e., $2n = 100$ in both fish. In this case, how can we be sure that the nucleus of the nucleo-cytoplasmic hybrid fish obtained from the combination of the above two kinds of fish really developed from the transplanted nucleus? Since there is no nuclear marker, the possibility that the recipient egg nucleus remained in the "enucleated" egg cytoplasm cannot be excluded, especially if the operations were done by unskillful experimenters. Fortunately, in our experiments we solved this question on the basis of the following facts: Whenever the morphological characteristics of the nucleus-donor-type fish and of the enucleated-recipient-type fish—i.e., common carp and crucian carp or vice versa— are distinctly different, we can be confident that the fish is a true nucleo-cytoplasmic hybrid if we can identify the morphological characteristics of the nucleo-cytoplasmic hybrid fish as mainly similar to those of nucleus-donor-type fish and the nucleus itself as diploid. Of course, if there are no remarkable morphological characteristic differences between the nucleus-donor-type fish and the enucleated-recipient-type fish used in the nuclear transplantation experiments, it is very difficult for the experimenters to confirm that the nucleus of the hybrid is truly from the transplanted nucleus. This is why we prefer to use remotely related species of fish with obviously different morphological characteristics as experimental materials in our nuclear transplantation experiments. The results we have obtained are quite clear.

2. *For how many generations can the "intermediate" characteristics displayed by some fish hybrids be perpetuated? Can you propose a mechanism to account for the stability of traits acquired from the egg cytoplasm?*

So far there is only one example of a nucleo-cytoplasmic hybrid fish (a combination of a common carp nucleus and crucian carp cytoplasm) which has produced offspring to four generations. The most obvious "intermediate" morphological characteristic that appeared in this nucleo-cytoplasmic hybrid (the number of scales along the lateral line) seems to be passed down. Therefore, such "intermediate" characteristics could be perpetuated. However, more generations of these nucleo-cytoplasmic hybrid fish should be observed before we can draw any final conclusions. It is rather difficult for the author to propose a proper mechanism to account for the stability of such a trait acquired from the egg cytoplasm right now. It is possible that the modified characteristics of the nucleo-cytoplasmic hybrid fish that we have obtained in our experiments are due to the interactions between the nucleus and the cytoplasm, rather than to a direct effect from the cytoplasm itself. As has been mentioned in the text of this chapter, we can only speculate that the modified characteristics of the nucleo-cytoplasmic hybrid fish are the result of the stimulation of the expression of "silent genes" in the nucleus and that this kind of gene expression can be observed only when nucleus and cytoplasm of different origins is combined. If this mechanism is correct, it should not be a problem to believe that the modified characteristics could become perpetuated.

CHAPTER 5

The Roles of Cell Lineage and Cell Interactions in the Determination of Cellular Fates in Vertebrate Embryos

Claudio D. Stern

Introduction

DURING THE DEVELOPMENT of any multicellular animal or plant, two types of tasks must be accomplished: the production of the correct cell types and their appropriate assembly in space. An organism or organ system can adopt one of two strategies to carry out these tasks (Fig. 1): (1) cells can become committed to their various fates irrespective of their position within the embryo (for example, by their previous lineage history), and then migrate to their correct sites, die, or sort out (Fig. 1a) or (2) cells may become determined at their final sites by local interactions, that is, as a result of their position within the embryo (Fig. 1b). There are advantages and disadvantages associated with both methods of generating cellular diversity. For example, while the second method requires less cell movement, no cell death, and little or no cell sorting, it is more difficult to produce the correct proportions of two or more different cell types in this manner.

If cell diversity is set up prior to and independently of cell interactions (in other words, if development is *mosaic*), there must be some extranuclear organization in the cytoplasm, cortex, or cell membrane that creates diversity upon cell division. It is this organization that constitutes the theme of this book.

To understand the relationship between cell diversity and pattern, it is important to recognize the difference between *fate* (the cell types *normally*

Cell diversity and pattern formation: two different strategies

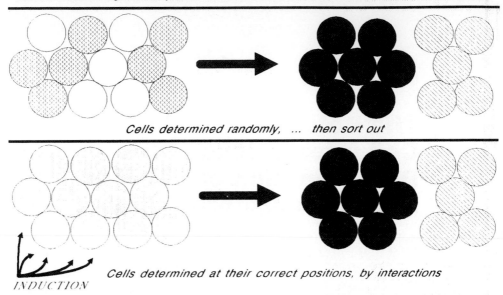

Cells determined randomly, ... then sort out

INDUCTION *Cells determined at their correct positions, by interactions*

FIGURE 1. Two strategies by which cell diversity can be generated in relation to the final pattern. (*a*) Cells are assigned different fates regardless of their position within the embryo, for example by their lineage history. The correct pattern is then generated by cell movements, cell sorting (differential adhesion?), or by differential survival of the two cell types at different locations. (*b*) All cells have the potential to become determined as either of the two cell types. Fates are allocated as a result of interactions of the cells with their microenvironment. The second strategy can be described as an "instructive induction," while the first may be viewed as a "permissive induction."

generated by a particular progenitor cell) and *developmental potential* (the *ability* of a progenitor cell to give rise to any given set of cellular phenotypes). The difference is illustrated by considering an experimental example: The *fate* of each of the two blastomeres of a 2-cell-stage amphibian embryo is to give rise to one-half of the adult body. However, each has the *potential* to give rise to a whole embryo. Each cell is able to realize its potential if the two blastomeres are separated carefully (see Refs. 28, 30). In fact, all four blastomeres generated by the second cleavage division are totipotent in terms of the cell types that they are able to generate. After the third cleavage, however, this is no longer the case: the *developmental potential* of the four animal blastomeres is no longer the same as that of the vegetal four. Commitment to a particular fate (or *determination*) occurs when the developmental potential of a cell no longer differs from its fate.

In this chapter I shall review the relationships between the generation of cell diversity and the formation of pattern during two important steps in the building of the vertebrate body plan: (1) gastrulation and (2) the laying down of segmental organization. Using these examples, I will argue that both the position-independent and local-interaction strategies for cell diversity and pattern formation are used, even in the same species and during the development of the same organ system.

Gastrulation

Lewis Wolpert (30) has stated the importance of gastrulation very succinctly: "it is not birth, marriage or death, but gastrulation which is truly the most important time in your life." Indeed, gastrulation is the period of early embryonic development in which the third germ layer, or *mesoderm*, arises as a distinct tissue. From the mesoderm will arise the skeleton, the muscle, and many of the internal organs of the adult organism. The segmental pattern of somites that develops in the mesoderm also dictates the pattern of some structures that do not derive from the mesoderm, such as the peripheral nervous system.

Like the rest of embryonic development, gastrulation consists of a series of processes that fall into two major categories: cytodifferentiation and morphogenesis. It is commonly assumed that morphogenesis precedes and is required for the allocation of cell fates (30). Thus, cell diversity is thought to result from geometry: the morphogenetic movements of gastrulation are required to bring certain tissues together in the embryo, and the interactions ("induction") between these tissues are believed to influence the fate of the "responding" or "competent" cells.

The process of gastrulation therefore represents an excellent model system in which the relationships between morphogenesis and cell diversification can be studied. In the first part of this chapter I will survey the classical views of chick gastrulation and present some data which may question the accepted beliefs. I will argue that the evidence in favor of geometry as the causative force for cell diversification is not as strong as is generally assumed, at least in birds.

Polarity and Symmetry-Breaking

In order for morphogenesis to take place in an orderly fashion, the geometry of the embryo must be organized prior to the onset of major morphogenetic movements. Just prior to gastrulation, the chick embryo is a flat disc, about 2 mm in diameter, with two concentric regions: a central *area pellucida*, and a peripheral *area opaca*, each of which is two-layered. The *epiblast*, a pseudostratified epithelium, is continuous over both regions of the embryonic disc. The ventral layer of both regions consists of large, yolky cells. In the *area pellucida* this layer is known as the hypoblast. In addition to generating a third germ layer (*mesoderm*), the process of gastrulation must break the initial radial symmetry of the disc and generate a bilaterally symmetrical embryo.

DORSOVENTRAL POLARITY OF THE EPIBLAST

The cells of the epiblast are polarized along their apical-basal (i.e., dorsoventral) axis. Sodium (43) and water (44) are transported from apical to basal aspects, and this unidirectional transport generates a transepithelial potential of some 25 mV (basal side positive; Refs. 14, 43). Like all transporting epithelia, the epiblast is also polarized morphologically. The polar features include apical intercellular junctions and some apical microvilli, basal nuclei, and a hyaluronate-rich basal lamina. The apical-basal polarity of the epiblast is labile; it can be reversed quickly by applying a transepithelial potential of opposite polarity to that measured across it (35 mV, apical side positive; 43). It can also be reversed by placing the epiblast in a pH gradient so that the apical side is about 3 pH units more acid than the basal side (unpublished observations).

How is the apical-basal polarity of the epiblast set up and maintained? In freshly laid eggs, the albumen that bathes the apical aspect of the epiblast is strongly alkaline (pH 9.5), while the subblastodermic fluid is slightly acidic (pH 6.5). Since a pH gradient of 3 units is sufficient to reverse the polarity of the epiblast experimentally, it seems likely that this asymmetry, set up by the mother, plays a role in determining the polarity of the epiblast. As the epiblast develops, the transepithelial potential generated could also serve to maintain the polarity of the epiblast. The primitive streak region appears to be a zone where the apical-basal polarity of the epiblast is reversed, or at least disturbed: ionic currents from the interior of the embryo escape through it (14, 43, 44).

CRANIOCAUDAL POLARITY OF THE EMBRYO: BREAKING RADIAL SYMMETRY

How is bilateral symmetry set up in the embryonic disc? It has been suggested that craniocaudal polarity is established during the descent of the egg in the oviduct, under the influence of gravity (21). The earliest manifestation of this polarity can be observed during the formation of the hypoblast, which coalesces into a continuous sheet of cells starting at the caudal end: the craniocaudal midline of the hypoblast sheet marks the future midline of the embryo. Waddington (49) was the first to suggest that it is the hypoblast that induces the formation of the primitive streak, since rotation of the hypoblast through 180° at the appropriate stage of development results in 180° reversal of the craniocaudal axis of the embryo. Other evidence (26), however, suggests that both the epiblast and the hypoblast have their own craniocaudal polarity. This view is based on the results of experiments of reaggregation of dissociated hypoblast and epiblast: When the hypoblast is dissociated and combined with an intact epiblast, the polarity of the epiblast dictates the orientation of the future craniocaudal axis of the embryo, and vice versa (26).

What determines the origin and the shape of the primitive streak? It has been suggested (1, 35) that the primitive streak arises preferentially at the margin between the *area pellucida* and *area opaca* (the region called the "marginal zone"; (1, 19); this region is undoubtedly special in some way. The reason for the rodlike appearance of the primitive streak is unclear, but it is likely that two major forces contribute to determine its shape: the tension generated by the expansion of the blastoderm on the vitelline membrane, and some change in the shape, arrangement, and rate of proliferation of the cells in the epiblast portion of the primitive streak (4, 36, 38).

Morphogenesis

To date, the information available about the origin, movements, and subsequent development of each of the three germ layers is based entirely on direct observation of embryos at the appropriate stages of development, a few simple grafting experiments, and a few observations using time-lapse cinephotomicrography.

Briefly, the stages in the development of the chick embryo in the 36 h after the egg is laid are as follows (Fig. 2): At stage X, which corresponds to the time of laying, the embryo is a flat disc, where the central *area pellucida* is essentially single-layered. The early hypoblast at this stage is no more than a series of separate islands of a few cells each. By stage XII, about 4 h later, the hypoblast is starting to form a sheet from the future caudal end of the embryo (bottom of the diagram). At stage XIV, about 7 h after laying, the hypoblast sheet is complete. Shortly afterward (stage 2) the primitive streak

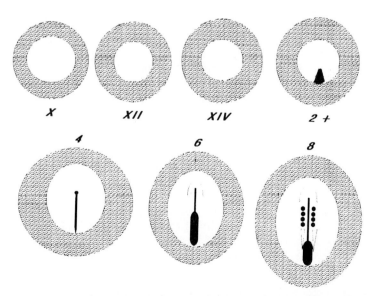

FIGURE 2. Diagram illustrating the stages of development of the chick embryo in the 36 h after the egg is laid. Stages denoted by Roman numerals (I–XIV) apply to the earlier stages of development, and follow the staging system described by Eyal-Giladi and Kochav (7), while those marked by Arabic numerals (2–45) correspond to the later stages of development and follow the system of Hamburger and Hamilton (11).

primordium becomes visible at the caudal end. This primitive streak then elongates until, by about 14 h of incubation, it reaches its full length, about two-thirds of the craniocaudal extent of the *area pellucida* (stage 4). The knob at the cranial end of the streak is "Hensen's node," which is responsible for inducing the neural plate in the overlying epiblast and which contains presumptive notochord cells. The notochord (head process) then elongates as a rod from the cranial end of the primitive streak, which has now started to shorten (regression). At stage 6, about 22 h into the incubation period, the regions around the cranial tip of the head process start to fold ventrally to produce the head fold, which starts the formation of the foregut. By about 26 to 30 h the first somites have started to appear on either side of the embryonic axis and thereafter continue to form toward the caudal end of the embryo.

MORPHOGENESIS OF THE LOWER LAYER

The earliest tissue to cover the lower layer is the hypoblast (Fig. 2). Its origin has been the subject of some controversy: one group of workers (8) argues that it arises by ingression of cells from the epiblast at many different sites, while others (40, 48) have suggested that it derives mainly from the caudal margin of the germ wall (lower layer of the *area opaca*). However, both views may be correct (see Ref. 28). The islands of hypoblast seen at very early stages (Fig. 2, stage X) appear to arise in situ, but the coalescence of the hypoblast into a sheet of cells takes place by two mechanisms: (1) the addition of cells derived from the caudal germ wall and (2) the spreading and joining of cells in the primitive islands. Thus, the primitive lower layer may be of mixed origin. The islands seen in young stages may constitute a *primary* hypoblast, while the *secondary* hypoblast that completes the primitive lower layer may be derived from the caudal margin of the germ wall (19, 48). This secondary hypoblast is said to be responsible for inducing the primitive streak (27).

During gastrulation, the hypoblast is gradually displaced by the appearance of the "definitive," or "gut," endoderm, which is derived from the cranial portion of the primitive streak, and by cells continuing to migrate centrally from the marginal germ wall (4, 40). The original hypoblast therefore becomes confined to a region close to the cranial *area pellucida–area opaca* margin by the end of gastrulation (full primitive streak stage; Fig. 2, stage 4; Ref. 11), forming a region known as the "germinal crescent" because the primordial germ cells are associated with it. The hypoblast itself does not contribute to the embryo proper; it gives rise only to the endoderm of the yolk sac stalk.

MORPHOGENESIS OF THE UPPER LAYER AND FORMATION OF THE PRIMITIVE STREAK

The upper layer (epiblast) undergoes complex cellular movements (see Refs. 28, 48 for reviews). Some of these movements are due to the expansion of the blastoderm on the vitelline membrane, while others are related to the change in shape that the blastoderm undergoes during gastrulation, from a circle to a pear shape. Around the periphery of the *area pellucida*, the epiblast cells move centrifugally, while near the primitive streak they move toward the midline. High-power time-lapse observations (unpublished) of the epiblast at stages X to XIV (7) (Fig. 2) show that there is considerable mixing of cells in all regions of the epiblast, which is perhaps surprising in a polarized epithelium containing intercellular tight junctions. The movement of the cells toward the axis of the primitive streak, however, does not appear to be related to the appearance of the mesoderm. Neither Vakaet's nor my own time-lapse observations (both unpublished, but see Ref. 48) lend any support to the widely accepted view that gastrulation consists of a convergence of cells to the primitive streak *accompanied* by a sheetlike involution of the epiblast into it to form the mesoderm. When a visible "groove" forms in the primitive streak, the cells lining it are elongated craniocaudally and do not move. Close to the end of gastrulation, while cells continue to converge toward the midline, no movement of cells into the primitive streak region can be seen.

MORPHOGENESIS OF THE MIDDLE LAYER

The appearance of the primitive streak is a remarkably rapid process. It is unusual to find a true stage 2 (11) embryo (Fig. 2), and in time-lapse films the formation of the primitive streak can be seen only "in retrospect," by projecting the film backward. This suggests that formation of the primitive streak is not a massive ingression of presumptive mesoderm cells but rather represents the coalescence of cells that were already under the surface of the epiblast (48).

The middle layer arises from the caudal portion of the primitive streak. *After* the primitive streak has formed, middle-layer cells migrate out of it to give rise to the lateral plate. Time-lapse films show that this happens at Hamburger and Hamilton's stage 3^+ (Fig. 2) at about the same time that the groove appears in the primitive streak. Before formation of the lateral plate, the mesoderm is packed densely at the primitive streak; as the lateral plate forms, it migrates massively away from the axis of the streak. The left and right halves of the lateral plate later become separated from each other by the regression (shortening) of the primitive streak that occurs after the end of gastrulation (38). The notochord is laid down as a rod of mesoderm by the cranial tip of the primitive streak (Hensen's node) and elongates as the primitive streak regresses. The mesoderm at the primitive streak displays an elevated level of hyaluronidase activity (37), and this enzyme may degrade the overlying basal lamina and, thereby, encourage more mesoderm cells to ingress (36).

Induction and Cytodifferentiation

From the foregoing discussion, it can be seen that interactions between cells and their environment play an important role in establishing the overall pattern of the early embryo. We must now consider the mechanisms that lead to the production of cell diversity. It has long been assumed that the mesoderm arises as the result of cell interactions. In the discussion that follows, I shall question whether current knowledge about gastrulation and induction supports this assumption.

THE APPEARANCE OF CELL DIVERSITY

By the time of laying (about stage X of Eyal-Giladi and Kochav, Ref. 7; Fig. 2), several distinct cell types are already recognizable by morphological criteria in the chick embryo. The epiblast consists mostly of small, columnar, polarized epithelial cells; the ventral surface of the embryo displays islands of hypoblast cells, which are larger and more yolky; and the germ wall has even larger and more yolky cells. In addition, the epiblast of the *area opaca* differs from that of the *area pellucida* in that the cells of the former are smaller and more cuboidal. During gastrulation, other cell types appear. The mesoderm of the primitive streak is a mesenchymal tissue, with small, fibroblastic, nonyolky cells. Initially these cells are packed tightly at the primitive streak, but later they migrate to give rise to the lateral and segmental plates and to the notochord. The notochord cells later become very vacuolated. The definitive (gut) endoderm also makes its appearance during gastrulation; it consists of flat cells that are more tightly adherent to one another than are those of the hypoblast.

Although these morphological differences help somewhat in understanding the origin and relations between the discernible cell types, they are not sufficiently well defined to represent good markers for these cell types. Moreover, lack of morphological differences does not necessarily indicate that the cells of tissues that look uniform are the same as one another.

MESODERM INDUCTION

According to several authors (28, 30) there are at least two distinct inductions that occur during the early development of amniote embryos. The first is induction of the mesoderm, which, in birds, is claimed to be the result of an interaction between the hypoblast (inducer) and the epiblast (competent ectoderm). The second is neural induction, which was first described in the amphibian embryo and for which Hans Spemann received the Nobel prize for Physiology and Medicine in 1935 (33). Here, the mesoderm is the inducer and the ectoderm the responding tissue. In birds, neural induction results from the interaction between the mesoderm of Hensen's node and notochord with the overlying epiblast to form the neural plate. In the rest of this discussion, we shall concern ourselves with induction of the mesoderm.

In *Xenopus laevis*, the cells that give rise to the mesoderm are already in a deep layer associated with the ectoderm (15), but the mesoderm is said to become determined as a result of an inductive interaction between the endoderm and ectoderm. This interaction can be made to occur in culture if an explant of ectoderm (containing both superficial and deep layers) is confronted with cells from the appropriate region of the endoderm (10, 30). Slack (30) has argued that this interaction is "instructive" rather than "permissive" because (1) there is no increase in volume in the explants or in the embryo during the relevant stages of development, (2) there is no visible cell death in the cultures, and (3) in confrontation cultures several markers characteristic of mesodermal derivatives are expressed (e.g., muscle actin). He states (p. 26): "It is not conceivable...that the[se] interactions...are permissive in character since they are clearly the foundation of the progressive regional subdivision and consequent increase in complexity of the body plan. If they are permissive then it means that some completely unknown process is responsible for generating the different types of cell...." Nevertheless, there is as yet no evidence for induction at the single cell level (see Ref. 10, p. 294 for a lucid, albeit brief, discussion of this problem). Recently, Slack and Smith and their collaborators (31, 32) have demonstrated that several substances are capable of inducing the expression of mesodermal markers in cultures of *Xenopus* ectoderm. They include fibroblast growth factor, an extract from chick embryos, and a protein secreted by an amphibian cell line, XTC, which may be identical to the transforming growth factor TGFβ2. All of these have relative molecular weights on the order of 16,000.

In the chick, our knowledge about mesodermal induction is even more limited. The avian equivalent of the inducing endoderm of the amphibian is the hypoblast. Since craniocaudal reversal of the chick hypoblast leads to reversal of the craniocaudal axis of the embryo (49), the hypoblast clearly plays a role in controlling the orientation of the primitive streak. Subsequently, it was shown that the result of this interaction depends critically upon the stage of the operation (2). It has never been shown, however, that the interaction between hypoblast and epiblast is truly an instructive induction in that it changes the fate of the cells of the epiblast. An alternative would be that it is only *permissive*, allowing the expression of cellular fates that otherwise do not become overt (see Fig. 1). Of course, virtually nothing is known about either the nature of the inductive signal or the nature of the response in bird embryos, and, unlike the case in amphibians, the mesoderm in birds cannot be induced easily by heterologous factors (28).

A FACTOR CAPABLE OF INDUCING SECONDARY AXES

Recently, a 50-kilodalton factor has been discovered (46) which is secreted by a human embryo cell line, MRC-5, and which is capable of "scattering" cultured epithelial cells. We (12) found that scatter factor can induce a secondary primitive streak to form in the epiblast: We grafted small pellets of MRC-5 cells into chick embryos of appropriate stages of development and found that secondary axes were formed, while grafts of cell lines that do not produce scatter factor had no effect. Primitive-streak-like structures formed in embryos grafted with MRC-5 cells, and supernumerary neural plates were found in about 80 percent of the grafted embryos (compared to some 50 to 60 percent after grafting the "natural" inducer, Hensen's node). Purified scatter factor applied locally is also effective (unpublished observations). We are investigating the possibility that chick cells with inducing ability, such as Hensen's node, primitive streak mesoderm, and hypoblast, produce scatter factor–like activity themselves.

The finding that cells that produce scatter factor can induce secondary axes in the chick could provide a useful tool for the study of mesoderm induction in these embryos. However, we still do not know whether mesoderm induction in amniotes is permissive or instructive. In order to address this question, it would be of great interest to identify stable cell-type-specific markers that could distinguish between different cell populations in the embryo during and prior to gastrulation.

Regional Markers

Embarking on a hunt for cell-type-specific markers is obviously a difficult task without some direction to guide the search. If regional differences were found, it would still have to be shown that these differences identify cell types with different fates and are not merely a reflection of momentary cell behavior within a diverse cell population. In our present state of ignorance, however, any regional differences at or before gastrulation will be helpful in the search for real markers. Recently, several such differences have been identified (Stern et al., in preparation). In the remainder of this section I shall discuss how one of these markers may help us to understand the relationships between cell diversification and morphogenesis.

Antibodies recognizing a complex sulfated carbohydrate epitope known as "L2," which is present on certain adhesion-related glycoproteins (N-CAM) show characteristic patterns of binding in early chick embryos (5). In stage 2 to 3 chick blastoderms, the L2 epitope identifies all of the primitive streak mesoderm, the hypoblast, and a few cells in the posterior (caudal) margin of the germ wall (*area opaca* endoderm) (Fig. 3). Western blot and immunohistochemical studies have shown that these regional differences are not due to differences in the distribution of N-CAM.

At stage XIII and earlier, long before the primitive streak makes its appearance, the hypoblast cells already bear the L2 epitope on their surface. We were surprised, however, to find that some cells in the epiblast are also labeled. These cells are distributed in an apparently random way in the epiblast, giving a "pepper and salt" appearance (Fig. 3). We do not yet know whether the L2-positive cells of the epiblast are mesodermal precursors, and we are investigating this possibility. If they were, the widely accepted notion that the mesoderm of amniote embryos results from an instructive induction will have to be revised. If it could be shown that L2-positive cells in the epiblast of the early embryo are indeed mesoderm precursors, this would imply that the mesoderm cells are determined as such prior to the formation of the primitive streak, and that they sort out from the nonmesodermal cells of the epiblast.

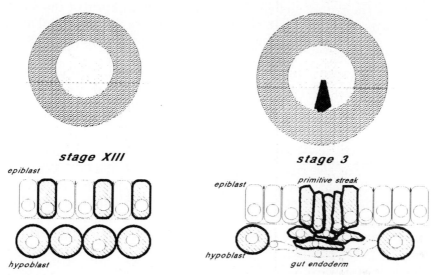

FIGURE 3. Diagram illustrating the distribution of the carbohydrate epitope L2 at two different stages of development in the early chick embryo. The upper diagrams show a general view of the embryo, while the lower diagrams show the distribution of L2 in a cross-section through the region marked by a dotted line in the corresponding upper diagram. At stage XIII, the hypoblast cells express L2 (heavy outline and hatching), while the epiblast contains a mixture of labeled and unlabeled cells. At stage 3, about the middle of gastrulation, the hypoblast is still positive, although it has become displaced to the periphery of the *area pellucida*. The cells of the forming primitive streak are also positive, but the rest of the epiblast no longer displays the epitope. Cartoon based on the results of Canning and Stern (5).

Summary

In summary, therefore, it seems that the dorsoventral and craniocaudal polarity of the early embryo are controlled as a result of dynamic interactions between cell populations and their environment. The production of the mesoderm as a new cell type during gastrulation has long been assumed also to be the result of cell interactions. However, careful consideration of the evidence in favor of this notion reveals that it is not sufficiently well founded to be accepted without further investigation, and it may yet turn out that the mesoderm and other cell types are allocated as a result of the lineage history of the progenitor cells. Cell lineage mapping, combined with transplantation and an analysis of new cell-type-specific markers as they appear should resolve this issue.

Setting up Segmental Organization

The segmental organization of vertebrate embryos is most obvious in the pattern of somites from which derive the vertebrae and ribs of the axial skeleton, the dermis of the trunk, and all the voluntary musculature of the adult are derived. The metameric pattern of somites determines the segmental arrangement of other structures in the embryo, such as the peripheral nervous system (16, 17, 41, 42, 45).

In the chick embryo, some 55 pairs of somites form, each somite being, at first, an epithelial sphere that buds off the rostral (anterior) end of each of the paired segmental plates of paraxial mesoderm that appear about 1.5 days after the egg is laid. Each pair of somites takes about 1.5 h to form. Some 6 to 8 h after its initial appearance, each somite splits up into two further components: (1) the dermomyotome dorsally, which retains some epithelial characteristics that give rise to the dermis of the trunk and to skeletal muscle and (2) the sclerotome ventromedially, which is a loose mesenchyme that gives rise, along with the notochord, to the axial skeleton. Each sclerotome is subdivided into a rostral (anterior) and a caudal (posterior) half (16, 17). Differences between the cells of the two halves determine the segmental organization of the peripheral nervous system: motor nerves and neural crest cells (16, 17, 29) are restricted to the rostral half of each sclerotome and are unable to colonize the caudal half.

During the formation of a somite, its progenitor cells have to make several decisions in order to realize their morphogenetic potential. Among them, we can distinguish: (1) When does a cell become committed to be part of a somite rather than part of any other mesodermal derivative? (2) When do sclerotome cells decide to become rostral or caudal? (3) When do cells choose to become dermomyotome as opposed to sclerotome? (4) When, if at all, are regional differences between somites determined?

Determination of Somitogenic Potential

Since the cranial portion of isolated segmental plates can form somites, while the caudal portion cannot (3, 34), the somitogenic potential of a cell may be determined during its sojourn in the plate.

A recent, albeit preliminary, experiment using lineage labels injected into single cells (39) may help us determine more precisely the time at which cells become restricted to a somitic fate. The results of this experiment, summarized in Fig. 4, show that as late as two cell divisions prior to segmentation, somite progenitor cells also contribute to other mesodermal tissues such as lateral plate and mesonephric kidney. A single injected cell at the caudal end of the segmental plate gives rise, 2 days later, to some 2^4 to 2^5 cells in a single somite and to some 2^8 cells scattered among other mesodermal tissues. Since the somite progenitor cells at this time appear to be scattered over a broad region at the caudal end of the segmental plate and primitive streak of the chick embryo, it seems unlikely that the decision between somite and nonsomite is made as a result of position. It is more likely that an unequal division of each progenitor cell gives rise to one daughter with somitogenic

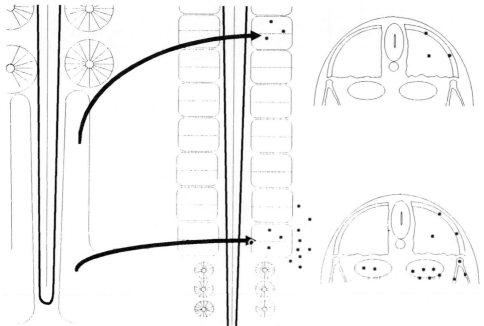

FIGURE 4. Summary of results from single-cell lineage experiments. The diagram on the left illustrates the embryo at the time of injection of rhodamine-lysine-dextran into a single cell, while the drawings in the center and right summarize the distribution of the progeny of the injected cell two days later in the whole embryo (center) and in transverse section (right). A cell injected anywhere within the cranial half of the segmental plate gives rise to a clone of about 16 to 64 cells which is restricted to one somite, but not to any particular portion of it. A cell injected in the caudal one-third of the segmental plate also gives rise to about the same cell number of labeled cells that are restricted to one somite, but a large number of labeled cells are also seen in the intermediate and lateral plate mesoderm, in the endothelium of the floor of the aorta, and in circulating blood. Based on the results of Stern et al. (39).

potential and a slow rate of cell division (9 to 10 h) and one with nonsomitic fate that divides faster (5 h) (Fig. 4). Thus, the commitment to a somitic fate is probably a consequence of *cell lineage history*, rather than of position (Fig. 5).

Rostral-Caudal Determination

When do sclerotome cells become committed to being rostral or caudal? If either half of a newly formed somite is excised and transplanted into any other site in the embryo, it always gives rise to sclerotome with the properties of the half of origin, irrespective of the position into which it is grafted (42). This rules out the possibility that the rostrocaudal fate of somite cells is determined after the time of overt somite formation.

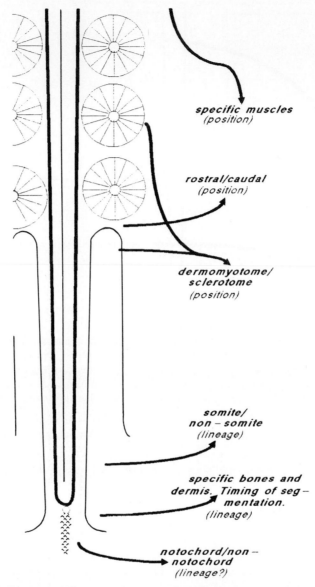

FIGURE 5. Diagram illustrating the positions at which the various cellular commitments involved in somite formation might take place. Since somite formation is a progression in time and space, such that "younger" cells are more caudally placed than more mature ones, the more caudal positions (lower portion of diagram) correspond to earlier points in the developmental sequence.

If rostrocaudal commitment is made before segmentation, there must be some mechanism to ensure that the correct cells end up in the correct half of the somite. There are two possibilities: (1) cells are fixed in position after determination or (2) cells are able to move in the segmental plate but can sort out according to their rostrocaudal nature (Fig. 1). There are arguments against both of these alternatives. An experiment by Menkes and Sandor (25) rules out the first possibility. They dissociated the segmental plate and found that it still produced normal somites. We have found that the rostrocaudal composition of the resulting sclerotomes is also normal. If cells are fixed in position in the segmental plate mesoderm, a restriction consistent with the existence of somitomeres (morphological presomite condensations in the segmental plate; Ref. 24), dissociation of the segmental plate should result in a loss of the rostrocaudal pattern. Moreover, cell movement has been reported to occur in the segmental plate mesoderm (41, 47, unpublished observations). If, on the other hand, cells move within the segmental plate, it would be difficult to maintain a fixed pattern of somitomeres. It would also be difficult to explain why rotation of the segmental plate results in reversal of the pattern (16), since cells should sort out to restore the original arrangement. Similarly, the experiment of Menkes and Sandor (25) should lead not to a normal repeated rostrocaudal sequence but rather to one large rostral and one large caudal half-sclerotome. Finally, in order for rostral and caudal cells to recognize each other for the purpose of sorting out, they would have to express some rostrocaudal differences in the segmental plate. There is no evidence for this at present.

Therefore, it seems likely that the rostrocaudal decision is made *during somite formation*, perhaps in relation to the length of time each cell spends adjacent to the developing segment border (41). One interesting consequence of this is that, since the determination of the rostral half of a newly forming somite as rostral would coincide with the determination of the caudal half of the preceding somite as caudal, rostrocaudal determination would be *para*segmental, as has been suggested for the epidermal segments of *Drosophila* (22). If this is the case, rostrocaudal determination would be an example of a developmental decision made in relation to *position*, rather than lineage history of the cells (Fig. 5).

Specification of Sclerotome and Dermomyotome

The dorsoventral polarity of the somites forming from the cranial tip of the segmental plate can still be reversed by inverting the plate (9, 13). This finding indicates that somite cells are specified as dermomyotome or sclerotome close to the time of somite formation, but before subdivision, some 6 to 8 h later. The commitment as dermomyotome or sclerotome therefore depends upon *position*, by interactions with the adjacent epiblast and endoderm (Fig. 5).

Regional Specification

Vertebrae in different regions of the spinal column are morphologically different from one another, suggesting that individual somites have defined regional identities. At what stage are somite cells determined to form particular skeletal elements? When thoracic segmental plate mesoderm is grafted into the cervical region, ribs develop in the neck (20). The same is true for the plumage pattern, which is derived from the dermatomes (23). The muscle pattern, on the other hand, does not behave in this way: when nonwing level somites are transplanted to the wing region, they give rise to normal wing muscles (6) and are innervated appropriately for their new position (18). These results could be interpreted to mean that skeletal and dermal derivatives of the somite are regionally determined in the segmental plate or earlier, while the voluntary muscles become determined much later. However, it is possible that regional specification for dermis and sclerotome does not take place until later; this alternative explanation requires that cells behave autonomously within the plate, being unable to take positional cues from other regions. Clearly, these transplantation experiments do not help us to determine when regional specification occurs.

Because somite pairs form sequentially, specification as cervical, thoracic, etc., could be linked to the time of formation of each somite pair. Heat-shock and other experiments (reviewed in Ref. 17) suggest that somite progenitor cells have an internal "clock" that makes them competent to segment at a particular time. This mechanism could control the size of somites by regulating the number of cells that segment at any one time. The experiments suggest that this clock is linked directly to the cell division cycle, and that this clock is already operating four division cycles before segmentation, that is, about two divisions *before* cells decide to become somitic (see above, "Determination of Somitogenic Potential"). It is possible that regional specification for sclerotome and dermal derivatives of the somite could also be linked to this clock. If this is the case, this would be another example of a *lineage*-related decision (Fig. 5). On the other hand, the myogenic cells, at least at limb levels, can become any muscle and be innervated by any motor nerve until they enter the limb, 24 h or more after the corresponding dermomyotomes form. The commitment to form a particular muscle, therefore, is an example of a *position*-related decision, as myoblasts would need to interpret positional cues from the rest of the limb (Fig. 5).

Summary

The preceding discussion illustrates that even within what appears to be a single process, that of somite formation, six of the developmental decisions that can be recognized are evenly divided in their use of the two strategies for creating cell diversity in relation to pattern: *somitogenic potential and the regional properties of the dermis and axial skeleton all appear to be determined by lineage history* (Fig. 1a), while it is likely that *the decisions between rostral and caudal sclerotome, between sclerotome and dermomyotome, and between different skeletal muscles all depend upon cell interactions* with neighboring tissues (Fig. 1*b*). In the case of the decisions we can identify during somite formation, therefore, it appears that those decisions that are made early (two cell cycles or more before overt segmentation) depend upon the lineage history of the cells, while those that are made late (during segmentation or later) depend upon cell interactions (Fig. 5).

Conclusions

During the development of vertebrate embryos, therefore, the fates of cells diverge both as a result of cell lineage history and as a result of interactions with their environment. Although it is easy to list advantages and disadvantages to each of these strategies, the mechanisms that cause cells to use one or the other method to generate diversity in relation to their final spatial arrangement are not always immediately obvious.

It is worth considering the possibility that different organisms use different strategies to generate similar structures. Because of the importance of cell diversity and pattern, it is possible that *form* is controlled by evolution more directly than the mechanisms used to produce it.

Acknowledgments

The research reported in this paper is funded by grants from the Wellcome Trust (gastrulation), the Medical Research Council (somite segmentation), and Action Research for the Crippled Child (neural segmentation).

General References

GURDON, J. B.: 1987. Embryonic induction—molecular prospects. *Development (Camb.)* **99**:285–306.

KEYNES, R. J. and STERN, C. D.: 1988. Mechanisms of vertebrate segmentation. *Development (Camb.)* **103**:413–429.

References

1. Azar, Y., and Eyal-Giladi, H. 1979. Marginal zone cells—the primitive streak-inducing component of the primary hypoblast in the chick. *J. Embryol. Exp. Morphol.* **52**:79–88.

2. Azar, Y., and Eyal-Giladi, H. 1981. Interaction of epiblast and hypoblast in the formation of the primitive streak and the embryonic axis in chick, as revealed by hypoblast-rotation experiments. *J. Embryol. Exp. Morphol.* **61**:133–144.

3. Bellairs, R. 1980. The segmentation of the somites in the chick embryo. *Boll. Zool.* **47**:245–252.

4. Bellairs, R. 1986. The primitive streak. *Anat. Embryol.* **174**:1–14.

5. Canning, D. R., and Stern, C. D. 1988. Changes in the expression of the carbohydrate epitope HNK-1 associated with mesoderm induction in the chick embryo. *Development (Camb.)* **104**:643–656.

6. Chevallier, A., Kieny, M., and Mauger, A. 1977. Limb-somite relationships: origin of the limb musculature. *J. Embryol. Exp. Morphol.* **79**:1–10.

7. Eyal-Giladi, H., and Kochav, S. 1976. From cleavage to primitive streak formation: a complementary normal table and a new look at the first stages of the development of the chick. *Dev. Biol.* **49**:321–337.

8. Eyal-Giladi, H., Kochav, S., and Yeroushalmi, S. 1975. The sorting out of thymidine-labelled chick hypoblast cells in mixed epiblast-hypoblast aggregates. *Differentiation* **4**:57–60.

9. Gallera, J. 1966. Mise en évidence du rôle de l'ectoblaste dans la différenciation des somites chez les oiseaux. *Rev. Suisse Zool.* **73**:492–503.

10. Gurdon, J. B. 1987. Embryonic induction—molecular prospects. *Development (Camb.)* **99**:285–306.

11. Hamburger, V., and Hamilton, H. L. 1951. A series of normal stages in the development of the chick. *J. Morphol.* **88**:49–92.

12. Ireland, G. W., Stern, C. D., and Stoker, M. 1987. Human MRC-5 cells induce a secondary primitive-streak when grafted into chick embryos. *J. Anat.* **152**:223–224.

13. Jacob, H. J., Christ, B., and Jacob, M. 1974. Die Somitogenese beim Hühnerembryo. Experimente zur Lageentwicklung des Myotom. *Verh. Anat. Ges.* **68**:581–589.

14. Jaffe, L. F., and Stern, C. D. 1979. Strong electrical currents leave the primitive streak region of chick embryos. *Science* **206**:569–571.

15. Keller, R. E. 1976. Vital dye mapping of the gastrula and neurula of *Xenopus laevis*. II. Prospective areas and morphogenetic movements of the deep layer. *Dev. Biol.* **51**:118–137.

16. Keynes, R. J., and Stern, C. D. 1984. Segmentation in the vertebrate nervous system. *Nature (Lond.)* **310**:786–789.

17. Keynes, R. J., and Stern, C. D. 1988. Mechanisms of vertebrate segmentation. *Development (Camb.)* **103**:413–429.

18. Keynes, R. J., Stirling, R. V., Stern, C. D., and Summerbell, D. 1987. The specificity of motor innervation of the chick wing does not depend upon the segmental origin of muscles. *Development (Camb.)* **99**:565–575.

19. Khaner, O., Mitrani, E., and Eyal-Giladi, H. 1985. Developmental potencies of area opaca and marginal zone areas of early chick blastoderms. *J. Embryol. Exp. Morphol.* **89**:235–241.

20. Kieny, M., Mauger, A., and Sengel, P. 1972. Early regionalization of the somitic mesoderm as studied by the development of the axial skeleton of the chick embryo. *Dev. Biol.* **28**:142–161.

21. Kochav, S., and Eyal-Giladi, H. 1971. Bilateral symmetry in chick embryo determination by gravity. *Science* **171**:1027–1029.

22. Martínez-Arias, A., and Lawrence, P. A. 1985. Parasegments and compartments in the *Drosophila* embryo. *Nature (Lond.)* **313**:639–642.

23. Mauger, A. 1972. Rôle du mésoderme somitique dans le dévelopement du plumage

dorsal chez l'embryon de poulet. II. Régionalisation du mésoderme plumigène. *J. Embryol. Exp. Morphol.* 28:343–366.

24. Meier, S., and Jacobson, A. G. 1982. Experimental studies of the origin and expression of metameric pattern in the chick embryo. *J. Exp. Zool.* 219:217–232.

25. Menkes, B., and Sandor, S. 1969. Researches on the development of axial organs. *Rev. Roum. Embryol. Cytol. Ser. Embryol.* 6:65–88.

26. Mitrani, E., and Eyal-Giladi, H. 1981. Hypoblastic cells can form a disk inducing an embryonic axis in chick epiblast. *Nature (Lond.)* 289:800–802.

27. Mitrani, E., Shimoni, Y., and Eyal-Giladi, H. 1983. Nature of the hypoblastic influence on the chick embryo epiblast. *J. Embryol. Exp. Morphol.* 75:11–20.

28. Nieuwkoop, P. D., Johnen, A. G., and Albers, B. 1985. *The Epigenetic Nature of Early Chordate Development. Inductive Interaction and Competence.* Cambridge University Press.

29. Rickmann, M., Fawcett, J. W., and Keynes, R. J. 1985. The migration of neural crest cells and the growth of motor axons through the rostral half of the chick somite. *J. Embryol. Exp. Morphol.* 90:437–455.

30. Slack, J. M. W. 1983. *From Egg to Embryo. Determinative Events in Early Development.* Cambridge University Press.

31. Smith, J. C. 1987. A mesoderm-inducing factor is produced by a *Xenopus* cell line. *Development (Camb.)* 99:3–14.

32. Smith, J. C., Dale, L., and Slack, J. M. W. 1985. Cell lineage specific labels and region-specific markers in the analysis of inductive interactions. *J. Embryol. Exp. Morphol.* 89(Suppl.):317–331.

33. Spemann, H. and Mangold, H. 1924. Über Induktion von Embryonalanlagen durch Implantation artfremder Organisatoren. *Wilhelm Roux' Arch. Entwicklungsmech. Org.* 100:599–638. [English translation. 1964. In: *Foundations of Experimental Embryology.* (B. H. Willier and J. M. Oppenheimer, eds.). Hafner Press, London.]

34. Spratt, N. T. Jr. 1955. Analysis of the organizer center in the early chick embryo. I. Localization of prospective notochord and somite cells. *J. Exp. Zool.* 128:121–164.

35. Spratt, N. T. Jr., and Haas, H. 1960. Integrative mechanisms in development of the early chick blastoderm. I. Regulative potentiality of separated parts. *J. Exp. Zool.* 145:97–137.

36. Stern, C. D. 1984. A simple model for early morphogenesis. *J. Theor. Biol.* 107:229–242.

37. Stern, C. D. 1984. Mini-review: Hyaluronidases in early embryonic development. *Cell Biol. Int. Rep.* 8:703–717.

38. Stern, C. D., and Bellairs, R. 1984. The roles of node regression and elongation of the area pellucida in the formation of somites in avian embryos. *J. Embryol. Exp. Morphol.* 81:75–92.

39. Stern, C. D., Fraser, S. E., Keynes, R. J., and Primmett, D. R. N. 1988. A cell lineage analysis of segmentation in the chick embryo. *Development (Camb.).* 104(Suppl.):231–244.

40. Stern, C. D., and Ireland, G. W. 1981. An integrated experimental study of endoderm formation in avian embryos. *Anat. Embryol.* 163:245–263.

41. Stern, C. D., and Keynes, R. J. 1986. Somites and neural development. In: *Somites in Developing Embryos.* (R. Bellairs, D. A. Ede, and J. W. Lash, eds.), pp. 147–159, Plenum Press, New York.

42. Stern, C. D., and Keynes, R. J. 1987. Interactions between somite cells: the formation and maintenance of segment boundaries in the chick embryo. *Development (Camb.)* **99**:261–273.

43. Stern, C. D., and MacKenzie, D. O. 1983. Sodium transport and the control of epiblast polarity in the early chick embryo. *J. Embryol. Exp. Morphol.* **77**:73–98.

44. Stern, C. D., Manning, S., and Gillespie, J. I. 1985. Fluid transport across the epiblast of the early chick embryo. *J. Embryol. Exp. Morphol.* **88**:365–384.

45. Stern, C. D., Sisodiya, S. M. and Keynes, R. J. 1986. Interactions between neurites and somite cells: inhibition and stimulation of nerve growth in the chick embryo. *J. Embryol. Exp. Morphol.* **91**:209–226.

46. Stoker, M., Gherardi, E., Perryman, M., and Gray, J. 1987. Scatter factor is a fibroblast-derived modulator of epithelial cell mobility. *Nature (Lond.)* **327**: 239–242.

47. Tam, P. P. L., and Beddington, R. S. P. 1987. The formation of mesodermal tissues in the mouse embryo during gastrulation and early organogenesis. *Development (Camb.)* **99**:109–126.

48. Vakaet, L. 1984. The initiation of gastrular ingression in the chick blastoderm. *Am. Zool.* **24**:555–562.

49. Waddington, C. H. 1933. Induction by the endoderm in birds. *Wilhelm Roux' Arch. Entwicklungsmech. Org.* **128**:502–521.

Questions for Discussion with the Editor

1. *Why do you suppose prevalent notions about organogenesis imply either/or phenomena? Your thesis is that perhaps both lineage and induction play major roles in specifying pattern formation events.*

I don't think that anybody will seriously defend the idea that any particular organism is either mosaic or regulative, although there may well be some organisms that are more regulative than others. However, there is indeed a prevalent view, which has found its way into many textbooks. To cite a popular one: *Molecular Biology of the Cell** states categorically that, "there are hardly any instances in normal development in which spatial order is created out of an initially random mixture of cell types." While this is clearly true, the converse is equally true. In fact, there are hardly any instances in which we *understand how* cell diversity is generated, so choosing one or the other method for a general statement is misleading.

Slack (Ref. 30; p. 26) implies that "instructive" inductions occur early in development, while "permissive" ones occur later. While this may be true in general terms, when looking broadly at the whole organism, it is not necessarily true when looking at the development of a particular organ system. Take, for example, the case of somite decisions discussed in this chapter: the early decisions seem to depend on lineage, while the later ones appear to depend on cell interactions—the reverse of Slack's general statement.

The view expressed in *Molecular Biology of the Cell* is clearly the one with the most followers among experimental embryologists, while the converse view (that

---

* Alberts, B. M., Bray, D., Lewis, J. H., Raff., M., Roberts, K., and Watson, J. D. 1983, *Molecular Biology of the Cell*. Garland, New York, 1st ed., p. 850.

cell lineage determines fate) tends to be favored by molecular biologists. Unlike those who study them, embryos seem to use all the mechanisms available to them to give rise to organized diversity.

2. *What are the prospects for superimposing a genetic approach on the vertebrate somite development problem?*

I am very optimistic. In the last 5 years, we have learned a huge amount about somite formation and about the subsequent development of vertebrate segments at the cellular level. This knowledge should help to bridge the gap between the pattern and the genetic approach. Many mouse mutations displaying abnormalities in various aspects of somite development have been known for many years (see Ref. 17 for review and references), but only a few have been studied in detail. This was probably due to a lack of understanding of the cellular processes involved in somite development. As this situation is changing, many more investigators are turning their attention to the study of these mutations. Another major advance will derive from vertebrates that have been introduced recently as developmental model systems. The most interesting recent addition is the zebrafish embryo, which allows a study of development at the molecular, genetic, and cellular levels.

To give a full answer to the editor's question, however, requires an understanding of what "the vertebrate somite problem" is. I see it as comprising many separate but related aspects. My aim in this chapter was to attempt to separate out some aspects of somite development that will probably have to be studied as independent issues. In order to confront the challenge of seeking the molecular bases of cellular decisions, we have to know the time and place at which these decisions are made.

But will meaningful results come from those approaching the problem from the genetic level or from those approaching it from the cellular-tissue level? Both are systematic, but while the latter is more elegant because experiments can be designed to follow each other logically, the former has proved to be more efficient. If asked for my own preference, I would choose logical elegance rather than efficiency. The indiscriminate molecular approach feels too much like solving a crossword puzzle by searching through the dictionary from "A" to "Z," looking for words that fit each clue.

CHAPTER **6**

Timing Mechanisms in Early Development

Noriyuki Satoh

Introduction

A FEW ASPECTS OF EARLY sea urchin embryogenesis will be briefly reviewed to provide examples of "timing in early development." A fertilized egg of *Strongylocentrotus purpuratus* begins its development with a series of synchronized cleavages. At the fourth cleavage, the blastomeres of the animal hemisphere divide equally, while the division of blastomeres of the vegetal hemisphere takes place unequally. The result of the fourth cleavage is an embryo composed of eight mesomeres, four larger macromeres, and four smaller micromeres. Further cleavages transform the embryo into a hollow blastula. Then, the embryonic blastula cells extend cilia from their apical surface, and they hatch out into the open sea. Soon after hatching, the cells comprising the vegetal pole region of the blastula elongate to form the vegetal plate. Several of the vegetal plate cells begin to migrate into a fluid-filled space termed *the blastocoel*. These cells, termed *primary mesenchyme cells*, are the descendants of the micromeres and are responsible for larval skeleton formation. Thereafter, gastrulation occurs by an ingression of the archenteron.

At this point, we can ask the following questions: Why does micromere formation always take place at the fourth cleavage and never at the third or fifth cleavage? Why does the embryo hatch prior to the morphogenetic movements that are associated with gastrulation? Why does the migration of the primary mesenchyme cells precede the formation of the archenteron? The answer to these questions is that every developmental event is *temporally* controlled within the embryo. In other words, embryonic cells must be provided with a "clock system" that can accurately measure time intervals for each developmental event from start to finish.

Is Exact Timing of Events Developmentally Significant?

The spatial patterning of early embryonic events is important for normal development. This is easily understood when we examine the bizarre appearance of the *Drosophila* mutant termed *Antennapedia*. In *Antennapedia* mutants, one of the antennae is replaced by a leg. Clearly, spatial patterning is of critical importance in early embryogenesis. Can the same be said of temporal regulation? The answer to that question is, yes, indeed. We will now begin to explore the temporal regulation of development using the following two examples: (1) the segregation of developmental fates in Ctenophora (12) and (2) the *grandchildless* mutation in *Drosophila* (23).

During the transition from the 4- to the 8-cell stage in ctenophore embryos, each blastomere produces one daughter cell with the potential to form comb plate cilia and one daughter cell that does not have this potential. If the second cleavage is blocked by cytochalasin B, the embryos, at the cleavage that coincides temporally with the third cleavage of untreated embryos, frequently form four blastomeres which resemble the blastomeres of a normal 8-cell embryo. At this division from two to four blastomeres, the comb plate–forming potential is segregated into one daughter cell. Alternatively, by compressing a 2-cell embryo in a plane perpendicular to the first plane of cleavage, it is possible to produce a 4-cell blastomere configuration that is identical to that produced following the inhibition of the second cleavage. However, under these circumstances the segregation of comb plate potential does not occur. These results suggest that the appropriate plane of cleavage must take place *at a precise time* for the correct spatial localization of developmental potential within blastomeres.

Another example that illustrates the significance of precise timing mechanisms is a temperature-sensitive maternal effect mutation in *Drosophila melanogaster*, termed *gs(1)N26*. During the very early development of *Drosophila* the nuclei divide in the central region of the egg cytoplasm, and cellularization of the embryo, termed *cellular blastoderm formation*, occurs after about 13 nuclear divisions. The cellularization of the posterior pole region begins first, resulting in the formation of pole cells which are the progenitors of the germ cells (see Chap. 11, this volume). The *gs(1)N26* mutant females lay eggs in which pole cell formation is lacking, thus their progeny are sterile. The defects in blastogenesis have been shown to be caused by the delay in the migration of the cleavage nuclei into the posterior pole region. Therefore, these observations indicate that a disharmony in the timing of two early events, blastoderm cellularization and nuclear migration, results in sterility.

Developmental Clocks

A "developmental clock" is the metaphor commonly used to describe the mechanisms within embryos which can measure the flow of time and determine the timing of initiation of specific kinds of developmental events. The developmental clock is not the same as the clock of circadian rhythms because the developmental clock is temperature-dependent. One example that illustrates the main characteristics of a developmental clock is shown in Fig. 1 (31). Acetylcholinesterase (AChE) is a muscle-specific enzyme expressed exclusively in the tail muscle cells of the ascidian larva. AChE activity is first detected in the neurulae, and its appearance is mediated by embryonic gene activity. As shown in Fig. 1, when embryos are allowed to develop at a lower than normal temperature (13°C), the first appearance of AChE activity is delayed by 8 h. In contrast, when embryos are allowed to develop more rapidly at a higher temperature (23°C), the time of the first appearance of AChE activity is accelerated. However, at both temperatures AChE activity is first detected in the presumptive muscle cells of the neurula. The total number of muscle cells in the tailbud-stage embryo that develops at 13, 18, and 23°C is always 36.

These experiments indicate that the timing of events during the early development of ascidian embryos is not mediated by chronological time or elapsed time since fertilization but is likely related to the number of cell division cycles. The driving force of the clock system may be attributed to cycles of cell division. There are several mechanisms that can possibly constitute the "alarm apparatus": (1) the number of cytokineses (a specific event that is initiated after a determined number of cytokineses), (2) nuclear divisions (DNA replications), and (3) the nucleo-cytoplasmic ratio.

One attractive idea is that the precise array of the developmental events move progressively through time, like the hands of a clock, by cellular events set in motion at fertilization. However, this idea does not seem plausible (17, 30–32). As will be described later, the beginning of asynchronous cleavage is determined by the nucleo-cytoplasmic ratio, and several developmental events, particularly those associated with transcription, require a certain number of DNA replications. This suggests that there is a close relationship

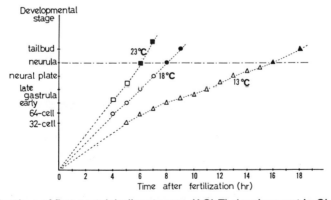

FIGURE 1. The time of first acetylcholinesterase (AChE) development in *Ciona intestinalis* (ascidian) embryos raised at 23°C (squares), 18°C (circles), and 13°C (triangles), respectively. Open symbols show embryos with no histochemically detectable AChE activity, and solid symbols show those with AChE activity (from Ref. 31).

between the number of cycles of DNA replication and the timing of certain developmental events. In addition, the initiation of some developmental events such as rRNA synthesis in *Xenopus* embryos is not timed by the number of cytokineses, nuclear divisions, DNA replications, or the nucleocytoplasmic ratio. This suggests that we should first study the timing mechanism for each developmental event and then try to organize the events into some kind of a hierarchy in order to provide an overall understanding of the temporal control of embryonic development.

The Number of Cytokineses and the Total Cell Number of the Embryo Are Not Critical Factors in Developmental Timing

Although we have just said that the cell cycle is the basis of the developmental timing mechanism, the number of cytokineses and, therefore, the number of cells which comprise an embryo are not critical factors in the initiation of developmental events. The issue is whether or not fertilized eggs must cleave before cellular differentiation and morphogenesis can occur. This issue was first elegantly elucidated by Lillie (19) in 1902 as the possibility of "differentiation without cleavage." Lillie showed that artificially activated *Chaetopterus* eggs can develop features of the trochophore larva in the absence of cytokinesis (Fig. 2a and b).

One of the best experiments which shows that cytokinesis is not always required for cellular differentiation is the expression of epidermis-specific

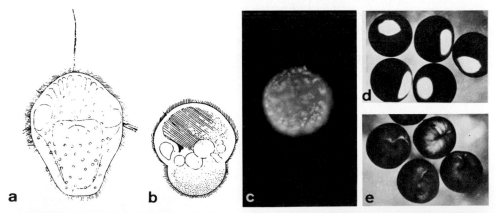

FIGURE 2. Differentiation and morphogenesis can occur without cleavage. (*a, b*) Differentiation without cleavage in *Chaetopterus* eggs (from Ref. 19). If this annelid egg is artificially activated with KC1-enriched seawater, the egg gives rise to a ciliated larva (*b*) that closely resembles a normal trochophore larva (*a*). (*c*) Development of epidermal-cell differentiation marker in cleavage-arrested *Halocynthia* (ascidian) egg, examined with a specific monoclonal antibody (from Ref. 24). (*d, e*) Pseudo-gastrulation of *Rana pipiens* eggs (from Ref. 36). Maturation of ovarian oocytes was induced by progesterone, and about 24 h thereafter, darkly pigmented surface of the animal hemisphere begins to cover the nonpigmented vegetal surface (*d*), and finally the surface invaginates (*e*).

markers in cleavage-arrested ascidian eggs. When fertilized ascidian eggs are immersed in seawater containing cytochalasin B, cytokinesis is completely blocked so that the eggs remain in the 1-cell state. If epidermal cell differentiation is examined using monoclonal antibodies that recognize epidermal cells, more than 90 percent of the cleavage-arrested eggs develop the epidermal marker (Fig. 2c; Ref. 24). A similar experimental result has been reported in the nematode *Caenorhabditis elegans*, in which cleavage-arrested fertilized eggs develop a hypodermis differentiation marker (7). That is, cytokinesis is not always required for cellular differentiation. How about morphogenesis? Can morphogenesis also take place without cleavage? The development of a trochophore-like larva from an activated *Chaetopterus* egg suggests that cleavage is not always an absolute requirement for morphogenesis. Figure 2d and *e* illustrates the phenomenon termed "pseudo-gastrulation" in *Rana pipiens* eggs (36). Oocytes were removed from the ovary, and maturation was induced by treating the oocytes with progesterone. Then the matured eggs were kept in Ringer's solution in the absence of sperm. These treated eggs and normally fertilized eggs undergo "pseudo-gastrulation" and gastrulation, respectively, after almost the same length of time. This is another example suggesting that morphogenesis can occur in the absence of cleavage.

Nucleo-cytoplasmic Ratio and the Onset of Asynchronous Cleavage

The fertilized eggs of most animals undergo several rapid synchronous cleavages followed by a transition to slower asynchronous cleavages. In amphibian eggs, at the time of the transition from synchronous to asynchronous divisions, distinct gaps between the time of DNA synthesis and mitosis, termed G1 and G2, occur in the cell cycle. The blastomeres actively synthesize RNA during the transition. This transition is, therefore, called the "midblastula transition (MBT)," which emphasizes the significant events that occur in the blastula. The onset of MBT is thought to be controlled by the attainment of a critical nucleus-to-cytoplasm volume ratio.

Kobayakawa and Kubota (18) obtained newt egg fragments composed of half or a quarter of the total egg volume by cutting a fertilized, uncleaved egg into two or four equal-sized parts. Asynchronous cleavage started one or two divisions earlier in the half or quarter embryos compared with the whole embryos. Newport and Kirschner (21) have shown that in *Xenopus* embryos there is likely a causal relationship between the onset of the MBT and the nucleo-cytoplasmic ratio. Experiments that support the nucleo-cytoplasmic ratio hypothesis are as follows. Half embryos with different cleavage schedules were obtained by partial constriction of a fertilized egg (Fig. 3a to f). However, both the advanced and the retarded half-embryos underwent the MBT after the same number of cleavages. The number of synchronous cleavages was not 12 but 11. The transition on the advanced side of the embryo occurred two cell cycles earlier (about 1 h) than on the retarded side, which received one of the nuclei of the advanced half after two cleavage cycles. Thus,

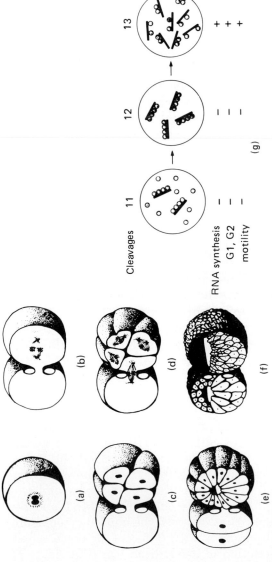

FIGURE 3. (a to f) Partial constriction experiment to test the dependence of the MBT on the nucleo-cytoplasmic ratio in *Xenopus* embryos. (b, c) During the first cleavage period, the egg is constricted preventing any of the daughter nuclei from migrating into one half of the egg. (d) At the third cleavage, one nucleus migrates across the constriction and (e to f) the nucleus in the left half proceeds to divide synchronously in both halves (from Ref. 21). (g) Titration model for the onset of the MBT. In this model a constant amount of cytoplasmic factor is titrated by an exponentially increasing amount of DNA. These factors are shown as putative inhibitors of transcription which become diluted out after the twelfth cleavage (from Ref. 22).

the MBT cannot be initiated by a counting mechanism based on the length of time since fertilization or by the number of rounds of DNA replication, but instead is likely initiated by the attainment of a critical nucleus-to-cytoplasm volume ratio. Moreover, Newport and Kirschner (22) have shown that when a plasmid containing a gene coding for yeast leucine tRNA was injected into cleaving eggs, the gene became transcriptionally active at the MBT. The suppression of transcription prior to the MBT could be reversed by the addition of vector DNA. The amount of DNA needed to induce premature transcription was equal to the amount of nuclear DNA present after 12 cleavages.

Newport and Kirschner (22) have proposed a titration model in order to explain the activation of transcription at the MBT (Fig. 3g). According to their model, the unfertilized egg contains a large cytoplasmic pool of a factor which is bound to chromatin and is capable of suppressing transcription. At the end of each synchronous round of DNA synthesis, the total amount of DNA in the whole egg is doubled and the new chromatin titrates a portion of the factor from the cytoplasmic pool. The depletion process continues until the completion of round 12 of DNA synthesis, at which time the cytoplasmic pool of the suppressor is depleted. This dilution process permits the activation of developmentally programmed transcription. The molecular analysis of the suppressor and the suppressor interaction with the chromatin are likely to be subjects of further studies.

The regulation of the timing of asynchronous cleavage by the nucleocytoplasmic ratio has been illustrated in other animals, including *Drosophila* (10), starfish (20), and sturgeon (6). Particularly, in both *Drosophila* and sturgeon, it has been shown that an artificial increase in the ratio of cytoplasm-to-nucleus volume can induce an additional synchronous cleavage compared with the normal number of synchronous cleavages. In the usual case, the timing of developmental events is rigidly programmed within embryos and cannot be experimentally altered. The MBT may be a unique example in which its timing can be altered.

Rounds of DNA Replication and the Initiation of Transcriptional Activity of Embryonic Genome

As mentioned above, cytokinesis is not always required for differentiation and morphogenesis. Is nuclear division a prerequisite for differentiation? The answer to this question appears to be no, because it has been shown in ascidian and *C. elegans* embryos that the differentiation of embryonic cells can occur when nuclear divisions are blocked by treating embryos with colchicine or colcemid (e.g., Ref. 39). Is DNA replication required for the differentiation of embryonic cells? Aphidicolin, a specific inhibitor of eukaryotic DNA polymerase-alpha, has been used in various kinds of experimental systems. Aphidicolin has been shown to block the development of trochophore-like larvae in artificially activated *Chaetopterus* eggs (2), to inhibit a heavy form of DNA ligase during early development of the axolotl embryo (35), to inhibit the transcription of the arylsulfatase gene in sea urchin embryos (Akasaka, personal communication), and to inhibit other developmental events. The synthesis of stage-specific polypeptides in early mouse embryos has been shown to be closely associated with the number of rounds of DNA replication (e.g., Ref. 26).

Cleavage-arrested ascidian embryos develop a specific type of AChE activity in the blastomeres of the muscle cell lineage. Figure 4a shows a 64-cell embryo in which cytokinesis and DNA synthesis were blocked with cytochalasin B and aphidicolin (33). All of the muscle lineage cells have completed the sixth round of DNA replication prior to the block, but none of the cells have developed AChE activity. In contrast, Fig. 4b shows a 110-cell embryo blocked with cytochalasin B and aphidicolin. Two distinct clusters of blastomeres, each composed of six presumptive muscle cells, are bilaterally positioned in the embryo and produce distinct AChE activity. Every AChE-producing cell in the embryo had completed the seventh round of DNA replication. These results suggest that AChE development in the presumptive muscle cells of the ascidian embryo requires a definite number of DNA replications.

Recently, aphidicolin has been used to investigate how DNA synthesis controls the expression of differentiation markers that appear in the gut, the hypodermis, and the muscle of *C. elegans* embryos (11). Indeed, the first several rounds of DNA replication are required for the expression of each differentiation marker, although the timing of marker expression is unlikely to be controlled by counting the normal number of rounds of DNA synthesis.

Holtzer et al. (15) have proposed the "quantal cell cycle" model, which states that cells can only alter their state of differentiation by passing through the S phase of a specific or quantal cell cycle. Recent studies of the differentiation of various types of cultured cells have shown that gene expression can occur in the complete absence of DNA synthesis (e.g., Ref. 5). However, as

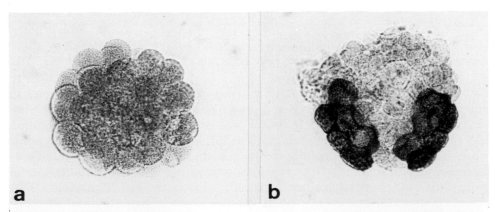

a **b**

FIGURE 4. AChE development in ascidian embryos in which cytokinesis and DNA synthesis are blocked with cytochalasin and aphidicolin, respectively. (a) Arrested 64-cell embryo showing no AChE development. (b) Arrested 110-cell embryo exhibiting AChE in two clusters of 12 presumptive muscle cells (from Ref. 33).

mentioned above, many early developmental events associated with gene activity are likely to require several rounds of DNA replication. In order to explain why DNA replication is necessary for the occurrence of some developmental events, at least two possible molecular mechanisms should be contemplated. The first is the possibility that some stepwise change takes place in the DNA itself during each round of DNA replication. DNA could count its replication number by some kind of modification such as DNA methylation at specific sites (14, 27, 31). Another possibility is that during a specific number of DNA replications there may be a gradual modification in the conformation of the chromatin.

Heterochronic Mutation

It has long been postulated that the temporal control of developmental events may be genetically controlled within embryos. In other words, there may be genes that are exclusively involved in the temporal regulation of development. Recently, several "heterochronic" mutants which affect only the temporal sequence of cell fates, but not the formation of spatial patterns, have been isolated in *C. elegans* (3, 4). Figure 5 shows two representative classes of the temporal alterations that are caused by mutations in the gene *lin-14*. The hypodermal cells of *C. elegans* larva have four cell division patterns, termed S1 to S4, each of which is characteristic of the respective larval stage L1 to L4. The X-linked semidominant mutation in the *lin-14* allele, termed *n536*, results in supernumerary molts. This mutant displays "retarded" development, in which the S1 division pattern that is specific to the L1 larva is repeated at the L2 larval stage. In addition, the S2 pattern occurs twice at both the L3 and the L4 larval stages. In contrast, a recessive mutation of *lin-14(n536n540)* causes a precocious development, in which the S1 pattern is skipped in the L1 larva and the precocious expression of S2, S3, and S4 cell lineage patterns occurs at the L1, L2, and L3 stages, respectively. Adult cuticle formation occurs twice: at the L3 molt and again at the normal time of the L4 molt.

The isolation of heterochronic mutants in *C. elegans* illustrates two important principles. First, there actually are genes which control the temporal order of developmental events. Second, these mutants tell us about the evolutionary significance of temporally altering embryological events. The sexual maturation of *lin-14(n536)* hermaphrodites occurs at the normal stage in spite of supernumerary molts occurring beyond the normal number of four molts observed in the wild type. Heterochronic mutations of this sort in *C. elegans* larvae thus result in neoteny, i.e., sexual maturation occurring in the larval body.

Developmental Events in Which Timing Is Not Directly Associated with Cell Cycles

There are many developmental events in which timing apparently is not controlled by the number of cell divisions, rounds of DNA replication, or the

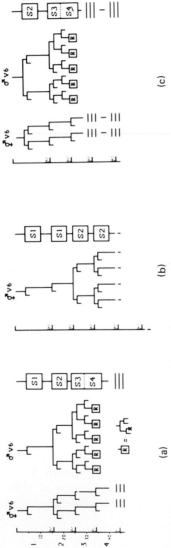

FIGURE 5. Heterochronic lateral hypodermal cell lineage patterns of different phenotypic classes of *lin-14* mutants. The vertical columns of boxes refer to the lineage patterns S1– S4 that comprise each lineage in the wild type. (*a*) Lineages generated by postembryonic blast cell V6 of the wild-type hermaphrodite and male. V6 is one of a set of similar lateral hypodermal blast cells positioned along the lateral line of the animal. (*b*) *lin-14(n536)*, class I phenotype; semidominant, retarded mutant. (*c*) *lin-14(n536n540)*, class II phenotype; recessive, precocious mutant (from Ref. 4).

nucleo-cytoplasmic ratio. Gastrulation, rRNA synthesis, and fibronectin synthesis in amphibian embryos, the formation of micromeres in sea urchin embryos, gastrulation in starfish and sturgeon embryos, and blastocyst formation in mouse embryos are examples of such events (for reference see Ref. 32).

The synthesis of rRNA in *Xenopus* embryos begins around the time of the MBT. If *Xenopus* embryos at early cleavage stages are treated with cytochalasin B or colchicine, cleavage is completely blocked. The amount of rRNA in colchicine-blocked embryos was less than $\frac{1}{25}$ that found in normal control embryos. However, rRNA synthesis in these cleavage-arrested embryos begins at almost the same time that it begins in normal embryos (38). In addition, the formation of nucleoli (which is a reflection of rRNA synthesis) in cleavage-arrested embryos begins at the normal time. Furthermore, the onset and rate of rRNA synthesis in embryos that developed from eggs in which about 40 to 50 percent of the cytoplasm had been extracted were not significantly different from that found in the control embryos.

Mouse blastocyst formation (cavitation) occurs on the fourth day of development at the morula stage when the embryo is comprised of approximately 32 cells. Several studies have examined the possible causal relationship between the "clock that signals" the time of blastocoel formation and the (1) absolute cell number, (2) total number of cytokineses, (3) nucleo-cytoplasmic ratio, and (4) number of DNA replication cycles. Artificial manipulation of embryos, including the inhibition of cleavage and the production of heteroploids, has eliminated the first three possible candidates (e.g., Ref. 37). In order to determine the role of the fourth candidate, preimplantation mouse embryos were treated with aphidicolin for 8 h during the S phase of the fourth cleavage division; therefore, DNA synthesis was inhibited by about 90 percent for the duration of the treatment. However, although aphidicolin treatment produced a delay in cell division, the onset of blastocyst formation was not affected. The treated embryos actually cavitated a few hours ahead of the control embryos at approximately half the normal cell number (9). This result indicates that the timing of blastocyst formation is not regulated by counting the number of DNA replication cycles completed since fertilization, but by some other intrinsic cellular clock.

Cytoplasmic Clocks

In 1904, Wilson (40) observed that when the polar lobe is isolated from trefoil-stage *Dentalium* (molluscan) eggs, they exhibit cyclic constrictions and changes in shape in parallel with the cleavage cycle of normal eggs. Similar morphological changes in the enucleated cytoplasmic fragments (merogons) have been reported in various kinds of invertebrate and vertebrate eggs, including mollusks, annelids, sea urchins, starfish, ascidians, amphibians, and mice (for reference see Ref. 31). These observations suggest that there are cyclical changes in egg cytoplasm that are common to most, if not all, animals.

One example of the cyclical activity of egg cytoplasm is shown in Fig. 6. The artificially activated *Xenopus* egg exhibits distinct cycles of "rounding up" and "relaxation." This behavior clearly indicates the oscillatory activities of the egg cytoplasm, which illustrates the principle of "cytoplasmic clocks" (16). The start of the clock is triggered in the absence of a nucleus or centriole and can be maintained for at least several cleavage cycles.

The rounding-up intervals of the nonnucleated fragments of *Xenopus* eggs are slightly, but significantly, longer than the cleavage intervals of the nucleated fragments and whole eggs. Transplantation of nuclei obtained from gastrula cells could accelerate the cytoplasmic cycle of nonnucleated *Xenopus* egg fragments. Furthermore, when sperm are treated with Triton X-100 and are injected into the rounding-up phase of nonnucleated *Xenopus* egg fragments, swollen vesicular nuclei are frequently observed, whereas swollen nuclei were never observed if sperm were injected into egg fragments in the relaxing phase (29). Therefore, the egg cytoplasm may play a more important role in regulating the phase and length of the cell cycle than any of the components that are associated with the mitotic apparatus.

The length of the cell cycle for synchronous cleavages is rigidly determined for each animal species. However, it has been shown that the duration of the cleavage cycle can be altered by reciprocal cytoplasmic transfers between a species with a fast (*Xenopus*) and a slow (*Pleurodeles*) cell cycle length (1). Injection of *Xenopus* egg cytoplasm, for example, induces precocious cleavage furrows in activated *Pleurodeles* eggs and leads to nearly a 30 percent shortening of the cell cycle compared with control eggs. This activity or the activity of the "cleavage timing system" (CTS) is found only in the clear supernatant that is obtained by centrifugation of *Xenopus* eggs at 120,000g for 1 h. The preliminary characterization of these fractions indicates that the factors are thermostable and resistant to RNase treatment but the CTS activity is abolished after protease treatment. These results suggest that there are cytoplasmic components with a proteinlike nature which determine the endogeneous cleavage-cycle length. In addition, Shinagawa (34) has demonstrated that the cytoplasmic factors that induce the periodic activities of the cell cycle are not distributed uniformly throughout the *Xenopus* egg but are primarily localized in the animal hemisphere.

There is evidence that suggests a causal relationship between a cytoplasmic clock and the timing of micromere formation in sea urchin eggs (e.g., Ref. 8). As previously mentioned, the fourth cleavage of the vegetal blastomeres occurs unequally, resulting in four macromeres and four micromeres. Experiments have shown that if an early cleavage is suppressed or retarded and the eggs are then returned to normal seawater, the treated eggs reach a delayed 8-cell stage, and the four cells on the animal side often become larger while those on the vegetal side are often smaller (25, 28). They simulate the micromere formation of the normal 16-cell stage. This phenomenon is designated "precocious" micromere formation. Rhythmic fluctuation of the sulfhydryl content of egg proteins that accompany the cleavage cycles has been demonstrated in sea urchin embryos. Dan and Ikeda (8) have shown:

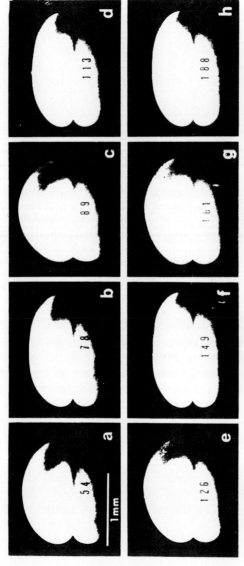

FIGURE 6. Eight sequential still pictures reproduced from a 16-mm time-lapse film, showing periodic changes in the height of an unfertilized *Xenopus* egg activated by pricking (the vitelline membrane removed). Numerals indicate time (min) after activation. Each picture includes reflection of the egg in the supporting glass surface (from Ref. 13).

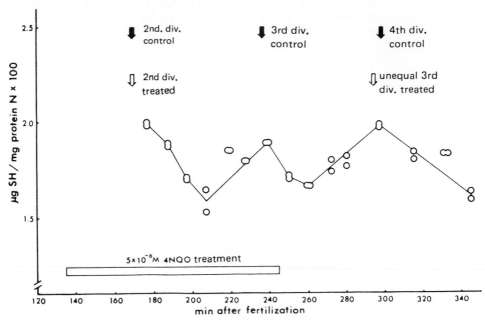

FIGURE 7. Relationship between periodic fluctuation of sulfhydryl content in the KC1-soluble fraction of egg proteins and timing of micromere formation in sea urchin eggs. Eggs are treated with 4NQO from 135 to 245 min after fertilization and later removed from the medium. The sulfhydryl content is at its greatest in spite of the absence of cleavage in the presence of 4NQO at the time corresponding to the third cleavage of the control embryos. The treated egg forms precocious micromeres at the third cleavage, the time of which almost coincides with the fourth cleavage of normal eggs (from Ref. 8).

1. If eggs are treated under conditions that stop nuclear division but preserve the sulfhydryl cycle, followed by a release from these conditions, precocious micromeres are formed (Fig. 7).
2. Conditions that freeze both the nuclear and cytoplasmic rhythms will not induce precocious micromere formation.
3. If conditions that leave nuclear activities intact but prevent cytoplasmic rhythms are followed by normal culture conditions, micromere formation does not occur at the 16-cell stage.

These experiments provide convincing evidence that the clock responsible for the timing of micromere formation is intimately related to the cyclic fluctuation of the sulfhydryl content in the cytoplasm and is not related to cytokinesis, nuclear division, or cycles of DNA replication.

Conclusions

Embryological development is a dynamic four-dimensional process. Time is an indispensable element in any interpretation of development. Many recent studies suggest that the emergence of events during early embryonic development is controlled by timing mechanisms or "developmental clocks" which are attributed to cell division cycles. Although it is an attractive idea that the development of an embryo is regulated by a single clock set in motion at fertilization, this is not the case. Several "alarm mechanisms" probably time the initiation of developmental events. Recently, a gene which only affects the temporal order of developmental events without affecting the spatial order has been isolated (3). Further research will no doubt elucidate more of these genes, which in turn will enhance our understanding of the molecular basis of timing mechanisms.

Acknowledgment

I am grateful to Dr. William R. Bates for his helpful and constructive suggestions on this manuscript.

General References

Satoh, N.: 1984. Cell division cycles as the basis for timing mechanisms in early embryonic development of animals. In: *Cell Cycle Clocks* (L. M. Edmunds, Jr., ed.) pp. 527–538, Marcel Dekker, New York.

Satoh, N.: 1985. Recent advances in our understanding of the temporal control of early embryonic development in amphibians. *J. Embryol. Exp. Morphol.* **89**:257–270.

References

1. Aimar, C., Delarue, M., and Vilain, C. 1981. Cytoplasmic regulation of the duration of cleavage in amphibian eggs. *J. Embryol. Exp. Morphol.* **64**:259–274.

2. Alexandre, H., De Petrocellis, B., and Brachet, J. 1982. Studies on differentiation without cleavage in *Chaetopterus*. Requirement for a definite number of DNA replication cycles shown by aphidicolin pulses. *Differentiation* **22**:132–135.

3. Ambros, V., and Horvitz, H. R. 1984. Heterochronic mutants of the nematode *Caenorhabditis elegans*. *Science* **226**:409–416.

4. Ambros, V., and Horvitz, H. R. 1987. The lin-*14* locus of *Caenorhabditis elegans* controls the time of expression of specific postembryonic developmental events. *Genes & Dev.* **1**:398–414.

5. Chiu, C. -P., and Blau, H. M. 1984. Reprogramming cell differentiation in the absence of DNA synthesis. *Cell* **37**:879–887.

6. Chulitskaia, E. V. 1970. Desynchronization of cell divisions in the course of egg cleavage and an attempt at experimental shift of its onset. *J. Embryol. Exp. Morphol.* **23**:359–374.

7. Cowan, A. E., and McIntosh, J. R. 1985. Mapping the distribution of differentiation

potential for intestine, muscle, and hypodermis during early development in *Caenorhabditis elegans. Cell* 41:923–932.

8. Dan, K., and Ikeda, M. 1971. On the system controlling the time of micromere formation in sea urchin embryos. *Dev. Growth & Differ.* 13:285–301.

9. Dean, W. L., and Rossant, J. 1984. Effect of delaying DNA replication on blastocyst formation in the mouse. *Differentiation* 26:134–137.

10. Edgar, B. A., Kiehle, C. P., and Schubiger, G. 1986. Cell cycle control by the nucleo-cytoplasmic ratio in early *Drosophila* development. *Cell* 44:365–372.

11. Edgar, L. G., and McGhee, J. D. 1988. DNA synthesis and the control of embryonic gene expression in *C. elegans. Cell* 53:589–599.

12. Freeman, G. 1976. The effects of altering the position of cleavage planes on the process of localization of developmental potential in *Ctenophore. Dev. Biol.* 51:332–337.

13. Hara, K., Tydeman, P., and Kirschner, M. 1980. A cytoplasmic clock with the same period as the division cycle in *Xenopus* eggs. *Proc. Nat. Acad. Sci. U.S.A.* 77:462–466.

14. Holliday, R., and Pugh, J. E. 1975. DNA modification mechanisms and gene activity during development. *Science* 187:226–232.

15. Holtzer, H., Rubinstein, N., Fellini, S., Yeoh, G., Chi, J., Birnbaum, J., and Okayama, M. 1975. Lineages, quantal cell cycles, and the generation of cell diversity. *Q. Rev. Biophys.* 8:523–557.

16. Kirschner, M., Gerhart, J. C., Hara, K., and Ubbels, G. A. 1980. Initiation of the cell cycle and establishment of bilateral symmetry in *Xenopus* eggs. In: *The Cell Surface: Mediator of Developmental Processes* (S. Subtelny and N. K. Wessells, eds.), pp. 187–215, Academic Press, New York.

17. Kirschner, M., Newport, J., and Gerhart, J. 1985. The timing of early developmental events in *Xenopus. Trends Genet.* 1:41–47.

18. Kobayakawa, Y., and Kubota, H. Y. 1981. Temporal pattern of cleavage and onset of gastrulation in amphibian embryos developed from eggs with the reduced cytoplasm. *J. Embryol. Exp. Morphol.* 62:83–94.

19. Lillie, F. R. 1902. Differentiation without cleavage in the egg of the annelid *Chaetopterus pergamentaceus. Arch. Entwicklungmech. Org. (Wilhelm Roux)* 14:477–499.

20. Mita, I. 1983. Studies on factors affecting the timing of early morphogenetic events during starfish embryogenesis. *J. Exp. Zool.* 225:293–299.

21. Newport, J., and Kirschner, M. 1982. A major developmental transition in early *Xenopus* embryos: I. Characterization and timing of cellular changes at the midblastula stage. *Cell* 30:675–686.

22. Newport, J., and Kirschner, M. 1982. A major developmental transition in early *Xenopus* embryos: II. Control of the onset of transcription. *Cell* 30:687–696.

23. Niki, Y. 1984. Developmental analysis of the *grandchildless (gs(1)N26)* mutation in *Drosophila melanogaster*: Abnormal cleavage patterns and defects in pole cell formation. *Dev. Biol.* 103:182–189.

24. Nishikata, T., Mita-Miyazawa, I., Deno, T., Takamura, K., and Satoh, N. 1987. Expression of epidermis-specific antigens during embryogenesis of the ascidian, *Halocynthia roretzi. Dev. Biol.* 121:408–416.

25. Painter, T. S. 1915. An experimental study in cleavage. *J. Exp. Zool.* 18:299–323.

26. Petzoldt, U., Illmensee, G. R., Burki, K., Hoppe, P. C., and Illmensee, K. 1981. Protein synthesis in microsurgically produced androgenetic and gynogenetic mouse embryos. *Mol. Gen. Genet.* **184**:11–16.

27. Razin, A., and Riggs, A. D. 1980. DNA methylation and gene function. *Science* **210**:604–610.

28. Rustad, R. C. 1960. Dissociation of the mitotic time-schedule from the micromere "clock" with X-rays. *Acta Embryol. Morphol. Exp.* **3**:155–158.

29. Sakai, M., and Shinagawa, A. 1983. Cyclic cytoplasmic activity of non-nucleate egg fragments of *Xenopus* controls the morphology of injected sperms. *J. Cell Sci.* **63**:69–76.

30. Satoh, N. 1982. Timing mechanisms in early embryonic development. *Differentiation* **22**:156–163.

31. Satoh, N. 1984. Cell division cycles as the basis for timing mechanisms in early embryonic development of animals. In: *Cell Cycle Clocks* (L. N. Edmunds, Jr., ed.), pp. 527–538, Marcel Dekker, New York.

32. Satoh, N. 1985. Recent advances in our understanding of the temporal control of early embryonic development in amphibians. *J. Embryol. Exp. Morphol.* **89**:257–270.

33. Satoh, N., and Ikegami, S. 1981. A definite number of aphidicolin-sensitive cell-cyclic events are required for acetylcholinesterase development in the presumptive muscle cells of ascidian embryos. *J. Embryol. Exp. Morphol.* **61**:1–13.

34. Shinagawa, A. 1985. Localization of the factors producing the periodic activities responsible for synchronous cleavage in *Xenopus* embryos. *J. Embryol. Exp. Morphol.* **85**:33–46.

35. Signoret, J., Lefresne, J., Vinson, D., and David, J. C. 1981. Enzymes involved in DNA replication in the axolotl. II. Control of DNA ligase activity during very early development. *Dev. Biol.* **87**:126–132.

36. Smith, L. D., and Ecker, R. E. 1970. Uterine suppression of biochemical and morphogenetic events in *Rana pipiens*. *Dev. Biol.* **22**:522–637.

37. Smith, R., and McLaren, A. 1977. Factors affecting the time of formation of the mouse blastocoele. *J. Embryol. Exp. Morphol.* **41**:79–92.

38. Takeichi, T., Satoh, N., Tashiro, K., and Shiokawa, K. 1985. Temporal control of rRNA synthesis in cleavage-arrested embryos of *Xenopus laevis*. *Dev. Biol.* **112**:443–450.

39. Whittaker, J. R. 1973. Segregation during ascidian embryogenesis of egg cytoplasmic information for tissue-specific enzyme development. *Proc. Nat. Acad. Sci. U.S.A.* **70**:2096–2100.

40. Wilson, E. B. 1904. Experimental studies on germinal localization. *J. Exp. Zool.* **1**:1–72.

Questions for Discussion with the Editor

1. *Do you suppose that some embryos, e.g., highly "mosaic" marine invertebrates, employ timing mechanisms more extensively than do others, e.g., highly "regulative" mammalian embryos?*

No, I don't think so. I suppose that, in principle, embryos of all animals employ similar timing mechanisms for each developmental process. However, timing mechanisms are not independent of other developmental phenomena but are very closely associated with them. The development of cellular patterns of embryos may reflect timing mechanisms and vice versa. In embryos having a mosaic type of developmental pattern, the timing mechanisms are also determinative, whereas in embryos having a regulative developmental pattern, the timing of developmental events can be changed by both internal and external environmental conditions. Therefore, my impression is that regulative embryos likely employ timing mechanisms more extensively than mosaic embryos. The former type of embryo appears to have a greater ability to adjust its clock, whereas in the latter type of embryo the clock is rigidly set.

2. *Which of the "timing systems" will, in your opinion, provide the most direct route to a complete understanding of its features at the molecular level?*

Investigation of heterochronic mutations would provide the most direct route. The content of my chapter might give readers the impression that each of the supposed timing mechanisms has not been explored thoroughly. That is true. From the standpoint of current genetic and molecular biology techniques, heterochronic mutants look like the most promising "timing system." Some of the questions to answer include: What kinds of alteration in the genes cause changes in the temporal pattern of cell divisions? What kinds of changes occur in the cytoplasm of cells and tissues? Are heterochronic genes common in every species? How do they function during normal embryogenesis? Such questions should be investigated in the future.

CHAPTER **7**

An Essay on Redundancy within Developmental Processes

George M. Malacinski and Anton W. Neff

*The cautious engineer makes a construction so
strong and durable that it will be able to stand a
load which in practice it will never have to bear.*
H. SPEMANN (1962)

Introduction

REDUNDANCY IN BIOLOGIC PROCESSES is prevalent and takes several forms. It operates in many situations, including those at the genetic, metabolic, cellular, and physiological levels. Although redundancy is widespread in developing and/or differentiating systems, many types are not particularly well acknowledged by either researchers or model builders. This chapter explains that in several simple systems it is becoming increasingly apparent that some gene products are dispensable. Multiple copies of various genes exist in many organisms. Deletion of one of the gene copies has, in many instances, no demonstrable effect on the organism's phenotype. At the cellular level, especially in vertebrates, cell death is a common feature of embryogenesis. Early in development there is an excess of motoneurons in relation to the number of muscle cell targets. Eventually, however, most motoneurons are eliminated. Even more extreme examples of redundancy exist at the level of cytoplasmic organization. Symmetry breaking in amphibian eggs, which normally is triggered by sperm entrance and usually involves complex cortical displacements, can be achieved entirely without participation of either of those events. Redundancy of that type most likely involves *alternative pathways* that under normal circumstances would not be employed but which can, should the need arise, be called into play.

This essay explores the extent to which redundancy exists in developing systems and discusses the implications of redundancy for understanding embryogenesis and for designing experimental strategies. Before considering examples of redundancy in more detail, however, a discussion of background information, especially concerning the context in which redundancy is best understood, is provided.

Two Extreme Views of Embryogenesis

"Perfectionist" Models

One view of the cytoplasmic organization and gene control systems which operate in developing systems is that they are optimally designed, highly ordered, and very efficient. This view holds that developmental events approach "perfection" in both overall design and execution. It is expected that the application of rational thought (e.g., human logic) will quickly lead in a straightforward manner to an understanding of, for example, the gene regulatory circuits which specify pattern in early embryogenesis. That is, the mechanisms which embryos employ to generate complex structures (morphogenesis) or tissue specializations (differentiation) are analogous to a modern assembly line in an electronics factory.

This view is often coherent and has emerged from at least two sources. Data on the molecular biology of the gene, which often are interpreted to emphasize economy of energy, size, and shape, provide one reservoir of thought. The other is the widespread use of computer programs to simulate various biologic phenomena.

Several examples of the types of models that this perfection oriented approach has generated are available. The Britten and Davidson (3) scheme for regulating embryonic gene expression through a simultaneous activation of batteries or groups of genes was one of the first and most prominent examples. More recent theories include those advanced by Stubblefield (48), which liken embryonic cells to "information management processors," or by Bailey (1), which employ sets and subsets of regulatory genes for producing neat cascades or "gene-activity trees." This view emphasizes the apparent efficient and streamlined logic displayed by embryonic systems. It is postulated that a set of regulatory genes is activated in a rigid, sequential fashion. Each developmental event is considered to be the consequence of a linear activation of control genes. The first control unit in the series is activated by factors that were generated during oogenesis.

Another contemporary example of these highly ordered models is the "regulator hypothesis" (8). A relatively small number of genes for cell adhesion molecules (CAMs) are thought to direct morphogenetic events in schedules that precede cytodifferentiation. A separate set of genes controls differentiation. This idea is global in scope (holistic) and attempts to explain how virtually all early morphogenetic events occur. It is also unifying, in that a diverse group of organisms is postulated to employ CAMs in key situations.

These perfectionist models are usually characterized by the implication that a single guiding principle or causal theme drives development in virtually all organisms. The perfectionist view does not necessarily disregard the impact of evolution. Rather, this view proposes that adaptive tendencies have met the challenge of selective forces head on and that perfection has resulted.

Other models strive to achieve simplicity based upon the use of a minimal number of assumptions or rules. Extensive discussions of such models, including the popular reaction-diffusion models, have been offered by Meinhardt (31) and Kaufman (24) in Vol. I of this series, *Primers in Developmental Biology*. A key feature of these models is their single-track nature. Pattern regulation, for example, is usually viewed more as involving a return to an earlier state of morphogenesis or differentiation and/or a reorganization of preexisting components. Seldom are alternative, surplus, or complex combinatorial mechanisms invoked to explain the reconstitution or redirection of a morphogenetic pattern.

In summary, the perfectionist view maintains that development is regulated by a set of streamlined control systems. A strict serial progression, with little need for alternative or fail-safe mechanisms, is considered sufficient to guide even the most complex developmental pathways.

A very different view of the cytoplasmic organization and gene control systems that regulate development has been derived from studies of comparative embryology which emphasize evolutionary relationships. This view holds that developmental systems by and large represent collections of mechanisms, each of which has been selected in evolution mainly because it works (e.g., see Ref. 4). In contrast to the perfection model, this "phylogenetic heritage" (i.e., "evolutionary hangover") model proposes that developmental mechanisms are not necessarily streamlined or highly efficient. Compared with the perfectionist view, this model emphasizes constraints as being more important than selective forces (16). That is, embryos are so complex, so cumbersome and, from their phylogenetic history, so specious, that, in reality, very few adaptive changes are possible. Hence, rather than streamlining, embryos keep adding on to preexisting structures. Major single-step restructuring events do not occur. Change occurs in small increments and modifies preexisting structures instead of generating entirely new ones.

An example of the product of a long history of evolution is the widely divergent patterns of gastrulation exhibited by various organisms. In marine invertebrate embryos the relocation of surface cells to the interior is often simple and direct. In vertebrates, however, cells destined to form the middle germ layer (mesoderm) arise at a remote location and need to be translocated a very long distance to their final site in the embryo's interior. Figure 1 illustrates the relatively streamlined cell relocation patterns of a marine protochordate (*Amphioxus*) embryo and the much more cumbersome movement patterns exhibited by prospective mesoderm cells of urodele amphibian and mammalian embryos. Prospective urodele mesoderm cells, for example, arise from a cell lineage that fails to position them close to their final destination. This apparent lack of streamlining no doubt reflects the evolutionary history of mesoderm cells. What is of supreme importance to the embryo is

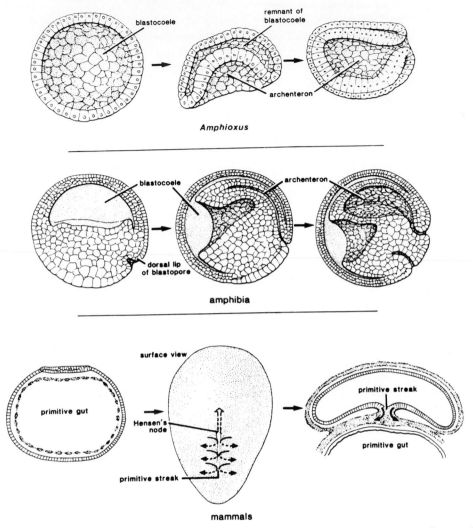

FIGURE 1. Gastrulation patterns in *Amphioxus*, amphibia, and mammals are very divergent.

that gastrulation succeeds in arranging the germ layers (especially the meso-derm) appropriately and in so doing establishes a body plan. The seemingly illogical scenario of moving cells over large distances is probably tolerated simply because it represents the only option for evolutionary change—the "add on" remodeling strategy.

"Phylogenetic Heritage"

The phylogenetic heritage view holds that superimposed on the triplet genetic code is a more complex information system. Some features of that information system are built into the genome, while others reside in the cell's cytoplasm. Furthermore, this view holds that the programs that specify pattern and regulate or control developmental events are *not* streamlined. This view acknowledges two circumstances.

1. Achieving order in developing systems is intrinsically more complex than simply expressing the triplet genetic code.
2. Cytoplasmic organizational systems are necessary, for they amplify the potential of the triplet genetic code.

That is, cytoplasmic systems provide many more opportunities for diversity and expansion of developmental programs than do simple nucleotide sequence changes. In combination, the cytoplasm and the genome expand the amount of information that can be stored in the limited confines of a single cell. A most striking example is inheritance of cortical patterning in ciliates (see Chap. 2, this volume).

An extreme view of the phylogenetic heritage model proposes that in higher eukaryotes, cytoplasmic information and organizational systems function much like bureaucracies. These systems are inclined to resist change, and, if anything, over time they tend to get larger rather than smaller. Most development (e.g., embryogenesis) occurs internally, either inside insulating membranes or embedded in a maternal environment (e.g., mammals). It is, therefore, not subject to the direct external environmental selection pressure routinely experienced by either lower organisms (e.g., prokaryotes) or various physiological processes in higher organisms (e.g., fertilization, sensory perception, etc.). Consequently, embryos, particularly the most recently evolved ones, often display vestigial tissues or organs. Table 1 provides representative examples drawn from several phyla of vestigial structures. Vestigial structures are, however, particularly abundant in higher organisms.

It has been estimated, for example, that there are at least 100 vestigial structures in a typical adult human. They have at least one common characteristic. They tend to show substantial individual variation. For example, the vermiform appendix can range from 2 to 20 cm in length. Bilateral structures show substantial left-right asymmetry. The palmaris longus muscle of the forearm, for example, shows substantial left-right variability and is absent from one or both sides in about 21 percent of all individuals. The muscle appears to be redundant since individuals with a diminished palmaris longus have normal wrist function. The muscle is phylogenetically degenerating and, therefore, is on its way to becoming vestigial. It has a short belly and long tendon, and the end of the tendon is expanded to form the palmar aponeurosis. The fact that the palmar aponeurosis originates from the distal tendon of the palmar longus muscle may be the reason that this muscle persists at all. The palmaris longus–palmer aponeurosis connection exemplifies one further point: the interlocking nature of developmental events is a major constraining force for evolutionary change. A mutation which deletes a tissue or organ is not easily accommodated because of the interdependence of developmental processes. Evolutionary change is more easily accomplished at the termini of

TABLE 1 Examples of Vestigial Structures (Tissues and/or Organs)*

Protozoa
  Flagella—*Nochtiluca miliaris*

Plants
  Extra nuclei of the embryo sac of angiosperms
  Leaves—*Spartium scoparium*

Arthropods
  Claw—*Uca*
  Back limbs—Hermit crab

Insects
  Wings—*Amalopteryx*
  Limbs—*Acerentomon*

Fish
  Fins—*Entropius laticeps*
  Eyes—*Anoptichthys jordani*

Amphibians
  Eyes—*Ichthyophis*
  Limbs—*Amphiuma*
  Lung—*Siphonops*

Reptiles
  Limbs—*Chalcides quentheri*
  Left lumg—snakes

Birds
  Wings—*Apteryx*
  Internal sex organs—right side in female birds

Mammals
  Allantois
  Vermiform appendix—humans
  Coccyx—primates
  Pelvic girdle—whales
  Ear muscles—humans seal
  Teeth—wisdom teeth in humans
  Eyes—blind mole
  Splint bones—horse
  Vestigial remnants of human embryonic urogenital structures†
    Male—gubernaculum testes, paradidymis, appendix of epididymis and testis, prostatic utricle, seminal colliculus.
    Female—rete ovarii, epoöphoron, paroöphoron, appendix of vesiculosa, duct of epoöphoron, duct of Gartner, hydatid, hymen

*Primarily from Ref. 25.
†From Ref. 34.

developmental sequences. Hence, as a consequence of their phylogenetic history, embryos have come to resemble "contraptions."

According to this view, development, especially embryogenesis, is primarily a utilitarian event. Compared to the length of the organism's entire life cycle, embryogenesis is of relatively short duration, and for most organisms, survival rates for embryos are exceedingly low.

Ironically, however, embryogenesis is the most effective place for change to occur. Even minor changes have the potential to cascade into major alterations in the morphology of the adult.

In summary, the phylogenetic heritage view portrays development as a rather imperfect event. It requires of most embryos only one thing: that they somehow manage to succeed in developing (usually in their insulated microenvironment). Even the burden of excess baggage (e.g., vestigial structures) is of secondary importance, as long as their survival rates are sufficiently high to ensure persistence of the species.

Significance of the Dichotomy of Views

It is certainly not necessary to portray embryogenesis in terms of only one or the other extreme type of mechanism. The development of some parts of an embryo, especially those which confer direct selective advantages (e.g., the sense organs such as eyes, ears, etc.), might be highly perfectionist. Others, such as those involved in digestion (e.g., exocrine and endocrine pancreas) might follow the phylogenetic heritage model more fully. Nevertheless, several significant points emerge from a discussion framed in terms of extremes:

1. *The perfectionist and phylogenetic heritage views generate divergent experimental strategies.* The perfectionist model favors a reductionist approach to experimental design: an understanding of details will permit a reconstruction and understanding of the whole embryo. Conversely, the phylogenetic heritage model proposes that only by understanding the whole embryo will the significance of its details become apparent. In the former model, molecular biology provides the key experimental approach. For the latter model, cell biology and experimental morphogenesis provide the main driving forces.

2. *Expectations for formulating rules or generalizations differ.* The perfectionist view predicts that details of developmental regulatory events will be similar in diverse organisms. For example, the genetic control of pattern specification events such as segmentation would be expected to share common features among many organisms. Hence, the discovery that the homeobox, a 180 base pair (bp) protein-coding sequence associated with several fruit fly genes that controls embryonic segmentation patterns, exists in the genomes of a wide variety of organisms has generated substantial excitement. It is even lauded as a possible breakthrough, a "Rosetta stone," in understanding the commonality of developmental regulation. Despite serious attempts at understanding its function, especially in vertebrates, no simple unifying statement has emerged. Rather, the converse has been the case. Despite the superficial similarities between segmentation in insect embryos and segmentation along the primary embryonic axis of vertebrates, no homeobox gene expression has yet been detected in the segmenting mesoderm of any of the half-dozen or more vertebrate embryos which have been thoroughly examined. Evolutionary considerations indicate that the common ancestors of vertebrates and arthropods were not segmented (21). Complicating matters further is the observation that the

genomes of unsegmented echinoderms contain homeobox sequences (7). Nevertheless, the perfectionist model proposes that "commonality" rather than "individuality" will be a prevalent feature of developmental processes.

The phylogenetic history view predicts that no general rules that are applicable across phyla will be generated to explain developmental phenomena in diverse species. It will likely be impossible to draw up a set of rules which govern, for example, the diverse gastrulation patterns in vertebrates (Fig. 1). At best, it will be possible to formulate a list of the key features of gastrulation in this or that organism. The role of model systems such as the nematode, fruit fly, or amphibian is to provide a guide for technical approaches and conceptual thinking rather than to serve as general information for dealing with more complex (e.g., mammalian) or novel situations.

3. *It may be necessary to concede that embryogenesis is intrinsically "illogical" (26).* This is not to imply that development is irrational. Instead, it implies that embryogenesis is controlled more by a bureaucratic program than by a well-designed constitutional authority. Acknowledging this fact may not be easy. As mentioned earlier, recognizing that the temporal and spatial organization of developmental events are more complex than the triplet code is not necessarily comforting. Even relatively recent accounts of pattern formation in the nematode *Caenorhabditis elegans* emphasize that pattern formation is directly specified by gene action. Interpretation of the latest data, however, finally recognizes the important role cell interactions might play in such key pattern formation events as vulval development (47) (also see Chap. 12, this volume).

4. *The phylogenetic heritage model but not the perfectionist model predicts that diligent searches will yield widespread redundancy at most levels of developmental control.* Contemporary analyses are slowly but steadily permitting answers to the general question "Is regulation of gene expression optimal?" Cavener (4), for example, recently paraphrased this question in the following terms: "Does expression always imply function?" The *Drosophila* enzyme glucose dehydrogenase (GLD) was used as a test system for assaying gene expression at all stages of development. The mutant phenotype of enzyme-deficient *Drosophila* is a tough and pliable puparium which cannot hatch in the absence of GLD. In wild-type individuals the enzyme is expressed at *all* stages of development, including the adult. However, mutant larvae manually removed from their tough puparium are perfectly viable and fertile, indicating that the enzyme's presence is dispensable at most stages. It appears that the seemingly "out of place" expression of genes such as GLD provides solid support for the phylogenetic heritage model. In this situation a battery of genes may code for disparate gene products. Controlled by common regulatory elements, an occasional mismatch or extravagance like GLD may have been left over from the evolutionary process.

From examples such as GLD it should be apparent that in order for biologists to construct a holistic model of embryogenesis in even a single organism, false starts, seemingly superfluous cell behavior patterns, and puzzling regulatory mechanisms will need to be incorporated to generate accurate and realistic features.

The phylogenetic heritage view, with its emphasis that redundancy exists at several levels of organization, including the genomic, cytoplasmic and cellular levels, will provide the focus of this essay. It will be emphasized that redundancy is more prevalent than is often acknowledged. It should be mentioned, however, that the perfectionist view, being easier to comprehend and more practical as an experimental foundation, often pervades contemporary (e.g., reductionist) studies. A more sober, and probably more realistic, approach is the theme of this essay.

Definitions of Redundancy

An accurate definition of the term "redundancy" requires reference to a specific context. In the present essay, redundancy will be defined in the context of levels of biologic organization. At the level of the genome, redundancy is defined as the existence of multiple, identical (or closely related) copies of a specific gene or nucleotide sequence (Fig. 2). The rRNA genes, for example, are generally regarded as being redundant (i.e., exist in multiple copies) in the genomes of most, if not all, higher eukaryotes. The *raison d'être* is the embryo's need to be able to transcribe copious quantities of gene product, i.e., rRNA, in a brief time span. Various other genes are also redundant, including the tRNA genes, histone genes, and 5 S RNA genes. Their copy number (so-called reiteration frequency) varies from approximately ten to several hundred depending upon both the specific gene and the organism.

Another example of redundancy at the genome level is the "gene family." In this case, the individual copies of the specific gene represented by the gene family differ from one another in subtle ways. For example, minor nucleotide sequence substitutions are usually revealed when the protein coding sequences of gene-family members are carefully compared. Gene families have a different *raison d'être* from the *identical* copy repeats mentioned above. Members of gene families often exhibit tissue- and/or stage-specific expression patterns. Individual members of some gene families are thought to display subtle physiological differences which enhance the performance of the cell or tissue in which they are expressed. For example, fetal hemoglobins have a higher oxygen affinity than adult hemoglobins, thus facilitating the transfer of oxygen across the placental membranes to the developing embryo. Another example is the acclimation of some fish (e.g., trout) to changes in ambient temperature. They respond, in part, by calling into play members of multigene families which code for metabolic enzymes (isozymes) that function more efficiently at the new temperature (20). In other cases, it is believed that individual members of a gene family are separately regulated. The existence of multiple copies of the structural gene permits the dedication of individual copies to specific regulatory modes. Fine tuning of spatial and temporal expression patterns is thereby achieved.

Figure 2 illustrates the usage patterns for redundant (i.e., repeated) genes. The situation is, however, even more complicated than depicted in Fig. 2. In some instances, as is documented below, mutants from which one of the

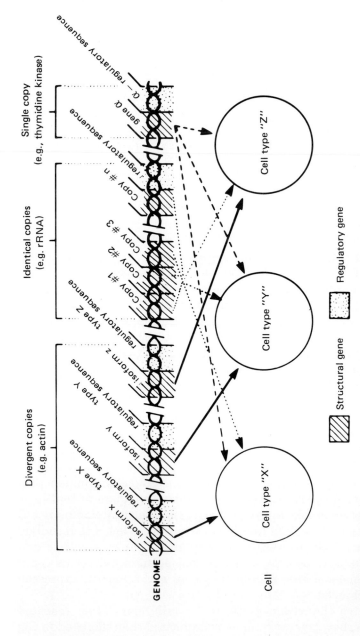

FIGURE 2. Genome redundancy for genes with various functions. Some gene families contain members (e.g., actin isoforms X, Y, and Z) which are expressed in different cell types. Other genes exist in multiple, identical copies and are expressed in all cells (e.g., rRNA genes). Finally, some single copy genes (e.g., thymidine kinase) are also expressed in all cells.

members of the gene family has been deleted display fully normal developmental and physiological properties. In those cases, *alternative, substitute, reserve*, or *surrogate* mechanisms are apparently called into play to carry out the function previously filled by the deleted gene. That is, *interchangeable components* exist in developing systems.

In this latter context, redundancy takes on a somewhat different meaning. Here it means that alternative ways of carrying out a function exist. In contrast to the widely accepted usage by molecular biologists of the term "redundancy" to describe identical, repeated copies of a gene, redundancy is also used by cellular and developmental biologists to mean that within a cell there is more than one way to carry out a function. A cell, or its subcellular components, normally carries out its duties in an optimal fashion. If that optimal route is, for whatever reason (e.g., deletion mutation, environmental stress, etc.), preempted, a substitute mechanism can be employed. The availability of alternative means of accomplishing a particular task represents, in the present essay, the *redundancy in the system. It is, therefore, always necessary to define the context or biologic level associated with each use of the term redundancy*. Molecular biologists equate redundancy with multiple copies of a gene or alternate triplet codes. In this chapter, however, the term is expanded to encompass not only molecular components but cellular and supracellular events as well.

Historical Perspective: *Double Assurance*

According to Spemann (45), the possibility that developing systems might be "overbuilt" in order to withstand perturbation was first proposed by Rhumbler in 1897 (37). He proposed the term "double assurance" to explain the role of astral rays during division of amphibian cells. That term was borrowed from engineering and denotes a condition under which a system is designed to accommodate substantially more stress than it normally encounters. More complex systems (e.g., tissues or organs) were likewise considered by Braus (2), as early as 1906, to employ similar double assurance mechanisms. Eruption of the amphibian limb was viewed as involving two separate processes to ensure that the limb bud perforates the tadpole skin. First, the growing limb bud generates sufficient force of its own to penetrate the skin, as demonstrated by grafting the bud under flank epidermis. Second, the operculum area, under which the limb bud normally develops, thins and develops a small perforation of its own, as shown by surgically removing the resident bud.

A parallel set of observations on lens development was made by Spemann himself. He was impressed by the fact that even in the absence of the inducing tissue (optic cup), a lens, albeit a small one, differentiates in surgically manipulated amphibian embryos (44). He invoked Rhumbler's term—double assurance—to account for that observation and further refined the "principle of double assurance" in the following terms (Ref 45, pp. 95–96):

This principle of double assurance seems to be of far-reaching importance, not only in development, but also in the functions of the adult organism. It may be understood in a twofold sense. In order to achieve a definite technical purpose two different arrangements may be made, and only when, or as far as, the first becomes deficient does the second, which at the outset remains in reserve, enter into action. Each of them is fully demanded at its time; but if the first has proved to be sufficient the second does not need to enter into action at all. On the other hand, there may be two arrangements working in the same sense from the beginning; either of them would be sufficient to do the whole work alone, but in normal conditions only half of the work is demanded of each. In the first case, one rope breaks after the other; in the second both hold or both break at once.

Other terms, such as "combinatorial achievement of unity" and "synergetic principle of development" were also employed by that earlier generation of embryologists to describe the concept of double assurance.

Examples of Redundancy

In recent years there has emerged a set of well-documented examples of redundancy at several levels in developmental processes which supplements the double assurance concept. These levels include gene expression (e.g., regulatory circuits), subcellular organization (e.g., symmetrization of various eggs), cell fates (e.g., superfluous immunoglobulin-producing cells), cell communication (e.g., migration cues), and physiology (e.g., homeostasis mechanisms). *The plethora of examples is due mainly to the increased resolution at which molecular genetics, developmental genetics, and cellular analyses are now routinely performed.* Prototype examples for gene expression and cell fates are illustrated in Fig. 3. Specific examples for each level of organization follows:

Gene Copy Redundancy

As mentioned previously, several genes, including the rRNA, 5 S RNA, tRNA, and histone genes, are usually represented in the genome as tandem repeats. These repeats serve the obvious function of generating vast quantities of gene product in a relatively short period of time. Although usually termed "gene redundancy," those cases do not fall in the same category as the alternative, surrogate, or substitute mechanisms being discussed here.

The genome of most eukaryotic organisms does, however, contain vast excesses of DNA. The amount of DNA required to code for all the proteins produced during a typical organism's life cycle has been calculated and accounts for only 1 to 5 percent of the total nucleotides in the genome. Furthermore, there exists no linear relationship between amount of DNA in an organism and its physiological-morphological complexity (one aspect of the so called "C-value paradox"). Typically, a eukaryotic genome contains only one essential gene per 30 kilobases (kb) of DNA (reviewed in Ref. 15).

This excess DNA has often been termed redundant. It might, however, represent either (1) structural DNA (e.g., centromeres), (2) regulatory DNA (e.g., enhancer sequences), or (3) vestigial DNA (i.e., serves no essential function—truly redundant or superfluous). In addition, some of this apparently excess DNA no doubt is part of multigene families, which may play important roles in cellular metabolism. Individual members of a multigene family can provide the subtle differences in protein function (or structure) which might be needed to change a reaction or to alter a metabolic pathway slightly. Earlier, several examples (e.g., globin) were mentioned. A computer search of the genetics literature will generate dozens, if not more, additional examples of gene families.

Only a few cases have been analyzed in sufficient detail to provide substantial insight into the possibility that much of the surplus DNA represents true vestigial or redundant DNA. Two cases will be summarized here.

In yeast, transforming DNA was integrated into the genome by homologous recombination to construct diploid strains that were heterozygous for random single disruptions. The percentage of those disruptions which were haploid lethal was surprisingly low: 12 percent were haploid lethal and another 16 percent had less dramatic phenotypes. It was therefore concluded that approximately 70 percent of the yeast genome is not essential for normal growth and differentiation of yeast cells (under laboratory conditions) (15). Thus these surplus sequences could be considered to represent vestigial DNA. A similar situation may exist in mammals, where the paucity of phenotypic effects resulting from attempts at insertional mutagenesis is striking (17).

Another example of putative vestigial DNA comes from the actin multigene family. Most eukaryotic genomes, with the notable exception of yeast, encode multigene families of actin. The situation has been especially well studied in the slime mold. In *Dictyostelium*, actin is a major protein in vegetative cells, for it makes up approximately 1 percent of a cell's newly synthesized protein (11). In addition, actin synthesis is developmentally regulated. As many as 20 actin genes are present in the slime mold's genome. Thus, the cellular and spatial patterning of actin expression is potentially very complex. In fact, the majority of actin genes are expressed only at a very low level (1 to 5 percent of the total actin mRNA) (39). They also appear to be independently regulated. That is, during development several of the actin genes are expressed in a stage-specific manner.

No precise, functional tests for whether any of these 20 actin genes are redundant are available, since our knowledge of *Dictyostelium* genetics is not yet sophisticated enough to develop such tests. Nevertheless, extrapolation from the genetically analyzed cases, such as the acetylcholinesterase, hsp 70, and *ras* genes, which are discussed below, allows one to easily imagine that the developmental life cycle of the slime mold would proceed normally even if deprived of several actin genes. The actin genes that are expressed at especially low levels represent prime candidates for vestigial DNA.

Two explanations for why vestigial DNA persists, rather than being quickly cast off during evolution, have been offered by Loomis and Gilpin (29). First, computer simulations of random duplications and deletions generate the notion that genome size will usually be large and that the number of copies of any individual gene will fluctuate. That is, as a genome evolves, it will

Prototype examples

Actual examples

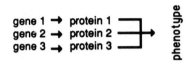

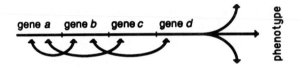

Either of two pathways generate the phenotype (e.g. heat shock response[1]).

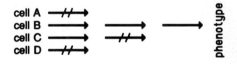

Multiple genes code for similar proteins (e.g. fruit fly yolk proteins[2]).

Many gene rearrangements are not productive (e.g. immunoglobins[3]).

Cell Fates

Prototype examples

Actual examples

Overproduction generates surplus cells which disappear (e.g. neurons[4]).

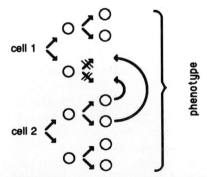

Cells of one type replace those of another type (e.g. Wolffian lens regeneration[5]).

FIGURE 3. Prototype examples of gene expression and cell fate redundancies. Reference 5 provides supplementary information for part 1, Ref. 49 for part 2, Ref. 52 for part 3, Ref. 9 for part 4, Ref. 53 for part 5, and Ref. 40 for part 6.

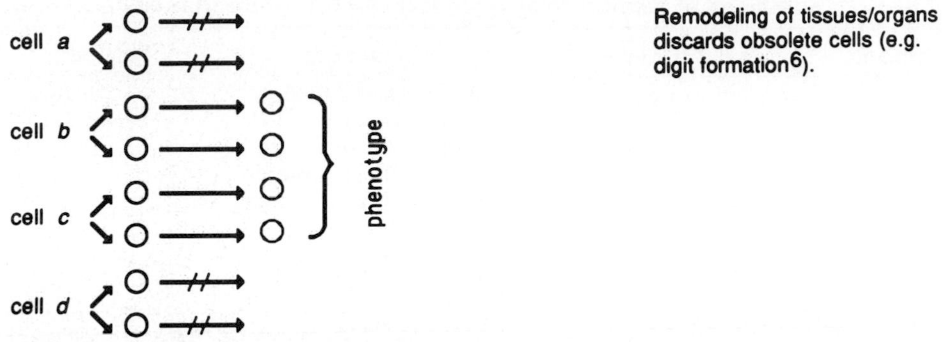

FIGURE 3. (Continued)

accumulate vestigial sequences. The genome's size will stabilize when it is sufficiently large to be able to tolerate occasional deletions without yielding catastrophic phenotypic effects (28). Second, the cost to a cell of maintaining vestigial DNA is estimated to be trivial. Compared with the energy consumption associated with normal housekeeping metabolism, far less than 1 percent of that amount is required to synthesize the vast quantities of vestigial DNA (28). Therefore, in eukaryotes no substantial selective advantage may be gained from eliminating that surplus DNA. Indeed, vestigial DNA may provide a valuable reservoir for facilitating evolutionary change. In a strict sense, therefore, those surplus DNA sequences may not be redundant after all! Perhaps they do serve a function: to provide an energetically inexpensive mechanism for expediting genetic adaptation, both for individuals as well as for whole populations or species.

Gene Expression Redundancy

Only from those organisms which are amenable to detailed genetic analysis can a thorough understanding of gene expression redundancy be achieved. Accordingly, virtually all examples are drawn from genetic studies and are confined to various invertebrates. Although relatively meager at present, the list of examples is growing rapidly. Table 2 summarizes several cases that are categorized with regard to the relevant organism from which the example is derived. Yeast, perhaps the most easily exploited of the genetic systems, offers the most examples. For example, yeast cells which carry a mutation in the topoisomerase I gene are viable. The activity of topoisomerase II can compensate for the defect (51). Furthermore, yeast contain two α-tubulin genes. By varying the functional copy number from zero to several extra copies of each gene, investigators observed that either gene alone is capable of providing all the functions normally ascribed to microtubules (42). These tubulins appear, therefore, to be interchangeable. One can compensate for the absence of the other, i.e., one of the tubulins is nonessential. Recently it has been observed

TABLE 2 Summary of Examples of Gene Expression Redundancy

Organism	Gene	Explanation	Reference
Yeast	Topoisomerase	Topoisomerase II can substitute for topoisomerase I	51
	α-Tubulin	Despite 10% divergence two tubulins are interchangeable	42
	Actin-binding protein	Deletion of one form yields no phenotypic effects	D. Botstein, personal communication
	Hexokinase	When both hexokinases deleted, shift to glucose-glucokinase occurs	30
	Heat shock proteins	Either of two genes confers heat resistance	5
	ras	Either of two *ras* genes is sufficient for growth	50
Nematode	Acetylcholinesterase	Class B enzyme can substitute for class A function	6
Silkworm	Vitellogenin	Vitellogenin-deficient eggs develop normally	54
Fruit fly	Vitellogenin	One of the three yolk proteins can be deleted	49
	bithorax complex	Overlapping regulatory systems substitute for one another	13
Mouse	Immunoglobin genes	Most rearrangements are not useful	52

that actin-binding proteins exist in multiple copies and that when one is genetically deleted, no apparent effects on cellular metabolism can be detected (D. Botstein, personal communication).

At the level of intermediary metabolism, various examples of metabolic pathway shifts, which occur according to the availability of exogenous nutrients, are well known in bacteria. Genetic studies have found similar shifts in yeast. A case in point is the role of the hexokinases. Two such enzymes, which are only 78 percent identical in their amino acid sequences, exist in *Saccharomyces cerevisiae*. Deletion of either enzyme fails to block fructose fermentation. Hexokinase I can functionally replace hexokinase II and vice versa. Furthermore, when both enzymes are deleted, cells can grow on glucose by using a third enzyme, glucokinase, which allows them to bypass the defect (30).

Other examples of dispensable gene products emerge from yeast genetic studies. When a variety of organisms, including both prokaryotes and eukaryotes, are grown at elevated temperatures, enhanced synthesis of a small number of well-defined proteins, the so-called heat shock proteins. Genetic analyses of the most abundant heat shock protein (hsp 70) in yeast have revealed the presence of two hsp 70 genes (*YG100* and *YG102*). These two genes are 97 percent identical to each other in their protein-coding regions but are separately regulated (5). Mutatations of those two genes have been prepared. No phenotypic effect of single mutations of either gene could be recognized. That is, the temperature-sensitive phenotype could be overcome by inserting either a functional *YG100* or *YG200* gene into the yeast's genome. Double mutants do, however, exhibit altered growth properties, demonstrating that heat shock function is indispensable.

It has been suggested that heat shock proteins play a role in normal metabolism as well as in protection of the organism against elevated temperature. The hsp 70 situation, therefore, is a candidate for the list of examples of "housekeeping" functions which employ interchangeable (i.e., in the sense of the present essay—redundant) components. One additional, perhaps conceptually similar, example is that of the yeast *ras* gene.

The yeast genome contains an evolutionary derivative of the cellular proto-oncogenes from the Harvey and Kirsten murine sarcoma virus, commonly known as *ras*. Null mutations in the two yeast *ras* genes have been constructed, and the phenotypes of the cells lacking those genes have been studied. Neither of the *ras* genes is separately essential for spore development, germination, growth, or any of the other normal cellular functions, such as meiosis or mitosis, which are easily examined. One wild-type allele is, however, required for normal function, since disruptions of both genes yield cells which fail to grow (50).

In the nematode *C. elegans* a striking example of interchangeability of enzyme activities is known. An analysis of acetylcholinesterase-deficient mutants has revealed the existence of one especially interesting gene—*ace*-1. Gene dosage experiments indicate that *ace*-1 represents a structural gene that contributes to the enzymic activity of the three different class A acetylcholinesterases. One of the enzyme deficient alleles of *ace*-1, *p1000*, is of special interest: animals which are homozygous for that mutation, and thus lack class A acetylcholinesterases, are behaviorally and developmentally indistinguishable from wild-type animals (6). The remarkable fact is that these class A acetylcholinesterases represent approximately 50 percent of the total *C. elegans* acetylcholinesterase activity. A substantial functional overlap with the remaining acetylcholinesterase (of class B) probably exists. That this is indeed the case is supported by both genetic and histological data. Animals which are homozygous for mutants in either class of enzyme are normal. In addition, histochemical staining for acetylcholinesterase reveals possible spatial overlap of the activities of the class A and B enzymes. It has even been suggested that both classes of acetylcholinesterase may be present in all functional cholinergic synapses, thereby accounting for the fact that deletion of either type of enzyme yields no obvious phenotypic effects.

Insects have provided several examples of genes which are either dispensable or are components of systems which exhibit overlapping functions. For example, vitellogenin makes up approximately 40 percent of the total protein of the silkworm egg. Yet vitellogenin-deficient eggs (prepared in male hosts) can be activated and develop through embryogenesis (54). Similarly, *Drosophila* yolk proteins can be at least partially deleted without deleterious effects. One mutant lacks one (Yp2) of the three yolk-protein genes, yet the females are normally fertile (49). Apparently, the three genes make up a redundant system, such that individual components are interchangable should one become nonfunctional.

The most difficult cases to analyze are those genes which code for regulatory functions rather than those which are translated into structural or enzymatic proteins. Nevertheless, detailed genetic analyses have revealed a particularly intriguing example in *Drosophila*. An instance was recently discovered in which the development of single central nervous system features can be controlled by more than one regulatory gene. A single gene cluster—the bithorax complex—controls, among other segmental features of the larva, the development of "presumptive leg neuromeres" (PLN). PLN are present in the internal thoracic ganglia of the larva. During metamorphosis the PLN develop into the leg neuromeres of the adult fly. Another internal function controlled by genes of the bithorax complex is development of "lateral dots" (LD), small structures in the first abdominal and thoracic ventral ganglia which are recognized by a monoclonal antibody (12).

Development of the segmental features controlled by the bithorax complex is believed to be regulated by derepression in any given segment of several specific genes. For external segmental features, i.e., those which develop from larval epidermal cells, considerable evidence indicates that a relatively straightforward expression of one (or more) bithorax complex genes is most likely involved in controlling the identity of a specific segment (27).

For internal tissue, such as trachea, information is limited, but sufficient data exist to suggest that a different type of regulatory circuitry probably is employed. Several bithorax complex genes are thought to be involved. The expression of each occurs in several consecutive segments rather than in single segments, as is the case for epidermal segments (27). That is, in various abdominal segments the expression of PLN and LD is actively suppressed by several of the same genes. More than one gene of the bithorax complex series, at least in part, appears to carry out the same function. Ghysen and Lewis (13) have in fact proposed that PLN suppression is reiteratively expressed by consecutive regions of the bithorax complex. Because of this apparent redundancy it is necessary to inactivate, or at least diminish, the function of more than one gene to permit either PLN or LD development.

It is of course not yet clear whether at the molecular level these redundant suppression functions are identical. What is clear, however, is that the final result—PLN or LD suppression—is the same regardless of which controlling gene is involved. The contrast between the redundant regulatory circuits involved in the control of internal features and the lack of similar redundancy associated with epidermal pattern specification is dramatic.

One final example of redundancy at the level of gene expression is the nonproductive gene rearrangements involved in immunoglobin production (reviewed in Ref. 52). According to current theory, B cells undergo a prolific differentiation of immunoglobin gene expression independent of antigen challenge. A collection of immunoglobin gene segments are rearranged to generate a broad spectrum of antibodies. Only a small proportion of those B cell—rearranged genes are, however, ultimately employed for combat with foreign antigens. This case represents redundancy on a massive scale, although it of course represents a highly specialized situation. Less widely accepted, however, is the contention that a significant fraction (up to one-third) of the gene recombination events involved in generating antibody diversity are nonproductive (reviewed in Ref. 52). These often defective immunoglobin rearrangements are far from ever approaching the perfectionist's dream of how a differentiation program should work. Recombination is many times imprecise, resulting in proteins that may not function as immunoglobulins, and even chromosomal translocations frequently occur. Indeed, published estimates of the frequency of nonproductive rearrangements are probably on the low side if one takes seriously the notion that repair-rectification mechanisms are at work to correct many of the mistakes.

Cytoplasmic Organization Redundancy

These examples, as well as the next series of cases, represent instances of either overbuilding or alternative strategies. They are more distantly separated, both in time and in space, from either the genome itself (e.g., gene copy redundancy) or gene expression (e.g., multigene families) than the previous cases. Overbuilding probably represents a "fringe" example of redundancy. No substitute or alternative mechanisms are yet known to be involved. Nevertheless, the striking excess capacity displayed in some subcellular systems, especially those associated with the cytoskeleton, warrant at least passing attention in the present essay.

Only a few examples have been reported, and they lack the precise definition at the functional level accorded our previous cases. That is due, in large part, to the present lack of the genetic approaches, which are rapidly becoming a prerequisite for an ultimate comprehension of the significance of a particular structure or process to a developmental event.

Approximately 20 to 35 percent of the protein synthesized in animal cells is devoted to the cytoskeleton. It consists of three main components—microtubules, microfilaments, and intermediate filaments (see Chap. 9, this volume). Contrary to the impression generated from viewing electron micrographs, the cytoskeleton is a very dynamic structure. Its subunits are continually being exchanged between the rigid, filamentous superstructure and the cell's soluble (i.e., dissociated) subunits. A rapid equilibrium between those two phases is thought to exist.

Microtubules are involved in a large number of cellular events, including mitosis, cytokinesis, and intracellular membrane traffic (see Chap. 9, this volume). It is not surprising, therefore, to learn that most cells are well endowed with tubulin subunits. What is intriguing is the extent to which some of the cytoskeletal systems are overbuilt. The mitotic spindles provide a case in point. The force required for normal chromosome movement during mitosis is very low. The spindle can, however, produce significantly greater force. In fact, it has been estimated that approximately 10,000 times more force can be generated than is actually required (36). It has even been suggested that the excess capacity could, if not properly controlled, lead to chromosome breakage and produce genetically defective cells.

Since the advent of experimental embryology at the turn of the century, biologists have tried to elucidate the mechanisms involved in egg polarization following fertilization. The amphibian egg, for example, is constructed during oogenesis with a cell surface (plasma membrane and underlying cortex) and internal cytoplasmic animal-vegetal polarity (e.g., yolk platelets are distributed by size and amount). It also displays a radial symmetry of the cell surface and internal cytoplasm about its animal-vegetal axis. Fertilization triggers a series of symmetry-breaking events that generate bilateral symmetry and lead to development of the dorsal-ventral axis. This dorsal-ventral polarity sets up the gastrulating embryo's primary embryonic axis. In monospermic amphibian eggs (e.g., *Xenopus*), the position of sperm entry in the animal hemisphere marks the future ventral midline of the emerging embryo (Fig. 4). About halfway in time between fertilization and the first cleavage there is an active polar movement (rotation) of the cell surface in relation to the underlying internal cytoplasm toward the future dorsal side (opposite the sperm entry point).

Despite over 100 years of experimental interest, identification of the subcellular components which specify amphibian egg dorsal-ventral polarity

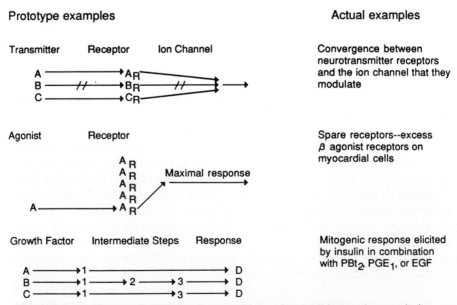

FIGURE 4. Multiple pathways for the regulation of amphibian embryo polarity.

have *not* been elucidated. We feel that the difficulty in finding the mechanism has been hampered by the apparent *multiplicity* (redundancy) of pathways involved in regulating amphibian egg polarity (Fig. 4). Even under natural conditions the role of the sperm is bypassed in a subpopulation of eggs (preprogrammed eggs) that appear to have inherited from oogenesis (35) a built-in cytoplasmic polarity. Under experimental conditions the role of the sperm can be bypassed either by tilting the egg or by mild centrifugation. The polarity of fertile eggs can be randomized with respect to the point of entry of the sperm by placing the eggs under simulated microgravity conditions or by transplanting a nucleus into an enucleated egg. The cytoplasmic movements that are associated with polarization of the fertile egg involve microtubules that move the cell surface relative to the underlying internal cytoplasm (10). This surface rotation correlates with the point of sperm entrance and appears to control egg polarization under normal conditions. If egg surface rotation is blocked by ultraviolet irradiation or by disruption of the microtubules through cold shock, the zygotes do not polarize. However, those potentially apolar zygotes can be *completely rescued* by a simple short-term tilt (41). These observations indicate that egg polarization can be achieved by a variety of mechanisms.

The polarization of the *Fucus* egg in response to different environmental cues provides another example of cytoplasmic redundancy. The unfertilized common brown seaweed (*Fucus*) egg is radially symmetrical. After fertilization there is a slow, progressive polarization of the symmetrical egg until, at about 16 h, a bulge forms on one end of the zygote. This bulge forms a rhizoid, which gives rise to the holdfast organ for anchoring the developing plant to a substrate. The other end of the developing zygote cleaves and gives rise to the thallus portion. Under normal conditions the sperm penetrates the egg surface randomly. In the dark, insulated from any environmental gradients, the sperm polarizes the *Fucus* egg. However, for a period of up to 9 to 12 h postfertilization, many environmental stimuli can *override* the polarization influence of the sperm. Polarization of rhizoid outgrowth can be cued by a wide and apparently disparate variety of environmental vectors (gradients) such as light, polarized light, temperature, pH, electric field, centrifugal field, and auxin (plant hormone) (23). There is clearly a redundancy in capacity to respond to stimuli for polarization of the *Fucus* egg. It is difficult to envision that stimuli like centrifugation and light would have a single receptor and a common coupling response. Current thinking concerning this problem does, however, envision a single receptor. Those environmental stimuli are thought to produce polar changes in the permeability of the egg membrane which lead to localized ion fluxes as well as to internal electric currents (38) which polarize the plasma membrane and the cytoplasm.

Why can amphibian and *Fucus* eggs achieve polarization by employing any one of a variety of cues? The establishment of a single primary embryonic axis, which follows from polarization of the egg, is such an important and essential event that redundant pathways serve as insurance that polarization will occur despite individual variability in egg construction and a variable and changing external environment. Amphibian eggs show considerable variability in such egg characteristics as pigmentation, size, viscosity of the internal cytoplasm, rate of rotation, degree of surface rotation, and biochemical characteristics.

For the *Fucus* egg "insurance" is necessary because the water that it finds itself in can vary in salinity, solute composition, temperature, available light, and wave action. Despite these variations it is imperative that the egg polarize if attachment to the appropriate substrate is to take place.

Surplus or Redundant Cells

Many examples of cell death or cell replacements are well known (e.g., Table 3). In many instances cell death occurs in large cells which are easily recognized (e.g., neurons—see Chap. 8, this volume), or in massive numbers (e.g., interdigital clefts). This situation contrasts sharply with the examples of cytoplasmic organization redundancy just discussed. Cytoplasmic redundancy requires elucidation at the ultrastructural level, which is often tedious. As a result, most cases probably remain undetected.

In a classic paper, Saunders (40) stated: "death of cells is the usual accompaniment of embryonic growth and differentiation." A catalog of instances

TABLE 3 Representative Examples of Production of Surplus Cells in Vertebrates

Example*	Circumstances	Presumed rationale†
Immune system	Immature lymphocytes proliferate, rearrange immunoglobin genes	Early random modifications of genes provide maximal antibody diversity later
Limb morphogenesis	Surplus cells of chondrogenic lineage display necrosis	Permits ornate shaping of appendage
Ovarian follices	Excess oocytes parceled out during menstral cycle	Provides accurate timing mechanism for follicle maturation
Neurons	Excess (nonfunctional) neurons degenerate	Accommodates individual variation in target organs as tissues
Embryogenesis	Almost all tissues and organs (see Ref. 14) shed surplus cells	Facilitates cell movements and organ separation
Pigmentation	White areas of the skin due to pigment cell death	Skin (dermal and epidermal layers) controls pigmentation pattern

*Abstracted from Ref. 33.

†These presumed rationales are retrospective at best. More likely the real reason for these circumstances is that during evolution various mistakes (mutations) which proved favorable to embryogenesis persisted.

of cell death in normal vertebrate ontogeny, containing several dozen entries, has been compiled by Glucksmann (14) and elaborated on by Michaelson (33). The widespread occurrence of cell death suggests that the production of surplus cells is a routine feature of embryogenesis. Some examples are: (1) As tissues are remodeled (e.g., cartilage, bone, etc.), cells are removed, die, and disappear, (2) as larval organs become obsolete (e.g., vertebrate pronephros), they degenerate, (3) as cells match up (e.g., neurons and muscle fibers), surplus cells (e.g., neurons) are eliminated, and (4) as appendages take form (e.g., hand-interdigital clefts), large numbers of cells are removed. In addition, for as yet unknown reasons, massive cell death can occur in an apparently non-specific fashion. An example that was mentioned earlier, is the necrosis and sloughing off of an apparently random number of cells during amphibian gastrulation (22). Events such as all those described above can probably be rationalized in one of two ways: (1) Sculpturing of embryonic tissues is most efficiently carried out by discarding cells, rather than by reprogramming them or (2) matching of cell types, such as those found in a neuron-muscle fiber complex, requires fewer genes when surplus neurons are discarded than when a precise, but cumbersome, feedback network inhibits neuron production.

Whether these examples represent cell suicide or cell assassination (40) is largely unclear. Mutations in *C. elegans* which affect programmed cell death have recently been identified (19). They reveal that in some cases, at least, death occurs by cell suicide. In the developing nervous system a natural selection which leads to the death and eventual removal of surplus neurons may occur. Edelman (9) has proposed a "Darwinian" type of phenomenon, i.e., neurons which become functionally active are selected from an enormous initial neuron population. The functionally active ones become permanent, while the others are believed to die and degenerate. Those that do not become functional, perhaps because they failed to receive a go-ahead stimulus for differentiation, die. They do not necessarily commit suicide. Perhaps they just fail to receive the necessary encouragement to keep on living.

Another phenomenon, cell replacement, reveals that not only are excess cells produced, but surplus developmental information for replacing cells which are inadvertently lost is maintained. In simple invertebrates, e.g., *C. elegans*, cells such as those belonging to the vulval cell lineage, if ablated, will be replaced by cells recruited from a nonvulval lineage (47). In various vertebrates, structures as diverse as the lens (53) or limb can be regenerated. These latter cases of regeneration have been attributed to redundancy of the information system for specification of the original tissue or organ.

Cell Communication Redundancy

Since both signal sender and signal receiver cells need to act in concert, multiple cues are employed to ensure that the receiver cells "get the message" (see Chap. 8, this volume). This is especially critical, since cell communication may (1) occur over relatively large (for a cell) distances, (2) occur on external surfaces (where environmental effects could erase a single cue), and (3) involve motile receiver cells, which may not make it to their rendezvous with signal molecules at exactly the right time.

In vertebrates, various types of cells either elongate over a long distance (e.g., neurons) or migrate over extracellular matrices [e.g., neural crest cells (see Chap. 17, this volume) and primordial germ cells (see Chap. 11, this volume)]. Neurons are perhaps the best studied of these examples. Evidence is available which indicates that neurons in a wide variety of animals, including invertebrates such as insects and most vertebrates, employ multiple, redundant mechanisms for selecting the appropriate pathway and/or target (43). When chick skin sensory innervation patterns were monitored, it was observed that no single mechanism (e.g., tracking motor axons) can fully account for the precision that characterizes skin sensory innervation patterns. Rather, a collection of cues (including (1) mechanical constraints provided by connective tissue, (2) homing tendencies which are genetically programmed into specific neurons (see Chap. 8, this volume), (3) extracellular matrix molecules, (4) adhesive differences along various migration routes, (5) target size, (6) diffusible attractant molecules (e.g., growth factors) which emanate from the target tissue, and (7) interactions between similar and/or competing neurons) are most likely employed. Hence, the remarkable specificity of innervation patterns is guaranteed, even though the number of neurons specified by a particular individual's genome might not match (during sexual recombination) the genetic programs which specify either target size or number. This case, as well as various others which involve cell migration [e.g., neural crest (see Chap. 17, this volume)], represents a persuasive example of the double assurance concept.

Regulatory Signal Redundancy

It is becoming clear that growth factors, mitogens, and other signal molecules play a role in early embryonic cell communication by controlling proliferation, differentiation, and morphogenesis (32). Communication and morphogenesis involving signal molecules may employ redundancy to ensure error-free communication in a noisy environment (46). Redundancy is generated by the large number of signal molecules, some of which may even have both stimulatory and inhibitory activity on the same cell. Coordination is, of course, necessary between the available signal molecules and the responding cell's competence to recognize them.

Studies of the roles played by signal molecules in developing systems are still in their infancy, so no bona fide examples of redundancy in the context of alternative pathways have yet emerged. Figure 5 illustrates prototype examples of redundancy in the response of a cell to signal molecules. A more realistic example is the modulation of gastrointestinal motility by regulatory signal molecules (hormones). The overall activity of motility in different regions of the gastrointestinal tract is regulated by both nerves and hormones. There is both an extrinsic autonomic nerve control and an intrinsic control (nerve cells within the wall of the gastrointestinal tract—the enteric nervous sys-

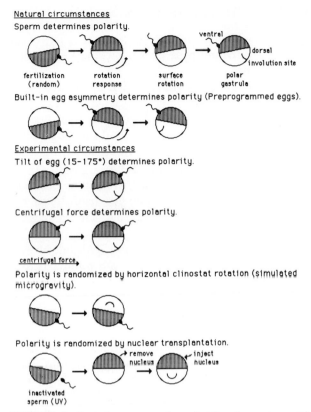

FIGURE 5. Prototype examples of redundancy in regulatory signaling.

tem). Motility is regulated by a balance between the extrinsic vagal and splanchnic nerves and the intrinsic enteric nervous system, with some redundancy: Recovery from inactivation of the extrinsic control occurs before reinnervation, which implies that a redundant control mechanism involving intrinsic enteric nerves exists. Intrinsic modulation of gastrointestinal motility involves regulatory peptides acting as both hormones and transmitters. The migrating motor complex (MMC) activity of the stomach and small intestine is modulated by hormones. Motilin, substance P, somatostatin, and neurotensin all increase MMC activity in fasted dogs. However, removal of one of them, e.g., motilin, has only transient effects (18). Also, motilin levels show considerable variation that ranges from 0 to 300 pmol/L. That degree of variability strongly suggests that redundancy exists in the regulatory circuit.

A Brief Concluding Discussion

A summary of several of the ideas and concepts presented in this chapter, as well as a few provocations, are provided by the following inquiries.

Why do embryos need redundancy? Several reasons can be listed, including the following:

1. To act as insurance against environmental perturbations—e.g., temperature shock, disorientation, etc.
2. To act as insurance against their own mistakes, which occur, occasionally, during organogenesis.
3. To accommodate variation. For example, physiological stress might be endured during the period when the organism, amphibian, fish, bird, etc., is constructing eggs (oogenesis). Redundancy in various processes will ensure that all eggs, even imperfect ones, survive.
4. To serve as a "tool kit" which substitutes for fine-tuned or well-designed changes in developmental mechanisms. Its usefulness becomes apparent when we consider the random nature of the occurrence of change during evolution.
5. To facilitate a broader range of evolutionary change. Redundant systems carry more information and therefore provide more opportunity for evolutionary change.

Why is the redundancy issue important? At least two reasons can be given.

1. For practical purposes it is important to understand whether a particular regulatory circuit or a specific cytoplasmic structure (e.g., cytoskeletal element), once characterized, represents the whole story or whether cytoplasmic information systems are so complex (e.g., contain redundant elements) that they will require exhaustive inventorying before any single developmental process can be considered fully understood. This might be called the "magnitude" or "complexity" problem. For example, in a mutant a function might not be deleted as expected. Likewise, injection of antisense mRNA or specific antibodies into an embryo may yield a negative result, which represents bypassing rather than lack of effect of the inhibitor.
2. For intellectual reasons, it is necessary to develop a coherent view of developmental events. It is important to know whether these events proceed—in an obligatory fashion—along a single, step-by-step pathway or whether they deviate in response, for example, to shortages of oogenetic stores (which might occur in overripe eggs). Why? Because this knowledge has substantial predictive value for the experimenter. The types of mutants that can be expected, and the kinds of regulatory circuits that might govern a particular developmental event, etc., are issues which depend on such knowledge.

How can the experimenter recognize instances of redundancy? This can be done in several ways.

1. Actively, by devising experiments that succeed in either bypassing, short circuiting, or overriding a known pathway but do not change the outcome of development.
2. Passively, by looking for instances in which the amount of an impor-

tant component in a regulatory system varies considerably from one embryo or condition to another.

3. By examining conflicting data which point toward two separate pathways yielding the same outcome.

Which experimental strategies most likely will be the most productive for discovering redundancy? The answer will, of course, depend on the type of redundancy (e.g., regulatory circuit, ultrastructure, cell communication, etc.). Since functional tests will be the most informative, mutant analyses will no doubt be especially valuable. That is, exhaustive searches for mutations which block a single pathway at the same step through different causal mechanisms will be the most useful approach. The most success will probably be achieved by focusing on a relatively simple phenotype—i.e., one which is *not* the product of a complex developmental cascade—and saturating the phenotype with as many mutant alleles as possible (see, e.g., Refs. 12 and 13).

For systems or processes which are not amenable to genetic analyses, e.g., the cytoskeleton, perhaps exhaustive ultrastructural studies accompanied by *stochiometric analyses* will be useful. Strict quantitation might provide clues regarding single or multiple routes.

What is the single most important problem which confounds assessments of redundancy? No single developmental event, with perhaps the exception of the fertilization reaction, is fully understood, so searches for flexibility or alternatives are not easily carried out.

At the levels which can be analyzed thoroughly, e.g., nucleic acids, extensive redundancy has been discovered. For example, repetitive sequences abound in the genomes of most organisms. Evolutionary processes would have cleared out these sequences, were it not for the beneficial effects these redundancies provide.

Redundancy as discussed in this chapter deals mainly with molecular, subcellular, and cellular events. Extensive examples of redundancy exist at the organismal level and include the following:

1. Multiple cues are employed by pigeons as they fly home. Some cues are dispensable.
2. Insects switch behavior patterns regularly, e.g., during prey searches. Various behavior-eliciting cues are employed.
3. Some species have redundancy of smell, taste, and vomeronasal organs.
4. Courtship behavior in many organisms involves multiple cues, not all of which are required.

Acknowledgments

The author's research on amphibian egg polarity, which spawned the present thesis on redundancy, is supported by a grant from NASA (NAG-2-323). Richard D. Campbell (Irvine) introduced GMM to the concept of redundancy. Rollin C. Richmond (Bloomington) provided valuable discussion, as well as a critical review of this manuscript. Susan Duhon kindly provided editorial assistance.

General References

MICHAELSON, J.: 1987. Cell selection in development. *Biol. Rev. Camb. Philos. Soc.* **62**: 115–139.

DAWKINS, R.: 1987. *The Blind Watchmaker*. Norton, New York.

References

1. Bailey, D. W. 1986. Genetic programming of development: A model. *Differentiation* 33:89–100.

2. Braus, H. 1906. Vordere Extremitat und Operculum bei Bombinator-Larven. Ein Beitrag zur Kenntnis morphogener Korrelation und Regulation. *Morphol. Jahrb.* 35:139–220.

3. Britten, P. J., and Davidson, E. H. 1969. Gene regulation for higher cells: A theory. *Science* **165**:349–356.

4. Cavener, D. R., and Douglas, R. 1987. Combinatorial control of structural genes in *Drosophila*: Solutions that work for the animal. *Bioessays* **7**:103–107.

5. Craig, E. A., and Jacobsen, K. 1984. Mutations of the heat inducible 70 kilodalton genes of yeast confer temperature sensitive growth. *Cell* **38**:841–849.

6. Culotti, J. G., Von Ehrenstein, G., Culotti, M. R., and Russell, R. L. 1981. A second class of acetylcholinesterase-deficient mutants of the nematode *Caenorhabditis elegans*. *Genetics* **97**:281–305.

7. Dolecki, G. J., Wannakrairoj, S., Lum, R., Wang, G., Riley, H. D., Carlos, R., Wang, A., and Humphreys, T. 1986. Stage-specific expression of a homeobox-containing gene in the non-segmenting sea urchin embryo. *EMBO J.* **5**:925–930.

8. Edelman, G. M. 1984. Cell adhesion and morphogenesis: The regulator hypothesis. *Proc. Natl. Acad. Sci. U.S.A.* **81**:1460–1464.

9. Edelman, G. M. 1987. *Neural Darwinism: The Theory of Neuronal Group Selection*. Basic Books, New York.

10. Elinson, R. P., and Rowning, B. 1988. A transient array of parallel microtubules in frog eggs: Potential tracks for cytoplasmic rotation that specify the dorso-ventral axis. *Dev. Biol.* **128**:185–197.

11. Firtel, R. A. 1981. Multigene families encoding actin and tubulin. *Cell* **24**:6–7.

12. Ghysen, A., Jan, L. Y., and Jan, Y. N. 1985. Segmental determination in *Drosophila* central nervous system. *Cell* **40**:943–948.

13. Ghysen, A., and Lewis, E. B. 1986. The function of *bithorax* genes in the abdominal central nervous system of *Drosophila*. *Roux's Arch. Dev. Biol.* **195**:203–209.

14. Glucksmann, A. 1950. Cell deaths in normal vertebrate ontogeny. *Biol. Rev. Camb. Philos. Soc.* **26**:59–86.

15. Goebl, M. G., and Petes, T. D. 1986. Most of the yeast genomic sequences are not essential for cell growth and division. *Cell* **46**:983–992.

16. Gould, S. J., and Lewontin, R. C. 1979. The spandrels of San Marro and the Panglossian paradigm: a critique of the adaptationist programme. *Proc. R. Soc. Lond. Biol. Sci.* **205**:581–598.

17. Gridley, T., Soriano, P., and Jaenisch, R. 1987. Insertional mutagenesis in mice. *Trends Genet.* **3**:162–166.

18. Grundy, D. 1985. *Gastrointestinal Motility*. MTP Press, Boston.

19. Hedgecock, E. M., Sulston, J. E., and Thomson, J. N. 1983. Mutations affecting programmed cell deaths in the nematode *Caenorhabditis elegans*. *Science* **220**: 1277–1279.

20. Hochachka, P. W., and Somero, G. N. 1984. *Biochemical Adaptation*. Princeton Univ. Press, Princeton.

21. Hogan, B., Holland, P., and Schofield, P. 1985. How is the mouse segmented? *Trends Genet.* **1**:67–74.

22. Imoh, H. 1986. Cell death during normal gastrulation in the newt, *Cynops pyrrhogaster*. *Cell Differ.* **19**:35–42.

23. Jaffe, L. F. 1968. Localization in the developing *Fucus* egg and the general role of localizing currents. *Adv. Morphog.* **7**:295–328.

24. Kaufman, S. A. 1984. Pattern generation and regeneration. In: *Pattern Formation* (G. M. Malacinski, ed.), pp. 73–102, Macmillan, New York.

25. Krumbiegel, J. 1960. *Die Rudimentation*. Gustav Fischer Verlag, Stuttgart.

26. Lewin, R. 1984. Why is development so illogical? *Science* **224**:1327–1329.

27. Lewis, E. B. 1978. A gene complex controlling segmentation in *Drosophila*. *Nature (Lond.)* **276**:565–570.

28. Loomis, W. F., Jr. 1973. Vestigial DNA? *Dev. Biol.* **30**:f.3–4.

29. Loomis, W. F., and Gilpin, M. E. 1986. Multigene families and vestigial sequences. *Proc. Natl. Acad. Sci. U.S.A.* **83**:2143–2147.

30. Ma, H., and Botstein, D. 1986. Effects of null mutations in the hexokinase genes of *Saccharomyces cerevisiae* on catabolite repression. *Mol. Cell. Biol.* **6**:4046–4052.

31. Meinhardt, H. 1984. Models for pattern formation during development of higher organisms. In: *Pattern Formation* (G. M. Malacinski, ed.), pp. 47–72, Macmillan, New York.

32. Meriola, M., and Stiles, C. D. 1988. Growth factor superfamilies and mammalian embryogenesis. *Development (Camb.)* **102**:451–460.

33. Michaelson, J. 1987. Cell selection in development. *Biol. Rev. Camb. Philos. Soc.* **62**:115–139.

34. Moore, K. L. 1977. *Clinically Oriented Embryology*, 2d ed. Sanders, Philadelphia.

35. Neff, A. W., Ritzenthaler, J. D., and Rosenbaum, J. F. 1988. Subcellular components of the amphibian egg: Insights provided by gravitational studies. *Adv. Space Res.* (in press).

36. Nicklas, R. B. 1983. Measurements of the force produced by the mitotic spindle in anaphase. *J. Cell Biol.* **97**:542–548.

37. Rhumbler, L. 1897. Stemmen die Strahlen der Astrosphare oder ziehen sie? *Arch. Entwicklungmech. Org. (Whilhelm Roux)* **4**:659–730.

38. Robinson, K. R., and Jafee, L. F. 1974. Polarizing fucoid eggs drive a calcium current themselves. *Science* **187**:70–72.

39. Romans, P., Firtel, R. A., and Saxe, C. L. III. 1985. Gene-specific expression of the actin multigene family of *Dictyostelium discoidium*. *J. Mol. Biol.* **186**:337–355.

40. Saunders, J. W. Jr. 1966. Death in embryonic systems. *Science* **154**:604–612.

41. Scharf, S. R., and Gerhart, J. C. 1980. Determination of the dorso-ventral axis in eggs by oblique orientation before first cleavage. *Dev. Biol.* **79**:181–198.

42. Schatz, P. J., Solomon, F., and Botstein, D. 1986. Genetically essential and nonessential α-tubulin genes specify functionally interchangeable proteins. *Mol. Cell. Biol.* **6**:3722–3733.

43. Scott, S. A. 1987. The development of skin sensory innervation patterns. *Trends Neurosci.* **10**:468–473.

44. Spemann, H. 1931. Ueber den Anteil von Implantat und Wirtskeim an der Orientierung und Beschaffenheit der induzierten Embryonalanlage. *Roux's Arch. Dev. Biol.* **123**:389–517.

45. Spemann, H. 1962. *Embryonic Development and Induction.* Hafner, New York.

46. Sporn, M. B., and Roberts, A. B. 1988. Peptide growth factors are multifunctional. *Nature (Lond.)* **332**:217–219.

47. Sternberg, P. W., and Horvitz, H. R. 1986. Pattern formation during vulval development in *C. elegans. Cell* **44**:761–772.

48. Stubblefield, E. 1986. A theory for developmental control by a program encoded in the genome. *J. Theor. Biol.* **118**:129–143.

49. Tamura, T., Kunert, C., and Postlethwait, J. 1985. Sex- and cell-specific regulation of yolk polypeptide genes introduced into *Drosophila* by P-element–mediated gene transfer. *Proc. Natl. Acad. Sci. U.S.A.* **82**:7000–7004.

50. Tatchell, K., Chaleff, D. T., DeFeo-Jones, D., and Scolnick, E. M. 1984. Requirement of either of a pair of *ras*-related genes of *Saccharomyces cerevisiae* for spore viability. *Nature (Lond.)* **309**:523–527.

51. Thrash, C., Bankier, A. T., Barrell, B. G., and Sternglanz, R. 1985. Cloning, characterization, and sequence of the yeast DNA topoisomerase I gene. *Proc. Natl. Acad. Sci. U.S.A.* **82**:4374–4378.

52. Van Ness, B. 1988. Immunoglobulin gene rearrangements. In: *Developmental Genetics of Higher Organisms* (G. M. Malacinski, ed.), pp. 29–51, Macmillan, New York.

53. Yamada, T. and McDevitt, D. S. 1974. Direct evidence for transformation of differentiated iris epithelial cells into lens cells. *Dev. Biol.* **38**:104–118.

54. Yamashita, O., and Irie, K. 1980. Larval hatching from vitellogenin-deficient eggs developed in male hosts of the silkworm. *Nature (Lond.)* **283**:385–386.

Questions for Discussion with the Editor

1. *What sort of molecular mechanisms might be responsible for the emergence and maintenance of redundancy within any given system?*

No doubt gene duplication provides the opportunity for redundancy to arise. For genome redundancy that is essentially a given. Multiple, identical copies of individual genes (reiteration) as well as multiple copies of similar genes (gene families) are generally understood to have emerged from ongoing gene duplication events. Additional redundancy probably arises as duplicated genes are mutated and generate protein products that exhibit slightly modified properties, so a newly evolved form can substitute, if necessary, for the original form.

Not all redundancy need be linked to structural genes, however. The lens crystallins provide an excellent example of a type of cell specialization which involves differential regulation of gene expression. Jiatigorsky [e.g., *Science* **236**:1554–1556 (1987)] and colleagues have identified several enzymes (e.g., argininosuccinate lyase, enolase, and lactate dehydrogenase) which are bifunctional. In various tissues and/or organs these enzymes are synthethized in relatively small quantities and perform metabolic functions. In the lens, however, they are synthethized in copious quantities and function as structural proteins (i.e., lens crystallins). The regulatory systems which operate in visceral organs are different from those which function in the eye. It is easy to imagine how similar examples of separate regulatory mechanisms, if intrinsically flexible (e.g., responsive to multiple cues), might provide substitution strategies at several levels of cytoplasmic organization.

Finally, redundant systems are probably maintained rather than cast off during evolution for four reasons: (1) There are not many environmental selective pressures at work on the embryo to screen them out. (2) Several of those redundant systems are probably called into play from time to time. Indeed, T. Sachs (*J. Theor. Biol.* **134**: 547–559 (1988)) even argues that epigenetic selection occurs within the embryo, such that individual developmental pathways are constructed from a broad menu, occasionaly on an ad hoc basis. (3) Maintenance, at the genome level, of extra nucleotide sequences is not an energetically expensive proposition. (4) An expansive genome provides increased opportunities for further change.

2. *Why is the "redundancy issue" so often overlooked?*

It flows counter to the widely held contemporary view that molecular analyses, especially at the level of gene regulation circuits, will generate a coherent understanding of development. That reductionist sentiment has as its historical antecedent the spectacular success achieved with attempts to understand bacteriophage morphogenesis.

Higher organisms, especially those which employ complex embryonic pattern specification mechanisms, do not function in nearly as streamlined a fashion as phage self-assembly. Evolution involves the modification, in minor ways, of preexisting genes. Lacking strong selective pressures, changes which are beneficial at one developmental stage are probably maintained, even if they require add-on modifications to function well at a later developmental stage. In our view, the typical vertebrate embryo is a "contraption." It is akin to something Rube Goldberg would design for moving the genome from the egg to the adult. This view contrasts with the traditional notion that development proceeds in as a series of stages or steps. Unlike insects, which exhibit clearcut break-points (e.g., instar larvae of *Drosophila*), vertebrates develop more continuously. Our view is that the process of evolution keeps development continuously moving toward the final destination— the sexually mature organism. Evolutionary processes do not reshape the embryo in a series of discrete steps that result in a streamlined adaptive form. The goal of embryogenesis is in part to function as a sort of delivery system which gets the genome to the right place at the correct time. In addition, the embryo serves as a scaffold for tissue and organ morphogenesis. Change occurs mainly through additions to preexisting structures, rather than the drastic redesign of old ones. From these viewpoints emerges the notion that developmental mechanisms are likely to be contrived, less than perfect, and highly redundant. No wonder that even "simple" organisms such as *C. elegans* exhibit phenomena such as the existence of 50 to 150 collagen genes [J. M. Kramer, G. N. Cox, and D. Hirsch, *J. Biol. Chem.* **260**:1945–1952 (1985)] in their genome.

CHAPTER 8

Multiple Cues in
Neuronal Patterning

Scott E. Fraser

Introduction

NEURAL DEVELOPMENT INVOLVES the elaboration of a highly stereotyped tissue from an apparently homogeneous sheet of cells—the neural plate. The generation of a complicated set of structures from a simple primordium is not unlike the development of other embryonic structures. The patterning of the cells within individual regions in the nervous system into recognizable structures results from morphogenetic movements, cell migrations, and cell differentiation, much like those found in other developing tissues. However, the nervous system has a second phase of patterning which depends on a uniquely neuronal feature, the axon. The axons grow out of the neurons to form a reliable pattern of connections both within and between regions of the nervous system. Because the axons are responsible for the flow of information from cell to cell, the complex and fascinating patterns they form play a central role in the functioning of the nervous system.

The most striking axonal patterns are formed by projection neurons, whose axons can traverse long distances to interconnect regions of the nervous system. Not only must the projection neurons find their target region, but they must also form a properly ordered set of connections within the region. In most cases, the fibers are organized in a topographical fashion, with a reproducible orientation. The simple, smooth pattern of topographical projections has led some researchers to propose that the order must be the product of simple patterning mechanisms. This optimism has resulted in many experiments that have yielded some important insights and offered at least circumstantial evidence for the involvement of specific mechanisms in nerve patterning. Supportive evidence exists for the involvement of mechanisms that range from cell surface chemical markers, on the one hand, to patterns of neuronal activity on the other. However, no concensus has emerged from the assembled evidence for a dominant mechanism that guides the assembly of nerve connections. Equally valid experiments have been taken as evidence for the sole operation of rather opposing mechanisms (cf. Ref. 10). In some experimental settings, interactions between neighboring neurons appear to guide the selection of a target site; in other settings, interactions of the projection neurons with the target sites themselves appear most important. *These disparate results suggest that multiple cues, both redundant and opposed, operate in parallel to guide the assembly of neural connections.* Thus, the challenge now becomes to understand the relative contributions of the many proposed cues in the formation of neuronal patterns.

As complex as the organization of the nervous system may seem, it offers some unique advantages for the study of embryonic patterning in general. The spatial separation of the projection neuron cell bodies and their termination sites permit independent manipulations of either group of cells. In contrast, the position and the pattern of cells within other tissue types are intimately associated. For example, it is difficult, if not impossible, to alter the position of a cell in the limb without also changing its position within the pattern of the limb and its set of neighboring cells. Similarly, because the recognition of the positional identity of a cell in the limb is based largely upon the pattern it forms with its neighbors, it is nearly impossible to assay the patterning of single, isolated limb cells. As is presented below, techniques to surmount these limitations are now becoming available for the developing nervous system. The unique data this is likely to generate should offer insights into the patterning of not only the nervous system but also of nonneural structures.

This chapter uses the lower vertebrate visual system to present an overview of the mechanisms that may play some role in neuronal patterning. Several articles have reviewed the basic phenomenology of the regenerating and developing lower vertebrate visual system (cf. Refs. 5, 7, 10, 27, and 33). Because these and other articles provide such fine reviews of the techniques and the data, this chapter does not attempt to fully describe the available data or to provide an all-inclusive review of the efforts of all laboratories. Instead, it will present representative data in schematic form to serve as a conceptual backdrop against which the possible mechanisms for neuronal patterning and some of the new approaches directed at testing these possibilities can be discussed.

The Lower Vertebrate Visual System

The visual system of the frog and other lower vertebrates offers several advantages that make it well suited to investigations of neuronal patterning. The projection neurons of each eye, the retinal ganglion cells, form a simple map, termed the retinotectal projection, on the surface of the contralateral optic tectum. The topographic projection formed by the eye onto the tectum displays a smooth and regular order. Ganglion cells from the dorsal portion of the retina project optic nerve fibers to the lateral (ventral) portion of the tectum; those from the ventral region of the retina project to the medial (dorsal) tectum; those from the nasal (anterior) region of the retina project to the caudal tectum; and those from the temporal (posterior) region of the retina project to the rostral tectum. The topography of the projections can be experimentally assayed by following the transport of tracers, applied to distinct regions of the retina, down the optic nerve and into the tectum. In addition, an electrode can be used to record activity from optic nerve terminals in the tectum following the presentation of visual stimuli, permitting the mapping of visual space (and hence the retina) onto the tectum (Fig. 1). In the example shown in Fig. 1, the correspondence of the sites in the visual field and the electrode positions demonstrates the two-dimensional organization of the retinotectal projection. It should be noted that the optics of the eye invert the image; therefore, stimuli in the nasal visual field excite the temporal retina, which projects to the rostral tectum.

A major advantage of the lower vertebrate visual system is that, unlike the higher vertebrate visual system, the optic nerve fibers can regenerate following trauma. Crushed optic nerve fibers can re-form a correctly ordered retinotectal projection within weeks. The regeneration is robust; a well-ordered projection re-forms reliably even when challenged by movement of the eye into an abnormal orientation or disruption of the normal pathways of the

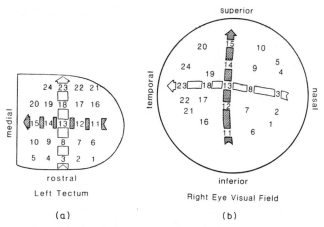

FIGURE 1. The order of the *Xenopus* retinotectal projection as assayed by electrophysiology. (*a*) Outline of the left optic tectum, showing the numbered sites at which an extracellular microelectrode was inserted. The electrode was lowered until its tip was within the neuropil where optic nerve fibers form their highly branched terminal arbors. The design of the electrode makes it preferentially record from the terminal arbors of the optic nerve fibers. (*b*) The visual field of the left eye is shown. Placement of a spot of light at the numbered sites evoked maximal electrical signals at the numbered electrode position (receptive fields). The full size of each receptive field was 2 to 3 times the size of the numbers. The shaded and open arrows connect a row and column of electrode sites and the corresponding receptive fields, emphasizing the two-dimensional order of the projection.

optic nerve (cf. Ref. 31). This has permitted experiments to be performed on the regenerating optic fibers of large and mature animals, which are easier to work with than developing embryos. The approach has been exploited by many laboratories in the hope that it may provide insights into the mechanisms that are important for the embryonic development of neural maps.

The retinotectal projection and some of the experiments performed on its regeneration are schematized in Figs. 2 and 3. The retinotectal projection in normal adult animals is topographical, but not perfect (Fig. 2a). Near neighbors in the retina do not terminate in a perfect, "crystalline" fashion at exactly neighboring sites in the tectum; instead, single optic axons reach *almost* the correct location where they elaborate terminal arbors that can fill more than 10 percent of the linear extent of the tectum (see reviews: Refs. 10 and 27). Following a nerve crush, larger and less precisely arranged terminal arbors reestablish contact with the tectum, and terminal arbors that would normally innervate distinct sites in the tectum overlap considerably (compare the terminals of retinal cells 1 and 2, Fig. 2b). With time, these terminal arbors are rearranged to yield near-normal topography (Fig. 2c). To gain some insight into the mechanisms guiding this regeneration, the pattern of connectivity has been assayed after manipulating the retina or the tectum (Fig. 3). The optic nerve fibers can find and synapse upon transposed or rotated pieces of tectal tissue (Fig. 3a). This result has been taken as evidence for some form of positional marker on the tectal surface. Conversely, the optic fibers can terminate at abnormal sites to reconstitute a somewhat normal overall projection pattern following regeneration of a whole eye into less than a complete tectum (Fig. 3b) or of a half eye into a whole tectum (Fig. 3c). These results have been taken as evidence against a dominant role for position-based cues and in favor of interactions among the regenerating optic fibers themselves. It should be noted that a possible shortcoming of any regeneration experiment is

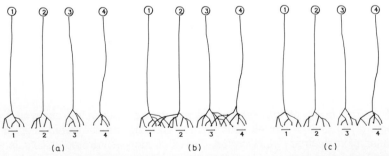

FIGURE 2. Schematic drawing of the normal pattern of the retinotectal projection and its regeneration. The four numbered circles along the top represent a row of retinal ganglion cell bodies projecting axons to the surface of the tectum. The sites they would be expected to occupy in the tectum of a normal animal are given by the numbered sites at the bottom of each panel. (a) The projection is well ordered, but not perfect. (b) Following a nerve crush, the optic nerve fibers regenerate terminal arbors that are larger and more crudely positioned. (c) With time, the arbors rearrange to form a nearly normal projection.

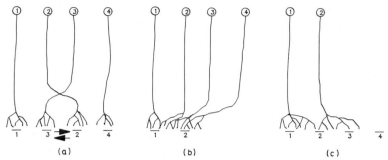

FIGURE 3. Schematic drawing of the regenerated retinotectal projection following experimental manipulations of the retina or tectum. (*a*) Optic nerve fibers can find and connect to the appropriate portion of the tectum, even after pieces of tectum have been interchanged. In this case, tectal regions 2 and 3 have been excised and moved to new locations. (*b*) The entire set of optic nerve fibers can compress to innervate the remaining portion of the tectum following deletion of a portion (regions 2 and 3). (*c*) The remaining optic nerve fibers expand to fill the entire tectum following removal of part of the neural retina. These classes of experiments have been taken as evidence that the regenerating optic nerve fibers prefer certain sites on the tectum, part *a*, but that they compete with one another for space in the tectum (parts *b* and *c*).

that the tectum had been previously innervated by optic nerve fibers. It remains possible that some aspect of the patterning observed during regeneration depends on the previous innervation of the tectum by an ordered set of optic nerve fibers. In addition, some cue that is present in the developing tectum or some property of the developing optic nerve fibers may have been lost by the stage at which the regeneration experiments were performed.

Because the outcome of the regeneration experiments might be the result of the previously formed retinotectal projection, it is important to perform similar experiments in the developing embryo. The lower vertebrate visual system is well suited for such experiments because embryonic development takes place in pond water after an external fertilization event. The nervous system of these animals forms rapidly from rudiments that can be experimentally manipulated. Much of classical experimental embryology was performed on amphibian embryos because of these advantages. This large and colorful literature offers both important insights and a variety of techniques that can be adapted for experiments on the development of the retinotectal projection.

Candidate Mechanisms for Retinotectal Patterning

The striking topography of the retinotectal projection immediately suggests that it might be formed by the selective growth of the optic nerve fibers in response to positional cues in the eye and brain. As compelling as this suggestion is, and as strong as the arguments that have been made for it are (cf. Ref. 30), decisive evidence has been difficult to obtain. This is mainly because the retinotectal projection does not arise in a random assemblage of cells but rather as an integral part of a developing and patterning embryo. Several processes under way in the embryo might be used as the dominant cues for retinotectal patterning, giving the false impression that position-based cues are guiding the axons. For example, regional differences in the timing of development of the retina and tectum might restrict the set of possible targets available to any given axon, thereby guiding the assembly of the retinotopic map (see below). A few of the possible schemes for establishing retinotopography are schematized in Fig. 4. Each of these simple mechanisms could result in the final topographic pattern of the projection. Because it is only through an understanding of the possible contribution of each mechanism that adequate experiments can be designed, we consider each of these mechanisms in detail below.

Selective Growth

One of the most commonly recurring, yet one of the most controversial, suggestions for the establishment of retinotopography is that the optic nerve fi-

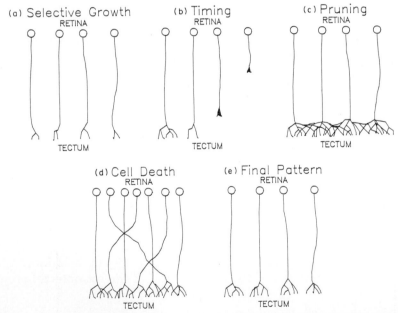

FIGURE 4. Schematic drawings of several different hypotheses for the establishment of the retinotectal projection during development. (*a*) The optic nerve fibers recognize sites on the tectum and selectively innervate them. Once there, they elaborate their terminal arbors to achieve the well-ordered final pattern (*e*). (*b*) A timing gradient in the retina leads to the outgrowth of some optic nerve fibers before the others. (*c*) A very crude or random set of optic nerve terminals is pruned to leave behind an ordered retinotectal projection. (*d*) A crudely ordered set of normal-sized terminal arbors are sorted into an orderly projection by killing those retinal ganglion cells that innervated the incorrect sites.

bers recognize their targets in the optic tectum. This suggestion, perhaps voiced most clearly in the chemoaffinity hypothesis of Sperry (30), is that optic nerve fibers from different regions of the retina recognize positional markers on the tectum and thereby choose their appropriate targets. Thus, the optic nerve fibers shown in Fig. 4a course directly to their correct target sites, where they then branch to form their terminal arbors (Fig. 4e). The appeal of this hypothesis may in part result from experiences in our everyday lives, in which we recognize roads, buildings, and rooms on the basis of their labels. However, the hypothesis raises many questions, such as: How many labels are needed and how unique are the labels? In the most extreme arguments, it has been suggested that there is insufficient information in the genome to completely specify each cell in even the relatively simple retinotectal projection. However, it must be remembered that the hypothesis does not depend on the wasteful scheme of unique chemical labels; as Sperry pointed out in his 1963 article (30), a few labels organized in a graded fashion could be used to specify position more economically than a large number of unique labels.

Experiments on the regenerating retinotectal projection of frogs and fish have been used to test the importance of position-based cues (cf. Ref. 10). Regenerating axons can take circuitous routes to their final target sites (12), perhaps suggesting that they are actively navigating over the tectal surface. In other experiments, optic nerve fibers were able to locate and synapse on their correct site in the tectum even after it had been surgically moved to another location in the tectum (Fig. 3a). However, in all regeneration experiments, the tectum has been previously innervated by both visual and nonvisual inputs. As a result, this class of experiment cannot offer definitive evidence concerning the role of positional cues in the development of the retinotectal projection.

Anatomical tracing of the pathways of developing optic nerve fibers in amphibian larvae has permitted some tests for the cues that are important in the initial formation of the retinotectal projection. Labeling subsets of the retinal ganglion cells with cobalt, tritiated amino acids, or horseradish peroxidase (HRP) results in an unambiguous assay of developing retinotopography. Using these techniques, it has been demonstrated that the dorsoventral topography of the projection forms as the fibers first reach the developing tectum (17, 23, 26). In experiments in which alternative cues, such as neuronal activity, are abolished or blocked, the fibers were still able to form a correct dorsoventral order (13). Such results have been used to argue, by a process of elimination, that the patterning of the retinotectal projection must be guided by positional cues (cf. Refs. 14 and 16).

Timing

Differences in the timing of development (see Chap. 6, this volume) could serve as an important determinant of neuronal connectivity. This hypothesis depends on cells within each neural center developing in a somewhat graded fashion, so that differentiation is well under way in one region while it is just beginning in another. As schematized in Fig. 4b, a gradient of maturation would result in a reliable order of outgrowth of the optic nerve fibers (more mature neurons sprouting axons first). Such a gradient of outgrowth could be used to establish retinotopography if either of two conditions are met: (1) The fibers grow into the tectum so that they first encounter the sites filled by the most mature optic nerve fibers, such that the fibers could fill the sites on a first-come, first-served basis or (2) the tectum matures in a graded fashion that matches the order of ingrowth, such that only the appropriate region of the tectum is receptive to innervation at the time any given set of axons enter the tectum. In the schematic of Fig. 4b, these two scenarios would correspond to the axons entering the tectum from left to right, or the tectum maturing from left to right, respectively. Interestingly, both of these conditions appear to be true in the developing retinotectal projection. The dorsal fibers are the first to reach the developing tectum, and they first reach the ventral tectum, which is both the most mature region of the tectum and the target region of the dorsal optic nerve fibers (16). The timing hypothesis is attractive because of its simplicity, and it may play a dominant role in the patterning of some invertebrate neural projections (20).

Although there are differences in the timing of both the arrival of optic fibers in the tectum and the maturation of the tectum, experimental results suggest that these differences are not critical for the formation of retinotopography. Reversing the natural order of arrival by grafting the dorsal part of the eyebud from a young animal into an old host does not alter the dorsoventral topography (16). These results argue that timing is not the dominant cue used by optic nerve fibers, but they cannot be taken as proof that timing plays no role. More complicated experimental designs, in which other potential cues are eliminated, will be required to test if timing plays any role.

Pruning

Another popular hypothesis is that topographical projections are formed by a somewhat nonselective growth of neuronal arbors followed by the pruning of "mistakes" from the terminal arbors. In this case an exuberant tangle of optic terminal arbors (Fig. 4c) is pruned to leave the topographical final pattern (Fig. 4e). In the most extreme version of this hypothesis, a topographical projection could form from a completely random initial distribution of terminal arbors. This hypothesis requires some method of recognizing a mistaken terminal branch so that it can be pruned. The recognition could be based on cell surface interactions either between branches or between a branch and the target tissue, but the most popular suggestion is that the recognition is based on neuronal activity. Activity could play an instructive role in the patterning of the projection because in some species there is a correlation in the activity of neighboring retinal ganglion cells that persists even in the dark (cf. Ref. 1). If the optic nerve terminals had some means of recognizing correlated activity, the smooth order of the projection could be the result of the fibers following the simple rule: "Fibers that fire together, synapse together."

The hypothesis is given some support by the refinement of the regenerating retinotectal projection (Fig. 2b and c), and by experiments that alter neuronal activity during regeneration. Abolition of activity in the optic nerve by intraocular injection of tetrodotoxin permits the regeneration of only a crudely ordered projection by unusually large terminal arbors (21, 28; for review see Ref. 27). Because this toxin is selective for the voltage-sensitive sodium channel, it has been argued that this effect demonstrates the importance of normal activity in the formation of patterned nerve connections. However, such data alone cannot separate possible instructive effects of neuronal activity from merely permissive effects. To determine if activity plays an instructive role, animals were reared in strobe light to disrupt locally correlated activity during regeneration. When all optic nerve fibers are thereby forced to fire somewhat synchronously, refinement of the projection is prevented, as was observed in the experiments with tetrodotoxin (29).

In apparent contrast to these regeneration experiments, a normally oriented projection can form even in the absence of neuronal activity (13), suggesting that activity is not required for the establishment of retinotopography. Unlike regenerating optic nerve fibers, the terminal arbors formed during development do not appear to be larger when they first form. The different effects of blocking activity and the differences in optic fiber anatomy may suggest that regenerating and developing optic axons follow different rules. Conversely, it remains possible that the differences in the assays employed heighten the apparent dichotomy.

Experiments on use-dependent changes in synaptic transmission suggest a mechanism for detecting correlated neuronal activity which might then be used to guide the pruning of inappropriate terminal branches. Two classes of glutamate receptors are found in the tectum and are classified by their response to the agonist, N-methyl-D-aspartate (NMDA). The NMDA class of receptors is blocked by extracellular magnesium ions unless the cell is depolarized; when opened, the receptors conduct calcium ions, unlike the non-NMDA-type glutamate receptors on the same cells (see review, in Ref. 2). As a result, the NMDA receptor channel on tectal cells would be expected to open only when the tectal cell is depolarized at the time the optic nerve fiber fires. Only those synapses that are active when the majority of the synapses on a given cell are active (so that the cell is depolarized) would be capable of opening the NMDA-type receptor. Thus, correlated activity would open this receptor channel, resulting in an influx of calcium ions and the triggering of intracellular second-messenger systems. This could lead to the selective strengthening or stabilization of the synapse (see review, Ref. 3), giving a competitive advantage to those synapses with correlated activity.

The possible role of the NMDA receptor has been tested with the specific NMDA-receptor blocking agent, aminophosphonovaleric acid (APV). When APV was applied to the tecta of three-eyed *Rana*, the ocular dominance stripes that normally form when more than one eye is forced to innervate a single tectum were totally disrupted in the treated region (4). Similarly, in kittens, APV has been shown to block the effects of monocular light deprivation (32) and to decrease the orientation selectivity of neurons in the primary visual cortex (19). The effects of activity and of the NMDA receptor on the initial development of retinotopography in the frog is discussed in a later section.

Cell Death

A final hypothesis worth noting is that the ordered retinotectal projection might be sculpted from a nearly random initial set of connections by killing those neurons that make mistakes (Fig. 4d). This hypothesis requires some means of recognizing mistakes, so that the appropriate cells can be targeted for death. The normal development of the amphibian visual system does not show the large number of mistakes predicted by this hypothesis (17), but some evidence indicates it may be active in the development of the rodent and chicken visual systems (for review see Ref. 33). Interestingly, the blockage of neuronal activity largely disrupts this mechanism, suggesting the presence of some activity-dependent mechanism, perhaps similar to that discussed above.

Tests for Redundant Cues in Retinotectal Development

A distinct limitation of almost all of the experimental designs discussed above is that they test only the involvement of a single cue in nerve patterning. If neuronal patterning is the product of several mechanisms acting in concert, it is possible that eliminating any one of them will not prevent the formation of retinotopography. Thus, simple experimental designs could, in the extreme, give the false impression that none of the involved mechanisms play a role (see Chap. 7, this volume for further explanation). For example, experiments that block activity with neurotoxin can show that neural activity is not absolutely required for the establishment of retinotopography during normal development, but they cannot test if activity is one of several cues. It is instructive to consider the parallels between research in pigeon homing and research in neuronal patterning (see Ref. 14). Experimental support for several different mechanisms of homing behavior had been obtained by different laboratories, yet it appeared that the pigeons could navigate in the absence of each of them. The issues were partially resolved only when test experiments were performed to eliminate some of the possible navigation cues in combination. In the case of the retinotectal projection, a few test experiments of combined cues have been performed (cf. Ref. 14); thus far, the results support the previous conclusions that positional markers are critical to the formation of the retinotopography. However, until all possible combinations of experimental conditions have been examined, it remains possible that some mechanisms that have been ruled out actually play a significant role. A more economical means to evaluate a system in which multiple, apparently redundant cues are employed is to refine the experimental system so that their relative roles might be assayed more directly.

A Vital Dye Assay of Retinotopography

Recently, we have developed a fiber-tracing technique which permits the topography of the retinotectal projection to be assayed as it forms in live *Xenopus* tadpoles (22). Because the technique permits the process to be observed directly, more detailed test experiments are possible. The assay employs the fluorescent vital dyes lysinated fluorescein dextran (LFD) and lysinated rhodamine dextran (LRD) injected into one-cell stage *Xenopus* embryos immediately after fertilization. These dyes are large and inert, and remain trapped within the injected cells and their progeny. Small groups of eyebud cells could then be excised from these embryos with the dye in all of their cells and transplanted to an unlabeled host. The fluorescent dextran labels not only the retinal ganglion cells but the optic nerve fibers that emanate from them as well. Therefore, the trajectories of the optic axons arising from the grafted cells can be followed in situ by placing the anesthetized host larva on the stage of an epifluorescence microscope and imaging the fibers through its transparent head. By using low-light-level video cameras and by protecting the animals from wavelengths of light that might bleach the dyes, possible phototoxic effects of the dyes are kept to a minimum and the labeled cells can be followed in a developing tadpole for a period of weeks.

An unambiguous assay of the topography of the developing retinotectal projection is made possible by performing homotopic grafts of fluorescein dextran (LFD)–labeled cells to one pole of the eyebud and of rhodamine dextran (LRD)–labeled cells to the opposite pole of the eyebud (22). By employing filter sets selective for rhodamine or fluorescein, the projection sites of both populations of labeled cells could be determined in the same animal. The relative positions of the two projection sites gives a straightforward assay of the topography of the projection; the overlap between the rhodamine- and fluorescein-labeled populations provides a measure of the precision of the topography. Dorsoventral topography was observed at early stages, shortly after the fibers had entered the tectal neuropil. The resolution of the projection was not highly refined; terminal arbors arising from ganglion cells at the poles of the eyebud often overlapped in the tectal neuropil. This overlap may indicate that another parallel process that has yet to act is required for the fibers to refine their connection sites. Conversely, it may be that within the small size of the developing tectum (about 100 μm), the fibers have adopted the most refined topography they can considering their relatively large terminal arbors (see Ref. 26). The observed topography and resolution are consistent with those of previous studies that traced fibers on fixed tissue by using autoradiography or cobalt (17, 26); however, in contrast to the previous studies, the topography could be assayed by a noninvasive technique in a living larva.

Nasotemporal topography was assayed by grafting labeled nasal and temporal cells to the host eyebud (22). In contrast to the animals with labeled dorsal and ventral cells, no topography was noticeable in the projection when assayed for the first few days of development (up to stage 45). That is, the LRD- and the LFD-labeled fibers overlapped in their projection to the tectum. The labeled fibers then gradually sorted out into a discernible anteroposterior topography over a period of days so that the topography was well established by stage 48. This gradual appearance of nasotemporal topography is consistent with previous studies that used cobalt to demonstrate that the arbors of individual optic nerve fibers fill most of the rostrocaudal extent of the tectum in young larvae (26). There are two obvious mechanisms by which retinotopography could slowly appear. It could be that the fibers form a widespread or "exuberant" projection to the entire rostrocaudal extent of the tectum that is later pruned back so that only the appropriate subregion is innervated. Conversely, it could be that both the nasal and temporal optic nerve fibers occupy a subregion of the optic tectum initially and that one of the two sets then sprouts selectively into a new region to establish retinotopography. Multiple observations of labeled optic fibers in the same animals at several stages indicate that the latter mechanism is correct (9, 22, 24). The nasal and temporal fibers both occupied only the more rostral portion of the developing tectum; later, the nasal fibers selectively grew new branches to the more caudal portion of the tectum and eliminated some of their rostral branches. The cause of this gradual refinement remains unclear. It could be that the positional cues that organize anteroposterior topography are weaker than those that guide dorsoventral topography, or that they are slower to appear. Additionally, the caudal tectum might not have matured sufficiently to support retinotectal synapses. Finally, the gradual refinement may indicate that some other process, such as neuronal activity, is important in nasotemporal topography.

A Role for Neuronal Activity?

The gradual appearance of anteroposterior topography occurs during a phase in which the visual system is becoming functional. This might suggest that neuronal activity is important for the establishment of nasotemporal topography, and that the delay results from the time required for neural activity to mature sufficiently. This suggestion is given some support by the superficial resemblance of the refinement process seen during development to the refinement of retinotectal topography during optic nerve regeneration. During regeneration, a crudely ordered projection forms and later refines only if locally correlated neuronal activity is permitted (cf. Ref. 27). However, the refinement of the regenerating projection appears to be the result of selective pruning of an exuberant projection, and the data presented above suggest that this is not the case for the developing projection. If this important difference is kept in mind, the approaches and lessons learned from regeneration experiments can be used to design experiments addressing the role of activity in development.

Direct test experiments of the role of neuronal activity in the formation of retinotopography are made possible by the knowledge that glutamate is the transmitter employed at the *Xenopus* retinotectal synapse (8, 9). Both NMDA-type and non-NMDA-type receptors are present and active in the developing tectum. To determine if an activity-dependent process plays some role in the development of anteroposterior topography, experiments were performed on animals with LRD- and LFD-labeled nasal and temporal eyebud cells. The NMDA-type glutamate receptor was blocked by bathing the larvae in APV. Bath-applied APV and many other drugs are taken up through the gills and are carried by the circulation to the tectum. As in normal animals, the terminals of the nasal and temporal fibers of APV-treated animals overlapped one another when assayed before stage 45 and were somewhat segregated into topographic order by stage 49/50 (several days later). However, the refinement of the topography was slowed by about 3 days in the APV-treated animals (9, 24). Presumably, by blocking the NMDA-type glutamate channel, the APV disrupted the mechanism for detecting coincident activity of the optic nerve terminals. To test this assertion, animals were raised in strobe light to force all of their nerve terminals to fire synchronously. The strobe-reared animals showed a decrease in the rate of refinement that was similar to that seen in APV animals (9, 24)

The eventual appearance of nasotemporal topography in both strobe and APV conditions suggests that neither locally correlated activity nor normal functioning of the NMDA class of glutamate receptor are strictly required. However, the retardation of the appearance of topography in both conditions suggests that neuronal activity is involved in some manner in the segregation of optic fibers along the anteroposterior axis. Thus, the results are consistent with the optic nerve fibers utilizing multiple, parallel mechanisms to establish retinotopography. For example, the proposal that the fibers pattern themselves in response to both positional cues and an activity-dependent cue is consistent with the data from these and other experiments (10). While the data suggest the presence of multiple cues, nothing in the experimental findings permits an analysis of whether activity is the strongest or the weakest of the cues that guide the optic fibers. Additional experiments would be required to determine the relative importance of each of these mechanisms. The in vivo fiber-tracing technique provides a means to perform these tests.

A Role for Positional Markers?

The results of experiments on the regeneration and development of the retinotectal projection suggest that some form of positional marker might play a role in the formation of retinotopography. One way to test for the presence of positional markers is to test the ability of the cells to find their correct target sites when grafted to ectopic sites in the embryo. If the eyebud as a whole is grafted to an ectopic site, the optic fibers turn and course toward the developing tectum, suggesting the presence of some positional cue(s) that can guide the optic fibers (15). To test if position-dependent properties within the cells of the eyebud guide the formation of the retinotectal projection, large fragments of the eyebud have been grafted to ectopic locations within the eyebud (cf. Ref 6). For example, dextran-labeled half-eyebuds were inverted along their dorsoventral axes and reimplanted into unlabeled hosts (23). When analyzed early in development, the labeled optic nerve fibers were found to project to the tectum as predicted by their position of origin in the donor embryo. This donor dependence is consistent with the presence of position-dependent cues in the eyebud. However, a large assemblage of eyebud cells were grafted in such experiments, leaving most neighbor interactions intact; therefore, guidance of the optic axons by mechanisms based on neighbor interactions, such as correlated neural activity, might also be contributing to the results.

A more direct test for the role of positional information in the formation of the ordered projection requires an experimental design that directly confronts possible position-dependent and neighbor-dependent cues. Figure 5*b* and *c* shows a schematic representation of such an experiment in which labeled eyebud cells are confronted with inappropriate neighbors (11). The experiment was performed by grafting small groups of eyebud cells from LRD- and LFD-labeled donors to nonequivalent sites in the eyebuds of unlabeled hosts. In Fig. 5, this is represented by the cells from positions 1 and 4 grafted to positions 4 and 1, respectively. If neighbor interactions (e.g. an activity-dependent mechanism) were dominant, the optic nerve fibers should project to

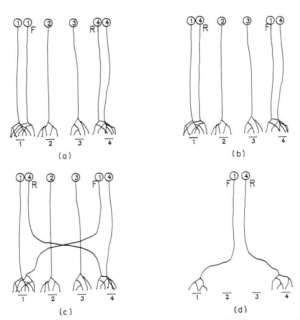

FIGURE 5. Schematic representation of grafting experiments in *Xenopus* tadpoles designed to test whether interactions between neighboring optic nerve fibers or interactions with the optic tectum dominate in the formation of the retinotectal projection. (*a*) Small grafts of eyebud tissue from fluorescein dextran (F)– and rhodamine dextran (R)–labeled donors project into the tectum in an ordered fashion. (*b*) The pattern expected following grafting to nonequivalent sites if interactions between neighboring optic nerve fibers were the major determinant of retinotopography. (*c*) The pattern expected if interactions between the optic nerve fibers and the tectum were the predominant interaction. (*d*) If interactions between the optic nerve fibers and the tectum are sufficient to order the projection, labeled optic nerve fibers should find their appropriate sites in the tectum in the absence of any neighbors.

the tectum in accordance with their new neighbors (Fig. 5b), i.e., the labeled region 4 cells should project together with their neighboring region 1 cells and vice versa. In contrast, if position-dependent cues inherent to the eyebud cells play a dominant role, the labeled fibers should terminate in the tectum as predicted by their position of origin (Fig. 5c), i.e., the labeled region 1 cell should project to region 1 in the tectum independent of the identity of its neighboring retinal cells. The experimental results were very consistent. In all cases, the heterotopically grafted cells projected to the tectum as schematized in Fig. 5c, suggesting that position-dependent cues are dominant (11).

Interestingly, while all the grafted cells projected to the tectum as shown in Fig. 5c, two different trajectories of the labeled fibers were observed. In some cases, the labeled optic nerve fibers grew into the tectum together with the host optic nerve fibers that surrounded them in the eyebud. These fibers adjusted their trajectory once within the tectum to arborize in the expected location, suggesting the presence of guidance cues on the surface of the tectum. In other cases, the labeled fibers corrected their trajectory within the optic nerve to enter the tectum in a fashion appropriate for its position of origin, suggesting that the fibers can recognize one another before reaching the tectum. These two distinct paths therefore support previous suggestions that fiber-fiber and fiber-tectum cues act in concert to guide optic nerve fibers (cf. Ref 10).

The grafts schematized in Fig. 5b and c demonstrate the presence of cell-autonomous cues in the eyebud, but they cannot elucidate the cues that the optic nerve fibers use to recognize their final position on the tectal surface. The fibers could be guided by positional cues on the tectum itself (cf. Ref. 10); conversely, the optic nerve fibers might recognize their correct tectal sites by directly or indirectly interacting with the large number of correctly positioned host optic axons that are present in the tectum (35). The operation shown in Fig. 5d has been used to distinguish between these alternatives. LRD- and LFD-labeled eyebud cells were grafted to the optic stalk of a host in which the eyebud, and hence the host optic nerve fibers, had been deleted. As expected if cues on the surface of the tectum guide the optic nerve fibers, the grafts innervated the tectum as predicted by their position of origin in the donor.

An interesting aspect of these results is that the cells seem to have a sense of their position within the eyebud at a stage when they remain unrestricted to any given phenotype. By grafting the cells of young embryos, it has become clear that the eyebud cells possess some form of positional information by stage 26. Injection of individual eyebud cells with lineage tracing substances indicates that stage 26 eyebud cells can give rise to a set of descendants made up of any combination of retinal cells, including both neurons and glia (18, 34). These findings suggest that cell position might be specified at an earlier stage than cell phenotype is specified. In fact, the cell lineage data could be taken to suggest that cell phenotype is assigned as late as the last mitosis of the precursor cells. This seems consistent with the radial unit hypothesis of cortical development (25), in which the local environment of a cell plays a large role in setting the phenotype but not the positional identity of a differentiating neuron.

Conclusion

The organization of neuronal interconnections is central to the correct functioning of the nervous system. The parallel operation of several patterning mechanisms would be one means to ensure that the correct organization is formed. When viewed in that light, it is perhaps not surprising that the recent experiments on the developing retinotectal projection suggest that several interactions play some role. The dominant mechanism appears to be the selective growth of the optic nerve fibers in response to positional cues in the retina and the tectum. Experimental evidence to date also suggests the involvement of neuronal activity in further refining the topography of the connections. As the resolution of the experiments improves, more direct tests will become possible to test the relative role(s) of timing, cell death, and other possible mechanisms. Thus, the mechanisms and experiments discussed above should not be viewed as a final statement of fact, but instead as a conceptual framework within which to design and interpret future experiments.

The experimental findings reviewed here point to the limitations of any experimental design that assumes that there is a single mechanism responsible for retinotopography. The results should also serve as a reminder that evidence favoring the presence or dominance of any one developmental mechanism cannot be taken as evidence against the involvement of any other possible mechanism. In fact, the operation of multiple mechanisms and the presence of relatively minor mechanisms may offer an explanation for the disparate results obtained from slightly different experiments performed in different laboratories. Nearly identical experimental protocols may lead to different cooperations or competitions between the mechanisms, thereby resulting in distinct outcomes. In the past, differing experimental results between laboratories were taken as a sign that one of them must be wrong. In light of the results discussed here, it is probably wiser to view such differences in results as signposts identifying interesting problems for future experiments. With the further refinement in experimental design and technique, it should become possible to turn today's points of contention into tomorrow's insights.

Acknowledgments

I thank my colleagues at the University of California, Irvine for the many stimulating discussions that helped to shape the viewpoints expressed in this article. In particular, I thank Drs. M. Bronner-Fraser and N. O'Rourke for their continued input and for their comments on this and related manuscripts. The research described was supported by a grant from the NSF, a gift from the Monsanto Corporation, and a McKnight Foundation Scholar Award.

General References

GAZE, R. M.: 1970. *The Formation of Nerve Connections*. Academic Press, London.

PURVES, D., and LICHTMAN, J. W.: 1985. *Principles of Neural Development*. Sinauer, Sunderland, Mass.

References

1. Arnett, D. W. 1978. Statistical dependence between neighboring retinal ganglion cells in goldfish. *Exp. Brain Res.* **32**:49–53.

2. Ascher, P., and Nowak, L. 1987. Electrophysiological studies of NMDA receptors. *Trends Neurosci.* **10**:284–288.

3. Collingridge, G. L., and Bliss, T. V. P. 1987. NMDA receptors—their role in long-term potentiation. *Trends Neurosci.* **10**:288–293.

4. Cline, H. T., Debski, E. A., and Constantine-Paton, M. 1987. *N*-Methyl-D-aspartate receptor antagonist desegregates eye-specific stripes. *Proc. Natl. Acad. Sci. U.S.A.* **84**:4342–4345.

5. Cowan, W. M., and Hunt, R. K. 1985. The development of the retinotectal projection: An overview. In: *Molecular Bases of Neural Development.* (G. M. Edelman, W. E. Gall, and W. M. Cowan, eds.), pp. 389–428, Wiley, New York.

6. Conway, K., Feiock, K., and Hunt, R. K. 1980. Polyclones and patterns in developing *Xenopus* larvae. *Curr. Top. Dev. Biol.* **15**:216–317.

7. Easter, S. S. 1985. The continuous formation of the retinotectal map in goldfish, with special attention to the role of the axonal pathway. In: *Molecular Bases of Neural Development.* (G. M. Edelman, W. E. Gall, and W. M. Cowan, eds.), pp. 429–452, Wiley, New York.

8. Fox, B. E. S., and Fraser, S. E. 1987. Excitatory amino acids in the retino-tectal system of *Xenopus laevis*. *Soc. Neurosci. Abstr.* **13**:766.

9. Fox, B. E. S., O'Rourke, N. A., Azterbaum, M., and Fraser, S. E. 1989. Neuronal activity plays a role in the initial formation of the retinotectal projection of *Xenopus*. *Proc. Natl. Acad. Sci. U.S.A.* (in press).

10. Fraser, S. E. 1985. Cell interactions involved in neuronal patterning: An experimental and theoretical approach. In: *Molecular Bases of Neural Development.* (G. M. Edelman, W. E. Gall, and W. M. Cowan, eds.), pp. 481–507, Wiley, New York.

11. Fraser, S. E. 1987. Intrinsic positional information guides the early formation of the retinotectal projection of *Xenopus*. *Soc. Neurosci. Abstr.* **13**:368.

12. Fujisawa, H., Tani, N., Watanabe, K., and Ibata, Y. 1982. Branching of regenerating retinal axons and preferential selection of appropriate branches for specific neuronal connections in the newt. *Dev. Biol.* **90**:43–57.

13. Harris, W. A. 1980. The effects of eliminating impulse activity on the development of the retinotectal projection in salamanders. *J. Comp. Neurol.* **194**:303–317.

14. Harris, W. A. 1984. Axonal pathfinding in the absence of normal pathways and impulse activity. *J. Neurosci.* **4**:1153–1162.

15. Harris, W. A. 1986. Homing behavior of axons in the embryonic vertebrate brain. *Nature (Lond.)* **320**:266–269.

16. Holt, C. E. 1984. Does timing of axon outgrowth influence initial retinotectal topography in *Xenopus? J. Neurosci.* **4**:1130–1152.

17. Holt, C. E., and Harris, W. A. 1983. Order in the initial retinotectal map in *Xenopus*: A new technique for labeling growing nerve fibers. *Nature (Lond.)* **301**:150–152.

18. Holt, C. E., Bertsch, T. W., Ellis, H. M., and Harris, W. A. 1988. Cellular determination in the *Xenopus* retina is independent of lineage and birthdate. *Neuron* **1**: 15–26.

19. Kleinschmidt, A., Bear, M. F., and Singer, W. 1987. Blockade of "NMDA" receptors disrupts experience-dependent plasticity of kitten striate cortex. *Science* **238**: 355–357.

20. Macagno, E. R. 1979. Cellular interactions and pattern formation in the development of the visual system of *Daphnia magna*. I. Interactions between embryonic reticular fibers and laminar neurons. *Dev. Biol.* **73**:206–238.

21. Meyer, R. L. 1983. Tetrodotoxin inhibits the formation of the refined retinotopography in goldfish. *Dev. Brain Res.* **6**:293–298.

22. O'Rourke, N. A., and Fraser, S. E. 1986. Dynamic aspects of retinotectal map formation revealed by a vital-dye fiber-tracing technique. *Dev. Biol.* **114**:265–276.

23. O'Rourke, N. A., and Fraser, S. E. 1986. Pattern regulation in the eyebud of *Xenopus* studied with a vital-dye fiber-tracing technique. *Dev. Biol.* **114**:277–288.

24. O'Rourke, N. A., Fox, B. E. S., and Fraser, S. E. 1987. Changes in optic fiber morphology during development. *Soc. Neurosci. Abstr.* **13**:368.

25. Rakic, P. 1988. Specification of cerebral cortical areas. *Science* **241**:170–176.

26. Sakaguchi, D. S., and Murphey, R. K. 1985. Map formation in the developing *Xenopus* retinotectal system: an examination of ganglion cell terminal arborizations. *J. Neurosci.* **5**:3228–3245.

27. Schmidt, J. T. 1985. Factors involved in retinotectal map formation: Complementary roles for membrane recognition and activity-dependent synaptic stabilization. In: *Molecular Bases of Neural Development*. (G. M. Edelman, W. E. Gall, and W. M. Cowan, eds.), pp. 453–480, Wiley, New York.

28. Schmidt, J. T., and Edwards, D. L. 1983. Activity sharpens the map during the regeneration of the retinotectal projection in goldfish. *Brain Res.* **269**:29–40.

29. Schmidt, J. T., and Eisele, L. E. 1985. Stroboscopic illumination and dark rearing block the sharpening of the regenerated retinotectal map in goldfish. *Neuroscience* **14**:535–546.

30. Sperry, R. W. 1963. Chemoaffinity in the orderly growth of nerve fibers and connections. *Proc. Natl. Acad. Sci. U.S.A.* **50**:703–710.

31. Sperry, R. W. 1965. Embryogenesis of behavioral nerve nets. In: *Organogenesis*. (R. L. DeHaan, and H. Ursprung, eds.), pp. 161–186, Saunders, Philadelphia.

32. Tsumoto, T., Hagihara, K., Sato, H., and Hata, Y. 1987. NMDA receptors in the visual cortex of young kittens are more effective than those of adult cats. *Nature (Lond.)* **327**:513–514.

33. Udin, S. B., and Fawcett, J. W. 1988. Formation of topographic maps. *Annu. Rev. Neurosci.* **11**:289–327.

34. Wetts, R., and Fraser, S. E. 1988. Multipotent precursors can give rise to all major cell types of the frog retina. *Science* **239**:1142–1145.

35. Willshaw, D. J., and von der Malsburg, C. 1979. A marker induction mechanism for the establishment of ordered neural mappings: Its application to the retinotectal problem. *Philos. Trans. R. Soc. Lond. B Biol. Sci.* **194**:431–445.

Questions for Discussion with the Editor

1. *Can the patterning of the vertebrate visual system serve as a useful paradigm for other cell patterning systems, or is it so specialized as to be a truly unique system?*

Although some aspects of the patterning of neuronal systems may be unique, most features are more general than they may appear at first glance: In the patterning of any tissue, there appears to be roles for interactions among the cells of the tissue as well as between those cells and other tissues. The nervous system is no exception in that the patterning of projection neurons results from interactions both among the nerve fibers and between the fibers and their targets. One feature that makes the patterning of neural projections seem unique is that these two classes of interaction can be separated from one another by great distances. This offers experimental advantages in permitting the independent manipulation of the interactions within and between tissues; however, this unique geometry cannot be taken as evidence for the operation of unique modes of interaction.

Perhaps the one unique aspect of the patterning of neuronal connection is the involvement of neuronal activity. This offers a means for a "fine-tuning" of the neuronal pattern based on its function. Most other tissues lack the action potential which is the basis of this interaction, so it seems unlikely that the exact mechanisms used by the nervous system will be shared with other tissues. However, even this may not be as unique as it appears. There are other time-variant features of the cells within any tissue, such as the cell cycle. It is possible that the patterning of cells depends in part on the coincidence of one of these features, just as the organization of projection neurons depends in part on the coincidence of neuronal activity. The ability to show that neuronal activity played a role in neuronal patterning relied on some well established electrophysiological techniques; tests of the roles of other cellular dynamics in tissue patterning may have to await the development of the appropriate technology to assay such dynamics within a developing tissue.

2. *Why didn't you invoke the term "hierarchy" when you discussed cue redundancy? Rather, you appear to prefer terms such as "parallel operation," "dominant" mechanism, etc.*

The eventual goal in dissecting a system with redundant cues is to understand the relative roles of each of the cues—to define the hierarchy of the cues. In discussing the patterning of the visual system, I have avoided the term hierarchy because it can overstate our present knowledge. Its use implies that we know which of the redundant cues is dominant in the normal patterning of the tissue. At present, some experiments show the operation of several different cellular interactions, and a few experiments indicate the dominance of one of the interactions in that setting. Without a more complete set of experiments, the experiments showing dominance could be assembled into a misleading hierarchy. For example, the grafting experiments performed in my laboratory (see Fig. 5) do not offer any evidence for a competition for space between optic nerve terminals. In contrast, experiments on the regeneration of the retinotectal projection suggest that a competition for terminal space plays a very strong role in the patterning (see Fig. 3*b* and *c*). The absence of evidence for competition in our experiments on developing connections cannot be taken as evidence for its absence. Competition, had it acted, would have generated results indistinguishable from those expected if it played no role.

In an attempt to construct a hierarchy of the cues, parallel work in my laboratory has used computer modeling (see Ref. 10). The goal of this work has been to determine if one set of cues, with fixed relative strengths, can generate the full range of experimental results now in hand. At present, it appears that a single model can fit most, if not all, of the regeneration experiments. Interestingly, in some experimental settings the predictions of the model are indistinguishable if one or more of the interactions is eliminated; in other settings, the same changes are disastrous. This finding has shaped my reticence to state a hierarchy, and it points to the importance of quantitative testing of hypotheses of developmental patterning.

Cytoplasmic Organization and Information Systems in Embryogenesis

CHAPTER 9

The Cytoskeleton in Development

David P. Hill, Susan Strome,
and Gary P. Radice

DEVELOPMENT OF METAZOANS consists of the proliferation, diversification, and morphogenetic movements of cells. This chapter considers the role of the cytoskeleton in these processes. In the first part of the chapter, the role of the cytoskeleton in functions common to all cells is described. These roles include maintenence of cytoplasmic organization and cell shape, cell division, and motility. Next, the manner in which these functions can be modified to serve specialized developmental roles, such as the distribution of developmental instructions to early embryonic cells, positioning of cells in early embryos, and movement of cells during morphogenesis is illustrated.

Introduction to Cytoskeletal Components

The cytoskeleton consists of at least three major filament systems and their associated proteins. The earliest identified and best characterized component is the microtubule system. It is composed of a framework of hollow tubes that are approximately 25 nm in diameter and are constructed from dimers of alpha and beta tubulin. The head-to-tail polymerization of the dimers generates polarity in the microtubules, and the two ends display different kinetics of assembly and disassembly (for review see Ref. 24). This component of the cytoskeleton also contains a plethora of proteins, termed "microtubule-associated proteins or MAPS," that bind to the core tubules and help to regulate their stability, arrangement, movement, and interactions with other cellular components (32). The arrangement of the microtubule system is dynamic, and it changes during the cell cycle from a complex and seemingly random cytoplasmic network radiating out from a microtubule organizing center or MTOC in interphase cells to a highly ordered spindle structure in mitotic cells (Fig. 1).

Another component of the cytoskeleton is the microfilament system. It is composed of core filaments of actin monomers polymerized end-to-end; two strings of actin monomers are intertwined into 6- to 8-nm-diameter microfilaments (Fig. 2A). These filaments are also polarized, and their two ends have different rates of polymerization and depolymerization. The arrangement of microfilaments varies widely both within a cell and from cell type to cell type. The most familiar, but highly specialized, microfilament system is seen in skeletal muscle fibers, where actin filaments are interdigitated with and slide along myosin filaments to contract the myofiber (for overview see Ref. 40). In nonmuscle cells, microfilaments form a cortical network of fibers, and they also run through the cytoplasm and often form bundles (Fig. 2B). Just as MAPs associate with microtubules, many proteins that interact with actin filaments, including a cytoplasmic myosin, have been isolated from various cell types. These proteins are probably involved in controlling the polymerization and crosslinking of actin filaments in vitro and may have the same functions in cells.

The third and least well understood component of the cytoskeleton is the

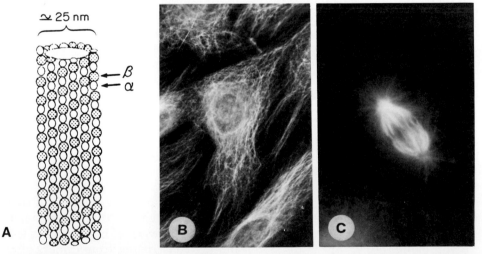

FIGURE 1. (a) Schematic representation of a microtubule. Dimers of alpha- and beta-tubulin polymerize head-to-tail into protofilaments, which associate side-to-side to form a hollow tube approximately 25 nm in diameter. Both subunits are approximately of the same molecular weight; they are shown as different sizes for clarity. (b) Immunofluorescence micrograph of a fibroblast cell in interphase. The cell was stained with antibodies that decorate the microtubules. (c) Immunofluorescence visualization of microtubules in the spindle of a fibroblast cell in mitosis. (Micrographs courtesy of M. Ladinsky and J.R. McIntosh.)

6-8 nm

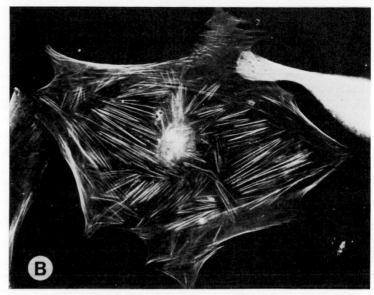

FIGURE 2. (*A*) Schematic representation of a microfilament. Two intertwined strands of actin subunits generate a 6- to 8-nm filament. (*B*) Fluorescence visualization of actin bundles in a spreading fibroblast cell. Cells were stained with the F-actin-specific stain rhodamine phalloidin. (Micrograph provided by E. Lazarides.)

intermediate filament network. It is composed of a variety of related proteins that polymerize to form a filamentous network that extends throughout the cytoplasm, often forming a "basket" around the nucleus. One likely role of intermediate filaments is in tension transmission in epithelial sheets. Epithelial cells are connected by specialized focal adhesions called desmosomes, and each cell's desmosomes are interconnected on the cytoplasmic side by the network of intermediate filaments. This allows tension in one cell to be transmitted to its neighbors and throughout the sheet. In addition, the primary structure of the proteins that make up intermediate filaments is closely related to that of nuclear lamins, and it has been hypothesized that the nuclear and cytoplasmic filaments interact, perhaps to position the nucleus. For the most part, however, it is not understood how intermediate filaments function and interact with one another and with other components of the cell. This mystery makes them and their behavior a popular topic for investigation.

Approaches to Analyzing Cytoskeletal Function

The oldest and most straightforward approach to analyzing the cytoskeleton and its components is direct observation. Originally, cells were fixed, sectioned, and stained for viewing by either light or electron microscopy. This type of analysis reveals global arrangements of the cytoskeleton and provides the basis for speculation about cytoskeletal function. However, because the three-dimensional distribution of filaments is often difficult to reconstruct from serial sections, this type of analysis has limitations. Moreover, not all cytoskeletal components can be visualized by this technique. This problem has in part been overcome by staining whole cells with fluorescently labeled antibodies directed against cytoskeletal components, then visualizing the components using fluorescence microscopy. In this way, entire cytoskeletons within cells can be visualized without sectioning. More recently, this technique has been dramatically improved by the development of confocal laser scanning microscopy, which allows the microscopist to optically section intact cytoskeletal arrays.

Another approach to the investigation of the cytoskeleton utilizes biochemical techniques. By analyzing cytoskeletal components in vitro, investigators hope to reconstruct cytoskeletal arrays and to mimic cytoskeletal functions observed in living cells.

There are drawbacks to both of the above approaches. In the first case cells are fixed. Since the cytoskeleton is dynamic, this approach may not provide an accurate picture of all of the changes that occur in the cytoskeletal array in vivo, not to mention possible alterations in cytoskeletal structure that may occur during fixation. An investigator may see two very different cytoskeletal arrangements in a cell at two different times but may not be able to tell how the transition took place. The second approach, analysis of components in vitro, may lead to an oversimplified or misleading view of the processes that occur in a cell, where the components may be under very different physical constraints. For example, two components found to interact in vitro may be sequestered to different compartments of a cell and may never contact one another in living cells.

Recently, technological breakthroughs in microinjection, molecular genetics, and microscopy have set the stage for high-resolution in vivo analysis of the cytoskeleton. Investigators can now microinject pharmacological agents, altered genes, presumed cytoskeletal components, and fluorescently tagged versions of cytoskeletal proteins. Once the molecules have been introduced, the physiological responses of the injected cells and resulting cytoskeletal changes can be monitored using video-enhanced microscopy. Results obtained with these technologies should provide detailed information about cytoskeletal dynamics and the effects of cytoskeletal disruption in living cells.

The Cytoskeleton Is Necessary for Many General Cellular Processes

Many of the activities of cells in a developing organism are universal to all cell types. This section discusses how these general processes depend on the cytoskeleton. The following section considers how these processes also serve specialized developmental roles.

Cell Shape

Without the application of force, cells would act as simple micelles in solution and adopt a spherical shape. Instead, living cells adopt a great variety of shapes and often change shape. Force must be applied to the inside or the outside of the cell membrane to cause the cell to deviate from a spherical shape. When these forces are applied internally, they are generated by the cytoskeleton.

One of the best-studied examples of the cytoskeleton's involvement in determining cell shape is the avian red blood cell. In this cell, all three of the major cytoskeletal filament systems appear to be involved in bringing about the cell's convex discoid shape (for review see Ref. 26; Fig. 3). Biochemical analyses have identified the critical membrane and cytoskeletal proteins and have defined connections between them. Analysis of the proteins' distributions has demonstrated that they are at the predicted locations in the cells. The current model that emerges from these studies is as follows: A bundle of microtubules, known as the marginal band, lies just beneath the equatorial membrane and forms a hoop that pushes the cell membrane out along the equator. This allows the cell to resist outside forces and participates in the reshaping of the cell after it has been perturbed (21). The force that pulls the apices of the cell together, thereby flattening it, seems to come from interactions of both intermediate filaments and microfilaments with an array of membrane proteins. Two membrane proteins, the ion transporter protein and glycophorin, appear to interact with the submembranous proteins, spectrin, protein 4.1, and ankyrin, which in turn interact with actin and intermediate filaments. Thus it appears that in red blood cells all of the known components of the cytoskeleton cooperate to maintain cell shape.

The red blood cell cytoskeleton is also dynamic. If ATP is not continuously supplied to the cells, they quickly lose their shape. Recent studies have implicated a complex set of phosphorylation pathways in the control of the state of the cytoskeleton; these pathways may ultimately regulate cell shape (17).

Cell Division

One of the basic tenets of the cell theory is that all cells arise from preexisting cells by the process of mitosis. The cytoskeleton plays a crucial role in the replication of cells, not only to ensure that chromosomes are distributed equally to the two daughter cells but also to ensure that cytokinesis takes place at the correct time and in the correct orientation relative to the dividing DNA (Fig. 4).

One of the most obvious structures in any mitotic cell is the mitotic spindle. It can be visualized in living cells by polarized light, Nomarski, or immunofluorescence microscopy (Fig. 2B). The spindle is composed of aligned microtubules that emanate from two poles at opposite ends of the cell. In the middle of the spindle, the condensed chromosomes line up along the

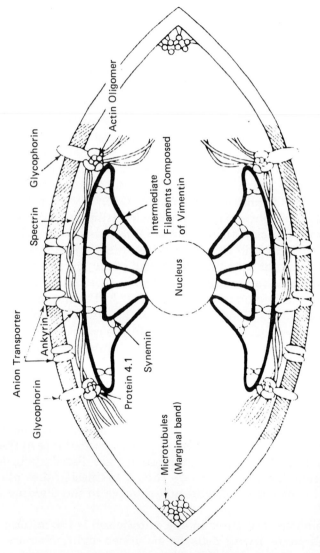

FIGURE 3. Diagram of the cytoskeletal elements that are responsible for the maintenence of cell shape in the avian red blood cell. (From Ref. 26.)

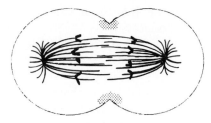

FIGURE 4. Diagram of a cell during the anaphase stage of mitosis. The spindle is composed of microtubules that emanate from two microtubule-organizing centers or spindle poles; chromosomes move toward the poles. As chromosome movement proceeds, microfilaments, shown as dots, align circumferentially at the cell cortex equidistant from the two spindle poles and initiate cytokinesis.

metaphase plate and are attached to spindle fibers at sites known as kinetochores. During anaphase the sister chromatids separate and move to opposite ends of the cell. The spindle microtubules attached to the chromosomes shorten during chromosome movement, but it is not known whether the microtubules themselves provide the motive force for chromosome movement or whether they simply serve as tracks along which chromosomes are moved. Current models include force generated by the chromosomes themselves, by microtubule-based motors, or by the depolymerization of microtubules behind the moving chromosomes (25).

After the chromosomes have been separated, cytokinesis divides the cell. This process is accomplished by the microfilament cytoskeleton. In animal cells, microfilaments accumulate as a band around the equator of the cell between the poles of the mitotic spindle (Fig. 4). The placement of the band is critical; it must be oriented so that the resulting cleavage distributes one daughter nucleus to each daughter cell. Experiments that alter the position of the mitotic spindle have shown that the microfilament band is formed equidistant from the two spindle poles (for review see Ref. 36). It is hypothesized that this placement results from the interactions between microtubules that emanate from the spindle poles and the cell cortex before cleavage begins. The actin filaments in the equatorial band are then thought to slide past one another, causing a contraction of the band that pulls the equatorial membrane "closed." The microfilament sliding is thought to be mediated by cytoplasmic myosin, since actin and myosin colocalize in the cleavage furrow during cell division and disruption of either actin or myosin by pharmacologic agents or by mutation inhibits cell division.

Cell Motility

Another characteristic that all cells share is their ability to locomote. Cells can move using motile forces from within or by being acted upon by external forces. When cells themselves provide the motile forces, the forces are generated by elements of the cytoskeleton.

Although the best understood form of cell motility is movement generated by the beating of cilia and flagella, this type of motility is only used to a limited extent during embryogenesis and is not considered here. Instead, we focus on the amoeboid type of cell motility that is observed in many embryonic cells.

Cell migration across surfaces is usually due to the extension of a protrusion in the direction of travel followed by movement of the cell body. The first step in this type of locomotion, extension of a protrusion from the cell body, involves a shape change. Next, the cytoplasm in the protrusion appears to "gel," and the protrusion becomes attached to the substrate. Finally, a contractile force somehow causes movement of the cell body toward the protrusion, either via streaming of the main body of cytoplasm into the protrusion or simply by dragging the rest of the cell body into the protrusion. It is not surprising to learn that this type of locomotion requires many complex cytoskeletal interactions. Furthermore, it appears that different amoeboid cell types may use different components of the cytoskeleton to accomplish each of the steps outlined above.

Amoeboid movement has been intensively analyzed in cells of simple slime molds such as *Dictyostelium discoidium* (for review see Ref. 13). The protrusion, called a lamellopodium or pseudopodium, contains a large amount of polymerized actin but is devoid of microtubules and myosin (Fig. 5). Microtubules and myosin are, however, seen in the cell bodies. In these cells, as in fibroblasts, it appears that the position of pseudopod formation may be directed by oriented arrays of microtubules and that microtubules may mediate attachment of the cell to the substrate (38). After the original protrusion or pseudopod is extended, polymerization of actin within the pseudopod causes the cytoplasm within to turn into a gel. The mechanism by which cytoplasm flows into the pseudopod has not yet been elucidated. It was originally hypothesized that contractions of actin-myosin cables in the cell body lead to cytoplasmic streaming. However, *Dictyostelium* mutants that lack the myosin

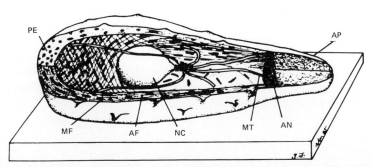

FIGURE 5. Schematic representation of a cell migrating across a surface. AF, actin filaments; AN, actin network; AP, anterior pseudopod; MF, myosin thick filaments; MT, microtubules; NC, nucleus; PE, posterior ectoplasm. Taken from Ref. 13.

heavy chain are able to locomote, although not efficiently (47). This suggests that actin-myosin contractions are not solely responsible for cytoplasmic streaming and that another mechanism may be utilized in this process. One interesting model suggests that the sol-gel transition in the pseudopod may create an osmotic pressure in the cell that results in cytoplasmic flow or that elastic forces in the gel itself may result in cytoplasmic streaming (33). Interestingly, the former mechanism may also be used to explain the formation of the pseudopod itself. If either of these hypotheses are correct, the microtubule and actin-myosin system may serve merely to fine tune the direction of streaming.

The mechanisms involved in generating protrusions in slime molds may differ significantly from the mechanisms used in other cells. For example, in certain algal cells, microtubules appear to be necessary for the formation of protrusions (44). Thus evolution may have adopted different strategies to achieve amoeboid motility.

Organization of Internal Components and Organelles

Cells are not merely bags of solution in which organelles and proteins float freely, governed by the laws of simple diffusion. The consistency and regularity with which structures are arranged inside cells indicates that internal components are anchored and transported in highly regulated ways. But these anchoring and supporting devices cannot be static structures, because as described earlier, cells undergo a number of dynamic changes. It has become clear in recent years that the organization of the cytoplasm is governed by the cytoskeleton.

The role of the microfilament cytoskeleton in altering cytoplasmic consistency was described in the discussion of cell motility. Recently, the intermediate filament network has been invoked in cytoplasmic organization, specifically in positioning the nucleus in many cell types (see Fig. 3 for a schematic view of this role in the red blood cell). Here we want to focus on the role of the microtubule cytoskeleton in intracellular organization and motility in interphase cells. The microtubule arrays found in interphase cells appear to function as tracks along which cellular vesicles and organelles move. Using model systems, such as the squid giant axon and the long slender processes from giant algal cells, investigators have identified several motors that can transport vesicles along single microtubules in vitro (46). These motors appear to be similar in shape to ciliary and flagellar dynein, which is known to be responsible for microtubule sliding in cilia and flagella. Moreover, like ciliary-flagellar dynein, they appear to be able to move objects only unidirectionally along microtubules. For example, the motor kinesin moves vesicles from the minus to the plus ends of microtubules, while the cytoplasmic dynein motor moves vesicles in the opposite direction. Both classes of motors require ATP for movement and are thought to cyclically attach to and slide along microtubules. In addition to the in vitro experiments described above, recent in vivo analysis suggests that vesicles and tubules of the endoplasmic reticulum move along a framework of microtubules within the cell, perhaps using the motors described above (43).

Cell Interactions with the Environment

Another characteristic of all cells is that they respond to environmental stimuli. Cells often move, change shape, divide, or even alter gene activity when an appropriate external stimulus is applied. It is clear, at least for movement, shape change, and cell division, that the cytoskeleton must respond to the external stimulus. To do this, some type of transducer must relay the signal from outside the cell to inside. These transducers are probably membrane-spanning proteins that change the physiological state of the cell either directly, such as gated ion channels, or indirectly, such as those that interact with G proteins. These changes in physiological state may then trigger changes in the cytoskeleton.

The microfilament cytoskeleton has also been shown to interact directly with membrane proteins whose external surfaces interact with the outside environment. One example is the integral membrane protein complex, integrin, which interacts with the microfilament cytoskeleton inside the cell and fibronectin outside the cell (see Chap. 18, this volume). These types of interactions may be very important inside embryos, where complex cell migrations may be directed by external signals, or where cells receive developmental cues by interacting with their neighbors.

The Cytoskeleton in Specialized Developmental Processes

Cells in developing organisms carry out many of the general processes described above. In addition some processes are adapted to specific developmental roles, such as (1) the differential partitioning of developmental information to early embryonic cells or blastomeres, (2) the generation of cells in correct positions relative to each other, (3) the migration of individual cells to new positions, (4) the migration of sheets of cells to form cell layers and tubes, and (5) the cell shape changes that accompany terminal differentiation. The next section of this chapter discusses some examples of the cytoskeleton's role in these developmental processes.

Segregation of Developmental Instructions

A central puzzle of early development is how blastomere nuclei, which contain apparently identical sets of genes, later transcribe different subsets of these genes and thus give rise to muscle, nerve, epidermis, and other specialized cell types. Somehow, each genetically identical nucleus acquires a unique set of instructions that gives the blastomere and its descendants their own identity. Classically, the adoption of different identities by early blastomeres has been thought to result from two mechanisms: (1) the segregation of cytoplasmic factors or "determinants" to different cells during the early cleavages and (2) the interpretation of external signals or positional cues from cell-cell interactions. It is now clear that most organisms use both of these mechanisms to generate cells with different identities in early embryos.

In some embryos, there is evidence that developmental "factors" or "instructions" are present in the egg and are segregated to specific blastomeres during the early cleavages (see Ref. 5 for a review of the evidence). There are a number of ways in which the cytoskeleton could participate in this segregation: (1) Cytoskeletal elements could act directly during segregation by attaching to determinants and moving them to subregions of the zygote's cytoplasm, analogous to kinesin moving vesicles along microtubule tracks. (2) Elements of the cytoskeleton could direct mass rearrangements of cytoplasm, for example, by modulating cytoplasmic streaming. (3) Cytoskeletal structures could anchor components in various regions of the cytoplasm. There is now evidence that different organisms use different elements of the cytoskeleton and different partitioning mechanisms to achieve a similar end result, the differential segregation of developmental instructions.

Many embryos that are thought to segregate developmental information during the early cell divisions undergo cytoplasmic rearrangements that are readily observable by light microscopy (see Chap. 13, this volume, for a description of this phenomenon in insects, and Chap. 10, this volume, for a discussion of the molluscan egg cortex). A graphic example of this phenomenon is seen in ascidian eggs (4). Over 80 years ago Conklin described a number of dramatic cytoplasmic shifts, referred to as ooplasmic segregation, that generated visibly different domains of cytoplasm in the egg before the first cell division (see Chap. 3, this volume). He hypothesized that the domains contained localized developmental potential. A number of experiments now support his hypothesis. Furthermore, recent evidence suggests that the microfilament system may be involved in the early phase of this cytoplasmic reorganization (Fig. 6). During the first phase of ooplasmic segregation, a vegetal contraction results in the localization of "cortical myoplasm" at the vegetal pole. During this period, a complex array of microfilaments can be visualized at the vegetal pole (39), and this phase of ooplasmic segregation is sensitive to inhibitors of microfilament function (T. Sawada and G. Schatten, personal communication).

Microfilaments have also been implicated in segregation events in embryos of the nematode *Caenorhabditis elegans*. Cell-fate specification in *C. elegans* embryos is controlled mainly by internally segregated developmental instructions. The *C. elegans* zygote undergoes a number of directed movements of internal components during the first cell cycle, then divides unequally into two daughter cells that differ in size and developmental fate. By using pharmacological agents, it has been shown that the reorganization events are sensitive to microfilament inhibitors (42). In the first cell cycle, there is a critical time interval during which disruption of microfilaments

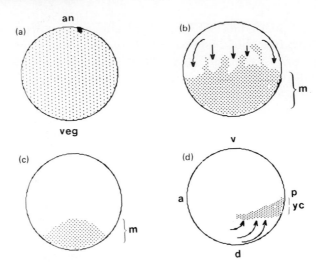

FIGURE 6. Diagram of ooplasmic segregation in ascidian embryos. (*a*) Newly fertilized egg. (*b*) The myoplasm, m, streams toward the vegetal end of the egg. This process is microfilament-mediated. (*c*) Myoplasm localized at the vegetal pole. (*d*) A microtubule-mediated shift moves the myoplasm into a crescent on the posterior side of the egg. The placement of this yellow crescent (yc) predicts the dorsal side of the embryo. a, anterior; p, posterior; d, dorsal; v, ventral.

leads to disruption of the asymmetry observed in that cell cycle as well as alteration of future development of the embryo (Ref. 16; Fig. 7). Two features of this disruption are notable. (1) actin disruption leads to the altered placement of cytoplasmic granules (P granules) that are normally segregated to the smaller daughter at the first division and eventually passed to the germ-line progenitor cell. Microfilaments but not microtubules appear to participate in positioning P granules to the posterior cortex of the zygote, after which neither cytoskeletal system is required to maintain its localization. Some other as yet unidentified element of the cytoskelton may function as the P-granule anchor. (2) Actin disruption alters pronuclear migration and the future placement of the mitotic spindle. Since both of these are microtubule-based events, we can infer that somehow the microfilament network is acting to modulate microtubule function during the critical time interval.

The mechanism by which microfilaments function in early *C. elegans* is not yet known. During the time interval when microfilaments are playing their critical role, cytoplasm streams toward the posterior end of of the cell (30). This streaming may carry P granules to the posterior cortex. Indeed microfilament disruption inhibits streaming. Interestingly, visualization of the microfilament cytoskeleton reveals that actin filaments become concentrated at the anterior end of the 1-cell embryo during the microfilament critical period (41). This is in contrast to amoeboid cells, in which actin is localized at the end of the cell toward which the cytoplasm streams. This would imply that streaming in *C. elegans* embryos is not the result of osmotic pres-

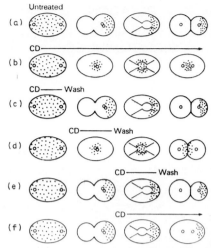

FIGURE 7. Schematic summary of the effects of treating *C. elegans* 1-cell embryos with the microfilament inhibitor cytochalasin D (CD) for brief periods during the first cell cycle. Each row shows four progressively later stages: pronuclear stage, pseudocleavage, and pronuclear migration, formation and asymmetric placement of the spindle, and first cleavage. Anterior is left; posterior is right. The small dots represent germ-line-specific P granules, and the ✕ symbols are spindles. (a) Normal embryo for reference. (b) Embryo treated continuously with CD. (c) Embryo treated briefly with CD during the pronuclear stage. (d) Embryo treated briefly with CD during the critical period from pronuclear migration until pronuclear meeting. (e) Embryo treated briefly with CD after pseudocleavage, pronuclear migration, and P-granule segregation until just before cytokinesis. (f) Embryo treated briefly with CD continuously after pseudocleavage, pronuclear migration, and P-granule segregation. (Taken from Ref. 16.)

sure changes caused by an actin gel. An alternative hypothesis is that streaming is induced as actin filaments "squeeze" the egg at the anterior end.

In some embryos, it appears that the microtubule cytoskeleton plays an active role in ooplasmic rearrangements. In ascidian embryos, the later stages of ooplasmic segregation appear to be mediated by microtubules (Fig. 6). During this phase, the myoplasm shifts from the vegetal pole of the egg to form a crescent-shaped area at the future posterior side of the embryo. T. Sawada and G. Shatten (personal communication) have shown that this phase of ooplasmic segregation requires the participation of the microtubule cytoskeleton and may result from growth of the sperm aster.

The microtubule system also appears to be involved in cytoplasmic reorganization in frogs. In frog eggs, a shift between the egg cortex and deeper cytoplasm occurs shortly after fertilization; this shift appears to play a role in determining the orientation of the dorsal-ventral axis of the embryo. During the shift, but not before or after, microtubules are found in the shear zone, oriented parallel to the direction of the shift, suggesting that they orient this movement (Ref. 7; Fig. 8). Also, microtubules have been implicated in the change in cytoplasmic consistency that accompanies the cytoplasmic shift (6).

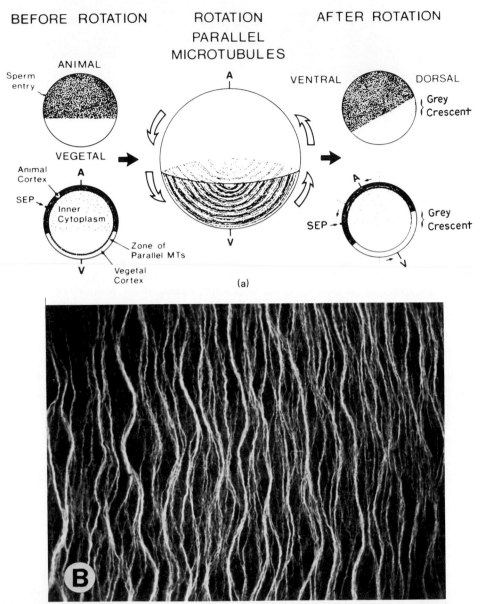

FIGURE 8. (*A*) Diagram of the cytoplasmic shift that occurs in frog eggs. (*B*) Immuno-fluorescence micrograph of the array of cortical microtubules aligned in the direction of the shift. (Taken from Ref. 7.)

Microtubules have also been shown to be important in the segregation of developmental information in the nemertean *Cerebratulus lacteus*. By manipulating the time of formation of the meiotic aster in fertilized eggs, Freeman (10) was able to alter the timing of the separation of gut and apical tuft potential. By removing the aster altogether, the separation was inhibited. These results suggest that the growing aster participates in the segregation of developmental potential.

It is clear from the examples that we have given and from the many more that exist that elements of the cytoskeleton can play a critical role in reorganization of ooplasmic components. Investigators are now primed to address localization mechanisms and to identify the factors that are being segregated. One exciting advance toward this end is the identification of localized maternal mRNA in embryos (see Chap. 3, this volume, for a discussion of localized RNA in *Xenopus*). Developmentally important mRNAs may be anchored or positioned by the cytoskeleton and thus become differentially segregated during embryogenesis. Jeffery and coworkers (20) have succeeded in isolating the myoplasmic cytoskeletal domain from ascidians, and interestingly, mRNA remains associated with the isolated cytoskeletons (19). However, as yet, no developmentally important RNAs have been shown to interact directly with the cytoskeleton.

One powerful approach that has been used to analyze localization mechanisms is genetics. An excellent example of this approach comes from *Drosophila*. The anterior-posterior and dorsal-ventral axes and the segmental patterning of *Drosophila* embryos are controlled by localized maternal mRNAs and proteins, many of which have been genetically defined (for reviews see Refs. 2 and 31). For example, anterior development is controlled by the *bicoid* gene, which encodes a mRNA that is tightly localized at the anterior end of oocytes and early embryos; as a result, the *bicoid* protein is distributed in a graded fashion from anterior to posterior (see Chap. 14, this volume). At least two genes, *swallow* and *exuperentia*, appear to be required for the anterior localization of *bicoid* mRNA. Identification of genes that encode proteins that are required for the localization of developmentally important factors is likely to target elements of the cytoskeleton that are directly responsible for partitioning events in embryos. Many of the cytoskeletal elements and mechanisms used are likely to be modifications of those used by all cells for general cellular functions (see the first section of this chapter). However, the possibility exists that embryos have evolved novel cytoskeletal mechanisms to serve these functions.

Cell-Cell Interactions in Embryos

As mentioned above, in addition to partitioned internal information, cell-cell interactions and external cues also play an important role in early development. One dramatic example of how the cytoskeleton can be influenced by the position of a cell in an embryo is discussed by Johnson (Chap. 15, this volume).

Many other examples exist in which embryonic cells recognize external signals and respond to them in a specific manner. In sea urchin embryos, primary mesenchyme cells migrate just before gastrulation into the blastocoel, where they form a ring of cells and eventually produce the larval skeleton. These cells need to recognize the correct target sites for the placement of the ring. Neurons in avian brains undergo complex migration patterns during neurogenesis; these migrations may be modulated as moving cells contact cues supplied by glial cells (1). Neuronal processes also undergo complex path-finding behaviors that result in the establishment of proper neuronal connections between cells. By generating and analyzing mutants in cell migration, cell signaling, and signal processing in such organisms as *Drosophila* and *C. elegans*, it may be possible to genetically dissect these processes and elucidate the involvement of the cytoskeleton in them.

Control of Cell-Division Patterns

Cell division plays a critical role not only in cell proliferation during embryogenesis, but also in ensuring passage of specific cytoplasmic domains to proper daughter cells and in generating daughter cells in correct positions relative to each other. When the internal components of a cell become polarized, it is ultimately the orientation of the cleavage furrow that determines whether partitioned elements are distributed to both or only a single daughter; an excellent example of this is discussed by Johnson in Chap. 15, this volume. Furthermore, divisions that generate developmentally different daughter cells are often unequal (i.e., generating daughter cells of different size). Since the position of the mitotic spindle determines the position of the cleavage furrow (see earlier discussion), the orientation and placement of the spindle dictates the division axis and daughter-cell sizes. The mechanism by which placement of the mitotic spindle is controlled is not known. In *C. elegans* embryos, the microfilament system appears to participate in both orienting and positioning the mitotic spindles during the early divisions (16, 18). How the microtubule and microfilament systems interact is still unclear, but mutations exist that result in the improper placement of the spindle; they may shed light on the mechanisms involved (23).

Morphogenetic Movements

Up to this point our discussion has focused on the activity of the cytoskeleton in individual cells. We now turn to its role in shaping of multicellular animals. Embryos acquire their shapes mostly through three processes: cell growth, cell death, and cell movements. Although the cytoskeleton may participate only indirectly in cell growth and death, it is critically involved in morphogenetic movements of individual cells, cell clusters, and cell sheets (see Ref. 45 for a general review of morphogenetic movements).

We have no reason to believe that cell movement in embryos differs in any fundamental way from movement of isolated cells in culture. That is, the mechanochemistry of actin, myosin, tubulin, kinesin, and their associated proteins probably does not change significantly when cells are removed from an embryo and grown in a petri dish. For example, cells migrating on flat glass or plastic tend to form broad, flat lamellae, whereas in vivo the same cells form fingerlike filopodia. Nevertheless, both structures are filled with microfilaments and may serve the same role in migration. We can determine precisely how the cytoskeleton acts during morphogenetic movements by carefully observing movements in vivo, by locating the cytoskeletal components in the cell, and by demonstrating that when a particular component is eliminated or its function defeated, the movement ceases.

We do not yet have such a complete analysis of how the cytoskeleton participates in a morphogenetic movement. However, we have learned a great deal about two important kinds of movements widespread in developing organisms: the folding of epithelial sheets of cells to form pockets and tubes and the elongation of cellular tubes.

FOLDING OF EPITHELIAL SHEETS

Formation of the optic cup, neural tube, thyroid placodes, and lung and salivary gland rudiments in higher vertebrates and of the archenteron in sea urchins and *Amphioxus* provide examples of morphogenesis by folding of a sheet of cells (see Ref. 9 for a more complete review). Those cases in which the invagination or involution begins on the surface of the embryo are more easily observed and manipulated, and it is those cases about which we know the most. We have combined the common features of several examples into a "generic" folding model, shown in Fig. 9. Invaginations are always preceded by cell elongation, termed "placode formation" or "palisading." Most often, elongation is followed by apical constriction. That is, the surface of the cell sheet that faces into the fold decreases in area and the cells become wedge-shaped. Invaginations are largely autonomous movements which arise from

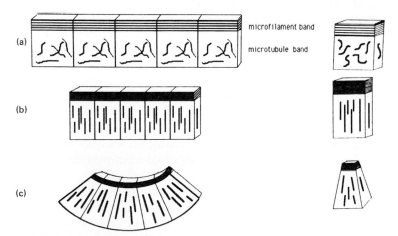

FIGURE 9. Schematic representation of a typical folding of an epithelial sheet. (*a*) Before folding, cells are cuboidal. (*b*) Cells elongate, and microtubules become aligned parallel to the direction of elongation. This is called "placode formation" or "palisading." (*c*) Cells contract at their apices, causing the sheet to fold.

the shape changes of the participating cells and not from buckling due to pressure from outside the area. Evidence for this comes from cell sheets cultured away from surrounding tissues. Sea urchin vegetal plates, amphibian neural plates, and fruit fly imaginal discs all fold more or less normally when cultured apart from their surrounding tissues (9).

One possible explanation for cell-shape changes and invagination is a change in adhesiveness between cells (e.g., Ref. 28). However, because we lack techniques to measure accurately the proposed changes in adhesiveness between cells, hypotheses that suggest differential adhesion as a mechanism, while potentially correct, have not been tested. However, an earlier hypothesis that elongation and apical constriction *cause* invagination, as originally proposed by Rhumbler (37) and Lewis (27), has been addressed. Ettensohn (9) has recently critically reviewed the evidence supporting this hypothesis, and his conclusions are summarized below.

1. Cell elongation in many cases depends on the presence of microtubules. The best-studied example of this is the amphibian neural plate, where cuboidal cells elongate as the plate folds into a tube. Before elongation, microtubules are randomly arranged, but then they become strikingly aligned parallel to the axis of elongation. When cuboidal cells are treated with inhibitors of microtubule function, the microtubules disappear and the cells fail to elongate. Elongated cells treated with the same inhibitors round up, demonstrating that intact microtubules are somehow necessary for maintaining elongation (3). How they do this is not known, but perhaps the orientation of the microtubules directs a lateral contraction of the microfilament system or an apical protrusion that stretches the cell.

2. There is always a high concentration of actin filaments arranged like a "purse string" at the apical surfaces of participating cells. However, there is also a high density of actin filaments at the apices of noninvaginating epithelial sheets, so these filaments may have other functions than in folding. For example, they may primarily stabilize adhesions, much like stress fibers stabilize focal contacts of cells in vitro.

3. The best evidence for contractile microfilaments in epithelia comes from the study of pigmented retinal epithelium (34). These cells have a dense hexagonal array of microfilaments only at their apical surfaces. When sheets of this epithelium are permeablized with glycerol and bathed in ATP, the apical surfaces constrict and the sheets break up into cup-shaped clumps. Moreover, contraction is inhibited by treatment with N-ethyl maleimide–inactivated myosin S1 fragment, evidence that actin-myosin interactions are required for apical contractions.

4. Interestingly, treatment with cytochalasins has not been as useful in elucidating the role of microfilaments in embryonic tissues as it has been with individual cells. Although cytochalasins do disrupt microfilaments in cells in epithelial sheets (22), a side effect is that the cells often lose their adhesions to each other. This effect, by itself, could halt invagination.

5. In some cases, folding can be regulated by changing calcium levels with papaverine, which reduces free intracellular calcium levels by closing membrane channels, or with the ionophore A23187, which raises cytoplasmic calcium by allowing influx from outside or release from internal stores.

None of the above observations by itself proves a role for the cytoskeleton in changing cell shape and hence the shape of the sheet. However, taken together, the observations suggest that microtubules orient and stabilize cell-shape changes, while actin-myosin interactions, powered by ATP and regulated by calcium, may be responsible for apical constrictions of some cells. This constriction can contribute to epithelial folding, but other forces probably also influence the final shape of the tissue. What is needed now is a genetic approach in which specific components of the cytoskeleton are deleted. Combined with ultrastructural and time-lapse observations, this would be a powerful tool in dissecting the cytoskeleton's role in epithelial folding.

ELONGATION OF A TUBE

Another widespread morphogenetic movement is the elongation of a tube. The cylinder is the most common shape in biology, and it is the basic body plan of most metazoans. How most organisms come to be cylindrical and not spherical is therefore a topic of considerable morphogenetic interest. Most of what we know about the mechanics of cylinder formation comes from studies of tissues in three widely divergent organisms: the leg imaginal disc of the fruit fly, the gut of a sea urchin, and the hypoderm of a nematode worm. In the first case, the *Drosophila* imaginal disc extends primarily because of changes in cell packing, without changes in cell shape (11). Unfortunately, we don't know how the cytoskeleton functions in these rearrangements, except that colchicine does not prevent cells from changing neighbors, suggesting that microtubules are not necessary (12). In the case of sea urchin archenteron elongation, the situation is much the same; elongation occurs primarily because of changes in cell packing (15). As in the imaginal disc, this movement does not require microtubules (14). Unfortunately, treatment of sea urchin tissues with cytochalasin D disrupts cell adhesions, so we do not know whether intact microfilaments are necessary for cell rearrangements (8). Further analysis of these two systems will probably exploit the unique biologic advantages of the different organisms. Studies of the fly leg disc may proceed rapidly if mutants defective in elongation can be produced. In sea urchins, genetic manipulations are impractical, but their transparency makes them ideal for microinjection of cytoskeletal analogs.

Both microtubules and microfilaments have been shown to participate in elongation of embryos of the nematode *C. elegans* into juvenile worms (35). At 5 to 6 h of development the ellipsoid embryo begins to elongate and over the next 2 h decreases in circumference threefold and increases in length fourfold. The mechanism of this elongation is quite different from that in the *Drosophila* imaginal disc or sea urchin archenteron. Changes in cell packing do not occur. Instead, five rows of hypodermal cells, joined to each other by belt desmosomes, surround and squeeze the embryo. That there is tension exerted is shown by inducing lesions in the hypodermis with a laser. When this is done, interior cells are squeezed out from the lesions. At the time of elongation, but not before or after, circumferentially oriented bundles of microfilaments appear at the apical surface of the hypodermal cells. Disruption of these microfilaments with cytochalasin D prevents generation of the tension and blocks elongation. The effect is reversible; when the inhibitor is washed out tension returns and embryos elongate. Microtubules are also present in hypodermal cells and are oriented circumferentially only during elongation. However, they apparently are not directly responsible for constricting the hypodermal cells, since embryos continue to elongate in the presence of the microtubule inhibitors colcemid, griseofulvin, or nocodozole. Treated embryos appear wrinkled, however, as if the microtubules normally function to help transmit the tensile forces generated by microfilaments.

Future Prospects

We have discussed some of the known and hypothesized roles of specific elements of the cytoskeleton, mainly the microtubule and microfilament systems, in general cellular as well as specialized developmental processes. The roles played by intermediate filaments are not well understood. Furthermore, cells may contain additional cytoskeletal systems that have not yet been discovered. In closing this chapter, it is important to reemphasize that the different elements of the cytoskeleton interact and cooperate to give cells their shape and to enable them to undergo cytoplasmic reorganizations, divide, move, and respond to their environment. A number of powerful new technologies will help biologists in their future investigations of cytoskeletal interactions and functions. The introduction of fluorescent analogs of cytoskeletal components into cells and improved microscopic techniques allow visualization of cytoskeletal arrays in living cells; cytoskeletal dynamics can be monitored as cells undergo normal events and respond to experimental manipulations. Molecular biologic alterations of gene products and gene disruption can be used to define the domains of cytoskeletal proteins that are responsible for specific interactions and functions. Genetic analysis provides a powerful means of mutationally manipulating known cytoskeletal genes and identifying additional factors involved in developmental processes. With these tools available to cell and developmental biologists, the next decade promises significant advances in our understanding of cytoskeletal functions in development.

General References

SCHLIWA, M. 1986. *The Cytoskeleton: An Introductory Survey*. Springer-Verlag, New York.

DARNELL, J. D., LODISH, H., and BALTIMORE, D. 1986. *Molecular Cell Biology*. Chaps. 18, 19, and 22, Scientific American Books, New York.

References

1. Alvarez-Buylla, A., and Nottebohm, F. 1988. Migration of young neurons in adult avian brain. *Nature (Lond.)* **335**:353–354.

2. Anderson, K. V. 1987. Dorsal-ventral embryonic pattern genes of *Drosophila*. *Trends in Genetics* **3**:91–97.

3. Burnside, B. 1971. Microfilaments and microtubules in amphibian neurulation. *Dev. Biol.* **26**:416–441.

4. Conklin, E. G. 1905. The organisation and cell lineage of the ascidian egg. *J. Acad. Natl. Sci. Phila.* **13**:1–43.

5. Davidson, E. H. 1986. *Gene Activity in Early Development*, 3d ed. Academic Press, Orlando.

6. Elinson, R. P. 1985. Changes in levels of polymeric tubulin associated with activation and dorsoventral polarization of the frog egg. *Dev. Biol.* **109**:224–233.

7. Elinson, R. P., and Browning, B. 1988. A transient array of parallel microtubules in frog eggs: potential tracks for a cytoplasmic rotation that specifies the dorsoventral axis. *Dev. Biol.* **128**:185–197.

8. Ettensohn, C. A. 1984. Primary invagination of the vegetal plate during sea urchin gastrulation. *Am. Zool.* **24**:571–588.

9. Ettensohn, C. A. 1985. Mechanisms of epithelial invagination. *Q. Rev. Biol.* **60**:289–307.

10. Freeman, G. 1978. The role of asters in the localization of the factors that specify the apical tuft and the gut of the nemertean *Cerebratulus lacteus*. *J. Exp. Zool.* **206**:81–108.

11. Fristrom, D. 1976. The mechanism of evagination of imaginal discs of *Drosophila melanogaster*. III. Evidence for cell rearrangement. *Dev. Biol.* **54**:163–171.

12. Fristrom, D., and Fristrom, J. N. 1975. The mechanism of evagination of imaginal discs of *Drosophila melanogaster*. I. General considerations. *Dev. Biol.* **43**:1–23.

13. Fukui, Y., and Yumura, S. 1986. Actomyosin dynamics in chemotactic amoeboid movement of *Dictyostelium*. *Cell Motil. Cytoskeleton*. **6**:662–673.

14. Hardin, J. D. 1987. Archenteron elongation in the sea urchin embryo is a microtubule-independent process. *Dev. Biol.* **121**:253–262.

15. Hardin, J. D., and Cheng, L. Y. 1986. The mechanisms and mechanics of archenteron elongation during sea urchin gastrulation. *Dev. Biol.* **115**:490–501.

16. Hill, D., and Strome, S. 1988. An analysis of the role of microfilaments in the establishment and maintenance of asymmetry in *Caenorhabditis elegans* zygotes. *Dev. Biol.* **125**:75–84.

17. Husain-Chishti, A., Levin, A., and Branton, D. 1988. Abolition of actin-bundling by phosphorylation of human erythrocyte protein 4.9. *Nature (Lond.)* **334**:718–721.

18. Hymen, A. A., and White, J. G. 1987. Determination of cell division axes in the early embryogenesis of *Caenorhabditis elegans*. *J. Cell Biol.* **105**:2123–2135.

19. Jeffery, W. R., 1984. Spatial distribution of messenger RNA in the cytoskeletal framework of ascidian eggs. *Dev. Biol.* **103**:482–492.

20. Jeffery, W. R., and Meier, S. 1983. A yellow crescent cytoskeletal domain in ascidian eggs and its role in early development. *Dev. Biol.* **96**:125–143.

21. Joseph-Silverstein, J., and Cohen, W. D. 1984. The cytoskeletal system of nucle-

ated erythrocytes. III. Marginal band function in mature cells. *J. Cell Biol.* **98**: 2118–2125.

22. Karfunkel, P. 1972. The mechanism of neural tube formation. *Int. Rev. Cytol.* **38**: 245–272.

23. Kemphues, K. 1988. Genetic analysis of embryogenesis in *Caenorhabditis elegans*. In: *Developmental Genetics of Higher Organisms*. (G. M. Malacinski, ed.), pp. 193–219, Macmillan, New York.

24. Kirschner, M., and Mitchison, T. 1986. Beyond self-assembly: from microtubules to morphogenesis. *Cell* **45**:329–342.

25. Koshland, D. E., Mitchison, T. J., and Kirschner, M. W. 1988. Polewards chromosome movement driven by microtubule depolymerization *in vitro*. *Nature (Lond.)* **331**:499–504.

26. Lazarides, E. 1987. From genes to structural morphogenesis: the genesis and epigenesis of a red blood cell. *Cell* **51**:345–356.

27. Lewis, W. H. 1947. Mechanics of invagination. *Anat. Rec.* **97**:139–156.

28. Mittenthal, J. E., and Mazo, R. M. 1983. A model for shape generation by strain and cell-cell adhesion in the epithelium of an arthropod leg segment. *J. Theor. Biol.* **100**:443–483.

29. Naccache, P. 1987. Cell movement, excitability, and contractility. *Int. Rev. Cytol.* **17**:457–492.

30. Nigon, V., Guerrier, P., and Monin, H., 1960. Du developpment chez quelque nematodes. *Biol. Bull. Fr. Belg.* **93**:131–202.

31. Nusslein-Volhard, C., Frohnhofer, G., and Lehmann, R. 1987. Determination of anteroposterior polarity in *Drosophila*. *Science* **238**:1675–1681.

32. Olmsted, J. B. 1986. Microtubule-associated proteins. *Annu. Rev. Cell Biol.* **2**:421–458.

33. Oster, G. 1988. Biophysics of the leading lamella. *Cell Motil. Cytoskeleton.* **10**:164–171.

34. Owaribe, K., Kodayama, R., and Eguchi, G. 1981. Demonstration of contractility of circumferential actin bundles and its morphogenetic significance in pigmented epithelium *in vitro* and *in vivo*. *J. Cell Biol.* **90**:507–514.

35. Priess, J. R., and Hirsh, D. I. 1986. *Caenorhabditis elegans* morphogenesis: the role of the cytoskeleton in elongation of the embryo. *Dev. Biol.* **11**:156–173.

36. Rappaport, R. 1986. Establishment of the mechanism of cytokinesis in animal cells. *Int. Rev. Cytol.* **105**:245–281.

37. Rhumbler, L. 1902. Zur mechanik des gastrulationsvorganges, insbesondere der invagination. Eine entwicklingsmechnische studie. *Wilhelm Roux' Arch. Entwicklungsmech. Org.* **14**:401–476.

38. Rinnerthaler, G., Geiger, B., and Small, J. V. 1988. Contact formation during fibroblast locomotion: involvement of membrane ruffles and microtubules. *J. Cell Biol.* **106**:747–760.

39. Sawada, T., and Osani, K. 1985. Distribution of actin filaments in fertilized egg of the ascidian *Ciona intestinalis*. *Dev. Biol.* **111**:260–265.

40. Squire, J. 1981. *The Structural Basis of Muscle Contraction*. Plenum Press, New York.

41. Strome, S. 1986. Fluorescence visualization of the distribution of microfilaments in gonads and early embryos of the nematode *Caenorhabditis elegans*. *J. Cell Biol.* **103**:2241–2252.

42. Strome, S., and Wood, W. 1983. Generation of asymmetry and segregation of germline granules in early *C. elegans* embryos. *Cell* **35**:15–25.

43. Terasaki, M., Chen, L. B., and Fujiwara, K. 1986. Microtubules and the endoplasmic reticulum are highly interdependent structures. *J. Cell Biol.* **103**: 1557–1568.

44. Travis, J. L., and Browser, S. S. 1986. Microtubule-dependent reticulopodial motility: is there a role for actin? *Cell Motil. Cytoskeleton.* **6**:146–152.

45. Trinkhaus, J. P. 1984. *Cells into Organs: The Forces That Shape the Embryo.* Prentice-Hall, Englewood Cliffs, NJ.

46. Vale, R. D., Schnapp, B. J., Mitchison, T., Steuer, T. S., Reese, T. S., and Sheetz, M. P. 1985. Different axoplasmic proteins generate movement in opposite directions along microtubules *in vitro. Cell* **43**:623–632.

47. Wessels, D., Soll, D. R., Knecht, D., Loomis, W. F., DeLozanne, A., and Spudich, J. 1988. Cell motility and chemotaxis in *Dictyostelium* amoebae lacking myosin heavy chain. *Dev. Biol.* **128**:164–177.

Questions for Discussion with the Editor

1. *Are most cells like red blood cells, in that they require ATP to maintain their cytoskeleton in its proper configuration and hence exhibit a stable cell shape? Do you expect that virtually all cytoskeletons are in a "dynamic" condition?*

It is likely that most cells require ATP to maintain proper cytoskeletal configurations and cell shape. Depletion of ATP using metabolic inhibitors causes disruption of cytoskeletal architecture and alters the shape of a variety of cell types (A. D. Bershadsky and V. C. Gelfand 1983. *Cell Biol. Int. Rep.* **7**:173–187). Since ATP is utilized both as an energy source and as a phosphate donor, it is likely to be involved with cytoskeletal function and stability at several levels. For example, ATP is required for those cytoskeletal functions that involve ATPases, such as dynein-microtubule and actin-myosin interactions. In addition, several microtubule- and microfilament-associated proteins are known to be phosphoproteins whose interactions with and effects on the cytoskeleton are modified by phosphorylation.

The dynamic nature of the cytoskeleton is evident in cells that are undergoing shape changes or division. However, even in terminally differentiated cells whose cytoskeletons appear static, there is evidence that the cytoskeletons are dynamic. For example, microtubules in living cells, when viewed using fluorescence analog cytochemistry, are constantly growing and shrinking. Furthermore, the treatment of cells with agents that stabilize microtubules or microfilaments usually has a detrimental effect on the cells, indicating that depolymerization of cytoskeletal arrays is as important to proper cell functioning as polymerization of the arrays. More precise answers to these types of questions will certainly come from application of the sophisticated techniques now available to cell biologists.

2. *Is the construction of a complex-ornate cytoskeleton akin to bacteriophage self-assembly? Or, does it rely on a preexisting scaffold inherited from a progenitor cell?*

This is a difficult question to answer or even to address. We know that purified cytoskeletal subunits, such as tubulin and actin, can polymerize into fibers in vitro, suggesting the potential for at least limited self-assembly. In living cells, observation of growth of cytoskeletal arrays after disruption of the cytoskeleton suggests that polymers grow from persistant nucleation centers. For example, microtubules grow out from centrosomes, and such "microtubule-organizing centers" can nucleate growth of spidery arrays of microtubules in vitro. However, what controls the pattern of fiber growth in cells is not understood. Wispy arrays of microtubules emanate from microtubule-organizing centers in interphase cells, whereas the microtubules in mitotic cells are organized into spindles. It may be that construction or alteration of the cytoskeleton in cells does resemble phage assembly in the sense that specific cytoskeletal components may be synthesized in temporally regulated fashions and that their successive interactions may lead to the final arrangement of the cytoskeleton (as discussed by E. Lazarides 1987. *Cell* **51**:345–356).

CHAPTER **10**

Organization and Function of the Molluscan Egg Cortex

M. R. Dohmen and J. E. Speksnijder

Introduction

THE EXISTENCE OF a distinct domain, termed the "cortex," at the periphery of an egg cell can be inferred from a number of experiments using micromanipulation, centrifugation, elasticity measurements, and isolation of the outer cytoplasmic layer adhering to the plasma membrane. There is considerable interest in the egg cortex because of its role in maturation and fertilization (reviewed in Ref. 27). In this paper the emphasis is on the role of the egg cortex in defining cellular asymmetries that are crucial to early development; special attention is given to the association of morphogenetic determinants with the egg cortex. Direct evidence for structural asymmetries in the cortex of molluscan eggs has been provided (6), but the relationship between the characteristic features of these cortical domains and the determinants of early development is restricted to a mere spatial correlation. Causal relationships are hard to demonstrate because of the poor characterization both of the cortical domains and of the determinants presumed to be localized in the same cortical or subcortical region.

We first briefly discuss the existence of cytoplasmic domains in molluscan egg cells and the evidence for a role of the cortex in establishing or maintaining these domains. In the second section the structure of the egg cortex is described, and an attempt is made to outline perspectives for future research, mainly by considering the data presently available on the organization of cortical domains in polarized epithelial and other cells.

Cytoplasmic Domains

In many molluscs, the two cells arising from the first cleavage division are already fundamentally different from each other in that they possess different developmental potential. This is seen, for example, in unequally cleaving eggs of the bivalves and in polar lobe-forming eggs of gastropods, bivalves, and scaphopods (see Ref. 40). In molluscs with equally cleaving eggs, the first difference between cells arises at the third cleavage, when micromeres and macromeres are formed with completely divergent developmental pathways. This early divergence in developmental pathways between blastomeres seems to be directed exclusively by maternal factors since no transcription of the zygotic genome can be detected during the first few cleavages. Besides informational molecules that are capable of directing early developmental events, the maternally inherited spatial organization of the egg cell is also indispensable for normal development.

Egg cells of molluscs as well as most other organisms possess a distinct polarity which may be expressed by the shape of the egg, position of the nucleus, distribution of cytoplasmic inclusions, surface architecture, or other characteristics. Several axes of polarity are distinguishable, such as the animal-vegetal, anterior-posterior, and dorsal-ventral axes. These axes may be established at different times. They may be imprinted during oogenesis, acquired through the action of external stimuli such as sperm entrance, or established as a result of internal events during cleavage (see below).

Animal-Vegetal and Anterior-Posterior Polarity

In molluscs, the animal-vegetal polarity of the egg cell is imprinted during oogenesis: it corresponds with the apical-basal axis of the developing oocyte. Evidence for this correspondence is provided by the eggs of *Bithynia* and *Nassarius*. In *Bithynia* the vegetal pole of the egg contains a conspicuous organelle, the so-called vegetal body (Ref. 5; Fig. 1). The origin of this organelle can be traced back to the previtellogenic oocyte (4). Its precursors consist of aggregates of small vesicles that are localized at the basal pole of the oocyte right from their first appearance during the previtellogenic stage. While more vesicles are added to the growing vegetal body, it becomes closely apposed to the basal cortex, finally resulting in a strong attachment such that after oviposition it cannot be displaced by centrifugal forces of up to 600*g*. The vegetal body thus provides a marker that allows one to retrace the origin of the polar axis of the egg. Another marker is found in eggs of *Nassarius*. The oocytes of this gastropod are attached to the ovarian wall by a region that is devoid of microvilli and also lacks the layer of extracellular material that covers the rest of the egg surface. These characteristics of the basal pole of the oocyte are retained after oviposition and become a characteristic feature of the vegetal pole of the egg (6).

It can generally be stated that the cytoplasm at the animal and vegetal pole of the egg confers specific developmental potential to the blastomeres

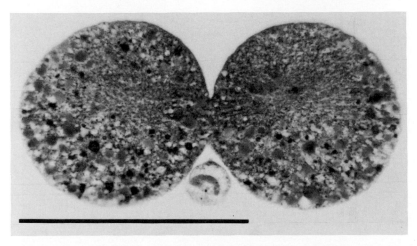

FIGURE 1. Section through the egg of *Bithynia* at first cleavage with polar lobe containing the cup-shaped vegetal body. Bar = 100 μm.

arising from each pole. Thus, the micromeres formed at the third cleavage will contain the cytoplasm of the animal pole of the egg and are thereby determined to become the stem cells of specific head structures. Similarly, the macromeres contain progressively less animally localized cytoplasm because of a series of unequal divisions that give rise to several tiers of micromeres. At the sixth cleavage micromeres are no longer formed and the macromeres now contain the vegetal part of the egg's cytoplasm, which determines that they become the stem cells of the endoderm. Thus the mechanisms that establish the animal-vegetal polarity finally determine the anteroposterior axis of the developing embryo.

The spatial organization of the egg cytoplasm can be severely disturbed by centrifugation. In spite of this disturbance, the original egg polarity appears to be maintained (see Ref. 37), and development proceeds normally in many cases. The explanation generally provided is that the position of most cytoplasmic inclusions is not crucial. What counts is the position of the morphogenetic determinants which apparently resist stratification. Fortunately, the vegetal body in the egg of *Bithynia* allows one to confirm directly the validity of this theory since the vegetal body resists displacement by moderate centrifugal forces which stratify all other visible cytoplasmic inclusions.

It should be mentioned that in certain eggs the animal-vegetal polarity can be overruled. Pressure experiments on eggs of *Limax* (see Ref. 37) show that before the second maturation division, the polar axis can be shifted to another position by experimentally shifting the position of the second maturation spindle perpendicular to the original animal-vegetal axis. As a result ooplasmic segregation and the cleavage pattern conform to the newly induced axis that is defined by the position of the second polar body. After the second maturation division it is no longer possible to change the animal-vegetal polarity of the egg.

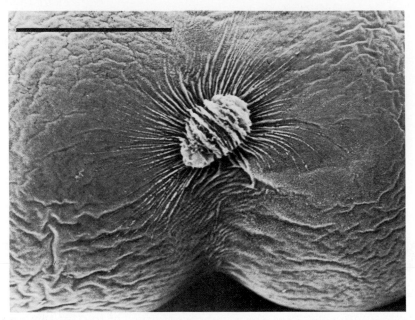

FIGURE 2. Egg of *Crepidula* at first cleavage. The cleavage plane is oriented at an angle of about 45° relative to the pattern of parallel ridges over the polar lobe area. Bar = 100 μm.

Dorsoventral Polarity of the Egg

The dorsoventral axis of the embryo is evidently not predetermined in the structure of molluscan eggs. In species whose eggs divide unequally, the dorsoventral axis is determined by the position of the first cleavage plane. It is not clear which mechanism determines the position of cleavage planes. The sperm entrance point may be a determinative factor, as in *Cumingia* (see Ref. 37) where the first mitotic spindle is always oriented perpendicularly to the path of the sperm to the center of the egg. In *Crepidula*, the first cleavage plane is always oriented at an angle of about 45° relative to the axis of symmetry of an array of surface folds (Fig. 2) present at the vegetal pole of the egg from the first maturation division onward (3, 4). These examples suggest that, in some species, the position of the first cleavage plane may be predetermined, possibly by factors localized in the egg cortex. In species whose eggs divide equally cellular interactions during the early cleavage stages appear to determine the dorsoventral axis of the embryo (37).

Cortical Domains

Extracellular Material

All molluscan eggs are surrounded by a layer of extracellular material (4). Although no specific directive influence on development has been attributed to

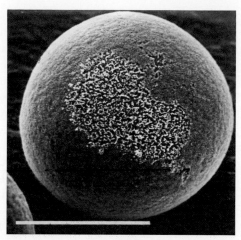

FIGURE 3. Uncleaved egg of *Nassarius* after second meiotic division, showing a patch of large microvilli at the vegetal pole. Bar = 100 μm.

this layer, the egg of *Nassarius* warrants closer attention because of the particular organization of its extracellular material. At the vegetal pole of this egg, the extracellular layer is virtually absent (6). Precisely in this area, extremely long microvilli of unknown function develop before the first cleavage (Fig. 3). The outgrowth of similar microvilli on the remaining egg surface might be repressed by the presence of the extracellular material.

Extracellular material may be a determinant of cell polarity, as shown for several epithelial cell lines (references in Ref. 39). The basal lamina in the ovarian wall may serve as the primary inducer of the apical-basal polarity of the oocyte. Later on, the vitelline membrane and other external substances that are closely apposed to the egg surface may help to maintain the acquired polarity through specific receptors in the plasma membrane. These extracellular components may also act to induce the inside-outside polarity of the blastomeres during the cleavage stages. Specific cell surface receptors for components of the extracellular material such as hyaluronate, laminin, fibronectin, and vitronectin have been detected in several cell types. In addition, transmembrane linkages between extracellular material and the cortical cytoskeleton have been demonstrated. Thus it seems plausible that the extracellular material exerts a considerable influence on the morphology and functioning of the egg as it does in other cells.

The Plasma Membrane

Specialized plasma membrane domains with distinct protein and lipid compositions seem to occur generally in polarized cells. Polarity has been studied most extensively in Madin-Darby canine kidney (MDCK) cells. In these cells an apical and a basolateral domain can be distinguished. These domains are characterized by an asymmetric distribution of membrane proteins such as aminopeptidase and $(Na^+ + K^+)$ ATPase and by the polarized budding of viruses from the cell surface (references in Ref. 39). There are also differences in lipid composition. The apical membrane is enriched in glycosphingolipids and sphingomyelin whereas phosphatidylcholine is concentrated in the basolateral membrane (see Ref. 38). The domains are separated from each other by a tight junction. Disruption of the epithelial organization, resulting in the disappearance of the tight junctions, abolishes the polarity of the cells. Therefore, the tight junction is thought to constitute a passive fence that prevents the diffusion of membrane components from one domain into the other, thereby maintaining the polarity of the plasma membrane. However, a recent report has shown that even in the absence of tight junctions, polarization of a certain apical protein does occur (39).

The maintenance of membrane domains in solitary cells such as egg cells, where tight junctions or comparable intramembrane barriers have never been observed, must depend on other mechanisms. Most probably the submembrane cytoskeleton is involved, as is discussed below, but other mechanisms may also operate, as illustrated by sperm cells. In mammalian sperm cells, plasma membrane domains without apparent structural boundaries have been described in which localized molecules are freely diffusing (see Ref. 2). Since free diffusion excludes trapping by the cytoskeleton, some other unknown mechanism must be responsible for domain formation in these cells.

In molluscs, the regionalization of the plasma membrane has been studied in eggs of the gastropod *Nassarius* (32). Freeze-fracture electron microscopy of four areas at different positions along the animal-vegetal axis has shown that there are quantitative differences in the distribution of intramembrane particles (Figs. 4 and 5; Table 1). The most prominent domain is found at the vegetal pole, where the density of intramembrane particles is much higher than in the other regions (Table 1). In the area adjacent to the vegetal pole region (area lll) the size distribution of the particles is quite dif-

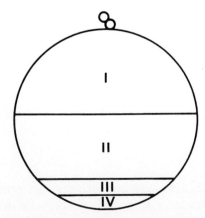

FIGURE 4. The four plasma membrane areas of the *Nassarius* egg that are distinguishable in the analysis of the pattern of intramembrane particles.

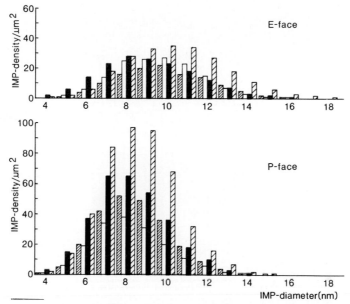

FIGURE 5. Frequency distribution (using 1-nm size classes) of intramembrane particles (IMPs) in the four plasma membrane areas shown in Fig. 4, on both E and P faces (see Table 1 for definitions of E and P faces). Ordinate: number of IMPs per μm^2; abscissa: IMP size in nm. Each bar represents the mean of the number of IMPs in a given size class. Finely hatched, area I; white, area II; black, area III; coarsely hatched, area IV.

TABLE 1   Density (per μm^2) of Intramembrane Particles in the Nassarius Egg Plasma Membrane*

Area	E face	P face	E/P
I	121.0 ± 4.7 (5)	238.2 ± 12.1 (3)	0.51
II	150.2 ± 7.4 (4)	170.1 ± 12.4 (4)	0.88
III	164.8 ± 20.4(5)	301.8 ± 51.7 (2)	0.55
IV	220.4 ± 19.9(3)	455.2 ± 144.3(2)	0.48

*The numbers represent mean ± SEM (standard error of the mean) and the numbers in brackets are the number of replicas analyzed. E face = exoplasmic fracture face; P face = protoplasmic fracture face.

ferent from the other regions since it contains predominantly small particles (Fig. 5).

In *Nassarius* the lateral mobility of plasma membrane lipids has been measured using the fluorescence photobleaching recovery method (33). During the first three cleavage cycles the lipid mobility is consistently greater at the vegetal pole. Superimposed upon this animal-vegetal polarity, there is a cell cycle–dependent modulation of lipid lateral diffusion in the vegetal plasma membrane with the highest values for the diffusion coefficient being reached during the S phase (Fig. 6).

The results of these two studies suggest a polar accumulation of mem-

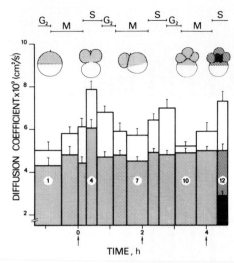

FIGURE 6. Lateral diffusion coefficient of plasma membrane lipids during early development of the *Nassarius* egg. Mean and SEM (vertical bars) are indicated. White, polar lobe; stippled, animal area; black, micromeres at the 8-cell stage.

brane proteins in the vegetal area and regional differences in lipid composition in the plasma membrane of the *Nassarius* egg. Numerous functions can be envisaged for a polar distribution of membrane proteins in egg cells. One example is the generation of steady polar ionic currents caused by the clustering of ion channels and pumps at either pole of the egg. Such transcellular currents have been measured in a large number of cells, including egg cells of various species. In molluscs, the only species investigated thus far is *Lymnaea*, whose eggs show cell cycle–dependent polar currents (7).

Regional differences in lipid composition can also result in topographical differences in the functional properties of the plasma membrane, since the lipid environment is known to modulate the function of various membrane proteins such as ion transport proteins, surface receptors, and membrane-bound enzymes. The membrane domains thus formed may play an important role in cell diversification during early development. As a result of the differential distribution of the egg plasma membrane during the early cleavages, blastomeres with different plasma membrane characteristics are formed, and these differences might be essential for the establishment of different cell lines within the embryo.

The Cortical Cytoskeleton

MICROTUBULES

Microtubules have an intrinsic structural polarity due to the polar arrangement of tubulin dimers in the microtubule lattice. Most microtubules are anchored with one pole at the centrosome, while the other pole extends toward the cell periphery. This microtubule distribution provides a radial array which can be used to organize the cytoplasm in an inward-outward fashion. Microtubules may also serve to create and maintain cell polarity. Both radial and polar organization probably require stable microtubules. The great majority of cytoplasmic microtubules are, however, labile polymers which exchange subunits rapidly with a soluble subunit pool. This dynamic instability is important for the rapid reorganization required by transitions between mitosis and interphase and by changes in morphology of the cells. In the model proposed by Kirschner and Mitchison (15) a restricted population of microtubules may become more stable, e.g., by detyrosination, and may thus provide a basis for stable conditions such as cellular polarity.

The localization of morphogenetic plasms in the polar subcortical regions of the eggs of molluscs and other organisms may in part be brought about by microtubules since these structures appear to be responsible for the routing of organelles in many cell types. In several instances it was found that a large number of microtubules are oriented parallel to the plasma membrane. The most striking example has been reported in the egg of *Xenopus*, where parallel tracks of microtubules cover the entire vegetal surface of the egg (8). They are thought to be involved in the rotation of the subcortical cytoplasm relative to the egg surface that occurs between 45 and 90 min after fertilization. This rotation seems to specify the embryo's axes. In cortices isolated from molluscan eggs, a dense network of microtubules is present during interphase (Fig. 7) but absent during mitosis (34, 34a). A similar cycle of assembly and disassembly of cortical microtubules has been described in starfish oocytes (29). In cortices of sea urchin eggs that were prepared in the same way as molluscan eggs, microtubules have not been detected (26). This discrepancy might be due

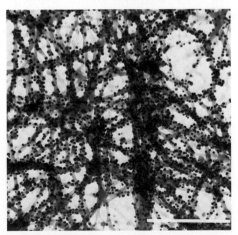

FIGURE 7. Electron micrograph of microtubules in the isolated cortex of an uncleaved egg of *Nassarius*, labeled with anti-tubulin using the protein A–gold method. Bar = 0.5 μm.

to subtle differences in the isolation procedure since cortical microtubules are easily washed off by the shearing forces used to remove the egg cytoplasm.

Specific interactions between astral microtubules and the cortex have been invoked to explain the positioning of the mitotic spindle which is strictly predetermined in cleaving eggs of molluscs and most other organisms. Such an interaction is particularly suggestive in extremely unequal cleavages (e.g., those leading to the formation of the polar bodies during meiosis). The shortening of the microtubules of the superficial aster, a process that allows the spindle to approach the cortex very closely, seems to be provoked by some cortical factor. Another type of interaction between microtubules and the cortex is probably necessary to induce the formation of a contractile ring of actin filaments as well as subsequent cleavage. Although both of these interactions are rather obvious, their mechanisms have not yet been elucidated. A local release or influx of calcium at the site of microtubule resorption or of cleavage furrow formation would probably suffice to explain these phenomena but has not yet been established.

ACTIN FILAMENTS, SPECTRIN, AND ASSOCIATED COMPONENTS

There are several reports suggesting that microfilaments may play a role in cytoplasmic localization in various cell types, including egg cells (references in Ref. 31). The cellular cortex, including that of egg cells, is particularly rich in microfilaments (34, 34a). The organization of this filament layer has been investigated in several egg species, and the results show correlations between the filament organization and ooplasmic segregation or other polar phenomena. For example, in *Tubifex* eggs the actin filament meshwork shows a characteristic organization at both poles of the egg that coincides with ooplasmic segregation after the second polar body formation (30). In the unfertilized ascidian egg, a network of actin filaments, present in the cortex, is involved in the segregation of cortical cytoplasm toward the vegetal pole following fertilization (28). In mouse eggs, cortical actin filaments form a dense aggregate at the animal pole during meiosis. This aggregate is thought to maintain the cortical location of the meiotic spindle (19). In eggs of the mollusc *Buccinum*, an extraordinarily thick layer of microfilaments is present at the vegetal pole, spatially correlated with the presence of morphogens in the subcortical cytoplasm in the same region (Fig. 8; Ref. 6).

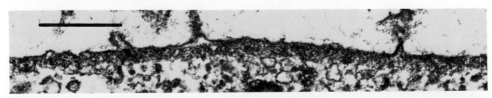

FIGURE 8. Electron micrograph of the cortex of the polar lobe in the egg of *Buccinum*, showing the thick layer of cortical microfilaments characterizing this area. Bar = 1 μm.

Experimental evidence that actin filaments may serve to bind peripherally located components to the cortex has been provided by experiments on sea urchin eggs. Cortical granules, which normally cannot be displaced by centrifugation, are dislodged from the cortex after treatment with the microfilament-disrupting agent cytochalasin B (22). The actin-filament layer itself is strongly bound to the egg surface as evidenced by the fact that it cannot be disrupted by centrifugation in *Tubifex* eggs (30) and by the observation that it resists the strong shearing forces used to remove the cytoplasm during the cortex-isolation procedure (26, 34, 34a).

This strong attachment may be due to an association with a membrane skeleton that consists of spectrin. In erythrocytes, which represent the best characterized model for cytoskeleton-membrane interactions, the membrane skeleton consists of a two-dimensional meshwork of spectrin tetramers and actin oligomers that are linked to integral membrane proteins by the intermediates ankyrin and band 4.1 protein.

Components analogous to the erythrocyte-membrane skeleton proteins have now been identified in a variety of cell types (see Ref. 21). They have also been found in egg cells from, for example, sea urchins, mice, and amphibians. Distinct spectrin variants (isoforms) may bind to specific membrane proteins and serve to immobilize these proteins in discrete regions of the cell, resulting in the differentiation of specialized membrane domains. Such a mechanism may be responsible for trapping the $(Na^+ + K^+)$ATPase in the basolateral membrane domain of MDCK cells since the development of this localization coincides with the formation of a dense layer of spectrin in the same region. Moreover, ankyrin has also been shown to bind to the ATPase (23). On the cytoplasmic side, spectrin and ankyrin are known to form linkages with microfilaments, microtubules, and intermediate filaments. Calpactins constitute another category of structural elements of the membrane skeleton (see Ref. 10). The two main forms, calpactin I and II, bind to actin and spectrin as well as to anionic phospholipids. They may thus provide a bridge between the membrane lipids and cytoskeletal proteins. Synapsin is still another interesting protein that we encounter in the study of the properties of the peripheral layer of cells. This phosphoprotein is associated with synaptic vesicles and interacts with spectrin, microtubules, and actin. It has been suggested that synapsin I is involved in the positioning of clusters of synaptic vesicles close to the plasma membrane by forming links with the cortical cytoskeleton (24). Since, in several molluscan eggs, morphogenetic plasms have been found to consist of aggregates of small vesicles that are closely associated with the egg cortex (3), synapsin-like proteins might serve to maintain the polar localization of morphogens.

INTERMEDIATE FILAMENTS AND FINE FILAMENTS

Although the structure and distribution of these filaments are known in great detail, their function is still not fully understood. Recently, it has been found that the carboxy-terminal region of vimentin binds specifically to the nuclear lamin B and that the amino-terminal region binds to ankyrin, thus allowing a network to be formed that interconnects the plasma membrane and the inner nuclear membrane (9). The involvement of intermediate filaments in establishing or maintaining specific cytoplasmic and membrane domains is strongly suggested by the occurrence, in lymphocytes, of a polarized submembranous aggregate of vimentin that coincides with a similarly polarized aggregate of spectrin (17).

There are only a few reports on the occurrence of intermediate filaments in egg cells. In *Xenopus*, both vimentin and cytokeratin are present in the egg. Vimentin filaments are evenly distributed throughout the cytoplasm of the mature egg, although during vitellogenesis they show a temporary polar distribution (11). Cytokeratin filaments, on the other hand, are localized in the egg cortex and their distribution is clearly polar (16). These observations indicate that intermediate filaments might be involved in polar phenomena in egg cells as well. In mouse eggs, cytokeratin proteins are present but only in a nonfibrillar form (18). Preliminary tests with anti-vimentin and anti-cytokeratin antibodies on isolated cortices from *Nassarius* eggs show that intermediate filaments are present in molluscan eggs (34a), but no data are available yet on distribution patterns.

A heterogeneous class of 2 to 5-nm fine filaments (see Ref. 25) might play a role in the functioning of the cortex, but the absence of data relevant to the present topic does not allow a meaningful discussion at this moment.

THE ENDOPLASMIC RETICULUM

The presence of a cortical endoplasmic reticulum has been demonstrated in eggs from many species. In deuterostomes, this reticulum is thought to serve as the internal source of calcium that supports the wave of free calcium found at egg activation (12). In addition, the cortical endoplasmic reticulum is thought to locally release and sequester calcium during various post-fertilization events such as waves of exocytosis and endocytosis (see Ref. 27). In protostomes, such as the molluscs, egg activation is dependent on the presence of external calcium, even though a cortical endoplasmic reticulum has been demonstrated in *Spisula solidissima* and *Nassarius reticulatus*. This suggests that the cortical endoplasmic reticulum in these species is not the primary source of calcium during egg activation. However, it may well act as a source and a sink of calcium for postactivation events such as cytokinesis.

Interestingly, a hexagonal pattern of the cortical reticulum is observed in isolated cortices of the sea urchin *Strongylocentrotus* (26) and the gastropod *Nassarius* (Fig. 9; see Ref. 34a). In the sea urchin egg the hexagonal pattern may be imposed by the presence of cortical granules. In the egg of *Nassarius*, however, no such cortical granules or comparable organelles are present in the cortex. Staining of the reticulum with the fluorescent dye 3, 3'-dihexyloxacarbocyanine iodide (35) shows the same hexagonal pattern to be present both in intact living or fixed eggs and in isolated unfixed or fixed isolated cortices, demonstrating that the pattern is not an artifact. The induction of this pattern may be due to an interaction between the endoplasmic reticulum and microtubules, since the reticulum seems to depend on

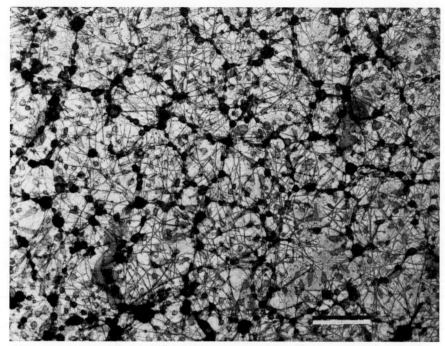

FIGURE 9. Isolated cortex of the uncleaved egg of *Nassarius*, showing a pentagonal-hexagonal pattern of cortical endoplasmic reticulum. Bar = 1 μm.

microtubules for its growth. This is inferred from the observation that the reticulum in the periphery of several cell types retracts with microtubule depolymerization and re-forms with the regrowth of microtubules (36).

Localization of Morphogenetic Determinants in the Egg Cortex

The vegetal body of *Bithynia* is a clear case of a morphogenetic plasm closely tied to the egg cortex. In the eggs of the gastropods *Crepidula fornicata, Buccinum undatum,* and *Littorina saxatilis* comparable, though less conspicuous, aggregates of vesicles have been observed that might have a similar developmental significance (3). The vegetal body of *Bithynia* contains a high concentration of RNA (5), whose characteristics remain to be established. If this RNA is the morphogenetic agent responsible for the effect of the polar lobe on development, it will likely be messenger RNA. The morphogenetic potential of maternal mRNA in establishing egg polarity and consequent embryonic axes has been demonstrated experimentally in *Drosophila*, where the defect of various dorsal mutants can be rescued by the injection of purified mRNA from wild-type eggs. Moreover, the site of injection determines the ventral side of the embryo (1). The localization of specific mRNAs in distinct domains of the egg cell has been shown in *Xenopus* (see Ref. 20) and in the ascidian *Styela* (13), and in both cases the cytoskeleton is thought to be responsible for this localization. In the annelid *Chaetopterus*, an association between maternal mRNA and a specific cortical domain has been demonstrated by Jeffery et al. (14). In this case the mRNA is bound to cytoskeletal elements that resist extraction by high ionic strength media, which suggests that intermediate filaments may be involved.

Conclusions and Perspectives

Investigations of a variety of cell types have revealed a large number of processes and structures operating in the cell periphery. Many of these will doubtlessly operate in a similar way in the cortex of egg cells. An important aspect of egg cells is their extremely stable polarity and, associated with this polarity, the regional localization of morphogenetic determinants. The available data strongly suggest that the egg cortex plays an important role both in establishing and maintaining egg polarity and in the positioning of morphogenetic determinants. Molluscan eggs seem to be particularly suited to study the mechanisms of cortical localization of morphogens because of the presence of well-defined morphogenetic plasms that are visibly associated with the egg cortex.

The data presently available for molluscan eggs show that the plasma membrane and the surface architecture of the egg are regionally differentiated into domains that coincide with subcortical localizations of morphogenetic determinants. The plasma membrane and the cortical cytoskeleton seem to be the structures that are most important in the localization process. A more detailed knowledge of their structure and its modulation during oogenesis and early development should reveal the mechanisms that govern the interaction of morphogenetic plasms with the egg cortex.

Acknowledgment

We thank Dr. Erwin Huebner for his constructive comments on the manuscript.

General References

DOHMEN, M. R.: 1983. The role of the egg surface in development. In: *Control of Embryonic Gene Expression*. (M. A. Q. Siddiqui, ed.), pp. 1–25, CRC Press, Boca Raton.

SARDET, C., and CHANG, P.: 1987. The egg cortex: from maturation through fertilization. *Cell Differ*. 21:1–19.

References

1. Anderson, K. V., and Nüsslein-Volhard, C. 1984. Information for the dorsal-ventral pattern of the *Drosophila* embryo is stored as maternal mRNA. *Nature (Lond.)* **311**:223–227.

2. Cowan, A. E., Myles, D. G., and Koppel, D. E. 1987. Lateral diffusion of the PH-20 protein on guinea pig sperm: Evidence that barriers to diffusion maintain plasma membrane domains in mammalian sperm. *J. Cell Biol.* **104**:917–923.

3. Dohmen, M. R. 1983. The polar lobe in eggs of molluscs and annelids: structure, composition, and function. In: *Time, Space and Pattern in Embryonic Development*, vol. 2. (W. R. Jeffery, and R. A. Raff, eds.), MBL Lectures in Biology, pp. 197–220, Liss, New York.

4. Dohmen, M. R. 1983. Gametogenesis. In: *The Mollusca*, vol. 3: *Development*. (N. H. Verdonk, J. A. M. van den Biggelaar, and A. S. Tompa, eds.), pp. 1–48, Academic Press, New York.

5. Dohmen, M. R., and Verdonk, N. H. 1974. The structure of a morphogenetic cytoplasm present in the polar lobe of *Bithynia tentaculata* (Gastropoda, Prosobranchia). *J. Embryol. Exp. Morphol.* **31**:423–433.

6. Dohmen, M. R., and van der Mey, J. C. A. 1977. Local surface differentiations at the vegetal pole of the eggs of *Nassarius reticulatus, Buccinum undatum,* and *Crepidula fornicata* (Gastropoda, Prosobranchia). *Dev. Biol.* **61**:104–113.

7. Dohmen, M. R., Arnolds, W. J. A., and Speksnijder, J. E. 1986. Ionic currents through the cleaving egg of *Lymnaea stagnalis* (Mollusca, Gastropoda, Pulmonata). In: *Ionic Currents in Development.* (R. Nuccitelli, ed.), pp. 181–187, Liss, New York.

8. Elinson, R. P. and Rowning, B. 1988. A transient array of parallel microtubules in frog eggs: Potential tracks for a cytoplasmic rotation that specifies the dorso-ventral axis. *Dev. Biol.* **128**:185–197.

9. Georgatos, S. D., and Blobel, G. 1987. Two distinct attachment sites for vimentin along the plasma membrane and the nuclear envelope in avian erythrocytes: A basis for a vectorial assembly of intermediate filaments. *J. Cell Biol.* **105**:105–115.

10. Glenney, J. R. Jr., 1987. Calpactins: calcium-regulated membrane-skeletal proteins. *Bioessays* **7**:173–175.

11. Godsave, S. F., Anderton, B. H., Heasman, J., and Wylie, C. C. 1984. Oocytes and early embryos of *Xenopus laevis* contain intermediate filaments which react with anti-mammalian vimentin antibodies. *J. Embryol. Exp. Morphol.* **83**:169–187.

12. Jaffe, L. F. 1985. The role of calcium explosions, waves, and pulses in activating eggs. In: *Biology of Fertilization*, vol. 3. (C. B. Metz, and A. Monroy, eds.), pp. 127–165, Academic Press, New York.

13. Jeffery, W. R., Tomlinson, C. R., and Brodeur, R. D. 1983. Localization of actin messenger RNA during early ascidian development. *Dev. Biol.* **99**:408–417.

14. Jeffery, W. R., Speksnijder, J. E., Swalla, B. J., and Venuti, J. M. 1986. Mechanism of maternal mRNA localization in *Chaetopterus* eggs. In: *Advances in Invertebrate Reproduction*, vol. 4. (M. Porchet, J. -C. Andries, and A. Dhainaut, eds.), pp. 229–240. Elsevier, Amsterdam.

15. Kirschner, M., and Mitchison, T. 1986. Beyond self-assembly: from microtubules to morphogenesis. *Cell* **45**:329–342.

16. Klymkowski, M. W., Maynell, L. A., and Polson, A. G. 1987. Polar asymmetry in the organization of the cortical cytokeratin system of *Xenopus laevis* oocytes and embryos. *Development* **100**:543–557.

17. Lee, J. K., and Repasky, E. A. 1987. Cytoskeletal polarity in mammalian lymphocytes *in situ*. *Cell Tissue Res.* **247**:195–202.

18. Lehtonen, E. 1985. A monoclonal antibody against mouse oocyte cytoskeleton recognizing cytokeratin-type filaments. *J. Embryol. Exp. Morphol.* **90**:197–209.

19. Longo, F. J. 1987. Actin–plasma membrane associations in mouse eggs and oocytes. *J. Exp. Zool.* **243**:299–309.

20. Melton, D. A. 1987. Translocation of a localized maternal mRNA to the vegetal pole of *Xenopus* oocytes. *Nature (Lond.)* **328**:80–82.

21. Moon, R. T., and McMahon, A. P. 1987. Composition and expression of spectrin-based membrane skeletons in non-erythroid cells. *Bioessays* **7**:159–164.

22. Morton, R. W., and Nishioka, D. 1983. Effects of cytochalasin B on the cortex of the unfertilized sea urchin egg. *Cell Biol. Int. Rep.* **7**:835–842.

23. Nelson, W. J., and Veshnock, P. J. 1987. Ankyrin binding to $(Na^+ + K^+)$ ATPase and implications for the organization of membrane domains in polarized cells. *Nature (Lond.)* **328**:533–536.

24. Petrucci, T. C., and Morrow, J. S. 1987. Synapsin I: an actin-bundling protein under phosphorylation control. *J. Cell Biol.* **105**:1355–1363.

25. Roberts, T. M. 1987. Fine (2-5nm) filaments: new types of cytoskeletal structures. *Cell Motil. Cytoskeleton* **8**:130–142.

26. Sardet, C. 1984. The ultrastructure of the sea urchin egg cortex isolated before and after fertilization. *Dev. Biol.* **105**:196–210.

27. Sardet, C., and Chang, P. 1987. The egg cortex: from maturation through fertilization. *Cell Differ.* **21**:1–19.

28. Sawada, T., and Osanai, K. 1985. Distribution of actin filaments in fertilized egg of the ascidian *Ciona intestinalis*. *Dev. Biol.* **111**:260–265.

29. Schroeder, T. E., and Otto, J. J. 1984. Cyclic assembly-disassembly of cortical microtubules during maturation and early development of starfish oocytes. *Dev. Biol.* **103**:493–503.

30. Shimizu, T. 1986. Bipolar segregation of mitochondria, actin network, and surface in the *Tubifex* egg: role of cortical polarity. *Dev. Biol.* **116**:241–251.

31. Speksnijder, J. E., and Dohmen, M. R. 1983. Local surface modulation correlated with ooplasmic segregation in eggs of *Sabellaria alveolata* (Annelida, Polychaeta). *Roux's Arch. Dev. Biol.* **192**:248–255.

32. Speksnijder, J. E., Mulder, M. M., Hage, W. J., Dohmen, M. R., and Bluemink, J. G. 1985. Animal-vegetal polarity in the plasma membrane of a molluscan egg: a quantitative freeze-fracture study. *Dev. Biol.* **108**:38–48.

33. Speksnijder, J. E., Dohmen, M. R., Tertoolen, L. G. J., and De Laat, S. W. 1985. Regional differences in the lateral mobility of plasma membrane lipids in a molluscan embryo. *Dev. Biol.* **110**:207–216.

34. Speksnijder, J. E., de Jong, K., Linnemans, W. A. M., and Dohmen, M. R. 1986. The ultrastructural organization of the isolated cortex of a molluscan egg. In: *Progress in Developmental Biology*, part B. (H. C. Slavkin, ed.), pp. 353–356, Liss, New York.

34a. Speksnijder, G. E., de Jong, K., Wisselaar, M. A., Linnemans, W. A. M., and

Dohmen, M. R. 1989. The ultrastructural organization of the isolated cortex in eggs of *Nassarius reticulatus* (Mollusca). *Roux's Arch. Dev. Biol.*, in press.

35. Terasaki, M., Song, J., Wong, J. R., Weiss, M. J., and Chen, L. B. 1984. Localization of endoplasmic reticulum in living and glutaraldehyde-fixed cells with fluorescent dyes. *Cell* **38**:101–108.

36. Terasaki, M., Chen, L. B., and Fujiwara, K. 1986. Microtubules and the endoplasmic reticulum are highly interdependent structures. *J. Cell Biol.* **103**: 1557–1568.

37. van den Biggelaar, J. A. M., and Guerrier, P. 1983. Origin of spatial organization. In: *The Mollusca*, vol. 3: *Development*. (N. H. Verdonk, J. A. M. van den Biggelaar, and A. S. Tompa, eds.), pp. 179–213, Academic Press, New York.

38. van Meer, G., Stelzer, E. H. K., Wijnaendts-van-Resandt, R. W., and Simons, K. 1987. Sorting of sphingolipids in epithelial (Madin-Darby canine kidney) cells. *J. Cell Biol.* **105**:1623–1635.

39. Vega-Salas, D. E., Salas, P. J. I., Gundersen, D., and Rodriguez-Boulan, E. 1987. Formation of the apical pole of epithelial (Madin-Darby canine kidney) cells: Polarity of an apical protein is independent of tight junctions while segregation of a basolateral marker requires cell-cell interactions. *J. Cell Biol.* **104**:905–916.

40. Verdonk, N. H., and Cather, J. N. 1983. Morphogenetic determinants and differentiation. In: *The Mollusca*, vol. 3: *Development*. (N. H. Verdonk, J. A. M. van den Biggelaar, and A. S. Tompa, eds.), pp. 215–252, Academic Press, New York.

Questions for Discussion with the Editor

1. *Please speculate on mechanisms whereby the extracellular matrix (ECM) might determine egg-cell polarity. For example, might the ECM generate regional metabolic differences in the cortex?*

There are many indications that the polarity of cells depends on an intact cytoskeletal system, suggesting that the cytoskeleton is the agent that is responsible for the spatial organization of the cytoplasm. It is widely believed, however, that the cytoskeleton does not act autonomously but that it depends on a signaling system that triggers its activity and provides spatial coordinates. The plasma membrane may contain such a signaling system, probably in combination with the ECM. Evidence for this view is the finding that specific receptors for ECM components, e.g., hyaluronate, laminin, and fibronectin, appear to be transmembrane glycoproteins that interact with actin filaments in the cortical cytoplasm. This suggests that ECM may influence the organization of the cytoskeleton and thereby regulate cell structure and behavior.

2. *Although an intact mollusc egg cortex can be mechanically isolated, is the cortex actually continuous with the internal cytoplasm or is there a distinct boundary? For example, is the internal cytoplasm free to move without regard to the egg cortex (as is the case with inverted amphibian eggs—see Chap. 7, this volume)?*

The mechanical links between the cortex and the internal cytoplasm should probably be thought of as relatively weak and dynamic structures. Both mechanical isolation of the cortex and centrifugation of intact eggs result in the separation of the internal cytoplasm from the cortex. When cortices are isolated, the amount of material adhering to the plasma membrane depends on the shearing force and on the isolation buffer used. It varies between a very thin microfilamentous lattice under stringent washing conditions and a thick layer of cytoplasm, including yolk granules, under gentle isolation conditions. The effects of artificial inversion on the egg cytoplasm has never been studied in molluscs as far as we know. Naturally occurring inversion can be observed in eggs of *Nassarius* in their capsule. Once isolated from the viscous capsule fluid, these eggs orient themselves with their animal pole pointing upward because of the presence of heavy yolk granules at the vegetal pole. Within the capsule, however, some eggs are usually forced into an inverted position for prolonged periods of time, and this inversion never leads to displacement of the yolk from the vegetal-pole area. This suggests that the links between the internal cytoplasm and the cortex are strong enough to resist unit gravity.

CHAPTER 11

Cytoplasmic Localization and Organization of Germ-Cell Determinants

Masami Wakahara

Introduction

SIGNIFICANT ADVANCES IN the molecular biology of the regulation of eukaryotic gene expression, especially regarding genes which are presumed to control pattern formation and morphogenesis, have recently been made (2, 29; see also Vol. III of this series). Several classically famous phenomena in developmental biology, however, await molecular approaches.

One of these phenomena is the enigma concerning the localization of germ plasm or of germ-cell determinants and their possible involvement in germ-cell formation during embryogenesis. Until recently, there has been a serious gap between several historically well known and cytologically recognizable phenomena (e.g., chromatin diminution in *Ascaris* and chromosome elimination in *Mayetiola*, which is discussed later) and knowledge of cellular and molecular biology in higher organisms.

One of the ultimate goals of modern developmental biology is to explain how the mechanisms which regulate cell differentiation in multicellular organisms, including germ-cell differentiation, function. This chapter attempts to combine the famous, classical examples of localization of the germ plasm or of germ-cell determinants in eggs and embryos with recent knowledge from cellular and molecular biology. Success with this attempt should provide further insights into not only the problems of germ plasm localization but also more generally into cytoplasmic information systems.

An Overview of Germ-Cell Determinants

It is approximately one century since Weismann formulated the germ-cell theory, stating that (1) germ-cell and somatic-cell lines segregate early and develop independently of each other and (2) the germ-cell line is continuous from one generation to the next by means of a substance transmitted directly from parental germ cells to the germ cells of the new individuals. He termed this substance *Keimplasma* (germ plasm) and this continuous route of transmission of the germ plasm *Keimbahn* (germ track) (for a detailed discussion of Weismann's germ-cell theory, see Refs. 4, 13, and 32). Although today that theory requires some modification, because *Keimplasma* referred to genetic materials of the nucleoplasm (later called genes), the general concept provides the basis of the present germ plasm hypothesis. It is also perhaps the most important and distinguished example of a cytoplasmic information system.

Historically, numerous modifications of the germ plasm hypothesis have been proposed. For instance, Nieuwkoop and Sutasurya (32) stated that, "in discussing germ-cell formation in the invertebrates, particularly the lower phyla, it has become clear that *the germ line concept is essentially inadequate*." Their monograph cites almost all the literature relating to germ-cell formation in all the animal phyla. From that database, they propose three modes for germ-cell formation in the animal kingdom: (1) epigenetic, (2) intermediate, and (3) preformistic.

1. In the epigenetic mode, sexual and asexual forms of reproduction are believed to alternate under the influence of environmental factors. During sexual reproduction, germ cells are formed from undifferentiated or dedifferentiated embryonic cells. During asexual reproduction, the same embryonic cells are thought to give rise to various somatic-cell types.

2. In the intermediate mode, germ cells, once formed, are considered to be no longer replaceable by other cells. They are, however, formed at a rather late stage of development from cells which earlier must have passed through a phase of somatic development.

3. In the preformistic mode, germ cells segregate from somatic cells at a very early stage of embryonic development. They are often predetermined by the presence of a germ-cell-specific germ plasm, so that either presumptive or true germ cells can be distinguished during most, if not all, of the life cycle.

It seems reasonable that all animals can be classified into one of these three groups with respect to their mode of germ-cell formation. From this author's point of view, however, such a classification system is not particularly useful for future studies designed to elucidate *basic mechanisms* which govern germ-cell formation. Until now, most investigators have employed widely different animals and have focused their efforts on elucidating the species-specific phenomena associated with their particular embryos. Some investigators have emphasized the indispensable role germ plasm or germ-cell determinants play in the formation of germ cells in certain species (22, 23, 25, 43). Others, however, have emphasized the epigenetic formation of germ cells or the induction of germ cells from pluripotent somatic cells in different species (24, 37). No unifying explanation is presently available for germ-cell formation throughout the animal kingdom. This author, however, proposes that it is now time to consider a comprehensive concept to explain germ-cell formation at the level of *modern cellular and molecular biology*, irrespective of the seemingly different modes of germ-cell formation exhibited by diverse species of animals.

It should be emphasized that all germ cells, once established, show virtually identical behavior. They undergo meiosis, which is specific to the germ cells, and they differentiate into female and male gametes, regardless of whether different apparent modes of germ-cell formation were employed. Furthermore, it seems reasonable to assume that the molecular mechanism which controls fundamental phenomena, such as the conversion of cell division from mitosis to meiosis and the bisexual differentiation of germ cells, is basically the same in all animals. One of the main discoveries of recent molecular studies of eukaryotic gene expression is that fundamental phenomena in cellular and developmental biology are controlled by similar mechanisms in different organisms. For instance, genes which might regulate pattern formation, including segmentation of embryos, are reported to share common sequences (homeobox) even in completely different animal groups such as insects and vertebrates (2, 29). Thus, it is tenable to assume that the formation of germ cells in all animals is regulated by basically the same mechanism, even though superficial differences in germ-cell formation are apparent from one animal to another. Germ-cell formation is one of the most fundamental events of animal morphogenesis and can be expected, therefore, to exhibit common features throughout the animal kingdom.

To date, the spatiotemporal regulation of eukaryotic gene expression during early organogenesis has been analyzed using marker genes such as a-actin (15) in amphibian embryos and acetylcholinesterase (30) and myosin heavy chain (33) in ascidian embryos. Likewise, insights into germ-cell formation will need to come initially from the analysis of the spatiotemporal expression of germ-cell-specific marker genes and, eventually, from the analysis of possible germ-cell-specific regulatory genes. It is important to note that the epigenetic mode of germ-cell formation is considered to be the result of relatively late expression of the relevant regulatory and germ-cell-specific marker genes, while the preformistic mode implies their much earlier expression. In this latter case, germ-cell-specific gene expression may be controlled by the presence in the egg cytoplasm of germ plasm or germ-cell determinants. Thus, the development of a comprehensive concept to explain mechanisms of germ-cell formation in the basically common terms of modern cellular and molecular biology must await knowledge of the regulatory mechanisms which underly germ-cell-specific gene expression patterns.

This chapter adopts the following strategy. First, a brief description of classical examples of germ plasm and germ-cell determinants in eggs and embryos of a variety of animals is provided. Second, modern biochemical and molecular approaches for solving germ plasm problems in several species are described. Third, future prospects for furthering our understanding of germ-cell determination are discussed. Finally, a model for germ plasm function is proposed in order to provide a foundation for a unified explanation for germ-cell formation throughout the animal kingdom.

Classical Examples of Germ Plasm or Germ-Cell Determinants

Nematodes: Chromatin Diminution

The most famous example of germ-cell determinants is that of parasitic nematodes (*Ascaris*). The selective loss of distal heterochromatin in somatic cells (so-called chromatin diminution) and its retention in the germ-line cells (Fig. 1) was first reported in *Parascaris equorum* (*Ascaris megalocephala*) in a classical paper by Boveri at the end of the nineteenth century. It was later confirmed in different species, e.g., in *Ascaris lumbricoides*. The relevant literature has been reviewed by Nieuwkoop and Sutasurya (32) and by Tobler et al. (41).

Several experimental approaches, including centrifugation of eggs, modification of the spindle orientation, ultraviolet (uv) irradiation, etc., have provided insight into possible mechanisms which guide chromatin diminution in *Ascaris* (41). At present, it is postulated that the polar organization of the egg (e.g., the differential distribution of animal and vegetal cytoplasmic substances) is responsible for chromatin diminution. It is widely accepted that the cytoplasm strongly influences chromosome behavior. That is, some cytoplasmic factor(s) are indispensable for the preservation of the full set of chromo-

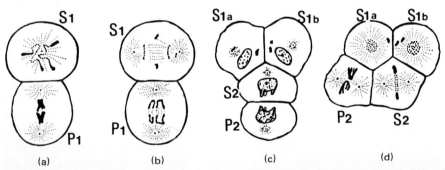

FIGURE 1. Development of an *Ascaris* egg from the 2-cell stage to the end of the 4-cell stage. (a) Two-cell stage, with the upper cell (S1) undergoing chromosome diminution and the lower cell (P1) showing normal chromosome behavior. (b) Two-cell stage, slightly later than in part a, S1 in side view. The ends of the diminished chromosomes remain in the center of the cell between the separating chromosome fragments. (c) Four-cell stage, resting nuclei. In the upper two cells (S1a and S1b), regenerating chromosome ends lie outside of nuclei. (d) Four-cell stage, nuclei dividing. S1a and S1b cells show the earlier diminished chromosomes. Lower right cell (S2) undergoes diminution but not in lower left cell (P2). The cells S1a, S1b, and S2 give rise exclusively to somatic cells, whereas the P2 cell represents the stem cell of the germ line. Redrawn from Nievwkoop and Sutasurya (32).

somes in the germ-cell line. However, little is known about differences in cytoplasmic substances in germ-cell and somatic-cell lines or about initial differences between the animal and vegetal halves of eggs. No clear cytoplasmic inclusions specific for the germ-cell line have been reported in *Ascaris*. Recently, germ-cell-specific cytoplasmic inclusions (P granules) have been reported in free-living nematodes, *Caenorhabditis elegans* (39, 40; discussed later). In *C. elegans* no chromatin diminution is, however, observed.

Dipteran Insect: Chromosome Elimination and Polar Plasm

Chromosome elimination similar to that seen in *Ascaris* is found in a different phylum, Arthropode, in dipteran families (3, 4, 32). The most famous example in this phylum is the gall midge, *Mayetiola destructor*, in which early cleavage is characterized by the elimination of chromosomes from presumptive somatic nuclei (only eight chromosomes are retained), whereas the full chromosome complement (about forty chromosomes) is kept by germ-line nuclei (3).

Several experiments involving uv irradiation and constriction and/or centrifugation of embryos show convincingly that the posterior polar plasm (or polar granules) prevent the nuclei from eliminating chromosomes (3). If the polar granules are dispersed by centrifugation or if nuclei are prevented by constriction from coming into contact with them before the fifth division, all nuclei divide with chromosome elimination. Thus, it is possible to obtain embryos that possess germ cells containing only eight chromosomes in their nuclei. Individuals whose germ-line nuclei have reduced chromosome numbers can develop to maturity. An ovary containing germ-cell nuclei with only eight chromosomes is, however, unable to form either oocytes or nurse cells. A testis containing germ-cell nuclei with only eight chromosomes is unable to form spermatocytes, but cells which come to resemble gametes are formed. Both males and females are sterile.

Among the dipteran species *Drosophila melanogaster* provides another outstanding example of the localization of germ plasm or germ-cell determinants, although neither chromosome elimination nor chromatin diminution is observed (25). Intensive experimental work using uv irradiation, cytoplasmic transplantation, and various mutants has demonstrated that the posterior polar plasm is indispensable for the formation of pole cells and primordial germ cells (PGCs) (22, 23). It is therefore widely accepted that polar plasm in *Drosophila* must contain germ-cell determinants (25). The characteristic organelles located exclusively in this polar plasm are called "polar granules" (Fig. 2). In *Drosophila* ultrastructural and cytochemical properties of the polar granules and their changes during development are well documented: polar granules originate during an early stage of vitellogenesis as small, electron-dense bodies which become attached to mitochondria during the growth phase of the oocyte. After fertilization the association of polar granules with mitochondria is lost. During cleavage and pole-cell formation, the polar granules fragment and become associated with mitochondria and polysomes, but after pole-cell formation, they coalesce. During pole-cell migration to the go-

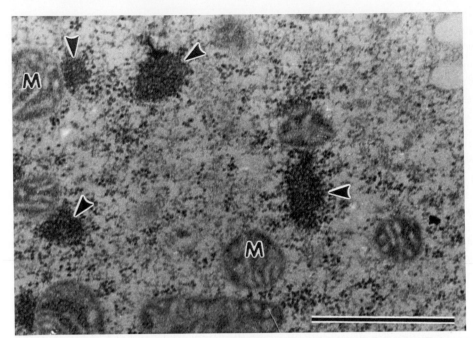

FIGURE 2. Ultrastructure of polar plasm of *Drosophila melanogaster* embryo. Polar granules (arrowheads) are attached to mitochondria (M) in the early embryo and have clusters of ribosomes associated with their periphery. Magnification bar = 1 μm. Courtesy Dr. M. Okada, Tsukuba University, Japan.

nads, they again fragment into fibrous bodies which become attached to the outer nuclear membrane. As described above, the polar granules undergo characteristic morphological changes before, during, and after pole-cell formation. These observations are consistent with the idea that polar granules carry the germ-cell determinants. At present, however, experiments still need to be done in order to determine whether the polar granules themselves actually contain the germ-cell determinants or whether cytoplasmic components which are only associated with the granules are the determinants.

Anuran Amphibians: Germ Plasm

The third example of localization of the germ plasm or germ-cell determinants is provided by anuran amphibians. Bounoure first recognized a special cytoplasmic localization in *Rana temporaria* eggs and embryos, the so-called germinal plasm or germ plasm, initially located in the subcortical layer of the vegetal region of the fertilized egg. Extensive studies have been carried out in different anuran species by many investigators, and the relevant literature has been reviewed by Eddy (13), Nieuwkoop and Sutasurya (31), Dixon (12), and Smith et al. (37).

In *Xenopus*, the germ plasm is reported to be continuously present from ovarian oocytes to fertilized eggs (11). In ovarian oocytes, the germ plasm is observed as very small islets dispersed among yolk platelets. After fertilization, the small islets of the germ plasm gather together to form larger cytoplasmic islands, which then combine together into much larger, morphologically distinct islands (Fig. 3A). During the first and second cleavages, the germ plasm is partitioned, more or less equally, between the first four blastomeres (12). During subsequent cleavages, only one daughter cell of each pair receives the germ plasm, generating the founder clone of four germ-plasm-containing cells (GPCC) or presumptive primordial germ cells (PGC) (12). During blastulation, the germ plasm moves from the peripheral cytoplasmic position it occupied during earlier stages to a position in contact with the nuclear membrane at the gastrula stage (6) (Fig. 3C).

The germ plasm contains numerous mitochondria and small electron-dense bodies (so-called germinal granules; Ref. 47) (Fig. 4). These germinal granules are composed of small electron-dense foci which appear to be embed-

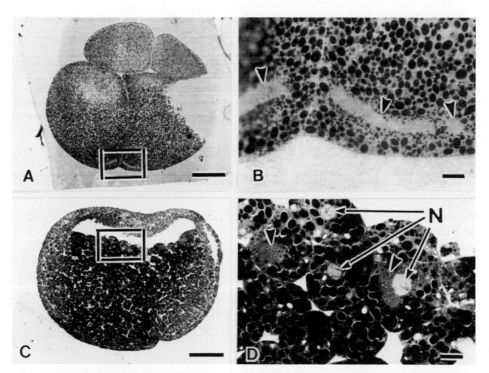

FIGURE 3. Light micrographs showing germ plasm in 8-cell (*A, B*) and early gastrula (*C, D*) stages of *Xenopus laevis*. (*A*) Large islands of germ plasm located in subcortical layer near vegetal pole of embryo are enclosed by a rectangle. (*B*) Enlargement of rectangular area in part *A*. Arrowheads indicate germ plasm. (*C*) Germ plasm containing cells embedded in endoderm are enclosed by a rectangle. (*D*) Enlargement of area in rectangle in part *C*. Arrowheads, germ plasm. N, nucleus. Magnification bars = 200 μm (*A, C*); 20 μm (*B, D*).

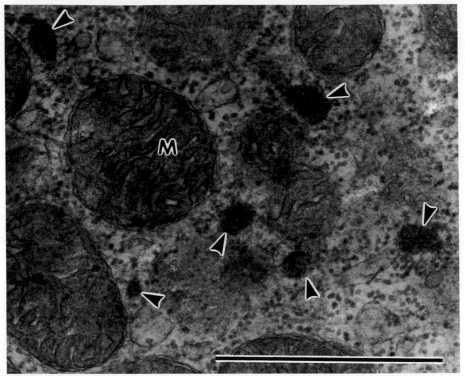

FIGURE 4. Ultrastructure of germ plasm in 8-cell embryo of *Xenopus laevis*. Major components of the germ plasm are mitochondria (M) and germinal granules (arrowheads). Magnification bar = 1 μm.

ded in a matrix of extremely fine fibrils. They are frequently found in contact with mitochondria, as are the polar granules in *Drosophila*. The germ plasm also contains ribosomes, pigment granules, glycogen, a few oil droplets, and a varying number of small yolk platelets. The origin of the germinal granules in *Xenopus* has been studied by Heasman et al. (16), who observed the germinal granules in the "mitochondrial cloud" of stage I oocytes of *Xenopus*. Ikenishi and Kotani (19) describe ultrastructural changes of the germinal granules during *Xenopus* development. At early stages the germinal granules in presumptive PGCs show a fibrillogranular structure, but they soon change, first, into irregular stringlike bodies, and, then, at the feeding tadpole stage, into granular material within the PGCs. Thus, the germ plasm can be traced from the stage I oocyte to the tadpole stage, where it resides in the germinal ridges.

Several experiments involving cytoplasmic withdrawal (9) and uv irradiation (18, 36) have convincingly shown that the intact germ plasm is indispensable for normal formation of the PGC in anuran amphibians. Akita and Wakahara (1) have demonstrated that the amount of germ plasm is directly related to the number of PGCs which develop: embryos which contain a larger amount of germ plasm develop a larger number of PGCs at tadpole stages.

Furthermore, the results of microinjection of vegetal cytoplasm (38), or of subcellular fractions of vegetal cytoplasm (44, 45), show that certain fractions contain the capacity to induce germ-cell formation. Recently, Ikenishi et al (20) have shown that a 20,000g fraction from the vegetal cytoplasm of fertilized *Xenopus* eggs is able to transform presumptive somatic cells to PGCs when microinjected into vegetal blastomeres that normally never contain germ plasm. Thus, it seems clear that the germ plasm or germinal granules have a PGC-inducing activity. At present, however, it remains unconfirmed whether the induced PGCs or PGCs that are transformed from presumptive somatic cells become functional germ cells. Therefore, key experiments still have to be done to determine whether the germ plasm or germinal granules contain true germ-cell determinant activity in anuran amphibians.

Other Animals

Other classical examples of germ plasm or germ-cell determinants are observed in Chaetognatha (*Sagitta* and *Spandella*) and Crustacea (*Cyclops*). In spite of the lack of experimental studies, both animals are famous for containing germ-line-specific structures called "special bodies" or "ectosomes" (13, 32).

Nuage

Many reports have demonstrated the presence of densely staining bodies in the cytoplasm of germ cells in numerous animals from coelenterates to mammals. The structure is generally called *nuage*, the French word for "cloud," because it is a descriptive and neutral term as far as composition or function are concerned. The relevant literature has been extensively discussed by Eddy (13) and by Nieuwkoop and Sutasurya (31, 32).

The nuage has attracted the attention of investigators because it is morphologically similar to polar granules in insects and to germinal granules in anuran amphibians. (1) It appears as a discrete, dense, fibrous, cytoplasmic organelle lacking a surrounding membrane, and (2) it is frequently seen in association with mitochondrial clusters or immediately adjacent to the nuclear envelope of germ cells (Fig. 5). The universal presence of the nuage in differentiated germ cells, such as oogonia and oocytes, in many species suggests that the nuage is only an indication of germ-cell differentiation rather than a causal factor for germ-cell determination. This concept is strengthened by the study of Noda and Kanai (34), who have reported changes in amount of the nuage during germ-cell differentiation in hydra. They found that interstitial cells and early nematocysts (somatic cells) of the asexual form, as well as the oogonia (germ cells) of the sexual form, contain varying amounts of nuage in their cytoplasm. The oogonia contain a large amount, interstitial cells a lesser amount, and young nematocysts only a small amount, which drops to zero as cellular differentiation proceeds.

Several authors have postulated that in many species a relationship ex-

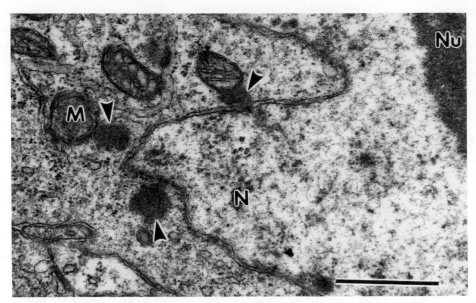

FIGURE 5. Electron micrograph showing nuage material (arrowheads) in cytoplasm of an oogonium of *Xenopus laevis*. Often, nuage is associated with mitochondria (M) and nuclear membrane. N, nucleus; Nu, nucleolus. Magnification bar = 1 μm.

ists between nuage and the germinal granules. It has been suggested that germinal granules gradually change into nuage, or, alternatively, that germinal granules originate from nuage (13). The biologic significance or the possible role of nuage during germ-cell development is still, however, unclear, because studies on nuage have been only descriptive or morphological; experimental studies that would determine its biologic role have not yet been performed. It seems, therefore, highly important to develop experimental systems to examine the role of nuage as well as to find useful probes for analyzing its molecular properties. For example, monoclonal antibodies that react with nuage would be very valuable.

Molecular Biologic Approaches to Understanding Germ Plasm

Genomic Differences between Germ-Line and Somatic Cells

At present virtually nothing is known about the molecular properties of either the germ plasm or of other germ-cell determinants. Furthermore, the mechanism of chromatin diminution or chromosome elimination, as well as the significance of germ-line-specific DNA sequences in species which exhibit chromatin or chromosome loss, remains unknown. One of the most important issues which should be addressed is Boveri's hypothesis that DNA eliminated from somatic cells contains genes essential for the germ line (32). Using molecular biologic techniques, it is now possible to examine this issue. For instance, a specific gene expressed only in the germ line could be cloned and its organization compared in germ-line and somatic-cell DNA. In the meantime, however, it will be informative to review several experiments which examined structural differences in DNA sequences between germ-line and somatic cells.

Ascaris

In *Ascaris lumbricoides*, about a quarter of the germ-line DNA is eliminated from presumptive somatic cells during chromatin diminution. Tobler et al. (41) have summarized what is known about differences in DNA sequences between germ-line and somatic cells in *Ascaris*: over 99 percent of all satellite DNA sequences present in the germ-line genome are expelled from the presumptive somatic cells by chromatin diminution. However, earlier methods, such as DNA hybridization-renaturation experiments, did not permit a demonstration of fine structural differences of DNA sequences or of changes at the level of gene function between germ-line and somatic cells.

Bennett and Ward (5) recently cloned a gene which encodes the major sperm protein (MSP) of *Ascaris lumbricoides*. The MSP appears to be expressed only in the testis of *Ascaris* and *Caenorhabditis*. Actin and α-tubulin genes served as somatically expressed gene controls. Their results demonstrate, surprisingly, that the MSP genes expressed only in germ-line cells are neither lost nor altered during chromatin diminution.

The constancy of the MSP gene organization appears to refute Boveri's hypothesis, but it is premature to discard the hypothesis since several technical and biologic limitations characterize those experiments. First, it is possible that MSP is, in fact, not germ-line specific but has some essential somatic functions at a level of expression below the present detection levels. Second, only a few genes have been tested, including MSP, actin, and α-tubulin. There may be other germ-line-specific genes, other than MSP, which are eliminated. Although at present we have only negative results from the cDNA-cloning studies on chromatin diminution in *Ascaris*, experiments that analyze qualitative differences as well as fine structural changes in DNA sequences between the germ line and somatic cell should be encouraged.

Tetrahymena

Tetrahymena provides well known examples of programmed elimination of specific DNA sequences and genomic rearrangement (7, 8). It does not, however, display a cytoplasmic localization similar to germ plasm. Knowledge of the molecular mechanisms which preserve germ-line-limited DNA sequences in *Tetrahymena* will, however, likely provide ideas concerning how the germ plasm acts in higher organisms.

Tetrahymena, like most ciliated protozoa, is distinctive in having within the same cell both a micronucleus and a macronucleus. From our point of view, the micronucleus, which is transcriptionally inactive during vegetative growth, is approximately equivalent to a *germ-line* nucleus, while the transcriptionally active macronucleus is the equivalent of a *somatic* nucleus. Both nuclei are formed from the descendants of parental micronuclei during sexual reproduction. The parental macronuclei are degraded. DNA elimination occurs during macronucleus development. Earlier studies of DNA renaturation kinetics have suggested that between 10 to 20 percent of the micronuclear genome is not present at all in the macronucleus (49). Many of the germ-line-limited DNA sequences belong to families of repeated sequences. The mechanism of DNA elimination in *Tetrahymena* has been extensively analyzed using a cloned micronuclear DNA fragment that is involved in elimination and in the formation of the macronuclear DNA fragment that results from this process (10). Recently, Stein-Gavens et al. (38) have reported that six members of a family of repeated micronuclear specific DNA sequences are homologous to a 1.5 kilobase poly(A)$^+$ RNA transcript present in starved mature cells but not in growing cells. Their presence suggests that at least one germ-line-specific DNA sequence is transcribed during a specific phase of the *Tetrahymena* life cycle. That transcript represents the first RNA of a germ-line-specific DNA sequence to be identified in any organism.

Because there is no evidence that there is any causal relationship between the micronucleus-specific DNA transcript and either the initiation or the development of sexual maturation, the function of the germ-line-limited DNA sequences remains unclear. Furthermore, there is no information on any cytoplasmic factor(s) which might be involved in conserving germ-line-limited DNA sequences in the micronucleus, although it is assumed that the ability to retain the germ-line-limited DNA sequences is not an intrinsic property of the germ-line nuclei. Rather, that ability is thought to be the result of the position of the micronucleus near cytoplasmically localized factors. Cloning germ-line-limited or micronucleus-specific sequences in *Tetrahymena* provides a promising model for examining Boveri's hypothesis because it looks at changes in nucleotide sequences of individual genes.

mRNA Which Might Be Responsible for Germ-Cell Determination

Polar plasm in *Drosophila* promotes pole-cell formation in any periplasmic region to which it is transplanted (22, 23) and restores fertility to embryos which have been uv-irradiated (35). It is therefore believed that molecular information is localized in the polar plasm (25). A subcellular fraction prepared from a homogenate of embryos at the early cleavage stage was demonstrated to restore pole-cell-forming ability to embryos when injected into the uv-irradiated posterior polar region (43). Similarly, a subcellular fraction from the vegetal-pole cytoplasm of anuran amphibian eggs has been shown to contain the germ-cell-forming activity (20, 44, 45).

Drosophila

Recently, Togashi et al. (42) demonstrated that maternal mRNA extracted from a subcellular fraction has an ability to restore pole cells to uv-irradiated embryos. The maternal mRNA-induced pole cells in uv-irradiated embryos exhibit a morphology similar to normal pole cells at the transmission electron microscope (TEM) and scanning electron microscope (SEM) levels. Experiments on pole-cell-inducing activity of poly(A)$^+$ RNA extracted from embryos at different developmental stages clearly shows that mRNA from the early embryos (20 or 90 min after egg laying, which corresponds to early cleavage and pole-cell-formation stages, respectively) has restoring activity while mRNA extracted from the later embryos (150 min after egg laying, which corresponds to the cellular blastoderm stage), however, does not. Thus, maternal mRNA sequences active in pole-cell formation seem to be inactivated on completion of pole-cell formation (42).

Those pole cells induced by mRNA were, however, demonstrated to be *unable* to form true germ cells: all the adults derived from embryos forming pole cells after irradiation and injection with mRNA were sterile. In contrast, about half of the adults derived from embryos irradiated and injected with intact polar plasm were fertile. Furthermore, injections of maternal mRNA in the anterior periplasm failed to induce ectopic pole-cell formation, while injections of intact polar plasm in the anterior periplasm did. Thus, it can be concluded that maternal mRNA extracted from a homogenate of oocytes or earlier embryos contributes to pole-cell formation but not necessarily to further germ-cell determination.

The importance of that experiment is twofold. First, it provides a molecular basis for understanding the segregation of the germ-cell lineage from the somatic-cell lineage: possibly a specific maternal mRNA is responsible for segregating pole cells from somatic cells. Second, it distinguishes between pole-cell formation and germ-cell determination at the molecular level: Pole cells may need additional information (possibly a second mRNA) before they will form germ cells. Further investigations will no doubt deal with issues such as when and where the messages are transcribed and what role their translation products play in pole-cell formation. This line of investigation, cDNA cloning for the mRNA involved in pole-cell formation and germ-cell determination, appears to be very promising.

ANURAN AMPHIBIANS

Histochemical or cytochemical procedures have been employed to understand the composition of anuran germ plasm (13, 31). Mahowald and Hennen (26) demonstrated the presence of RNA in the germinal granules of the egg and early embryo in *Rana pipiens* by electron microscopic histochemistry. Thus, it can be expected that amphibian germ plasm might have the mRNA needed for germ-cell formation, as was discussed above for *Drosophila*. However, recent experiments using in situ hybridization with poly (U) failed to detect poly (A)$^+$ RNA in the *Xenopus* germ plasm, even in fertilized eggs and early embryos (M. Wakahara, unpublished observations). Conversely, this author has speculated that germ plasm might possess an activity to suppress transcription in PGCs; poly (A)$^+$ RNA does not accumulate in developing PGCs in germinal ridges, but most somatic-cell nuclei possess poly (A)$^+$ RNA (46).

As noted, little is known concerning the molecular or chemical properties of amphibian germ plasm since, except for histochemical or cytochemical procedures, useful molecular probes which can be used to analyze them do not yet exist. A novel technique for screening DNA sequences with monoclonal antibodies directed against protein gene products has been developed (50). Thus, monoclonal antibodies specific for germ plasm or for components that are active in germ-cell formation can be prepared and employed for cloning genes.

URODELE AMPHIBIANS

Since urodeles have been reported to show an epigenetic mode of germ-cell formation, that is, germ cells are formed by a cell-cell interaction or by induction (24, 31, 37), we cannot easily propose a molecular basis for germ-cell determination. Although Ikenishi and Nieuwkoop (21) have found a germinal granule–like structure in developing germ cells in *Ambystoma*, neither its chemical components nor biologic significance is yet known.

Monoclonal Antibodies against Germ Plasm

Caenorhabditis elegans

The free-living soil nematode, *Caenorhabditis elegans*, exhibits germ-line-specific cytoplasmic granules (P granules), which display asymmetric segregation during early cleavage (39, 40). P granules were found accidentally by staining particulate cytoplasmic components of germ-line cells using fluorescein isothiocyanate-conjugated rabbit anti-mouse IgG (39). Later, a more specific monoclonal antibody to P granules was made (40, 48). Although the biologic functions of P granules remain to be discovered, the P granule monoclonal antibody has proven to be a powerful probe for analyzing the mechanism responsible for asymmetric segregation of P granules. The ability to employ mutations (17) in *C. elegans* (see Chaps. 8 to 10 in Vol. III of this series) provides us with exciting possibilities for analysis of germ-cell determination which do not exist in higher animals.

Drosophila melanogaster

In order to obtain probes, not only for studying germ-cell determination but also more generally for detecting maternal factors active in regulating early *Drosophila* development, a library of monoclonal antibodies against ovarian antigens has recently been established (28). A monoclonal antibody has been identified which reacts with granules abundant in the posterior region of early embryos. These granules are found in the posterior polar cyto-plasm at the cleavage stage and in pole cells at the blastoderm stage. Some granules, however, were also found in the somatic region of embryos, suggest-ing that they are not true polar granules. To date, there have been no reports of success in obtaining monoclonal antibodies that react exclusively with the polar granules or that recognize specific molecules localized in the polar plasm in *Drosophila*. Obtaining specific monoclonal antibodies against polar gran-ules seems especially important for further progress. Since many maternal-effect mutations are available in *Drosophila* (25), studies that combine monoclonal antibodies and mutations might be fruitful, albeit laborious.

Prospects for Future Germ Plasm Research

Species Specificity of Germ Plasm Function

The morphological similarity of germ plasm between species suggests a com-mon role for it in specifying germ cells. Thus, it might be interesting to ex-amine the species specificity of germ plasm function. Although interspecific transplantation of polar plasm from *Drosophila immigrans* to *D. melanogaster* embryos has been successfully performed (27), no systematic analyses have been done with a wide variety of donors and recipients. These experiments should be done not only between similar species within either dipteran insects or anuran amphibians, but also between completely different organisms such as dipterans and anurans. For example, the 20,000g fraction of vegetal cyto-plasm from *Xenopus* eggs (20, 45) might be injected into *Drosophila* embryos. It is relatively easy to prepare a large amount of an active fraction from the vegetal-pole cytoplasm of *Xenopus* eggs (44, 45). It is, however, difficult to as-say the germ-cell-forming activity by microinjecting it into somatic blastomeres of developing embryos (20). Conversely, pole-cell forming activity or the ability to rescue uv-induced sterility is assayed more easily in *Drosophila* than in *Xenopus*. Preparation of *Drosophila* cytoplasmic fractions responsible for pole-cell formation is unfortunately rather difficult (M. Okada, personal communication). Transplanting *Xenopus* germ plasm to *Drosophila* embryos may be more practical.

Furthermore, experiments should not be limited to animals with the preformistic mode of germ-cell formation but should be extended to animals with the epigenetic and intermediate modes (32). Microinjection of the germ plasm (or cytoplasmic fractions which have been proven to have a germ-cell-forming activity; e.g., Refs. 43–45) into embryos or somatic blastomeres of an-imals with the epigenetic or intermediate mode will provide further insight not only into germ-plasm function but also into the general mechanisms un-derlying germ-cell formation. The significance of this type of experiment is twofold. First, it is relatively easy to do, and the information obtained could be of considerable value. Second, even though the molecular properties of germ plasm or germinal granules remain unknown, it would still be an informative exercise. If success is achieved in transforming the urodele somatic blastomere to germ cells by introducing anuran germ plasm, a new level of understanding of the mechanisms of germ-cell differentiation would be achieved, i.e., no substantial differences between the preformistic mode and the intermediate or the epigenetic modes of germ-cell formation may exist!

Regulatory Genes for Germ-Cell Differentiation

Nieuwkoop and Sutasurya (32) have suggested that germ-cell formation should be considered a specific type of cellular differentiation, in principle no different from that of somatic cells, except that germ cells retain the potential for supporting total development. From that point of view, it should be possible to analyze the molecular basis of germ-cell formation with the same procedures which have been applied to the investigation of the spatiotemporal expression of several structural genes during early development (14, 15, 30, 33). Recent progress in the molecular biology of regulatory genes involved in pattern formation or morphogenesis (e.g., homeotic and segmentation genes in *Drosophila*, Ref. 29) provides a possible model for the molecular basis of regulatory gene function in germ-cell differentiation. In this regard, this author supports the point of view that germ plasm contains products of regulatory genes and/or germ-cell-specific marker genes (Fig. 6). At present, however, there is no direct evidence to demonstrate that the control of regulatory genes is a prerequisite for germ-plasm action. Furthermore, no probes for detecting and analyzing regulatory gene expression in germ-cell differentiation are available.

It has often been observed that in molecular biology progress on a difficult problem has been achieved by selecting particularly favorable biologic systems for analysis (14). In this regard, perhaps the restricted DNA loss from somatic nuclei, but full preservation of it in germ-line nuclei, known to occur in a wide variety of animals (3, 7, 8, 41), will provide an excellent system for analyzing germ-line versus somatic differentiation. *Tetrahymena* appears to be especially favorable for analyzing changes in gene function between germ-line and somatic nuclei (38).

Because such investigations have just started, the data are insufficient to demonstrate whether genes indispensable for germ-cell differentiation exist. However, the successful demonstration of micronucleus-specific DNA transcription in *Tetrahymena* is consistent with the idea that regulatory genes for germ-cell differentiation are a common feature of virtually all animals. Once regulatory genes specific for germ-cell differentiation are recognized and cloned, the strategy that has been employed to detect and analyze the homeobox-containing genes in higher animals (2) will no doubt be employed.

A Model for Germ Plasm Function

Assuming that germ-cell formation requires the expression of both regulatory genes and structural genes, the three different modes of germ-cell formation

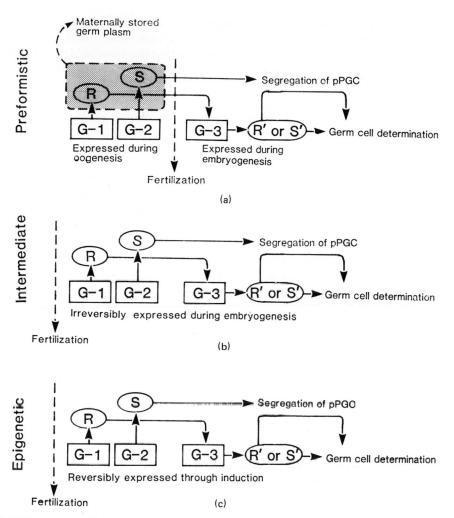

FIGURE 6. A model for the role of gene expression in three different modes of germ-cell formation: (*a*) preformistic, (*b*) intermediate, and (*c*) epigenetic.

which have been proposed (epigenetic, intermediate, and preformistic; Ref. 32) can be explained in terms of temporal gene expression. The epigenetic mode of germ-cell formation is considered to result from a later expression of those genes, while the preformistic mode implies a much earlier, possibly even maternal, expression. The localization of germ plasm is considered a cytoplasmic manifestation of factors preformed during oogenesis and stored in oocytes until fertilization, which exerts instructive influences on the nucleus when presumptive germ-line nuclei are exposed to germ plasm. Therefore, germ plasm can be considered to consist of specific gene products from the regulatory and structural genes which govern germ-cell formation.

As shown by Togashi et al. (42) in *Drosophila*, a specific maternal mRNA (or possibly its translation product) works in segregating pole cells from the somatic cells. However, they have also shown that maternal mRNA is not sufficient for final germ-cell determination: pole cells may need additional information to be able to form germ cells. These observations indicate that pole-cell formation can be distinguished from germ-cell determination and that germ-cell formation proceeds stepwise. At least two different mechanisms are necessary for complete germ-cell formation from the initial segregation of presumptive PGCs (or pole cells) to the final determination of germ cells. Combining these recent results with previously known facts produces a model of the type illustrated in Fig. 6.

The simplest model employs only three genes: G-1 is a regulatory gene, whose product R regulates the expression of another gene (G-3); G-2, a structural gene, whose product S works in segregating presumptive PGCs (pPGCs) (or pole cells); and G-3, either a structural or a regulatory gene which is indispensable for germ-cell determination.

In animals with the preformistic mode of germ-cell formation (and which also show the localization of germ plasm) (Fig. 6a), a regulatory gene (G-1) and a structural gene (G-2) are both expressed during oogenesis. Either their transcripts or the gene products (R = regulator and S = structural) are stored in the germ plasm. In the case of *Drosophila*, the products of G-2 might be bound to a specific cytoplasmic fraction (42, 43) or to the polar granules, whereas the regulatory gene (G-1) products are stored in a different compartment of polar plasm from the polar granules. After fertilization and cleavage, the G-2 product S is involved in segregating pPGC from the somatic cells. When the nuclei of the pole cells are eventually exposed to polar plasm, the G-1 product R exerts its regulatory role by instructing the expression of G-3, which governs germ-cell determination. The G-3 product (R' or S') governs germ-cell determination either directly (R') or indirectly (S'). In amphibians, it is believed that the actual determination of germ cells occurs at least as early as the gastrula stage, when germ plasm moves from the cell periphery and becomes associated with the nucleus (6).

In animals showing the intermediate mode of germ-cell formation, neither the regulatory gene (G-1) nor the structural gene (G-2) is expressed before fertilization (Fig. 6b). Instead, they are first expressed during early embryogenesis. Once G-2 is expressed in certain cells, the gene products instruct the cells to form PGCs. In contrast, gene products of G-1 turn on G-3 expression, resulting in germ-cell determination. Because the expression of both G-1 and G-2 never occurs before fertilization, preformed germ plasm or germ-cell determinants can be neither synthesized nor stored during oogenesis by animals with this intermediate mode. However, the gene products of G-1 from animals that show the intermediate and preformistic modes of germ-cell formation exert a similar regulatory role in the expression of G-3. Once the germ cells are determined as a result of G-3 expression, provided that this expression is irreversible during normal development, they will fulfill their destiny as germ cells.

Finally, in animals showing the epigenetic mode of germ cell formation (Fig. 6c), the regulation of G-1 and G-2 differs from their regulation in animals showing the preformistic and intermediate modes. The epigenetic mode might be controlled by a cell-cell interaction or by induction (31). As in embryonic induction (14), certain inducer molecules instruct G-1 expression and the gene products (regulatory molecules) switch on G-3. As a result, cells which have been exposed to inducer molecules or to signals from a cell-cell interaction become germ-line cells. However, the expression of G-3 is reversible in this group of animals: If G-1 is not expressed, i.e., if the inducer molecules or some cues from cell-cell interaction are lacking, G-3 might be inactive and thereby, the cells are led to differentiate into somatic cells.

This model proposes that germ plasm or germinal granules are morphological manifestations of the maternal storage of gene products of at least two different genes: (1) a regulatory gene (G-1) whose transcript or translate regulates the expression of another regulatory or structural gene (G-3) which is indispensable for germ-cell determination and (2) a structural gene (G-2) involved in the initial step of the segregation of germ cells. The idea that germ plasm contains important information for germ-cell development which is synthesized from the maternal genome and stored during oogenesis has been proposed by many investigators (25). Until now, however, the importance of stored regulatory gene products in the germ plasm has been neglected.

Acknowledgments

I would like to thank G. M. Malacinski for his critical reading of the manuscript and for valuable discussions, and Susan Duhon for editorial assistance. Research presented here was supported in part by Grants-in-Aid for Scientific Research from the Ministry of Education, Science and Culture of Japan (60440100, 60540452).

General References

Nieuwkoop, P. D., and Sutasurya, L. A.: 1979. *Primordial Germ Cells in the Chordates. Embryogenesis and Phylogenesis.* Cambridge University Press, Cambridge.

Nieuwkoop, P. D., and Sutasurya, L. A.: 1981. *Primordial Germ Cells in the Invertebrates. From Epigenesis to Preformation.* Cambridge University Press, Cambridge.

Mahowald, A. P., and Boswell, R. E.: 1983. Germ plasm and germ cell development in invertebrates. In: *Current Problems in Germ Cell Differentiation.* (A. McLaren, and C. C. Wylie, eds.), pp. 3–17, Cambridge University Press, Cambridge.

References

1. Akita, Y., and Wakahara, M. 1985. Cytological analyses of factors which determine the number of primordial germ cells (PGCs) in *Xenopus laevis. J. Embryol. Exp. Morphol.* **90**:251–265.

2. Awgulewitsch, A., Utset, M. F., Hart, C. P., McGinnis, W., and Ruddle, F. H. 1986. Spatial restriction in expression of a mouse homeo box locus within the central nervous system. *Nature (Lond.)* **320**:328–335.

3. Bantock, C. 1970. Experiments on chromosome elimination in the gall midge *Mayetiola destructor. J. Embryol. Exp. Morphol.* **24**:257–286.

4. Beams, H. W., and Kessel, R. G. 1974. The problem of germ cell determinants. *Int. Rev. Cytol.* **39**:413–479.

5. Bennett, K. L., and Ward, S. 1986. Neither a germ line–specific nor several somatically expressed genes are lost or rearranged during embryonic chromatin diminution in the nematode *Ascaris lumbricoides* var. *suum. Dev. Biol.* **118**:141–147.

6. Blackler, A. W. 1970. The integrity of the reproductive cell line in the Amphibia. *Curr. Top. Devel. Biol.* **5**:71–87.

7. Bostock, C. 1984. Chromosomal changes associated with changes in development. *J. Embryol. Exp. Morphol.* (Suppl.) **83**:7–30.

8. Brunk, C. F. 1986. Genome reorganization in *Tetrahymena. Int. Rev. Cytol.* **99**:49–83.

9. Buehr, M., and Blackler, A. W. 1970. Sterility and partial sterility in the South African clawed toad following the pricking of the egg. *J. Embryol. Exp. Morphol.* **23**:375–384.

10. Callahan, R. C., Shalke, G., and Gorovsky, M. A. 1984. Developmental rearrangements associated with a single type of expressed a-tubulin gene in *Tetrahymena. Cell* **36**:441–445.

11. Czolowska, R. 1969. Observations on the origin of the 'germinal cytoplasm' in *Xenopus laevis. J. Embryol. Exp. Morphol.* **22**:229–251.

12. Dixon, K. E. 1981. The origin of the primordial germ cells in amphibia. *Neth. J. Zool.* **31**:5–37.

13. Eddy, E. M. 1975. Germ plasm and differentiation of the germ cell line. *Int. Rev. Cytol.* **43**:229–280.

14. Gurdon, J. B. 1987. Embryonic induction—molecular prospects. *Development* **99**: 285–306.

15. Gurdon, J. B., Mohum, T. J., Fairman, S., and Brennan, S. 1985. All components required for the eventual activation of muscle-specific actin genes are localized in the subequatorial region of an uncleaved amphibian egg. *Proc. Nat. Acad. Sci. U.S.A.* **82**:139–143.

16. Heasman, J., Quarmby, J., and Wylie, C. C. 1984. The mitochondrial cloud of *Xenopus* oocytes: the source of germinal granule material. *Dev. Biol.* **105**:458–469.

17. Hirsch, D., Kempheus, K. J., Stinchcomb, D. T., and Jefferson, R. 1985. Gene affecting early development in *Caenorhabditis elegans. Cold Spring Harbor Symp. Quant. Biol.* **50**:69–78.

18. Ijiri, K. 1976. Stage-sensitivity and dose-response curve of UV effect on germ cell formation in embryos of *Xenopus laevis. J. Embryol. Exp. Morphol.* **35**:617–623.

19. Ikenishi, K., and Kotani, M. 1975. Ultrastructure of the 'germinal plasm' in *Xenopus* embryos after cleavage. *Dev. Growth & Differ.* **17**:101–110.

20. Ikenishi, K., Nakazato, S., and Okuda, T. 1986. Direct evidence for the presence of germ cell determinant in vegetal pole cytoplasm of *Xenopus laevis* and in a subcellular fraction of it. *Dev. Growth & Differ.* **28**:563–568.

21. Ikenishi, K., and Nieuwkoop, P. D. 1978. Location and ultrastructure of primordial germ cells (PGCs) in *Ambystoma mexicanum*. *Dev. Growth & Differ.* **20**:1–9.

22. Illmensee, K., and Mahowald, A. P. 1974. Transplantation of posterior polar plasm in *Drosophila*. Induction of germ cells at the anterior pole of the egg. *Proc. Nat. Acad. Sci. U.S.A.* **71**:1016–1020.

23. Illmensee, K., and Mahowald, A. P. 1976. The autonomous function of germ plasm in a somatic region of the *Drosophila* egg. *Exp. Cell Res.* **97**:127–140.

24. Kotani, M. 1957. On the formation of the primordial germ cells from the presumptive ectoderm of *Triturus* gastrulae. *J. Inst. Polytech. Osaka City Univ. Ser. D.* **8**: 145–159.

25. Mahowald, A. P., and Boswell, R. E. 1983. Germ plasm and germ cell development in invertebrates. In: *Current Problems in Germ Cell Differentiation.* (A. McLaren, and C. C. Wylie, eds.), pp. 3–17, Cambridge University Press, Cambridge.

26. Mahowald, A. P., and Hennen, S. 1971. Ultrastructure of the "germ plasm" in eggs and embryos of *Rana pipiens. Dev. Biol.* **24**:37–53.

27. Mahowald, A. P., Illmensee, K., and Turner, F. R. 1976. Interspecific transplantation of polar plasm between *Drosophila* embryos. *J. Cell Biol.* **70**:358–373.

28. Maruo, F., and Okada, M. 1987. Monoclonal antibodies against *Drosophila* ovaries: their reaction with ovarian and embryonic antigens. *Cell Differ.* **20**:45–54.

29. McGinnis, W., Levine, M. S., Hafen, E., Kuroiwa, A., and Gehring, W. J. 1984. A conserved DNA sequence in homeotic genes of the *Drosophila* antennapedia and bithorax complexes. *Nature (Lond.)* **308**:428–433.

30. Meedel, T. H., and Whittaker, J. R. 1984. Lineage segregation and developmental autonomy in expression of functional muscle acetylcholinesterase mRNA in ascidian embryos. *Dev. Biol.* **105**:479–487.

31. Nieuwkoop, P. D., and Sutasurya, L. A. 1979. *Primordial Germ Cells in the Chordates. Embryogenesis and Phylogenesis.* Cambridge University Press, Cambridge.

32. Nieuwkoop, P. D., and Sutasurya, L. A. 1981. *Primordial Germ Cells in the Invertebrates. From Epigenesis to Preformation.* Cambridge University Press, Cambridge.

33. Nishikata, T., Mita-Miyazawa, I., Deno, T., and Satoh, N. 1987. Muscle cell differentiation in ascidian embryos analyzed with a tissue-specific monoclonal antibody. *Development* **99**:163–171.

34. Noda, K., and Kanai, C. 1977. An ultrastructural observation on *Pelmatohydra robusta* at sexual and asexual stages, with a special reference to 'germinal plasm.' *Ultrastruct. Res.* **61**:284–294.

35. Okada, M., Kleinman, I. A., and Schneiderman, H. A. 1974. Restoration of fertility in sterilized *Drosophila* eggs by transplantation of polar cytoplasm. *Dev. Biol.* **37**: 43–54.

36. Smith, L. D. 1966. The role of a 'germinal plasm' in the formation of primordial germ cells in *Rana pipiens. Dev. Biol.* **14**:330–347.

37. Smith, L. D., Michael, P., and Williams, M. A. 1983. Does a predetermined germ line exist in amphibians? In: *Current Problems in Germ Cell Differentiation.* (A. McLaren and C. C. Wylie, eds.), pp. 19–39, Cambridge University Press, Cambridge.

38. Stein-Gavens, S., Wells, J. M., and Karrer, K. M. 1987. A germ line specific DNA sequence is transcribed in *Tetrahymena. Dev. Biol.* **120**:259–269.

39. Strome, S., and Wood, W. B. 1982. Immunofluorescence visualization of germ-line-specific cytoplasmic granules in embryos, larvae, and adults of *Caenorhabditis elegans*. *Proc. Nat. Acad. Sci. U.S.A.* **79**:1558–1562.

40. Strome, S., and Wood, W. B. 1983. Generation of asymmetry and segregation of germ-line granules in early *C. elegans* embryos. *Cell* **35**:15–25.

41. Tobler, H., Muller, F., Back, E., and Aeby, P. 1985. Germ line—soma differentiation in *Ascaris*: a molecular approach. *Experientia (Basel)* **41**:1311–1319.

42. Togashi, S., Kobayashi, S., and Okada, M. 1986. Functions of maternal mRNA as a cytoplasmic factor responsible for pole cell formation in *Drosophila* embryos. *Dev. Biol.* **118**:352–360.

43. Ueda, R., and Okada, M. 1982. Induction of pole cells in sterilized *Drosophila* embryos by injection of subcellular fraction of them. *Proc. Nat. Acad. Sci. U.S.A.* **79**: 6946–6950.

44. Wakahara, M. 1977. Partial characterization of 'primordial germ cell–forming activity' localized in vegetal pole cytoplasm in anuran egg. *J. Embryol. Exp. Morphol.* **39**:221–233.

45. Wakahara, M. 1978. Induction of supernumerary primordial germ cells by injecting vegetal pole cytoplasm into *Xenopus* eggs. *J. Exp. Zool.* **203**:159–164.

46. Wakahara, M. 1982. Chronological changes in the accumulation of poly (A) + RNA in developing cells of *Xenopus laevis*, with special reference to primordial germ cells. *Dev. Growth & Differ.* **24**:311–318.

47. Williams, M. A., and Smith, L. D. 1971. Ultrastructure of the 'germinal plasm' during maturation and early cleavage in *Rana pipiens*. *Dev. Biol.* **25**:568–580.

48. Yamaguchi, Y., Murakami, K., Furusawa, M., and Miwa, J. 1983. Germline-specific antigens indentified by monoclonal antibodies in the nematode *Caenorhabditis elegans*. *Dev. Growth & Differ.* **25**:121–131.

49. Yao, M.-C., and Golovsky, M. A. 1974. Comparison of the sequences of macro- and micronuclear DNA of *Tetrahymena pyriformis*. *Chromosoma (Berl.)* **48**:1–18.

50. Zipursky, S. L., Venkatish, T. R., and Benzer, S. 1985. From monoclonal antibody to gene for a neuron-specific glycoprotein in *Drosophila*. *Proc. Nat. Acad. Sci. U.S.A.* **82**:1855–1859.

Questions for Discussion with the Editor

1. *Do you agree that the lack of practical genetics in most vertebrate systems is going to continue to limit the rate of progress on the molecular biology of germ plasm? Could techniques for the production of transgenic animals provide a useful experimental approach?*

Undoubtedly yes. In *Drosophila*, a lot of maternal-effect mutations affecting pole-cell formation and function have been separated as a series of *grandchildless* mutants. It has generally been accepted that an association of the genetic and the developmental biologic approaches provides a basis for substantial progress in the molecular biology of germ-cell formation and germ-cell determination in *Drosophila*. In most vertebrate systems, however, germ-cell formation has been studied only by morphological and embryological approaches as described in this chapter. Neither practical genetics nor developmental genetics have been developed in the vertebrate systems, probably because it is very tedious to separate mutations showing defects of germ-cell development, especially in animals which have a relatively long generation time. Thus, even though some mutations may occur which affect germ-cell formation and function, it is difficult to separate these mutants from the wild type because regardless of whether the mutants are recessive homozygous or dominant heterozygous, the mutants theoretically never produce progeny. Temperature-sensitive mutations affecting germ-plasm function may overcome this difficulty.

Transgenic animals will probably be used mainly to study the spatial and temporal regulation of genes that have already been cloned, and to clone genes that have already been mapped. At present, however, genes (or cloned DNA) that affect germ plasm function or that cause germ-cell segregation (e.g., "master" genes for germ-cell development) are not available in vertebrate systems. The production of transgenic animals and the transformation of somatic cells to germ cells using the master genes for germ-cell development will be next-generation experiments. Maybe there is no single general tool for progress on the molecular biology of germ plasm, but rather combinations of techniques, such as (1) the production of germ plasm–free eggs or germ-cell-free embryos, whether of genetic or experimental origin, (2) the injection of antibodies against the germ plasm into eggs and embryos, (3) the introduction of well defined, cloned genes (e.g., for actin, vitellogenin, crystallin, etc.) into the germ plasm-containing cells or primordial germ cells, or (4) the introduction of inducible promoters into the differentiating germ cells by appropriate vectors.

2. *Do you suppose that even somatic cells of vertebrates undergo chromosome diminution, perhaps on an ultramicro scale? That would explain, for example, why transplantation of somatic-cell nuclei fail to promote development of enucleated eggs beyond the swimming tadpole stage.*

It has generally been accepted that genomes appear to be stable from one generation to the next at a *very coarse sequence level* and that all cells of the adult organism are "clones" of the fertilized egg. That is, they all contain exact copies of the genes donated to the zygote by the male and female parents. On the one hand, there is quite good evidence to suggest some specific somatic cells of vertebrates undergo DNA rearrangements at the level of terminal differentiation, for example, immunoglobulin genes in B lymphocytes and cell surface receptor genes in T lymphocytes in mammals. Furthermore, an insertion of transposable genetic elements, infection of retroviruses, or somatic recombination have been reported to cause diversification of genomes and modification of gene structure and function in several systems, including vertebrates. These phenomena possibly show a rather "plastic" structure of genomes, and they are consistent with the concept that even somatic cells of vertebrates undergo genomic changes on an ultramicro level. At present, however, I feel that speculation on the rearrangement, modification, and/or elimination of chromosomes in higher organisms seems a bit premature, because we know little about the comprehensive structure of genomes; problems such as what kinds of genes, how many genes, what sort of hierarchy of genes, what is the significance of noncoding sequences etc., await solving.

Irreversible changes to genomes during cell differentiation or possible chromosome elimination at any level might favor the inability of somatic-cell nuclei to support full development when transplanted into enucleated eggs. Chromosome elimination and the potency to promote development will raise different issues as exemplified by terminally differentiated lymphocyte nuclei (the genome is thought to have been rearranged to synthesize a specific antigen) which are reported to support full development after serial transplantation in *Xenopus*. Thus, nuclear transplantation can provide evidence on the stability of genomic structures at the very coarse level but not at the ultramicro level. Although there is no available experimental evidence I suppose that elimination of noncoding sequences and even of a part of evolutionarily "new" genes from amphibian genomes could not alter the potency for promoting development of eggs beyond the swimming tadpole stage if a series of more important genes (e.g., "master" genes for development) are left intact. Evolutionarily "old" and "sleeping" genes (which are believed to be accumulated in genomes but not to be expressed at all during normal development and differentiation) would possibly be expressed if several conditions, including tentative interactions of genes, cytoplasmic environments, etc., exist. Nuclear transplantation experiments using *Ascaris* or *Mayetiola* somatic-cell nuclei, if technically doable, will address, in part, this issue.

CHAPTER **12**

Role of Cytoplasmic Factors That Affect Cleavage in Leech Embryos

David A. Weisblat, Stephanie H. Astrow,
Shirley T. Bissen, Robert K. Ho,
and Katharine Liu

Introduction

UNDERSTANDING EMBRYONIC DEVELOPMENT requires an explanation of cellular, molecular, biophysical, and mechanical changes that occur as an egg gives rise to an adult organism. Development can be thought of as a quasi-historical process, in which changes at each level of organization can affect the course of events at the other levels. The critical difference between development and truly historical processes (such as the evolution of species or the life of Abraham Lincoln) is that, whereas historical events are essentially unique occurrences (despite common themes), embryonic development of a given species repeats itself, down to very fine details, generation after generation. Precise descriptions of the process can therefore be obtained by observing it many times using different techniques, and hypotheses as to the underlying mechanisms can be tested by perturbing the process and comparing the resultant embryo with unperturbed controls. By contrast, the details of history are never repeated and thus are not subject to this sort of experimental analysis.

We study the development of a leech, *Helobdella triserialis*, whose embryos offer relatively large, identifiable, and experimentally accessible cells that are amenable to most of the experimental techniques needed to obtain a truly satisfying explanation of this complex process. In this chapter, we describe studies on (1) the formation of distinct domains of cytoplasm, called teloplasm in the leech zygote, (2) the role that teloplasm plays in determining cleavage patterns in *Helobdella*, (3) the segregation of teloplasm into a particular cell line in the early cleavages, and (4) the analyses of cell cycle composition. The latter study provides further information on the differences that appear between various cell lines during cleavage. But first we briefly describe leech development so that the material presented subsequently can be viewed in the context of the long-range goal of understanding how the mature leech arises from the zygote.

Much of the work on leech development, beginning with that of C. O. Whitman in 1878 (16–18), has focused on cell lineage. As members of the phylum Annelida, which contains also the polychaetes and oligochaetes, leeches (hirudinids) are segmented worms that develop via a highly predictable sequence of cell divisions. Compared to arthropods, differences between segments are slight and the cellular organization is simple. Many cells, especially neurons, can be uniquely identified in any given segment, and homologues to identified cells occur from segment to segment and individual to individual. Although the complete lineages for the definitive neurons and other cells have yet to be established, it seems likely that they arise as particular products of a cell-lineage pattern that is invariant in normal development, as in the nematode *Caenorhabditis elegans* (10, 14, 19).

Morphologically defined segments arise as the products of interdigitating clones of small numbers of segmental founder cells called *blast cells* (15). Blast cells are generated in a longitudinal array from five bilateral pairs of embryonic stem cells called *teloblasts* (Fig. 1). In *Helobdella*, the teloblasts range in size from 50 to 100 μm in diameter. Among the teloblasts is one pair of mesodermal precursors, the *M* teloblasts, and four pairs of ectodermal precursors, the *N, O/P, O/P,* and *Q* teloblasts. The teloblasts arise by a stereotyped

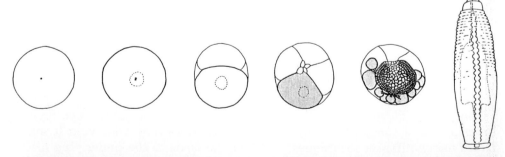

FIGURE 1. Summary of glossiphoniid leech development. Diagrammatic views of *Helobdella* embryos at progressively later stages from left to right, viewed from the animal pole (left-hand four drawings) or from the dorsal aspect (right-hand two drawings). Polar bodies are shown as small circles (left-hand two drawings); teloplasm (dashed circle) is segregated into cell CD of the 2-cell embryo and thence into macromere D', the five pairs of teloblasts and their bandlets of blast cells are stippled. The bandlets on each side have joined to form the germinal bands. The small cells in the 8-cell embryo are the first four micromeres. Drawings are roughly to scale; the diameter of the uncleaved egg is about 400 μm.

sequence of cleavage divisions from a cell called the *D' macromere* in the 8-cell embryo. (A number of much smaller cells called *micromeres* are also formed during this period by highly unequal divisions.) The D' macromere is identifiable as the largest cell in the embryo at this stage of development. It is also unique in that it contains pools of yolk-deficient cytoplasm at the animal and vegetal ends of the cell. These cytoplasmic domains, generally referred to as *pole plasm* in annelids but as *teloplasm* in leeches, arise by cytoplasmic rearrangements that occur prior to the first cleavage.

Distribution of Teloplasm Affects Cleavage Patterns

In normal development, only the D' macromere of the 8-cell embryo cleaves to generate teloblasts, and hence the segmental tissues of the leech. The D' macromere also inherits the bulk of the teloplasm during the first three cleavages. Is the correlation between cytoplasmic inheritance and future cleavage pattern merely a coincidental outcome of the fact that cell D' is also the largest cell in the embryo, or is it indicative of a determinative role for the yolk-deficient cytoplasm? It is important to realize the fundamental impossibility of testing any such hypothesis merely by observing normal development. Rather, it is necessary to perturb the process under investigation and observe the results.

Centrifugation has often been used to disrupt the cytoplasmic organization of cells. The rationale for this procedure is that cytoplasmic components of varying densities are normally held in nonequilibrium positions through relatively weak forces that can be overcome by increasing the effective gravitational force. The force used must be strong enough to reposition cytoplasmic components, but gentle enough to allow subsequent cleavages to occur. For *Helobdella*, it was found that $350g$ applied roughly parallel to the animal-vegetal axis of 2-cell embryos shortly before the second cleavage was sufficient to partially stratify the cytoplasm without blocking further cell divisions (2). Yolk-deficient cytoplasm, most of which is normally contained within the animal and vegetal pools of teloplasm in cell CD at this stage, accumulated in a layer at the centripetal ends of the cells (Fig. 2).

Thus, when the second cleavage occurred, yolk-deficient cytoplasm was often partitioned into both cells C and D more or less equally, whereas normally it is inherited almost entirely by cell D. The basic result of these experiments was that if the teloplasm was evenly distributed between cells C and D, both cleaved further to generate embryos with supernumerary teloblasts and bandlets of blast cells. Moreover, the location of the second cleavage plane relative to the teloplasm and to the cell as a whole varied in these experiments, presumably due to unavoidable variability in the orientation or exact developmental state of the embryos during centrifugation. Thus, the relative size of cells C and D also varied, and varied independently from the relative partitioning of the teloplasm. This made it possible to rule out the possibility that size, rather than inheritance of teloplasm, was the critical factor in determining the future cleavage pattern of cells C and D, because it was ob-

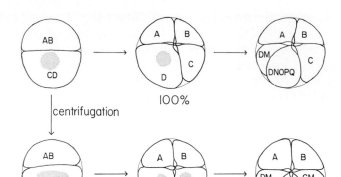

FIGURE 2. Centrifugation redistributes yolk-deficient cytoplasm and induces abnormal cleavages in *Helobdella*. Embryos are drawn as viewed from the animal pole. During centrifugation, the embryos orient so that the gravitational vector is roughly parallel to the animal-vegetal axis and directed into the plane of the figure. Stippling indicates the distribution of yolk-deficient teloplasm in normal and centrifuged 2- and 8-cell embryos.

served that cell C would cleave to form teloblasts when it received appreciable amounts of teloplasm, even if it was still smaller than cell D. The fates of the extra teloblasts and blast cells generated in centrifuged embryos cannot be studied because such embryos fail to gastrulate normally and die before segments form.

From these experiments, it was concluded that teloplasm inherited by blastomeres during early cleavage does play a role in determining the subsequent cleavage pattern of those cells. Of course this finding raises new questions as well as answering the original one. For example, does the teloplasm act qualitatively, so that any amount above some threshold unleashes a fixed program of cell divisions, or quantitatively, so that the number of cleavage divisions in a particular cell line is proportional to the amount of teloplasm in the precursor cell? The fact that redistributing teloplasm to both cell C and cell D results in the formation of supernumerary teloblasts suggests that the former alternative may be closer to the truth, but since the number of teloblasts arising in centrifuged embryos is variable, this conclusion is only tentative. A clearer answer might be gained by transferring cytoplasm directly from one cell to another, but to date this has not been successfully accomplished in *Helobdella* embryos; the cytoplasm of this species is too viscous to be drawn up into any micropipet that is fine enough to successfully reimpale blastomeres.

Other questions that follow from this work are: How does teloplasm form? How does it come to be segregated preferentially into one cell line? What is there in teloplasm that influences cell division patterns and by what mechanism(s) do such factors operate? Progress in addressing these questions is outlined in subsequent sections.

Animal and Vegetal Teloplasms Intermingle during Normal Development

Although the centrifuged embryos in which both cells C and D make teloblasts fail to complete development, it seemed clear nonetheless that distinct ectodermal (CNOPQ and DNOPQ) and mesodermal (CM and DM) precursor blastomeres were formed, as judged by their subsequent cleavages and by the mitotic patterns of their progeny (2). It was previously assumed that the animal and vegetal teloplasms contain distinct ectodermal and mesodermal determinants, respectively, but this belief was called into question after these centrifugation experiments, in which animal and vegetal teloplasms seem to mix at the animal end of the cells, and yet distinct mesodermal and ectodermal precursors still arise.

Further circumstantial evidence against the notion that animal and vegetal teloplasms contain distinct cytoplasmic determinants comes from the discovery that the two domains of teloplasm are not kept separate even during *normal* development, but rather they intermingle in the D' macromere of the 8-cell embryo shortly before it cleaves to form the mesodermal (DM) and ectodermal (DNOPQ) precursors (9). Teloplasm movements can be inferred by examining sections prepared from synchronously developing embryos fixed at sequential time points, or can be seen in living embryos using fluorescence microscopy to follow the movements of mitochondria labeled with rhodamine 123. With both techniques it was concluded that soon after the D blastomere cleaves to form macromere D' and micromere d', most, though not necessarily all, of the vegetal teloplasm is translocated toward the animal end of the cell. So by the time cell D' cleaves to DM and DNOPQ, there appears to be but a single pool of teloplasm located near the micromeres at the animal end of the cell. Figure 3 presents cutaway views of embryos showing the distribution of teloplasm (stippling) in cells D and D' at progressively later times between the 4-cell stage and the cleavage of macromere D' to cells DM and DNOPQ. In each drawing cell D is flanked by cell A or A' at left and C or C' at right. Each of these cells is designated with the prime (') after it has generated a micromere en route from the 4-cell to the 8-cell stage. In this figure, micromeres d' and c' are indicated in the three right-hand panels. As macromere D' cleaves, the merged teloplasms are once again divided by the oblique orientation of the furrow, so DM and DNOPQ receive some teloplasm.

It should be noted that, although these results argue against the localization of distinct mesodermal and ectodermal determinants in vegetal and ani-

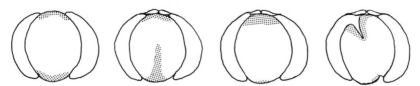

FIGURE 3. Translocation of vegetal teloplasm after the third cleavage.

mal teloplasm, there could still be important differences between animal and vegetal teloplasm. One might imagine, for example, that the two teloplasms, like the A and B components of epoxy cement, must be kept separate until shortly before "use," and the mixing we observe in the 8-cell embryo brings them together to initiate a new type of cleavage pattern, by which the DM and DNOPQ teloblast precursors generate teloblasts.

Teloplasm Formation in *Helobdella* Is a Microtubule-Dependent Process

Yolk-deficient domains of cytoplasm also appear in the precleavage embryos of other leech species and in more distantly related annelids. This process has been best studied in *Tubifex hattai*, an oligochaete worm. Oligochaetes, including the familiar earthworms, have many features in common with leeches, and the two classes are widely assumed to have arisen from a common ancestor. Indeed, some scholars classify the two groups together in a single class or subphylum, Clitellata. Therefore, it seemed likely that the mechanisms underlying the formation of the domains of yolk-deficient cytoplasm might be the same in *Helobdella* and *Tubifex*, and we refer to the yolk-deficient cytoplasm as teloplasm for both organisms, even though this term is normally applied only to leeches.

Both *Helobdella* and *Tubifex* eggs are fertilized internally but remain in meiotic arrest until they are laid. Teloplasm formation becomes evident after the formation of the second polar body. Figure 4 represents the process of teloplasm formation in *Helobdella* via drawings of meridional sections through embryos at progressively later times during the first cell cycle. This

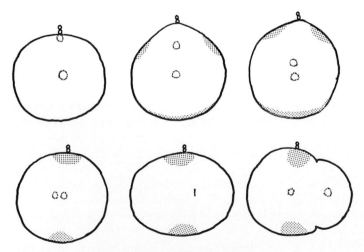

FIGURE 4. Teloplasm formation in *Helobdella*.

process is very similar to that described previously for *Tubifex* by Shimuzu (11–13). The animal pole is at the top in each panel; the polar bodies are represented by small circles at the top of each embryo. Small zones of clear cytoplasm (indicated by dashed circles) contain the pronuclei. The onset of teloplasm accumulation is accompanied by a slight latitudinal contraction about two-thirds of the way from the equator to the animal pole, giving the animal hemisphere a slightly domed shape. Teloplasm is evident as two ringed domains of clear cytoplasm (stippling) only parts of which are visible in these meridional profiles. Compact domains of teloplasm are formed as the rings of teloplasm move poleward. During this process, the female pronucleus moves from the animal pole toward the male pronucleus at the center of the zygote; additional yolk-deficient cytoplasm accumulates at the center of the egg around the nucleus. The embryo then elongates along the future dorsal-ventral axis; by metaphase, the chromatin (black bar) is located eccentrically with respect to the animal-vegetal axis. The first cleavage furrow is unequal, segregating most of both pools of teloplasm into the larger daughter cell CD (left).

Regarding the mechanism of teloplasm formation, two lines of evidence suggest that the rings of teloplasm arise from cytoplasmic rearrangements rather than from metabolic conversion of yolky to nonyolky cytoplasm. First, as the latitudinal rings of teloplasm form in *Tubifex*, longitudinally oriented strands of teloplasm extend perpendicular to the ring in both directions, as if cytoplasm were streaming into the location of the ring, and coincident longitudinal deformations appear in the surface of the embryo, somewhat like the lines on a pumpkin. Second, rhodamine 123 has been used to visualize mitochondria moving into the rings of teloplasm in the living embryo in both *Tubifex* and *Helobdella*.

Using transmission electron microscopy of zygotes and isolated cortices and fluorescence microscopy of *Tubifex* embryos stained with rhodamine-phalloidin conjugates, Shimizu (11, 12) observed cortical actin microfilaments, which suggested that a cytoskeletal actin network might be involved in generating the forces required to form the rings of teloplasm and subsequently, translocating them to the poles. Evidence for this hypothesis was also obtained from pharmacological studies. Bath application of cytochalasin B, which binds to actin monomers and thereby blocks their polymerization, prevented teloplasm formation. In contrast, colchicine, which blocks the polymerization of microtubules, the other major component of the cytoskeleton, did not prevent teloplasm formation. Colchicine blocked polar body formation, which is known to depend on microtubules. Thus it appears that teloplasm formation and movement in *Tubifex* is mediated by microfilament networks in the cytoskeleton.

As for the leech, Fernandez and coworkers (7) carried out similar studies on zygotes of *Theromyzon rude*, a species of leech in the same family (Glossiphoniidae) as *Helobdella*, and found similar longitudinal deformations of the embryo during teloplasm formation and similar evidence (but as of yet without pharmacological studies) for the presence of actin microfilaments. Thus, when we started to examine teloplasm formation in *Helobdella*, we ex-

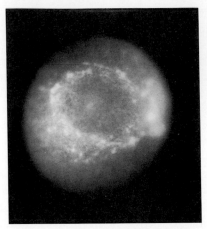

FIGURE 5. Fluorescence photomicrograph of a *Helobdella* zygote that had been injected with rhodamine 123. Viewed from the animal pole, midway through teloplasm formation, mitochondria are clustered into a ring at the surface of the embryo and converge toward the animal pole (cf. Fig. 4, top right panel).

pected simply to confirm and extend the *Tubifex* story for another species (Figs. 4 and 5); it was surprising to find that cytochalasins failed to block teloplasm formation in *Helobdella* (2a). Neither bath-applied [using 0.1% dimethyl sulfoxide (DMSO) to increase solubility] nor microinjected cytochalasin blocked teloplasm formation, although these treatments did block polar body formation and first cleavage, suggesting that access of the drug to the cytoplasm was not a problem. By contrast, colchicine, the microtubule polymerization inhibitor, did partially prevent teloplasm formation, and nocodazole, another inhibitor of microtubule function, clearly blocked teloplasm formation at concentrations as low as 0.05 μ*M*. Further evidence for the specificity of the effect was obtained with stereoisomeric forms of a third compound, tubulazole. *cis*-Tubulazole, which binds to tubulin monomers and blocks polymerization, also blocked teloplasm formation, but *trans*-tubulazole, which has no specificity for tubulin, was without effect.

Thus, the results obtained with pharmacological agents on teloplasm formation in *Helobdella* are apparently in conflict with those obtained for what is almost certainly a homologous embryological process in another clitellate annelid, *Tubifex*. The paradox deepened when we repeated the *Tubifex* experiments and found the same sensitivity to cytochalasin and the expected lack of sensitivity to nocodazole and tubulazole. These results suggest that perhaps both classes of cytoskeletal elements can be used to achieve teloplasm formation in annelids but that their relative contributions differ between *Tubifex* and *Helobdella*. A less attractive hypothesis is that the process called teloplasm formation in oligochaetes and leeches is not really a homologous process at all.

Unequal Cleavages Segregate Teloplasm into the D Cell Line

Teloplasm forms at the animal and vegetal poles of the zygote, which roughly correspond to the future anterior and posterior ends of the organism, respectively. All else being equal, the teloplasm would be bisected by a truly meridional first cleavage plane and, on the basis of the centrifugation experiments, we would expect both cells (AB and CD) of the 2-cell embryo to generate teloblasts. Thus, it is of considerable significance that the first cleavage plane lies parallel to, but does not include, the animal-vegetal axis. Consequently, the first cleavage is unequal, with the larger cell (CD) inheriting most of the teloplasm. Since the teloblasts arise from cell CD in a stereotyped manner and since the definitive segmental tissues arise from the teloblasts in an equally predictable manner, it can be said that the unequal first cleavage, in destroying the radial symmetry of the late zygote, also serves to define the dorsoventral axis of the embryo. How does this unequal first cleavage come about? Is the dorsoventral axis present but cryptic in the zygote, or is it determined at random by the orientation of the first cleavage?

The orientation of the cleavage furrow is determined by the orientation of the mitotic spindle at metaphase. Accordingly, an analysis of unequal divisions in *Helobdella* embryos showed that, by metaphase, the mitotic apparatus is eccentrically located with respect to the animal-vegetal axis (S. Settle, personal communication). Thus, the question of how the unequal first division occurs reduces to how the nucleus or mitotic apparatus moves during prophase or early metaphase and whether this motion is along a preexisting but cryptic dorsoventral axis. Again, either microtubule or microfilament networks in the cytoskeleton could be responsible for this motion, although work in many other systems suggests that microtubules are the most likely candidate. In principle, microtubule-based motion could arise either by the generation of pulling forces (traction) arising from tubule depolymerization on one side of the embryo or from pushing forces (compression) arising from tubule polymerization on the other side of the embryo (4).

Figure 6 shows the results of experiments designed to gain information for the eventual resolution of this issue. The left side of Fig. 6a shows embryos that had just extruded the second polar body and which had been placed in wells in an agarose disk about 0.5 cm (middle circle) or 1.0 cm (bottom circle) from a source (S) of nocodazole dissolved in DMSO. Animal-vegetal axes of the embryos are oriented roughly normal to the plane of diffusion. After several hours (shown on the right side of Fig. 6a), when control embryos have cleaved, the embryos closer to the source of nocodazole are still at the 1-cell stage, indicating that nocodazole has diffused from the source and has blocked cleavage. Embryos further from the source do cleave and the angle (theta) between the AB/CD axis and the nocodazole source can be measured. Because the nocodazole concentration was decreasing exponentially with distance from the source and because the diffusing wave affected the proximal side of the zygote first, it was assumed that microtubule function would be disrupted preferen-

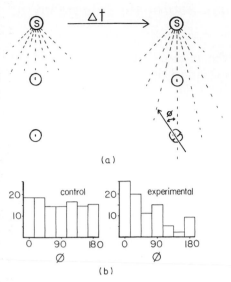

FIGURE 6. Orientation of the first cleavage plane is influenced by exposure to a nocodazole gradient.

tially at the proximal side of the embryo. If the eccentric metaphase results from pushing forces generated by microtubule polymerization, then preventing polymerization at the end closest to the source should have resulted in (1) a greater push from the other side, (2) a movement of the nucleus toward the source of the gradient, and (3) the formation of the CD cell at the distal end of the embryo. In control embryos (left side, Fig. 6b), which were exposed to a gradient of only dilute DMSO and marker dye, the orientation of the first cleavage furrow was random, as expected, but when nocodazole was present (right side, Fig. 6b), the cleavages were oriented such that cell CD tended to form *proximal* to the source. From this it seems more likely that the nucleus or mitotic apparatus is being pulled toward the cortex at one side of the embryo, rather than being pushed away from the other side of the embryo. These observations are in accord with Dan's studies (5, 6) on the unequal cleavage in echinoderms.

With this experiment, we can also address the question of whether the dorsoventral axis is preestablished, but cryptic, in *Helobdella* or whether it is established at random by the segregation of the teloplasm at the unequal first cleavage. If there is a preset dorsoventral axis, then forcing the first cleavage furrow to assume a particular orientation by using the nocodazole gradient should disrupt normal development, by displacing teloplasm relative to the dorsal pole. But when embryos were removed from the nocodazole gradient after first cleavage, development continued normally. Thus, although we cannot exclude the possibility that a preferred dorsoventral axis is present in the early zygote, we *can* conclude that any such preference can easily be overridden, and moreover, that it is the segregation of teloplasm, not the "preferred dorsoventral axis," that is the final arbiter of dorsoventral polarity.

Polyadenylated RNAs Are Localized in Teloplasm

Once we know that teloplasm plays a role in determining the future cleavage patterns of the early blastomeres, the question arises as to what component or components of teloplasm are responsible for this effect. Initially, C and D blastomeres, from both normal and centrifuged embryos, were examined by two-dimensional gel electrophoresis in the hope of finding proteins that were unique to one cell or the other (2). No unique proteins were found, although this negative result must not be given much weight, especially since this technique is limited to the detection of relatively abundant proteins. What *was* learned in these experiments is that cell D contains certain proteins in greater relative abundance than does cell C, that these proteins are associated with the teloplasm, and that in embryos centrifuged at the 2-cell stage to redistribute teloplasm, these proteins are distributed more uniformly between cells C and D at the 4-cell stage.

Another way of approaching the issue of what factors in teloplasm influence cell cleavage was suggested by the observation that acridine orange, a fluorescent dye that binds to nucleic acids, stains teloplasm. This observation has been confirmed through in situ hybridization experiments, in which tritiated polyuridylic acid, [³H]polyU, was used as the probe, to localize polyadenylated mRNAs in the early embryo (9). Figure 7 shows a *Helobdella* embryo that had been fixed at the 2-cell stage, incubated with [³H]polyU, and then embedded in glycol methacrylate, sectioned, and processed for autoradiography. Cell AB is at left, cell CD is at right, and the animal pole is toward the top. Figure 7a shows a bright field view of yolk platelets, visible as dark circles after toluidine blue staining, which are absent from cytoplasm

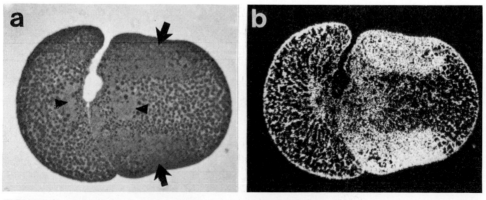

FIGURE 7. Teloplasm is enriched for polyadenylated RNAs.

around the nuclei (arrowheads) and in the animal and vegetal domains of teloplasm (arrows). Figure 7*b* shows a dark field view of the same section. Silver grains (which appear as white dots), indicative of the distribution of the bound [^{3}H]polyU, are concentrated over the two domains of teloplasm. Control embryos incubated with tritiated polyadenylic acid exhibit no binding above background levels. From these experiments, we concluded that polyadenylated RNAs, like mitochondria and granular cytoplasm, accumulate in the teloplasm as it forms.

Polyadenylic Acid Induces the Formation of Supernumerary Teloblasts

Of course, a priori, there is no more reason to suspect that mRNAs are more important than mitochondria or other cytoplasmic constituents in determining future cleavage patterns for a cell. But RNAs have been shown to act as cytoplasmic determinants in certain cases in other organisms, most notably *Drosophila* (1), and another set of experiments with *Helobdella* offers circumstantial evidence that some aspect of RNA metabolism or localization may affect the early cleavage patterns. These experiments began with the discovery that microinjecting polyadenylic acid (polyA) into blastomeres or newly cleaved teloblasts of the *Helobdella* embryo results in additional equal or almost equal cell divisions (8). In Fig. 8, the drawing at left represents a *Helobdella* embryo just after the ectodermal precursor (DNOPQ‴) has cleaved to form left (stippled) and right (striped) ectodermal precursors (left and right NOPQ). Normally, cell NOPQ cleaves further to generate four teloblasts, N, O/P, O/P, and Q, represented in the drawing on the right as the striped contours, plus some micromeres. But if an NOPQ cell is microinjected with polyA, it generates more and slightly smaller teloblasts, represented in the drawing at right by the stippled contours. As a result, supernumerary teloblasts can be produced which, presumably after the injected polyA has been degraded, generate blast cells and definitive progeny of a particular lineage.

This effect is quite specific for polyA. Extra cleavages can be induced by, for example, overinjection with regular lineage tracers, but only rarely, and the resultant cells usually die. Microinjection of polyG or polyC has quite the opposite effect; the blastomere fails to divide further and eventually dies.

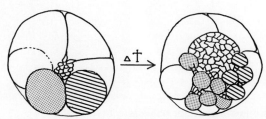

FIGURE 8. Supernumerary teloblasts arise from blastomeres microinjected with polyA.

Yeast RNA, tRNA, and plasmid DNA have no effect on cleavage pattern. Polydeoxyadenylic acid (polydA) does induce extra equal divisions, however, with the important difference that no blast cells are produced. The progeny of the injected cell continued dividing equally for as long as we were able to follow them. Presumably the persistence of the effect of polydA relative to that of polyA is a result of the resistance of polydA to degradation.

Whether these supernumerary teloblasts arise through *new divisions* that would not normally have occurred or through a *change in the symmetry of divisions* that would have normally generated other cells, for example micromeres, remains to be determined. Moreover, it is important to realize that the effect of injecting polyA might not have anything to do with its structural relationship to polyadenylated mRNA. [The various experiments in which other injected polymers do *not* show duplication of teloblasts argues against this however, as does the observation that microinjecting early embryos with cordycepin (3'-deoxyadenosine, an inhibitor of mRNA polyadenylation) results in the generation of supernumerary large blastomeres.] Such caveats notwithstanding, however, this discovery, in addition to its utility as a technique for altering the cellular composition of the embryo, is consistent with the notion that some aspect of RNA localization or metabolism is important in controlling the pattern of early divisions and the transition between cleavage divisions in early embryogenesis and stem-cell divisions (that generate blast cells) in the middle period of development. For example, one of several possibilities is that the injected polyA increases the lifetime of maternal RNAs that maintain the D cell line in a "cleavage mode" of cell divisions, and that when these are exhausted, the transition to stem-cell divisions takes place. PolyA might affect the metabolism or translation of endogenous polyadenylated RNAs by serving as an alternate substrate for RNases, by occupying binding sites in the cytoplasm or by affecting the synthesis of new RNAs.

Cell Cycle Composition During Cleavage Varies in a Cell-Specific Manner

A major question, and one that is perhaps the most difficult to approach, is how the teloplasm exerts its influence on the fates of cells. The polyA injection experiments are one line of investigation that may prove relevant to this question. Another important approach is to refine our description of this period of development, with the goal of observing cellular and molecular phenomena in detail sufficient to permit the formulation and testing of reasonable hypotheses regarding causal relationships among them. Accordingly, one approach that we have followed is to study the composition of the cell cycles in the early embryo. In the lineage tree used to represent cleavage divisions, time is measured along the vertical axis and each vertical line segment represents the life of an individual cell, from the time it is created through the cytokinesis of a parent cell until the time that it divides or dies later in development. Further analysis has enabled us to break down the lifespan of each cell into its various phases, namely, mitosis (M), DNA synthesis (S), and the gap phases that usually follow mitosis (G1) and synthesis (G2) (Fig. 9; Ref. 3).

We entered into the area of the analysis of the cell cycle in early *Helobdella* embryo laboring under two common assumptions. The first was that during early divisions, cells cycle directly between the S phase and the M phase without going through G1 or G2, presumably because most of the proteins needed for mitosis are available as part of the maternal inheritance. The second assumption was that changes in duration of the cell cycle would reflect changes in the length of the G1 phase, because once a cell has, during G1, supplemented preexisting stores with the rest of the machinery needed for a new round of replication and division, it would proceed through the rest of the cell cycle (S, G2, and M) at a rate characteristic of the species.

In fact, a detailed analysis of the cell cycles of identified blastomeres in *Helobdella* has shown that this embryo exhibits the same degree of cell-specific differences in cell cycle composition as in cell cycle duration and mitotic symmetry; moreover, neither of the starting assumptions holds up to

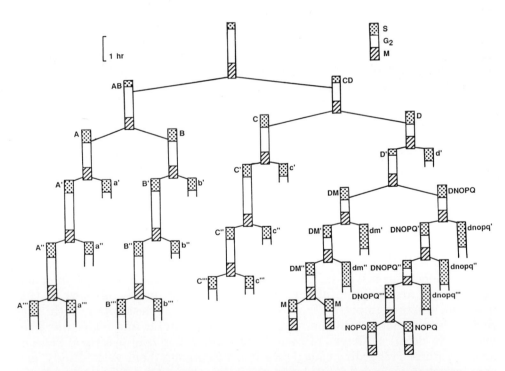

FIGURE 9. Cell cycles of the early cells in the *Helobdella* embryo. Each vertical line segment represents the total life cycle of a cell, which is composed of phases of DNA synthesis, (S, stippled), gap (G2, white), and mitosis (M, striped). The sloping horizontal lines indicate cytokineses. Macromeres and blastomeres are designated by capital letters and micromeres by lowercase letters. Open-ended cell cycles (e.g., A''', dnopq', etc.) indicate that the rest of the cell cycle is not known.

scrutiny. First, the early blastomeres do not cycle directly between M and S phases of the cell cycle but rather exhibit G2 phases of significant duration (up to 1.5 h or 60 percent of the total cell cycle) (3). Slightly later though, the teloblasts exhibit very short G phases (<15 percent of the cell cycle) as they undergo iterated, highly unequal divisions, generating blast cells at the rate of about one per hour. Second, it was found that variation in the duration of the cell cycle arises not so much from differences in the length of the G1 phase as from differences in the length of G2. Although this is evident among the early blastomeres, it is even more pronounced among the various classes of ectodermal blast cells, whose cell cycle times range from 21 to 33 h. For all classes of blast cells, the G1 phase was not discernible, the S phase lasted for 4 to 5 h, and the entire difference in the cell cycle duration was accounted for by differences in the length of the G2 phase. From these results, it appears that there is considerably more complexity in the regulation of cell cycles in vivo than is apparent from the study of cells in vitro.

Regarding the possible influence of cytoplasm on the cell cycle, it is of interest to note that cell cycles in the D cell line are considerably shorter than those in the A, B, and C cell lines. This could simply reflect the fact that the teloplasm inherited by these cells is enriched in mitochondria and ribosomes, and therefore, they should be able to manufacture all the machinery needed for cell division at a faster rate than the cells that are impoverished for yolk-deficient cytoplasm. It is not immediately obvious how this difference alone would account for all the differences in the cleavage pattern between the D cell line and the others.

Summary: Directions for Future Research

The work summarized above encompasses various experimental techniques and observations that pertain to phenomena observed at different stages of development. They are presented here together because a central problem for developmental biologists is to understand how such more or less distinct phenomena are *connected* to bring about the development of a particular type of organism. Although neither the connections nor the phenomena themselves are well understood, it is important to consider the issues, focusing on what is *not* known, so that future studies may be well directed.

Teloplasm formation entails cytoplasmic rearrangements that might be thought of as occurring in two stages. In the first, mitochondria, ribosomes, granular cytoplasm, and polyadenylated RNAs move relative to yolk platelets to accumulate in two rings just below the surface of the embryo in the animal and vegetal hemispheres. These movements are temporally correlated with the completion of the meiotic divisions of the female pronucleus and oriented so that the planes of the rings are perpendicular to the animal-vegetal axis, but the actual mechanism(s) by which teloplasm formation is triggered and oriented is not known. Another question is whether certain components of the cytoplasm are passively translated as a result of the active movements of other components. In the second stage of teloplasm formation, the rings of teloplasm converge at their respective poles. At the moment the only information on the mechanism(s) of these two cytoplasmic rearrangements is that they appear to require microtubules in *Helobdella*. Whether the second movement is simply a continuation of the first, and the extent to which both these movements operate according to basic mechanical constraints of the cytoskeleton without any prepattern are important questions for future investigation.

The first cleavage is unequal, with the result that teloplasm is segregated to the larger cell, CD, rather than being divided equally. It is clear that this unequal division has important consequences for future development (by effectively determining the dorsal axis), but by contrast, the position of the cleavage plane does not seem important. Whatever the cellular cue that causes the nucleus to move in a particular direction, it can be overridden and a normal dorsal-ventral axis forms in accord with the inheritance of teloplasm. The most attractive interpretation of the observations for us is that (1) the nucleus and/or the mitotic apparatus is constrained to move off center by metaphase, but the *direction* of the movement is not predetermined; (2) inheritance of teloplasm alone is sufficient to determine the future dorsal pole of the embryo. Perhaps the same mechanism operates in cell CD as in the zygote to force the cell into an unequal division. Perhaps it is the teloplasm itself that is partly responsible. In any case, yet another fascinating problem that remains to be explained is what causes the nuclei of all four cells to move toward the animal pole for the third cleavage and why the D' nucleus drops back to the center afterward for an equal fourth cleavage, while those of A', B', and C' stay near the animal pole and generate more micromeres.

The nature of the factor or factors in teloplasm that affect the cleavage pattern of cells that inherit it is unknown. It is suggestive that RNAs segregate with teloplasm, and the results obtained with polyA injections provide circumstantial evidence in favor of the notion that RNA localization or metabolism is somehow affecting cleavage divisions. But it must be remembered that ribosomes, mitochondria, cytoskeletal elements, and soluble proteins are also associated with teloplasm. Indeed, far from being a "specialized" cytoplasm, it might be that teloplasm is closer to being "normal" cytoplasm and that it is the remainder of the cytoplasm that is specialized by virtue of its yolk content. Thus, the question becomes whether teloplasm acts instructively, by virtue of one or more particular factors, or permissively, by virtue of its being enriched for all the normal cytoplasmic constituents.

Since so little is known about what factor(s) in teloplasm influence cleavage patterns, it is no surprise that the mechanism of action is also unknown. A strategy for pursuing this problem is to work backward from the observed differences in cleavage pattern with the aim of finding differences in more fundamental cellular processes that precede division. The analysis of the cell cycle composition falls under this strategy; another important undertaking will be to carefully plot the movements of the nuclei and mitotic apparatuses in the various cell lines. From such analyses, it may be possible to formulate or eliminate hypotheses regarding the role of teloplasm in determining cleavage patterns. For example, if speed and direction of nuclear movements in two cells are the same, but one cell proceeds through its cycle more rapidly, then differences in the symmetry of cell division would simply be a consequence of the fact that one cell enters its mitotic cycle earlier than the other. Note that this strategy relies heavily on what are often mistakenly disdained in these times as "descriptive" studies. But in fact such studies are a cornerstone of developmental biology. For although the development of each *individual* species is merely *quasi*-historical, i.e., reproducible and subject to experimental analysis, the rare accidents that *change* developmental processes, ultimately generating *new* species, are part of a truly historical process, evolution. Thus, the success with which *generalizations* about developmental mechanisms can be made seems limited, and the ultimate answer to the question of how an organism develops may be little less cumbersome than a high-resolution description of the events leading from the egg to the adult.

General References

STENT, G. S., and WEISBLAT, D. A.: 1982. The development of a simple nervous system. *Sci. Am.* **246**(1):136–146.

WEISBLAT, D. A., ASTROW, S. H., BISSEN, S. T., HO, R. K., HOLTON, B., and SETTLE, S. A.: 1987. Early events associated with determination of cell fate in leech embryos. In: *Genetic Regulation of Development*. (W. F. Loomis, ed.), pp. 265–285, Liss, New York.

References

1. Anderson, K. V., and Nusslein-Volhard, C. 1984. Information for the dorsal-ventral pattern of the *Drosophila* embryo is stored as maternal mRNA. *Nature (Lond.)* **311**:223–227.
2. Astrow, S. H., Holton, B., and Weisblat, D. A. 1987. Centrifugation redistributes factors determining cleavage patterns in leech embryos. *Dev. Biol.* **120**:270–283.
2a. Astrow, S. H., Holton, B., and Weisblat, D. A. 1989. Teloplasm formation in a leech *Helobdella triserialis*, is a microtuble-dependent process. *Dev. Biol.* (in press).
3. Bissen, S. T., and Weisblat, D. A. 1989. The durations and composition of cell cycles in embryos of the leech, *Helobdella triserialis, Development* **105**:105–118.
4. Bjerknes, M. 1986. Physical theory of the orientation of astral mitotic spindles. *Science* **234**:1413–1416.
5. Dan, K. 1979. Studies on unequal cleavage in sea urchin. I. Migration of the nuclei to the vegetal pole. *Dev. Growth & Differen.* **21**:527–535.
6. Dan, K. 1978. Unequal division: It's cause and significance. In: *Cell Reproduction: In Honor of Daniel Mazia* (E. R. Dickson, D. M. Prescott, and C. F. Fox, eds.), pp. 557–562, Academic Press, New York.
7. Fernandez, J., Olea, N., and Matte, C. 1987. Structure and development of the egg of the glossiphoniid leech *Theromyzon rude*: characterization of developmental stages and structure of the early uncleaved egg. *Development* **100**:211–225.

8. Ho, R. K., and Weisblat, D. A. 1987. Replication of cell lineages by intracellular injection of polyadenylic acid (polyA) into blastomeres of leech embryos. In: *Molecular Biology of Invertebrate Development*. (J. D. O'Connor, ed.), pp. 117–131, Liss, New York.

9. Holton, B., Astrow, S. H., and Weisblat, D. A. 1989. Animal and vegetal teloplasms in the early embryo of the leech, *Helobdella triserialis*. *Dev. Biol*. **131**: 182–188.

10. Kramer, A. P., and Weisblat, D. A. 1985. Developmental neural kinship groups in the leech. *J. Neurosci*. **5**:388–407.

11. Shimizu, T. 1982. Ooplasmic segregation in the *Tubifex* egg. Mode of pole plasm accumulation and possible involvement of microfilaments. *Roux's Arch. Dev. Biol*. **191**:246–256.

12. Shimizu, T. 1984. Dynamics of the actin microfilament system in *Tubifex* during ooplasmic segregation. *Dev. Biol*. **106**:414–426.

13. Shimizu, T. 1986. Bipolar segregation of mitochondria, actin network and surface in the *Tubifex* embryo: Role of cortical polarity. *Dev. Biol*. **116**:241–251.

14. Weisblat, D. A., Kim, S. Y., and Stent, G. S. 1984. Embryonic origins of cells in the leech *Helobdella triserialis*. *Dev. Biol*. **104**:65–85.

15. Weisblat, D. A., and Shankland, M. 1985. Cell lineage and segmentation in the leech. *Philos. Trans. R. Soc. Lond. Biol. Sci*. **312**:39–56.

16. Whitman, C. O. 1878. The embryology of *Clepsine*. *Q. J. Microsc. Sci*. **18**:215–315.

17. Whitman, C. O. 1887. A contribution to the history of the germ layers in *Clepsine*. *J. Morphol*. **1**:105–182.

18. Whitman, C. O. 1892. The metamerism of *Clepsine*. Festschrift zum 70, Geburtstage R. Leuckarts, pp. 385–395.

19. Zackson, S. L. 1984. Cell lineage, cell-cell interaction, and segment formation in the ectoderm of a glossiphoniid leech embryo. *Dev. Biol*. **104**:143–160.

Questions for Discussion with the Editor

1. *When, following fertilization, is zygote nuclear gene expression initiated—early (first few cleavages) or late (blastula-gastrula)? How far into embryogenesis do maternal controls dominate before zygote nuclear activity takes over?*

The answers to these important questions are just beginning to emerge from studies in which RNA synthesis is assayed by incorporation of tritiated UTP or an immunohistochemically detectable analog, BrUTP (S. T. Bissen and D. A. Weisblat, in preparation). At this point one general conclusion is that it is important to look at such issues at the level of individual cells and not exclusively in homogenized embryos, because, at least in the leech, different cell types may initiate transcription at different times or at different levels. Somewhat analogous to the midblastula transition in amphibian embryos, smaller cells seem triggered to begin transcription; thus, it seems that blast cells and micromeres initiate RNA synthesis soon after they arise. We hope that the effects of microinjecting polyadenylic acid into early blastomeres may be explained by effects on localization, synthesis, or stability of mRNAs, so that further studies of this phenomenon may give new insights into the developmental transition between maternal and zygotic controls.

2. *Readers might be interested in learning what the prospects are for expanding the research potential of this marvelous organism by building a developmental genetics system out of it. Please comment.*

Of the various kinds of leeches that are now used for scientific purposes, some species of *Helobdella* may be the best suited for developmental genetics. Annelids are hermaphroditic, but in most species, cross-fertilization is required for reproduction. But *H. triserialis* is capable of both cross- and self-fertilization. Its generation time is about 9 weeks, and one individual can produce as many as several hundred progeny over the course of about 18 weeks. Because of its small size (1 to 2 cm for adults), large numbers of individuals can be maintained in a relatively small space. The genome size is equal to that of *Drosophila*, and it has more than a dozen pairs of chromosomes (13a). These properties make *Helobdella* extremely attractive as a subject for genetic analysis compared with the simplest vertebrate systems, such as mouse or fish. However, compared with the nematode *Caenorhabditis elegans* or the insect *Drosophila melanogaster*, the slow generation time, large chromosome number, and the large amount of time and resources that would be necessary to establish the genetics of *any* "new" organism lead us to question the strategic wisdom of attempting to do classical genetics on leech. However, there are many possibilities in *Helobdella* for doing what is sometimes called "reverse" genetics, taking advantage of the many advances that have been made in molecular biology in the past decade. For example, it should be possible to clone certain genes of developmental interest on the basis of sequence similarity to genes characterized in other organisms and to study their expression and function in the leech embryo. We have begun some such experiments (13a).

CHAPTER 13

Control of Mitosis and Ooplasmic Movements in Insect Eggs

Jitse M. van der Meer

Introduction

THE CONTROL OF MITOSIS and of cytoplasmic movements is not usually studied in early insect embryos. The mitotic waves that traverse early insect eggs used to be studied because they were believed to establish the segmentation of the body pattern. This turned out to be coincidental at best (34). In this chapter I suggest that early insect embryos are interesting because they may provide evidence for a common metabolic control of mitosis and cytoplasmic movement. I present a classification of patterns of mitosis and early ooplasmic movements. I propose that a calcium pulse removes metabolic arrest in the unfertilized egg and triggers a metabolic cycle which drives subsequent nuclear and ooplasmic cycles.

Early insect eggs have the advantage that cell size, cell division, and nuclear division are uncoupled. The first 12 to 13 mitoses occur in the absence of cell division in a cell of roughly constant size. The result is a plasmodium. Mitosis is synchronized, and in the fruit fly *Drosophila* nuclear cycles 1 to 10 may be simplified in that there are two, rather than the normal four, phases (34).

Mitotic Patterns

In early embryonic development of many insects, nuclear division is not accompanied by cytokinesis. Nuclear multiplication begins in the yolk-rich center of the egg. As the nuclei multiply, they gradually move toward the periphery of the egg, where they finally enter the periplasm, a yolk-free layer of peripheral cytoplasm. Nuclear multiplication continues in the periplasm, resulting in thousands of nuclei within a single plasmodial cell. The plasmodial blastoderm develops into a cellular blastoderm when the peripheral nuclei are surrounded by incomplete cell membranes. This is followed by gastrulation. The division of nuclei in the early plasmodial phase and of cells in the later cellular phase is synchronized. When the nuclei or cells are out of phase, the mitotic phases occur in bands or waves that move along the egg (metachrony).

Patterns of mitosis are best studied with in vivo time-lapse film and analysis of quick-heated or quick-frozen embryos. Movements of nuclei or ooplasm in one optical section of the egg can be recorded using objectives that have a small depth of focus (Figs. 1 and 2). This reveals local patterns of movements such as the opposite flows of ooplasm. Time-lapse observation is suitable to reveal mitotic waves in vivo, measuring the speed of those waves, and correlating individual mitotic phases with the various types of ooplasmic movement. However, the magnification required to see the details of mitosis and local ooplasmic movements precludes observation of the complete mitotic pattern. The latter requires (1) using lower magnification, (2) freezing the dynamics of the pattern at different points in the process by fixation, (3) enhancing the contrast of the chromosomes by staining, and (4) mapping the stages of mitosis in the egg. This requires a fixative which minimizes artificial ooplasmic flow and associated distortion of mitotic patterns. A combination of these methods in a number of species (34) has shown that intravitelline and superficial mitoses are coordinated. Details such as the site of initiation and the speed and spatial pattern of mitotic waves may vary within a species (34), or between genera or higher taxa (Fig. 5). Mitotic waves may start in the anterior cleavage center (*Wachtliella*, a gall midge, and *Callosobruchus*, the pea beetle), or there may be an additional posterior origin (*Drosophila*).

Movements of Ooplasm and Nuclei: Description

Wachtliella

Figure 1 shows the normal development in the gall midge *Wachtliella* from the synkaryon stage (left egg) to the eight-nuclei stage (right egg) showing all the movements which occur along the longitudinal egg axis during the division cycles 1 to 3. These cycles are associated with waves 2 to 4 of randomly oriented saltation (WROS) of individual yolk particles. (See p. 275 for a more detailed description of WROS.) Every 20 min a WROS starts between 50 and 60 percent of the egg length (E.L.), moves at a constant speed of 25 ± 5 μm/min toward the two egg poles, and occupies about 10 percent of the E.L. (densely hatched area in Fig. 1b). Cyclic ooplasmic flows cause a "passive" shifting of individual yolk particles (dashed lines) and of the cleavage nuclei. Arrows mark the tracks of nuclei that move opposite to the surrounding ooplasmic flow. *PC* marks the formation of the pole cell.

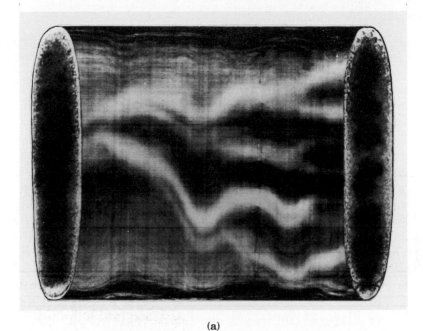

(a)

(b)

FIGURE 1. Kymographic record (a) and diagram (b) of normal development in the gall midge *Wachtliella*. Abbreviations: *p*, prophase, ending with the breakdown of the nuclear membrane (indicated by a sudden reduction in nuclear volume); *m*, prometaphase and metaphase; *a*, anaphase; *t*, telophase followed by the phase of "active" nuclear migration (location and timing marked by black-rimmed hatched areas). Courtesy of Dr. R. Wolf, Ref. 44.

Pimpla

Figure 2 shows yolk movements (thin lines) and nuclear migration (thick lines) in the egg of the ichneumonid wasp *Pimpla* from maturation until the formation of pole cells. Figure 2*a* shows a kymographic record of ooplasmic movements along the curved long axis of the egg (see left side). (See p. 272 for an explanation of kymography.) During the accumulation of periplasm, a mixing motion (MM) of ooplasm appears throughout the egg. This continues during meiosis I as a weak initial flow of yolk (I) in the anterior of the egg. This is followed by two unipolar flows (U1 and U2). U1 coincides with meiosis II. The transfer flow (T) begins in the cleavage center (85 to 89 percent of the egg length), shows three pulses (a, b, c) which coincide with mitoses 1 to 3, and then continues as two poleward flows (bipolar fountain flow, F). The latter show five pulses (dots 1 to 5) coinciding with anaphase and telophase of subsequent mitoses. During the transfer flow and the fountain flow, yolk in the posterior one-third of the egg moves slightly in an anterior direction (R). This, together with the transfer flow causes the egg to shorten, creating two polar perivitelline pockets (PPP) between the vitelline membrane and the oolemma. Interrupted lines in the diagram indicate a trace of the fourth pulse of the two fountain flows. Figure 2*b* shows that the curved long axis of the egg has been straightened so that it is parallel to the ordinate. Horizontal lines trace the movement of the central yolk mass. Interrupted vertical lines indicate the onset of flow pulses. Thick lines indicate the front of the transfer and fountain flows which coincide with the moving edge of the expanding nuclear sphere. Three more smaller pulses (N, I, and II; see Fig. 1*a*) follow the fountain flow and help move the nuclei into the posterior pole region where pole cells form during pulse II.

Drosophila

Figure 3 shows the yolk and cytoplasmic movements associated with mitoses 11 and 12 in *Drosophila melanogaster*. In Fig. 3*a* the embryonic nuclei are all in interphase. Figure 3*b* shows that just after the nuclei pass from interphase into mitosis, the yolk mass retracts from both poles (white arrows). Simultaneously, the periplasm begins to flow from the equator toward the polar regions (black arrows). The periplasm nearest to the poles moves first, followed by a progressive recruitment of more and more equatorial material into the stream of poleward flow. The zone where periplasm is being recruited into the stream, which progressively shifts from each pole to the equator, seems to coincide with the region where nuclei enter anaphase. Note that the direction of periplasmic flow here is counter to both the direction of the shifting front of cytoplasmic recruitment and the direction of the mitotic wave. In Fig. 3*c*, the nuclei are all in a late phase of mitosis (anaphase or telophase). In Fig. 3*d*, the yolk mass relaxes and reenters the regions near the poles. The initiation of this yolk movement appears to trigger a chain of periplasmic movement as in Fig. 3*b*, except that this time the periplasm flows from each pole back toward the equator. The front of the periplasmic movement coincides with passage of

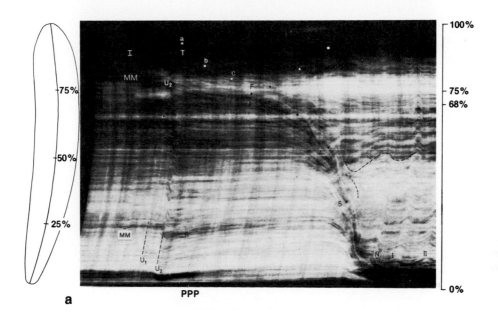

a

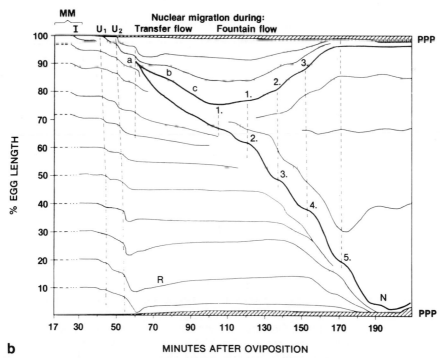

b

FIGURE 2. Yolk movements (thick lines) and nuclear migration (thick lines) in the egg of the ichneumonid wasp *Pimpla* from maturation until the formation of pole cells. Modified from Ref. 4.

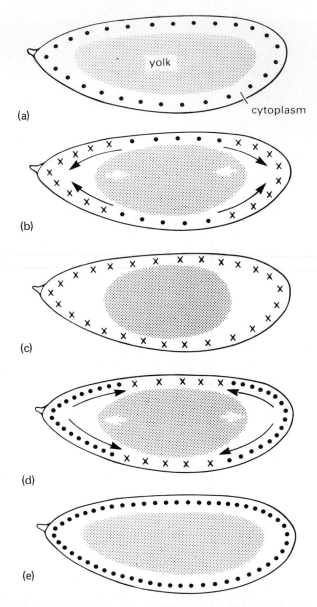

FIGURE 3. Yolk and cytoplasmic movements associated with mitoses 11 and 12 in *Drosophila melanogaster*. Central yolk-rich ooplasm is stippled; yolk-free peripheral cytoplasm (periplasm) is untextured. (●) Nuclei in interphase, prophase, or metaphase; (x) nuclei in anaphase or telophase. Only the peripherally located nuclei in one optical section are depicted; pole cells are not shown. The relative size of the yolk mass is reduced and the scale of the yolk movements is exaggerated for clarity. Courtesy of Dr. V. Foe, Ref. 8.

the nuclei into their next interphase and with the start of the budding cycle. (See pp. 276 to 277 for a more detailed description of the budding cycle.) Note that the direction of periplasmic flow, the front of periplasmic recruitment, and the mitotic wave are all traveling in the same direction. While the initial yolk retraction in Fig. 3b is rapid, the yolk relaxation in Fig. 3d is slow and persists for much of interphase. In Fig. 3e, mitosis is over, and all of the nuclei are in their next interphase.

Nuclear and Ooplasmic Cycles in Wachtliella

Figure 4 shows the nuclear and ooplasmic cycles in the gall midge *Wachtliella persicariae* during intravitelline nuclear multiplication. Timing is with reference to the one marked nucleus (egg 1, *), since different egg regions are in-

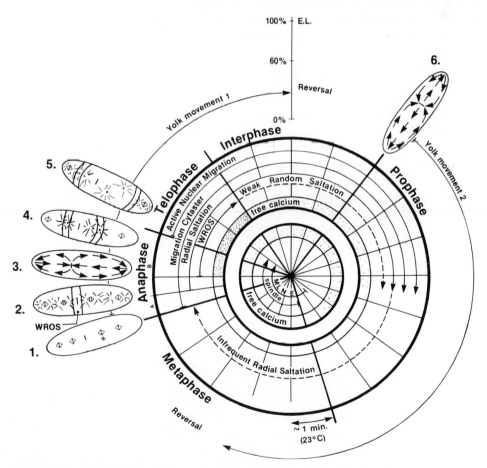

FIGURE 4. Nuclear and ooplasmic cycles in the gall midge *Wachtliella persicariae* during intravitelline nuclear multiplication. Abbreviations: E.L.: egg length (0 percent E.L. = posterior pole). From Ref. 34.

volved in the various cycles at different times. Nuclear cycles (inner circle) include a multilayered nuclear envelope (MLNE), a mitotic spindle, and nuclear free calcium. Nuclear free calcium is assumed to be persistently low because the MLNE and the migration cytaster act as calcium sinks (for details see p. 284). The outer circle shows the ooplasmic cycles. Nuclei are in metaphase when a WROS leaves the cleavage center at 50 to 60 percent of the egg length (egg 2). A nucleus enters anaphase and radial saltation (RS) starts when the WROS arrives at the outskirts of the cytaster (egg 2). (RS is described more fully on pp. 275 to 276.) Telophase and active migration (see p. 273 for a discussion of active versus passive nuclear migration) begin when the WROS reaches the nucleus (egg 4). RS and weak random yolk saltation continue in its wake. Note that a nucleus enters anaphase A and RS starts before arrival of the WROS (egg 2). This would be expected if the WROS were preceded by a putative wave front of free calcium (for details see p. 286). RS begins at anaphase B and ends when the migration cytaster disassembles (see p. 273), i.e., during late prophase of the next mitosis. The presence of the migration cytaster, RS, and active nuclear migration are strictly correlated. Active nuclear migration begins at telophase. The separation of the daughter nuclei during anaphase B is due to spindle elongation and continues as active nuclear migration to early prophase of the next cycle. The timing of passive nuclear migration coincides with that of the two yolk movements. Both movements and passive nuclear migration overlap with active nuclear migration. Therefore, it is difficult to give the exact timing of passive nuclear migration. Eggs 3 and 6 show the reversal of yolk flow during late interphase and late metaphase. One arrow represents the total extent of yolk displacement (about one-sixth of the length of the egg). Movements 1 and 2 correspond to *yolk contraction* in *Wachtliella* and *relaxation* in *Drosophila* and *Callosobruchus*. In *Callosobruchus*, the anaphase waves for superficial mitoses 9 to 13 coincide with the flow of periplasm that compensates for the yolk contraction (25). In *Drosophila*, yolk contraction and relaxation during superficial mitoses 10 to 13 (Fig. 3) are similar to those in *Callosobruchus*. However, yolk movement 1 (contraction) begins in early prophase, and the recruitment of periplasm for the compensatory poleward flow rapidly follows at the onset of anaphase with which it coincides in timing and direction (Fig. 3*b*). Yolk movement 2 (relaxation) is correlated with late telophase–early interphase (8).

Comparison of Patterns of Mitosis and Central
Yolk Movement

Figure 5 shows a comparison of patterns of mitosis and of central yolk movement. Simultaneous compensatory surface movements always occur in the opposite direction (Fig. 3) but are not indicated. Movements during maturation and after completion of the cellular blastoderm are not included for lack of comparative data. When the ooplasm moves back and forth without net displacement, only the first reverse movement is indicated. (See the interval between mitoses 1 and 2 in *Wachtliella*; 10 and 11 in *Pimpla*; and 1 and 2 in *Callosobruchus*.) In *Acheta* (Orthoptera) no random yolk saltations were re-

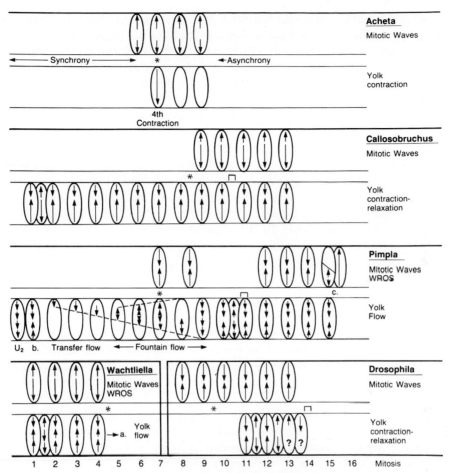

FIGURE 5. Comparison of patterns of mitosis and of central yolk movement. Eggs have the anterior pole up and the ventral side to the left. Empty eggs, no yolk movements occur; no eggs, no data available. *, nuclei arrive in periplasm; ⊓ , cell membranes begin to form. Length of arrow indicates extent of displacement. WROS = wave of randomly oriented yolk saltation. From Ref. 34.

ported to accompany mitosis. In *Callosobruchus* (Coleoptera) each mitosis is accompanied by random yolk saltation, but it is uncertain whether it is a wave. Therefore, the arrows in the top row indicate mitotic waves only (25). In *Drosophila melanogaster* (Diptera), either the previous pattern is repeated (not shown) or relaxation starts toward the anterior pole and then reverses (both shown) during mitosis 13. This contraction does not relax during the next two mitoses. A question mark indicates no data reported. No (W)ROS was reported (8). (See p. 275 for a discussion of WROS and ROS.) Mitoses 14 to 16 have complex patterns. In *D. montana*, Wilson (40) observed 12 cycles of yolk movement (see Ref. 34). In *Pimpla* (Hymenoptera), U_2 indicates second unipolar flow and b shows the first bipolar flow; an interrupted line indicates the front of expanding nuclear sphere. The transfer flow encompasses a band of about 30 percent of the egg length which moves rhythmically toward the posterior pole. The four pulses probably coincide with telophase of mitoses 2 through 5 (45). Yolk particles in the wake of this pulsating wave swing back and forth in the rhythm of mitosis. The fountain flow may span a variable number of mitoses. It has four to five pulses, each of which coincides with anaphase of mitoses 5 through 8 or 9. On its way to the posterior pole, the posterior component of the fountain flow is accompanied by a moving, ring-shaped contraction of the ooplasm. Moreover, there is a stationary contraction at the origin of the fountain flow. Starting with mitosis 10 a period of very irregular cyclic yolk saltations follows. Mitotic waves occur occasionally during mitoses 7 and 8. Their direction was inferred from Ref. 4, Fig. 11d. Part c shows that mitotic wave 15 sometimes stops around the equator or proceeds to the anterior pole, where no mitotic wave originates. WROSs have been seen to accompany mitotic waves (44). In *Wachtliella* (Diptera), part a shows that this pattern of yolk flow weakens but continues for an undetermined period (41).

Movements of Ooplasm and Nuclei: Analysis

Ooplasmic movements are studied by time-lapse film and kymography. A kymogram (e.g., Figs. 1a and 2a) is made by projecting a time-lapse film in a dark room on a narrow slit which is shaped to expose the egg region to be studied. Photographic paper is attached to a kymograph drum that is rotating at constant speed behind the slit. Moving nuclei and yolk particles will create light traces on the photographic paper. The rotation speed is set such that the traces are slanted at 45°. Light vertical time lines (isochrones) are projected onto the paper by periodically blackening individual frames of the film. It is necessary to observe the film in order to identify the recorded movement as ooplasmic flow, waves of contraction, or changes of color since these will all produce the same kind of record on the kymogram. By projecting the slit appropriately, the kymogram will show movements in the center or at the periphery of the ooplasm.

Flow and Contraction of Yolk

My focus is on the coordination of ooplasmic movements with plasmodial mitosis. Considerable differences of type and pattern of movement exist. Two basic types are "flow" and "contraction." In some species the endoplasm and the yolk particles suspended in it are incoherent and periodically flow along the long axis of the egg.

Note that in *Pimpla* a net displacement of yolk particles takes place (Fig. 2), but in *Wachtliella* the ooplasm just swings back and forth (Fig. 1). In other species there are series of contractions and relaxations of a more coherent mass of yolk (Figs. 3 and 5). The contractions shorten the egg, such that in the poles periplasm accumulates under the oolemma and a fluid-filled pocket appears between the oolemma and vitelline membrane (8).

The flow or contraction of yolk in the center of the egg is compensated for by an opposite movement at the surface (Figs. 3 and 4). This *surface flow* may affect the periplasm alone, or also the yolk underneath. Localized yolk contraction occurs at the surface and in various directions, resulting in the local constriction of an egg (34).

Flow and contraction are distinguished on the basis of the different dynamics of their movements, but this distinction is not supported by their respective reactions to drugs. Both cytochalasin and colchicine inhibit flow in one species and contraction in another. The effect is sometimes stage-dependent (34). This suggests that flow or contraction is not associated with one type of cytoskeletal component but with an association of microtubules, microfilaments, and various proteins that modulate their reaction. An inventory of cytoskeletal components is presently being developed by using standard biochemical procedures, drugs (Table 1), and electron and immuno-fluorescence microscopy (reviewed in Ref. 34). However, identification of cytoskeletal components deep in the egg is hampered by nonspecific absorption of fluorescent antibody to the yolk or by absorption of fluorescent light by the yolk mass.

Nuclear Migration

Nuclei are both passively distributed by the flow of ooplasm and actively pulled by microtubules (44) or by contractile cytoplasmic strands of unidentified type (42). Active migration should not be confused with the process of spindle elongation during anaphase B, which widens the distance between two daughter nuclei and changes into nuclear migration. Active nuclear migration is sometimes superimposed upon passive movement. The passive component is revealed when active migration is blocked by colchicine (42). Conversely, active nuclear migration is revealed when cytochalasin is used to stop yolk flow or when nuclei are seen moving upstream. The latter event happens in *Wachtliella* when nuclei start moving from the cleavage center in a posterior direction (Fig. 1b).

The microtubules that pull a nucleus through the ooplasm are organized in a migration cytaster, i.e., the microtubules radiate outward from a center to which a nucleus is attached (43). Migrating nuclei of *Drosophila* and of the pteromalid wasp *Nasonia* also have a cytaster facing the oolemma, but a migratory function has yet to be established (34). In the migration cytaster of *Wachtliella* the microtubules are associated with membranous vesicles (43) and presumably microfilaments. These temporarily adhere to the egg surface while the nucleus is being pulled through the ooplasm. Wolf (42) suggests that treadmilling accompanied by a net disassembly of the microtubules pulls the associated nucleus toward the egg periphery. However, the estimated rate of treadmilling (0.014 μm/s, Ref. 23) is between 9 and 30 times too slow to account for nuclear migration rates that range from 0.13 to 0.43 μm/s (41, 43), especially since the nuclear migration rates result from forces that are exerted on the nucleus from different directions. Shortening of microtubules at both ends could ac-

TABLE 1 Inhibition of Nuclear and Ooplasmic Dynamics by Colchicine and Cytochalasin[a]

	Colchicine, μg/mL[b]	Cytochalasin, μg/mL[b]
Wachtliella (gall midge)	0.8–2	1
Mitosis	+ [c]	–
Active nuclear migration	+	–
Yolk flow	–	+
WROS[d]	+	–
RS	+	–
Drosophila	40	1–10
Mitosis	+ [e]	–
Nuclear migration	+	+
Yolk contraction	+	+ [f]
Periplasmic saltation	+ [g]	– [h]
Membrane furrowing	–	+
Accumulation of periplasm	+	
Bombyx (silk moth)	2	1
Mitosis	+	
Nuclear migration	+	+
Membrane furrowing	–	+
Heteropeza (gall midge)		
Mitosis	+ 5.6[i]	– 17[j]
ROS	+ 8[i]	– 1–192[j]

[a]From Ref. 34.

[b]Minimum intracellular concentrations are shown, except for *Bombyx* and *Heteropeza* where minimum external concentrations are given. The effects of colchicine and cytochalasin are opposite except for the inhibition of yolk contraction in *Drosophila* by colchicine.

[c] + , inhibition; –, no inhibition; blank, no data.

[d]Abbreviations: RS: radial saltation; (W)ROS: (wave of) randomly oriented yolk saltation; saltation is a bouncing movement of individual yolk globules.

[e]Inhibition is often concentration dependent. At 1 μg/mL two more abnormal mitoses occur while at 10 μg/mL mitosis is blocked.

[f]Inhibition is often concentration dependent. Yolk contraction is normal at 2 μg/mL, abnormal at 20 μg/mL, and blocked at 200 μg/mL.

[g]Inhibition is often concentration-dependent; 8 μg/mL.

[h]Inhibition is often concentration-dependent; 200 μg/mL.

[i]The same results were obtained with 7.3 μg/mL of vinblastine.

[j]Concentrations without effect.

count for the rates of nuclear migration measured, but no published shortening rates are available yet. Cytaster-confined membranes are not likely to generate the force (14) since (1) they are radially oriented and, therefore, would not provide direction to movement and (2) they are not anchored to other egg structures. Since cytochalasin does not block active nuclear migration in *Wachtliella*, the filaments connecting the nucleus with the cytaster must be insensitive to it (42, 43). This raises the possibility that similarly insensitive microfilaments are also involved in active nuclear migration as appears to be the case in *Wachtliella* (Table 1). Obviously, the role of microfilaments in nuclear migration needs clarification.

Randomly Oriented Saltation

Periodically, randomly oriented saltation (ROS), a bouncing movement of yolk particles, encompasses the whole egg. In some species the ROS appears simultaneously throughout the whole egg. In others it starts near the cleavage center and expands as a wave of randomly oriented saltation (WROS) toward both egg poles (Figs. 1 and 5). The (W)ROS cannot be observed without the use of time-lapse cinematography, i.e., the saltation is too slow to be classified as Brownian motion. Resolution of the wave shape of a WROS depends on the exposure frequency of the time-lapse used. The (W)ROS is inhibited by colchicine but not by cytochalasin B (Table 1). In the paedogenetic gall midge *Heteropeza*, 1 μg/mL of cytochalasin B blocks morphogenesis of the germ band, but 200 μg/mL does not block random yolk saltation in the same egg and at the same stage (J. Kaiser, personal communication). This suggests a role for microtubules. However, these experiments do not exclude a role for microfilaments because the latter may have been insensitive to cytochalasin. Microfilaments would have to be excluded only if, for instance, anti-actin or anti-myosin antibody failed to block the (W)ROS.

Kaiser (personal communication) suggests that the (W)ROSs are caused by treadmilling microtubules to which yolk particles temporarily adhere. A particle would ride on a subunit of a treadmilling microtubule until the subunit detaches, joins another microtubule, and rides in a different direction. The velocities of saltation and of treadmilling (50 μm/h; Ref. 23) agree. Treadmilling represents one of a number of possible states for cytoplasmic microtubules (19). There may be shifts from a slow-growing to a fast-shrinking set of microtubules or from an unstable dynamic state of predominantly slow-growing microtubules to a stable equilibrium of treadmilling ones. Whether a (W)ROS is associated with any of these shifts in the dynamic behavior of the microtubule population is not known. If the (W)ROS is caused by a shift to the treadmilling state, then treadmilling will not occur immediately before the next anaphase, at which time there is no (W)ROS. Also, since the yolk particles are attached to microtubules then a (W)ROS should be sensitive to anti-tubulin antibody but not sensitive to anti-actin or anti-myosin antibody. However, if the yolk particles are connected to microfilaments, then contraction of the microfilaments could move adhering yolk particles if the microfilaments were attached to a microtubular scaffold. Connections between microtubules and microfilaments are known to occur in other systems. The inhibition of the (W)ROS by colchicine would then be due to disruption of the microtubular scaffold.

We can test for possible connections between the (W)ROS and the cytoskeletal network by exploiting the observation that ROS in *Callosobruchus* fails to continue outside the expanding sphere of nuclei (the cytoskeletal components may be preferentially located inside this sphere). Likewise, the expanding sphere of nuclei which triggers the saltation of periplasmic particles in *Drosophila* (8) may do so by supplying contractile elements.

Radial Saltation

When a WROS approaches a migration cytaster, it triggers a radial saltation (RS) of yolk particles within it either in a single direction toward the center of the cytaster or oscillating around a fixed position. RS is not to be confused with the saltation of periplasmic particles at right angles to the oolemma of the *Drosophila* egg (8). RS is visible only with time-lapse cinematography and is, therefore, too slow to qualify as Brownian motion. Inhibition of RS with colchicine (42) suggests involvement of microtubules. Assuming a network of microfilaments and microtubules in the cytaster, RS might be explained in one of three ways.

1. RS may reveal the jumpy movements of a microtubule that is shortening at both ends (19) and that contains adhering yolk particles. One end is attached to the cytaster center and the other to peripheral egg structures.

2. Jumps could be transmitted to yolk particles if they were attached to microfilaments. Also, if the net movement of microtubules is toward the cytaster center because of unequal shortening rates at opposite ends of the microtubule or because of treadmilling, microfilaments would be centripetally stretched. A microfilament would periodically detach from a microtubule resulting in radial saltation of the attached yolk particles. Rates of RS [0.22 to 0.43 μm/s (41); 0.4 to 0.6 μm/s (42)] are much higher than the estimated rate of treadmilling (0.014 μm/s). This suggests that the microtubules of the cytaster may be shortening at both ends and that the rate of RS may be closer to this shortening rate.

3. RS can be caused by direct contraction of microfilaments radially constrained by the microtubular scaffold of the cytaster.

The last two explanations require testing the lack of sensitivity of microfilaments to cytochalasin B because it fails to inhibit RS.

Changes of Periplasm

Addition or loss of periplasm is characterized by variations in the thickness of a layer of yolk-free cytoplasm under the oolemma caused by an unknown mechanism. In *Drosophila* it is blocked by colchicine (47), suggesting that microtubules in the periplasm may exclude yolk particles from the egg's periphery (17). These particles may displace cytoplasm from the interior toward the egg surface.

Surface Changes and Cytokinesis

Cell division that is caused by a contractile ring of actin filaments which pulls the cell membrane into a furrow is termed "cytokinesis." This may occur only in some insect eggs. Cytokinesis is preceded by a variety of *surface changes* that accompany superficial mitosis. I use the term "cytokinesis" or "cellularization" to indicate the complete and permanent separation of the surface nuclei from each other and from the underlying yolk by cell membranes. Surface changes preceding cytokinesis and occurring in the presence of nuclei are termed "cyclic budding" and "cyclic furrowing." *Pseudocleavage* indicates surface changes or cytokinesis in the absence of nuclei. Budding is the temporary protrusion from the egg surface of the oolemma and the underlying periplasm above the nuclei. Cyclic budding precedes cellularization in *Callosobruchus* and *Drosophila*, but in eggs of the silk worm *Bombyx*, cytokinesis occurs by means of budding (34).

Temporary furrowing is the infolding of the oolemma away from the egg surface and down between the nuclei; this is followed by retraction. It accompanies superficial mitosis in *Wachtliella* and is followed by lateral extension of the membranes underneath the nuclei. The oolemma has the potential to develop transverse furrows as early as intravitelline mitosis (41). In eggs of the blowfly *Calliphora* temporary furrows appear only once, viz., during cycle 13, while cytokinesis occurs during the subsequent interphase (22). In *Drosophila*, mitotic cycles 10 through 14 are accompanied by five budding cycles, the last of which precedes cytokinesis in stage 14. Furrowing is not caused by a contractile ring or band of microfilaments, but F-actin and myosin are associated with the furrow canals formed during cellularization (34). Contrary to the impression given above, there is confusion about whether the membranes that separate the nuclei form by the protrusion of buds upward from the egg surface followed by their lateral expansion or by the infolding of the oolemma from the surface down into the egg (34).

Ooplasmic and Nuclear Cycles Are Correlated

Despite differences in the types and spatial patterns of yolk movement (Figs. 1, 2, and 5), the movements are correlated with mitosis. The following description of this correlation in *Wachtliella* (Diptera—intravitelline mitoses 1 to 4) also summarizes events in *D. melanogaster* (Diptera—superficial mitoses 11 and 12), *Heteropeza* (Diptera), *Pimpla* (Hymenoptera), and *Callosobruchus* (Coleoptera), while leaving out some species-specific variations (see section "Nuclear and Ooplasmic Cycles in *Wachtliella*").

Basically, each nuclear cycle is accompanied by a cycle of ooplasmic movement. Figures 1 and 4 show how the nuclear and the ooplasmic cycles are related. The ooplasmic cycle begins with radial saltation of yolk particles around the zygote nucleus which is in anaphase A. Next a WROS leaves the area of the zygote nucleus in the direction of the two poles. As the WROS leaves, the first nuclear cycle proceeds through late anaphase and telophase. A few minutes before the WROS appears, the yolk endoplasm begins to flow along the longitudinal axis of the egg, away from the egg's poles and toward a region anterior to the zygote nucleus (Figs. 1 and 5). The zygote nucleus completes mitosis while it is caught in the anteriorly directed flow in the posterior two-thirds of the egg. One daughter nucleus is shifted anteriorly along with this flow, whereas the other daughter nucleus migrates actively upstream, i.e., toward the posterior pole (Fig. 1*b*). The two flows are compensated for by oppositely directed surface flows (not visible in Fig. 1). As the daughter nuclei enter interphase, the flow of yolk endoplasm that is directed anteriorly reaches its greatest extent and subsequently reverses so that there is no net displacement of yolk particles (Fig. 1b). By this time the nuclear and the cytoplasmic cycle are completed and will be repeated between 12 to 13 times, depending upon the species, until the cellular blastoderm has been formed.

All subsequent nuclear divisions are accompanied by a WROS initiated at the cleavage center. As a WROS approaches, RS begins and the nucleus proceeds from metaphase to anaphase A, and these events are highly coincidental. When a WROS is experimentally desynchronized with the nuclear cycle, the WROS may engulf a nucleus unprepared for anaphase. Such a nucleus does not complete anaphase until it is engulfed by the next WROS (44). Whether it is arrested in metaphase remains to be confirmed (R. Wolf, personal communication). A weaker continuation of the ROS in *Heteropeza* and of both the WROS and the RS in *Wachtliella* (Fig. 4) fills the period between anaphase and the next prophase. Synchronization among the central yolk flow or yolk contraction, the WROS, and the anaphase waves has been observed in vivo in seven different species representing three orders (34). Once the nuclei have arrived in the periplasm, superficial mitosis is synchronized with budding of the cell surface above the nuclei. However, the timing of budding relative to the nuclear cycle differs between species and also varies somewhat among individuals of the same species (34). These correlations suggest that flow and contraction of ooplasm, mitosis, budding and furrowing of the egg surface, and cellularization are under common control.

Ooplasmic and Nuclear Cycles Are Relatively Independent

(W)ROSs, RS, cytaster migration, and cell surface changes are independent of the nuclear cycle. In the unfertilized egg of *Wachtliella*, anucleate cytasters exist which react normally to the approach of a WROS, i.e., they show RS and divide. The WROSs appear with lower frequency but in the normal pattern (44). Such "pseudocleavage" has been observed in a variety of insect eggs and sometimes continues as late as the (pseudo)blastoderm stage. In *Wachtliella*, for instance, the oolemma is able to invaginate in the absence of nuclei and anucleated cells are formed (34).

Additional evidence has come from the manipulation of the frequency, speed, and direction of WROSs in *Wachtliella* by local reduction of gas exchange or by imposing a temperature gradient, within the physiological range, along the egg (44). WROSs start in the region that is the warmest or that has free gas exchange, and they appear more frequently than normal. Impaired gas exchange or low temperature slows down WROSs, nuclear division, and nuclear migration. As a result, the next WROS starts in the warmer egg region before the previous one has disappeared and two or more WROSs traverse the egg simultaneously. Nuclei that are reached by the first WROS will enter anaphase, complete mitosis, and start migrating; they are, however, unprepared for the remaining WROSs. Slowed nuclei that are not in late metaphase do not respond to a WROS, omit two or more divisions, become desynchronized with the WROSs and, therefore, have to wait for the next WROS before entering anaphase. While waiting, sometimes three instead of two cytasters appear, indicating that one of the two cytasters has divided autonomously. After equilibration of the temperature in eggs that have both normally directed and reversed WROSs and mitotic waves, retarded egg regions catch up with the development in the rest of the egg and subsequent development is normal. In summary,

1. Cytasters show RS and divide in the absence of a nucleus.
2. WROSs can be induced anywhere in the egg and in regions without a nucleus or cytaster.
3. WROSs and nuclear division can be desynchronized.
4. Nuclei react to a WROS with division only when they are in late metaphase.

These observations can be used to distinguish between a nuclear cycle (division) and an ooplasmic cycle (indicated by WROS, RS, and migration cytaster), and they suggest that the nuclear cycle is driven by the ooplasmic cycle.

The cyclic flow of yolk is another component of the ooplasmic cycle that can be dissociated from the nuclear cycle. Experimentally activated anucleate eggs of *Pimpla* display a variety of flow patterns ranging from almost normal to incomplete, and they also show pseudocleavage (38). Colchicine and cytochalasin have opposite effects on nuclear and ooplasmic dynamics in *Wachtliella* and in *Drosophila* (Table 1). Ooplasmic flow in *Wachtliella* continues at the same frequency when WROSs are inhibited (42). In the house cricket *Acheta* meiotic yolk contractions proceed in the absence of nuclear division (31). In the ant *Camponotus* accumulation and differentiation of periplasm starts before meiosis when females are prevented from oviposition (K. H. Bier, unpublished observation). Thus, the different components of the ooplasmic and nuclear cycles can be dissociated without affecting each other. This suggests that the ooplasmic timer is not affected by colchicine or cytochalasin and that it continues to trigger those parts of the system not disabled by these drugs. However, in *Pimpla* absence of nuclei and/or experimental activation results in abnormal yolk flow. As these examples demonstrate, more comparative studies on the effects of colchicine and cytochalasin are needed, as well as studies on the effects of concentration (Table 1). In *Drosophila* the use of mutants would seem to be the best method for elucidating the relationships between the ooplasmic timer and the various ooplasmic and nuclear cycles.

The evidence from insect embryos is in agreement with the observations made in a wide variety of eukaryotic cells that show that many components of the cell cycle (such as the behavior of monasters or chromosomes as well as cytokinesis, mitosis, cell surface properties, cell morphology, DNA synthesis, and levels of maturation promotion factor and protein phosphorylation) are mutually independent events and cycle parallel to each other (34).

A Metabolic Cycle Model for Regulating Nuclear and Ooplasmic Cycles

I suggest that removal of metabolic arrest in oocytes can initiate a metabolic cycle, such as an oscillating glycolysis and/or respiration (e.g., Ref. 15). The model (Figs. 6 and 7) has the following minimum assumptions:

1. Calcium stimulates metabolism.
2. Metabolic activity increases because of a fast calcium-stimulated rise of inositol triphosphate (IP_3) and an IP_3-stimulated rise of free calcium.
3. This increase in metabolic activity is followed by a decline in activity that is a result of calcium influx into the endoplasmic reticulum (ER), mitochondria, or perivitelline space via ATP-stimulated calcium pumps and of various negative feedback pathways that curtail the calcium signal. This creates a cycle of free cytoplasmic calcium and of metabolic activity.
4. Metabolism, once triggered, cycles independently of calcium.
5. Cycles of calcium and, possibly, of other metabolites drive the nuclear and ooplasmic cycles.

Direct measurements of free calcium in insect eggs are not available. However, it is likely that ooplasmic movements, mitosis, budding, furrowing, and cellularization in insect eggs are controlled by fluctuations in ooplasmic free calcium. First, free cytoplasmic calcium triggers meiosis, mitosis, as well as cytokinesis in various cell types (34). Second, the cytoskeletal elements involved in ooplasmic movement are sensitive to calcium (34) and inositol lipids (2). Third, the behavior of the WROS in *Wachtliella* and of the calcium-induced wave of cortical vesicle exocytosis in the fish medaka are similar. Fourth, all these processes are highly correlated.

I also hypothesize that the calcium cycle mediates between the metabolic cycle and the other cycles. Metabolic control of both meiosis and mitosis is suggested by the concurrence of metabolic and meiotic activation and by the similarities between meiotic and mitotic activation. The transformation of interphase chromosomes into metaphase chromosomes in oxygen-deprived

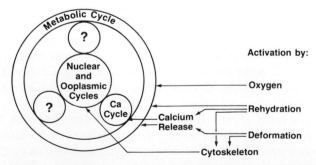

FIGURE 6. Simplified model of the putative interactions between cycles of metabolism, cytoplasmic free calcium, mitosis, and ooplasmic movements during activation and early development of insect eggs. From Ref. 34.

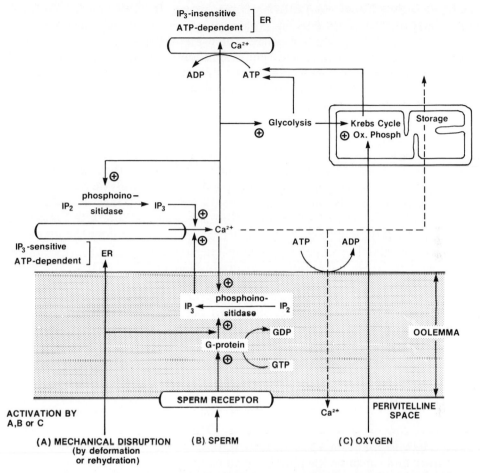

FIGURE 7. Proposed mechanism for the initiation and maintenance of a cytoplasmic calcium cycle. Fast effects, _____; slow effects, ----, G-protein, GTP-binding protein.

Drosophila eggs (9) supports the hypothesis that metaphase arrest corresponds to a low point in the metabolic cycle. The predictable changes in the patterns of mitosis and ooplasmic movements that are caused by local oxygen deprivation of *Wachtliella* eggs suggest that both mitotic and ooplasmic cycles are under metabolic control.

Activation: The First Calcium Pulse Triggers Metabolism and Meiosis

The pulse of free calcium elicited by sperm entry triggers the resumption of meiosis and of the metabolic cycle in the oocytes of fish, echinoderms, worms, mollusks, amphibians, and mammals (34). Evidence in insects is scarce. In *Pimpla*, several of the explanted oocytes that were injected with the calcium ionophore A23187 developed into hatching larvae (R. Wolf and D. Wolf, personal communication). It may be possible to demonstrate directly a role for calcium by injecting calcium-sensitive fluorescent probes either into eggs or into the hemolymph for incorporation into oocytes (35). Monitoring of the intensity of fluorescence during activation should reveal sources and sinks of calcium. Mitochondria, perivitelline space, and ER are potential sources and sinks. Mitochondria control large and long-term calcium fluctuations. ER controls small and rapid calcium fluctuations.

Activation may start when a sperm-receptor complex activates G-protein (GTP-binding) in the oolemma (Fig. 7). This in turn activates the phosphoinositidase that catalyzes GTP-dependent IP_3 production (2). The GTP-dependent IP_3 pulse triggers a GTP-independent, autocatalytic calcium cycle. The rising phase of this calcium cycle would consist of a fast IP_3-stimulated release of calcium from IP_3-sensitive ER and a calcium-stimulated release of IP_3 from the ER (Ref. 39; Fig. 7). The calcium pulse may trigger the glycolytic oscillator via the effect of calcium-mediating proteins on glycolytic enzymes. Consequently, it also stimulates the production of ATP which is used by (Ca^{2+})ATPase for fast uptake of calcium into ER (Fig. 7). Thus a decline in the level of calcium depends directly upon metabolism. Various other negative feedback pathways also cooperate to curtail the calcium signal (2).

In insect eggs, mechanical release of calcium from internal sources is probably common. Deformation of an egg in the oviduct or swelling of dehydrated oocytes after oviposition may disrupt the ER. For instance, mechanical disruption of the ER in parthenogenetically developing male-yielding eggs of *Pimpla* (24) causes resumption of meiosis, and this includes ooplasmic flow (37). Oocytes of many insects are slightly dehydrated, and this may be associated with the metabolic arrest of oocytes. This correlation is known to occur in eggs, larvae, and imagos of insects and other organisms that enter a dormant state and consume very little or no oxygen under harsh environmental conditions. If dehydration imposes metabolic dormancy on mature oocytes, rehydration would seem to be a necessary part of activation since rehydration would provide the physiological environment for metabolic reactions. Calcium may be released from ER by permeabilizing its membranes or by triggering the production of IP_3, first in the oolemma and subsequently in the ER (Fig. 7). IP_3 activates the oocytes of echinoderms and amphibia (e.g., *Discoglossus*) (20) by causing the release of calcium (reviewed in Ref. 2). Release of calcium by disruption of internal sources is also consistent with the ubiquity of parthenogenesis in insects and with the movements of ooplasm that occur in unfertilized eggs (36) or in eggs that are activated mechanically (38). In oxygen-deprived oocytes of *Carausius* metabolic arrest may be directly lifted when oxygen enters the egg's micropyle after oviposition (28).

Coupling the Metabolic Cycle to the Nuclear and Ooplasmic Cycles

The fundamental timekeeper in the egg is envisioned to be the glycolytic oscillator, which causes oscillations in the Krebs cycle and oxidative phosphorylation, and in the concentrations of their metabolites such as GTP, ADP, and H^+ (e.g., Ref. 15). The calcium cycle, which is an integral part of the feedback control of the inositol lipid pathway (2), may be causally independent of, but entrained by, the glycolytic cycle. Alternatively, the glycolytic cycle may cause the calcium fluctuations.

If the calcium cycle is coupled to the metabolic oscillator, their frequencies must be similar. Mitochondrial calcium homeostasis would be too slow to control the cycle. A calcium pulse can act quickly via calcium-modulating proteins which stimulate enzymes at strategic positions in the metabolic network, such as phosphorylase kinase b (46). This enzyme stimulates glycogenolysis and subsequent extramitochondrial glycolytic production of ATP. It also stimulates the Krebs cycle and oxidative phosphorylation by raising the levels of pyruvate and glycerol 3-phosphate, both of which easily diffuse into mitochondria. Consequently, a calcium pulse raises the level of ATP and stimulates fast calcium uptake into IP_3-insensitive ER by calcium-stimulated ATPase (Fig. 7). This, in turn, reduces the calcium level outside of the ER.

The metabolic cycle may control the nuclear and cytoplasmic cycles via Ca^{2+}, H^+, and metabolites whose concentrations cycle, such as ATP, GTP, cAMP, IP_3, and others. For instance, gel-to-sol transitions and ooplasmic movements may involve changes in H^+ concentration, which controls the assembly of cortical actin (1), and the inositol lipids which control actin polymerization and attachment of actin filaments to the cell membrane (2). Some metabolites may trigger cycles of sensitivity to calcium in contractile proteins (5).

Calcium Pulses May Trigger Mitosis

In *Xenopus*, meiosis and mitosis appear to be under similar control. Mitosis in metaphase is blocked when cytoplasm from oocytes arrested in meiosis is injected into fertilized eggs. This block is lifted by subsequent injection of calcium (27). Continuous monitoring of free cytoplasmic calcium in a sea urchin egg has revealed that calcium peaks at fertilization, pronuclear migration, nuclear envelope breakdown, the metaphase-anaphase transition, and cleavage (29). Similar measurements show a coincidence of calcium signals and of the metaphase-anaphase transition in plant cells (18) and mammalian cells (30).

There is a large body of evidence that calcium also triggers mitosis in a wide variety of cells (34). Such evidence usually consists of showing that mitosis can be blocked by compounds that block mediators and regulators of calcium-dependent effects, such as microtubules and calmodulin. Calcium has been found in the mitotic apparatus of different cell types together with a calcium-sensitive calcium pump in ER vesicles and calmodulin. Calmodulin mediates the effect of calcium on the spindle by binding to the calcium pump which is located in membraneous vesicles in the mitotic spindle.

Evidence for the control of mitosis in plasmodial insect eggs by calcium may be obtained in desynchronized *Wachtliella* eggs. In such eggs the WROS engulfs a nucleus unprepared for anaphase (44). I suggest that such a nucleus remains arrested in metaphase and that the arrest is lifted when the next calcium wave arrives at the nucleus. In normal eggs mitotic synchrony is a result of a simultaneous rise in calcium concentration throughout the egg, whereas metachrony (mitotic waves) and asynchrony result from temporal differences in the onset of such a rise. Genetic dissection of the components of, and prerequisites for, meiosis and mitosis in *Drosophila* appears to be a promising approach to the study of the nuclear cycle, especially when combined with injection of factors known to control the cell cycle.

Compartmentalization of Calcium

It must be assumed that regional control of the concentration and effects of free cytoplasmic calcium explain the nuclear and ooplasmic cycles. The sensitivity of organelles to calcium may be modulated by proteins associated with the organelles (3, 32). Regional differences in calcium concentrations may be caused by an nonhomogeneous distribution of the components of the IP_3 pathway, receptors, stores, and enzymes (2).

In cell types other than insect eggs, membraneous vesicles act as sources and sinks in the local control of free calcium levels (14). The location and behavior of membranes around nuclei of early insect embryos suggests the existence of similar compartments. Mitotic spindles and interphase nuclei are enclosed by several layers of smooth ER (34). In *Wachtliella, Heteropeza,* and *Drosophila* the spindle envelope is connected to the ER and surrounds the nucleus only during the period of assembly of spindle microtubules, viz., from late prophase until late anaphase or early telophase (34). A calcium wave passing a nucleus would activate putative calcium-sensitive calcium pumps in the spindle envelope, thereby, temporarily reducing the amount of free calcium inside the spindle. Outside the spindle the calcium level would remain high after the passage of the wave front. This has been observed in PtK_1 cells (30), and it would explain the simultaneity of the assembly of spindle microtubules inside the nucleus and the absence of the microtubules of the migration cytaster outside the nucleus during metaphase in *Wachtliella* (Fig. 5 in Ref. 43). At metaphase, ER vesicles of the spindle envelope in *Nasonia* extend into the polar regions and assume an astral arrangement in the cytaster (48). In *Wachtliella,* as the chromosomes move toward the spindle poles, membraneous vesicles aggregate around the poles where the migration cytaster starts to grow. Later, mitochondria and ER are lined up radially between the microtubules of the cytaster (43), and this may allow the calcium concentration in the spindle to rise to that found in ooplasm. The rise would progress from the spindle equator toward the poles, while the free calcium level in the cytaster would decrease. Therefore, microtubules would disappear from the spindle and reappear in the cytaster. Storage of cytoplasmic free calcium in ER and mitochondria may permit the extension of the cytaster microtubules toward the cell surface in the wake of the calcium wave.

Simultaneous disassembly of spindle microtubules and the growth of cytaster microtubules also occurs in sea urchin eggs (12) and in *Drosophila* during nuclear cycles 10 and 11 (17). In contrast to *Wachtliella*, cytaster microtubules and membraneous vesicles in *Drosophila* are also present before anaphase and the vesicles are not relocated (17, 33). This serves as a reminder that redistribution of free calcium does not require redistribution of membraneous vesicles.

Control of Ooplasmic Movements via Cytoskeletal Elements

Yolk movements may require gel-to-sol transitions which are controlled by free calcium and mediated via various gelation factors. The cytoskeletal ingredients for gel-to-sol transitions are present in insect eggs (34). In *Drosophila* the compensatory surface flow is accompanied by a change in consistency of the periplasm (8). Deep puncture causes yolk to flow out of the egg rapidly, but superficial puncture does not result in loss of periplasm, except at the poles (40). This egg may consist of a periplasmic gel, an endoplasmic sol, and regions of gel-to-sol transition in the poles that are similar to those found in acellular slime molds.

Whether the ooplasm contracts or flows may depend on the kinds, arrangement, and calcium sensitivity of the components of the cytoskeleton. In crude cell extracts of *Amoeba proteus* the sol state is characterized by an absence of actin filaments, the gel state by a random network of filaments, and the contracted state by bundles of actin filaments (26). Colchicine inhibits contraction in *Drosophila*, but does not inhibit flow in *Wachtliella* (Table 1). In eggs that undergo contraction, microtubules may provide a scaffold for microfilaments to generate force, whereas flow may occur in eggs without a scaffold. The reaction to a pulse or wave of calcium would consist of a gel-to-sol transition and a contraction or flow, depending on the organization of the cytoskeleton. For instance, in *Drosophila* a calcium wave would cause both a contraction of the internal sol and a gel-to-sol transition in the periplasm.

The mechanical and chemical bases of ooplasmic movement need clarification. How general are gel-to-sol transitions, and how are the forces generated by them transmitted through the yolk? How does one explain the interspecific differences in spatiotemporal patterns of movement (see Fig. 5)? Regional differences in patterns of movement also require explanation. Why are the ROSs confined to the expanding sphere of nuclei in *Callosobruchus*? In *Wachtliella* yolk flow and the WROS can occupy different ooplasmic domains (Fig. 1*b*). How can a gel-to-sol transition be restricted to a WROS, whereas the associated yolk flow requires the solation of a much larger territory of yolk? The explanation of such spatiotemporal patterns of yolk movement probably requires additional assumptions that involve compartmentalization of calcium and the existence of spatiotemporal patterns of sensitivity to calcium. A comparative analysis of egg cytoskeletons and associated gelation factors is needed.

A calcium wave can cause a wave of gel-to-sol transition to traverse an egg. Within this wave, yolk particles saltate randomly resulting in a WROS. Two models for this saltation can be tested. If yolk particles are attached to microtubules, calcium may shift the dynamic equilibrium between various microtubular states (19) toward treadmilling. Since the rates of treadmilling and saltation are similar, treadmilling may explain the (W)ROS. If the (W)ROS cannot be blocked by different microfilament inhibitors, it must be caused directly by microtubules. On the other hand, if some of these inhibitors do block the (W)ROS, it is likely to be a manifestation of a calcium-induced contraction of microfilaments to which yolk particles can adhere. It should be noted that colchicine treatment may not discriminate between the two models since it may remove the scaffold that microfilaments need for contraction and generation of the (W)ROS.

As a putative calcium wave induces a temporary gel-to-sol transition, nuclei are passively shifted downstream with the resulting yolk flow. The calcium pulse may shift the state of the microtubules of the migration cytaster toward one where the microtubules are shortening at both ends. Shortening rates should, therefore, be in the range of rates of active nuclear migration [0.13 to 0.43 μm/s minimally (34)].

Assuming a network of microfilaments and microtubules in the cytaster, RS might be triggered by a calcium pulse by one of the three ways suggested earlier. RS continues until midprophase of the next nuclear cycle (Fig. 4), i.e., long after the putative ooplasmic calcium wave has passed. RS cannot be due to Brownian motion because it is inhibited by colchicine (43) and it is too slow. RS may reflect a calcium-initiated calcium cycle confined to the cytaster, such as may underlie caffeine-induced monaster cycling in sea urchin eggs (13) or monaster cycling in parthenogenetically activated molluskan eggs (21). It is not clear how the disassembly of the cytaster in *Wachtliella* is initiated.

In *Drosophila* a transient rise in the level of free calcium may trigger (1) the division of buds by acting upon the caps of myosin and F-actin located between the nuclei and the protruding oolemma and (2) the protrusion of the newly formed surface caps by acting upon the same myosin-actin caps as well as upon the hexagonal network of myosin-actin which is located below the furrows of the oolemma (34). Involvement of actin filaments in cytokinesis is also suggested by the collision of anaphases and the disturbances of cellularization that follow cytochalasin treatment (47).

Oxygen Deprivation Affects Mitosis and Ooplasmic Movements

Reduction of gas exchange or temperature at one egg pole causes WROSs to be triggered from the nonsuppressed pole in *Wachtliella* (44). Likewise, when eggs of *Drosophila* are submitted to a temperature gradient, mitotic waves always start in the warm pole (L. Boring and G. Schubiger, personal communication). These treatments may create a gradient of metabolic cycle frequencies with the highest frequency in the nonsuppressed pole, from which various cycles would spread over the egg.

Under natural conditions the metabolic oscillator would be faster in areas with higher respiratory intensity. Local respiratory intensity in insect eggs may depend on the local rate of gas exchange, which in turn depends on the local surface-to-volume ratio. Since this ratio is largest at the egg poles, the implication is, all other things being equal, that waves of calcium and mitosis would have to start from polar egg regions. However, wave patterns might also be affected by unequal surface-to-volume ratios between the poles, differences in abundance and activity of mitochondria, or differences in the turnover rate of ATP (16). The variability in spatial pattern, speed, and direction of mitotic waves that has been observed in studies of patterns of mitosis and ooplasmic movement (34) may be explained by inadequate control of factors that limit metabolism.

Barriers to the propagation of calcium waves may exist. In *Pimpla*, the last blastodermal mitotic wave is unable to cross a slanted line that is located at about the middle of the egg (Fig. 5). In fixed eggs of *Drosophila*, mitotic waves which originate in both egg poles during plasmodial and cellular blastoderm stages 12 through 14 are unable to cross the middle of the egg when one of the two waves is delayed (8). These barriers may be composed of substances that render cells insensitive to a mitotic stimulus.

Similarities between the Activating Calcium Wave in the Fish *Oryzias* (Medaka) and the WROS

The existence of similarities between the WROS in *Wachtliella* and the activating calcium wave in medaka also supports the hypothesis that metabolic control of ooplasmic movements is mediated by calcium. In the medaka egg, when two ionophore-induced calcium waves meet, they fail to propagate through one another (11). Likewise, when two WROSs arise simultaneously in the egg of *Wachtliella* (44) or when the first or second postblastoderm mitotic wave arises at two or more points simultaneously in *Drosophila*, the two waves annihilate each other upon contact (J. A. Campos-Ortega, personal communication). Both the calcium wave in medaka and the WROS in *Wachtliella* can be made to appear anywhere in the egg, and the wave and the WROS can have a bandwidth of about one-tenth of the egg's diameter or egg length, respectively.

One difference between the two systems is that the velocity of mitotic waves in four different insect species is about 8- to 30-fold lower than the speed of the calcium wave elicited after fertilization of medaka eggs and of the supposedly calcium-induced waves of cortical vesicle fusion in a small marine invertebrate and larger freshwater vertebrate eggs (5 to 15 μm/s, Ref. 34). However, this may be due to ooplasm-specific differences in pH, pCa, or calcium sensitivity. In medaka the calcium wave can be quickened by raising the pH above 7.1. It slows when the pH is set below 7.1, or when the pCa is between the resting level and the concentration found at the threshold for a self-sustaining calcium wave (10). In *Wachtliella*, WROSs slow down or are extinguished when they pass through an area with reduced gas exchange or when they are immediately preceded by another WROS; both processes presumably deplete calcium stores. Likewise, the fivefold variation in speed of mitotic waves in *Drosophila* (8) appears to be related to the availability of oxygen (V. Foe, personal communication). Finally, among noninsect eggs there is a considerable variation in the speed of calcium waves (34). Such rate differences probably reflect the dynamics of the underlying control mechanism rather than a different nature of the waves.

Cell Cycle Models in Light of Data on Early Insect Embryogenesis

So far, four models have been proposed for the cell cycle. These are as follows:

1. The cell cycle is a dependent and partially branched sequence of events which closes on itself. There are simple causal relations between the events.
2. Cell size, expressed as the ratio of cytoplasmic to nuclear mass, is held constant. Deviant ratios trigger cell and nuclear division. Cytoplasmic mass could be monitored as cell surface area, protein or RNA content, etc.
3. There is no cell cycle. The transition of a cell from one state to another is random.
4. Cells have an internal clock in the form of biochemical oscillations. Cell cycle events are triggered independently and in the correct sequence when the oscillations reach given threshold values (for reviews see Chaps. 14 to 16 in Ref. 7).

At present, there is not enough information on the nuclear cycle in early insect development for a rigorous assessment of all four models. However, based on the observations that the various component cycles of the cell cycle in *Acheta* (*Gryllus*), *Pimpla*, *Drosophila*, and *Wachtliella* are independent, it is unlikely that the cell cycle is a simple, causally related sequence of events. This is further supported by the observation that a cytoplasmic factor in the plasmodial egg of *Drosophila* is required for mitosis (6). This factor is translated from maternal messenger RNA which is in limited supply. The need for this factor, however, increases with the number of nuclei until the translational capacity of the egg is exhausted. At this point (cycle 14) mitosis stops. Thus, whereas calcium pulses may control the timing of mitosis, the total number of mitoses may depend on the nucleo-cytoplasmic ratio.

Acknowledgments

My sincere appreciation goes to Drs. Victoria Foe, Ulf-Rüdiger Heinrich, Klaus Sander, Helmut Sauer, Helmut Vollmar, Richard Warn, and Rainer

Wolf for comments on a previous version of this paper, to Drs. Landin Boring, Victoria Foe, Johannes Kaiser, Dirk Went, and Rainer Wolf for permission to use unpublished data, and to Justin Cooper and John Ensing for relief from teaching.

General References

COUNCE, S. J.: 1973. The causal analysis of insect embryogenesis. In: *Developmental Systems: Insects 2*. (S. J. Counce, and C. H. Waddington, eds.), Academic Press, London.

SANDER, K.: 1976. Morphogenetic movements in insect embryogenesis. In: *Insect Development*. (P. Lawrence, ed.), Symposium of the Royal Entomological Society, London, No. 8, pp. 35–52, Blackwell, Oxford.

VAN DER MEER, J. M.: 1988. The role of metabolism and calcium in the control of mitosis and ooplasmic movements in insect eggs: a working hypothesis. *Biol. Rev. Camb. Philos. Soc.* **63**:109–157.

References

1. Begg, D. H., and Rebhun, L. I. 1979. pH regulates the polymerization of actin in the sea urchin egg cortex. *J. Cell. Biol.* **83**:241–248.
2. Berridge, M. J. 1987. Inositol triphosphate and diacylglycerol: two interacting second messengers. *Annu. Rev. Biochem.* **56**:159–193.
3. Brady, R. C., Cabral, F., and Dedman, J. R. 1986. Identification of a 52-kD calmodulin-binding protein associated with the mitotic spindle apparatus in mammalian cells. *J. Cell Biol.* **103**:1855–1861.
4. Bruhns, E. 1974. Analyse der Ooplasmaströmungen und ihrer strukturellen Grundlagen während der Furchung bei *Pimpla turionellae* (Hymenoptera). *Wilhelm Roux's Arch. Dev. Biol.* **179**:55–89.
5. Cobbold, P. H. 1980. Cytoplasmic free calcium and amoeboid movement. *Nature (Lond.)* **285**:441–446.
6. Edgar, B. A., Kiehle, C. P., and Schubiger, G. 1986. Cell cycle control by the nucleo-cytoplasmic ratio in early *Drosophila* development. *Cell* **44**:365–372.
7. Edmunds, L. N. Jr. 1984. *Cell Cycle Clocks*. Dekker, New York.
8. Foe, V. E., and Alberts, B. M. 1983. Studies of nuclear and cytoplasmic behaviour during the five mitotic cycles that precede gastrulation in *Drosophila* embryogenesis. *J. Cell Sci.* **61**:31–70.
9. Foe, V. E., and Alberts, B. M. 1985. Reversible chromosome condensation induced in *Drosophila* embryos by anoxia: visualization of interphase nuclear organization. *J. Cell Biol.* **100**:1623–1636.
10. Gilkey, J. C. 1981. Mechanisms of fertilization in fishes. *Am. Zool.* **21**:259–375.
11. Gilkey, J. C., Jaffe, L. F., Ridgway, E. B., and Reynolds, G. T. 1978. A free calcium wave traverses the activating egg of the medaka, *Oryzias latipes*. *J. Cell Biol.* **76**:448–466.
12. Harris, P. 1975. The role of membranes in the organization of the mitotic apparatus. *Exp. Cell Res.* **94**:409–425.

13. Harris, P. 1983. Caffeine-induced monaster cycling in fertilized eggs of the sea urchin *Strongylocentrotus purpuratus*. *Dev. Biol.* **96**:277–284.

14. Hepler, P. K., and Wolniak, S. M. 1984. Membranes in the mitotic apparatus: their structure and function. *Int. Rev. Cytol.* **90**:169–238.

15. Hess, B. 1979. The glycolytic oscillator. *J. Exp. Biol.* **81**:7–14.

16. Kalthoff, K., Kandler-Singer, I., Schmidt, O., Zissler, D., and Versen, G. 1975. Mitochondria and polarity in the egg of *Smittia* spec. (Diptera, Chironomidae): UV irradiation, respiration measurements, ATP determinations and application of inhibitors. *Wilhelm Roux's Arch. Dev. Biol.* **178**:9–121.

17. Karr, T. L., and Alberts, B. M. 1986. Organization of the cytoskeleton in early *Drosophila* embryos. *J. Cell Biol.* **102**:1494–1509.

18. Keith, C. H., Ratan, R., Maxfield, F. R., Bajer, A., and Shelanski, M. L. 1985. Local cytoplasmic calcium gradients in living mitotic cells. *Nature (Lond.)* **316**:848–850.

19. Kirschner, M., and Mitchison, T. 1986. Beyond self-assembly: from microtubules to morphogenesis. *Cell* **45**:329–342.

20. Kline, D., and Nuccitelli, R. 1985. The wave of activation current in the *Xenopus* egg. *Dev. Biol.* **111**:471–487.

21. Kuriyama, R., Borisy, G. G., and Masui, Y. 1986. Microtubule cycles in oocytes of the surf clam, *Spisula solidissima*: An immunofluorescence study. *Dev. Biol.* **114**: 151–160.

22. Lundquist, A., and Lowkvist, B. 1984. Cell-surface changes during cytokinesis in a dipteran egg. *Differentiation* **28**:101–108.

23. Margolis, R. L., and Wilson, L. 1981. Microtubule treadmills—possible molecular machinery. *Nature (Lond.)* **293**:705–711.

24. Middeldorf, J. 1983. *Ultrastrukturelle Untersuchungen zur Oogenese von Pimpla turionellae (L. Ichneumonidae, Hymenoptera).* Diplomarbeit, Ruhr-Universität, Bochum.

25. Miyamoto, D. M., and Van der Meer, J. M. 1982. Early egg contractions and patterned parasynchronous cleavage in a living insect egg. *Wilhelm Roux's Arch. Dev. Biol.* **191**:95–102.

26. Nagata, K., and Ichikawa, Y. 1984. Changes in actin during cell differentiation. In: *Cell and Muscle Motility*, vol. 5: *The Cytoskeleton*. (J. W. Shay, ed.), pp. 171–193, Plenum Press, New York.

27. Newport, J. W., and Kirschner, M. W. 1984. Regulation of the cell cycle during early *Xenopus* development. *Cell* **37**:731–742.

28. Pijnacker, L. P., and Ferwerda, M. A. 1976. Experiments on blocking and unblocking of first meiotic metaphase in eggs of the parthenogenetic stick insect *Carausius morosus* Br. (Phasmida, Insecta). *J. Embryol. Exp. Morphol.* **36**:383–394.

29. Poenie, M., Alderton, J., Tsien, R. Y., and Steinhardt, R. A. 1985. Changes of free calcium levels with stages of the cell division cycle. *Nature (Lond.)* **315**:147–149.

30. Poenie, M., Alderton, J., Steinhardt, R., and Tsien, R. 1986. Calcium rises abruptly and briefly throughout the cell at the onset of anaphase. *Science* **233**:886–889.

31. Sauer, H. W. 1966. Zeitraffer-Mikro-Film-Analyse embryonaler Differenzierungsphasen von *Gryllus domesticus. Z. Morphol. Oekol. Tiere* **56**:143–251.

32. Schliwa, M., Euteneuer, U., Bulinski, J. C., and Izant, I. G. 1981. Calcium lability of cytoplasmic microtubules and its modulation by microtubule-associated proteins. *Proc. Nat. Acad. Sci. U.S.A.* **78**:1037–1041.

33. Stafstrom, J. P., and Staehelin, L. A. 1984. Dynamics of the nuclear envelope and of nuclear pore complexes during mitosis in the *Drosophila* embryo. *Eur. J. Cell Biol.* **34**:179–189.

34. Van der Meer, J. M. 1988. The role of metabolism and calcium in the control of mitosis and ooplasmic movements in insect eggs: a working hypothesis. *Biol. Rev. Camb. Philos. Soc.* **63**:109–157.

35. Vollmar, H. 1972. Frühembryonale Gestaltungsbewegungen im vitalgefarbten Dotter-Entoplasma-System intakter und fragmentierter Eier von *Acheta domesticus* (Orthopteroidea). *Wilhelm Roux's Arch. Dev. Biol.* **171**:228–243.

36. Went, D. F. 1972. Zeitrafferfilmanalyse der Embryonalentwicklung *in vitro* der vivipar paedogenetischen Gallmücke *Heteropeza pygmaea*. *Wilhelm Roux's Arch. Dev. Biol.* **170**:13–47.

37. Went, D. F., and Krause, G. 1974. Alteration of egg architecture and egg activation in an endoparasitic hymenopteran as a result of natural or imitated oviposition. *Wilhelm Roux's Arch. Dev. Biol.* **175**:173–184.

38. Went, D. F., and Nuss, E. 1976. Das Bewegungsmuster während der Furchung im künstlich aktivierten Ei von *Pimpla turionellae* (Hym.). *Wilhelm Roux's Arch. Dev. Biol.* **180**:257–286.

39. Whitaker, M., and Irvine, R. F. 1984. Inositol 1,4,5-triphosphate microinjection activates sea urchin eggs. *Nature (Lond.)* **312**:636–639.

40. Wilson, J. C. 1970. Analysis of pre-blastoderm cytoplasmic and nuclear movements in *Drosophila montana*. M.A. Thesis. University of Texas at Austin.

41. Wolf, R. 1973. Kausalmechanismen der Kernbewegung und -teilung während der frühen Furchung im Ei der Gallmücke *Wachtliella persicariae*. I. Kinematische Darstellung des "Migrationsasters" wandernder Energide und der Steuerung seiner Aktivität durch den Initialbereich der Furchung. *Wilhelm Roux's Arch. Dev. Biol.* **172**:28–57.

42. Wolf, R. 1978. The cytaster, a colchicine-sensitive migration organelle of cleavage nuclei in an insect egg. *Dev. Biol.* **62**:464–472.

43. Wolf, R. 1980. Migration and division of cleavage nuclei in the gall midge, *Wachtliella persicariae*. II. Origin and ultrastructure of the migration cytaster. *Wilhelm Roux's Arch. Dev. Biol.* **188**:65–73.

44. Wolf, R. 1985. Migration and division of cleavage nuclei in the gall midge, *Wachtliella persicariae*. III. Pattern of anaphase-triggering waves altered by temperature gradients and local gas exchange. *Wilhelm Roux's Arch. Dev. Biol.* **194**:257–270.

45. Wolf, R., and Krause, G. 1971. Die Ooplasmabewegungen während der Furchung von *Pimpla turionellae* (Hymenoptera), eine Zeitrafferfilmanalyse. *Wilhelm Roux's Arch. Dev. Biol.* **167**:266–287.

46. Woods, N. M., Cuthbertson, K. S. R., and Cobbold, P. H. 1987. Agonist-induced oscillations in cytoplasmic free calcium concentration in single rat hepatocytes. *Cell Calcium* **8**:79–100.

47. Zalokar, M., and Erk, I. 1976. Division and migration of nuclei during early embryogenesis of *Drosophila melanogaster*. *J. Microsc. Biol. Cell.* **25**:97–106.

48. Zissler, D., and Sander, K. 1982. The cytoplasmic architecture of the insect egg cell, In: *Insect Ultrastructure*, vol. I. (R. C. King, and H. Akai, eds.), pp. 189–221, Plenum, New York.

Questions for Discussion with the Editor

1. *Recalling the old adage "a model is only as good as it is testable," please offer a few direct experimental tests for your metabolic cycle–calcium cycle model.*

The highest priority should be given to direct visualization of putative calcium pulses by high-resolution monitoring of light emitted by a calcium-sensitive probe. It is possible to avoid the pitfall of absorption of light by the yolk because meiosis and the later stages of plasmodial mitosis occur at the surface of the egg. If calcium pulses do occur, their relationship to other cytoplasmic and nuclear cycles, and to a possible metabolic cycle, needs clarification. This requires monitoring of calcium pulses in eggs that have been manipulated to change various cycles. This can be done in eggs of *Wachtliella* in which the WROS can be reversed by local low temperature or by oxygen deprivation. *Drosophila* eggs have the advantages of artificial activation, manipulation of mitotic waves by oxygen deprivation, and mutational dissection. Other species (*Pimpla, Venturia, Smittia,* and *Psychoda*) allow control only over the timing of activation, but they have a much higher percentage of activation than *Drosophila*. *Pimpla* eggs are useful because they can be activated by injection of A23187; however, the exact activation percentage remains to be established. Eggs of *Heteropeza* develop parthenogenetically in vitro and can easily be monitored throughout early development. We also need bioassays for substances that control the various nuclear and ooplasmic cycles. We can study these cycles by blocking them by injection of cytoplasm from arrested oocytes into fertilized eggs (see Ref. 27), and by mutants that exhibit blocks at the appropriate points in the cycle. Removal of the block can be attempted with calcium, IP_3, and various other ooplasmic substances.

Local fluctuations in sensitivity to calcium should be tested even if calcium pulses do occur. Calcium sensitivity of organelles such as microtubules is mediated by calcium-binding proteins (CBPs). CBPs should be isolated and used to prepare antibodies. Fluorescent anti-CBP antibodies may reveal changes in the quantity, distribution, and association of CBPs with target organelles.

To investigate the role of metabolism, continuous spectrophotometric and potentiometric recording of concentrations of metabolic intermediates and of enzyme activity (see Ref. 15) has to be applied to insect eggs. For instance, one could monitor the fluorescent light emitted by a probe sensitive to the ooplasmic pH or to changes in membrane potential.

2. *Do you suppose that the observations on ooplasmic movements represent only the "tip of the iceberg?" That is, do you expect that increased resolution will reveal even more ooplasmic traffic?*

Better temporal resolution (through the use of longer time lapses) at constant spatial resolution may reveal movements that are slower than those observed so far. However, they may be difficult to detect if they overlap with simultaneously occurring ooplasmic movements or with the later postblastodermal morphogenetic movements. For instance, two simultaneous movements, such as the WROS and ooplasmic flow, can both be detected (Fig. 1). This is because their frequencies are different enough that they do not interfere but similar enough so as not to render one of them too slow for detection. It is difficult to detect the WROS at later stages because of the presence of more nuclei whose cytasters display RS simultaneously with the WROS; the similarity of the frequencies of the WROS and RS leads to interference.

Better spatial resolution, using light-emitting or other molecular probes, will reveal successively smaller objects such as organelles, supramolecular structures (e.g., protein filaments, enzymes), and macromolecules. Their freedom of movement will depend on the state of assembly of the cytoskeleton, whether they are attached to the cytoskeleton, and whether their cytoplasmic environment is in a gel or sol state.

The term ooplasmic movement should exclude movements that are not uniquely cellular, such as thermal motion of molecules and Brownian motion. Thus, by increasing the power of spatial resolution one will meet a limit below which there is no ooplasmic movement.

CHAPTER **14**

Cytoplasmic Localization in Insect Eggs

Klaus Kalthoff and Michael Rebagliati

*Dedicated to Klaus Sander on the occasion of his
60th birthday*

Introduction

AN INSECT EGG CONTAINS not only genetic information encoded in DNA but also structural asymmetries which arise during oogenesis and which may be laid down in the eggshell (chorion), plasma membrane, or cytoplasmic components. We can usually predict the polarity axes of the developing embryo and the site of primordial germ-cell (pole-cell) formation from asymmetrical features in the egg of a particular species. In *Drosophila*, for instance, the head of the embryo develops where the chorion has an opening for the sperm (micropyle) and a hatch for the larva. Similarly, the ventral side of the embryo, as defined by the furrow formed during gastrulation, develops along the most convex meridian of the eggshell. Finally, primordial germ cells located near the posterior pole form at the site of distinctive cytoplasmic organelles known as polar granules. Sometimes, but not always, correlated with the overt structural asymmetries are localizations of cytoplasmic factors that actually determine the fates of embryonic cells and the spatial coordinates of the developing organism. The biologic activities of such determinants are generally shown by transplantation experiments, and their distributions are made visible by specific molecular probes. In this chapter, we discuss the localization of cytoplasmic determinants in *Drosophila* and two other dipterans, *Chironomus* and *Smittia*.

Bioassays for Localized Cytoplasmic Determinants

The term "localized cytoplasmic determinants" is often used to refer to physiologically active components that are associated with visible markers and are present in sharply defined egg regions. This view tends to set localized determinants apart from the idea of morphogen gradients. The latter are usually thought of as small diffusible molecules emanating from an invisible source and specifying different cell fates in a concentration-dependent way. We propose to drop these connotations for several reasons. First, the association of some localized cytoplasmic determinants with visible markers is incidental. Second, the formation of concentration gradients is not limited to small molecules, especially not in insect embryos, which become cellularized only after many nuclear divisions and can, therefore, utilize macromolecules as signals between nuclei. Third, some cytoplasmic determinants are distributed in a gradient fashion, but whether different concentrations activate different genes remains to be seen (see below).

We define "localized determinants" as asymmetrically distributed (or asymmetrically activated) molecules that determine certain cell lineages and/or specify part of the embryonic body pattern. One strategy to prove the existence of localized cytoplasmic determinants is heterotopic transplantation. This strategy was used in the pioneering experiments of Sander who pushed, by means of a blunt needle, posterior pole plasm into the middle of the egg of the leaf hopper, *Euscelis plebejus*. Such translocations, especially when followed by transverse ligations, dramatically altered the polarity and segment patterns formed in the separated egg fragments (51, 52). Similarly, posterior pole plasm from *Drosophila* eggs, when injected anteriorly or ventrally into host eggs, caused the ectopic formation of pole cells that could give rise to functional gametes (Chap. 11, this volume). Both transplantation experiments were aided by the presence of visible markers in the transplanted cytoplasm, which consisted of a ball of symbiontic bacteria in the case of *Euscelis* eggs and of basophilic fibrous granules (polar granules) in *Drosophila* eggs. The association of the visible markers with the physiologically active components has proven to be rather loose in *Euscelis* but appears to be close in the case of the polar granules in *Drosophila*. Attempts to find visible structures associated with other localized determinants in insect egg cytoplasm have not been successful (73).

Another operational criterion for localized cytoplasmic determinants is the rescue procedure. For instance, *Chironomus samoensis* embryos can be programmed, by uv inactivation of anterior cytoplasmic components, to develop the abnormal segment pattern "double abdomen," where head and thorax are replaced with a mirror-image duplication of abdominal segments (72). Such embryos were "rescued," i.e., restored to normal development, by microinjection of anterior cytoplasm or RNA from unirradiated donors (Fig. 1; see also Ref. 13). Similar rescue experiments were carried out using as recipients *Drosophila* embryos that developed abnormally because of maternal-effect mutations (3, 44, 54).

Both heterotopic transplantation and the rescue procedure can be used as

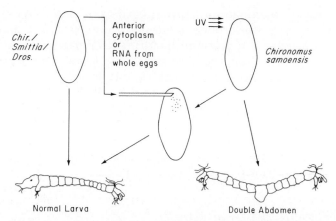

FIGURE 1. Rescue bioassay for anterior determinant activity. Recipient eggs from *Chironomus samoensis* are programmed for double abdomen development by uv irradiation of the anterior pole region. Such embryos can be "rescued," i.e., restored to normal development, by injecting anterior cytoplasm or RNA from unirradiated donors. Rescuing activity is measured as an increase in the percentage of normal embryos among the surviving embryos.

bioassays for characterizing cytoplasmic determinants. By testing cytoplasm from different donor regions and developmental stages, one can determine where and when the activity is present. In the case of the *Chironomus* egg shown in Fig. 1, the rescuing activity decreases from the anterior to the posterior pole and is present from egg deposition to gastrulation. At any given stage, the nature of the inducing or rescuing activity can be determined by transplanting subcellular or molecular fractions instead of whole cytoplasm. In several cases, the activity was found to be associated with poly(A)$^+$ RNA, suggesting that the active cellular components are messenger ribonucleoprotein (RNP) particles (Refs 3 and 13; Chap. 11, this volume).

Three Sets of Cytoplasmic Determinants in Insect Eggs

Cytoplasmic determinants control three major features of dipteran embryos: germ-cell formation, the anteroposterior polarity, and the dorsoventral polarity. The development of each feature is affected by several maternally expressed gene products, some of which are known to be localized while others are not.

Germ-Cell Determinants

Germ-cell determinants in *Drosophila* eggs become localized in the posterior pole plasm during oogenesis. Transplanted posterior pole plasm induces ectopic formation of functional primordial germ cells and completely rescues embryos sterilized by posterior uv irradiation. However, only limited rescue is observed after injection of a postmitochondrial RNP particle fraction or with poly(A)$^+$ RNA extracted from this fraction. The "rescued" embryos form pole cells which become incorporated into gonads but do not give rise to gametes. Likewise, ectopic pole cells are not induced by RNP particles or poly(A)$^+$ RNA. Apparently, these fractions contain some, but not all, of the components required for germ-cell development (Chap. 11, this volume). In accordance with this conclusion, several maternal-effect mutants interfere with different steps of germ-cell development, including polar granule formation, nuclear migration to the posterior pole, and gametogenesis.

Determinants of Anteroposterior Polarity

Localized cytoplasmic determinants of the anteroposterior axis have been analyzed by experimental and genetic means (26, 44, 51, 52). After uv irradiation, centrifugation, or other experimental manipulations, *Chironomus* and *Smittia* eggs form one of four basic body patterns: double abdomens (Fig. 2), double cephalons, inverted embryos, or normal embryos (26, 72). Clearly, anterior as well as posterior embryonic halves can produce either anterior or posterior segments. The basic decision between these two developmental pathways appears to be controlled by mutually repressive anterior and posterior cytoplasmic determinants. Anterior determinants have been characterized as localized cytoplasmic RNP particles in *Smittia* (26). This conclusion has been confirmed and extended with *Chironomus* eggs (13) using the bioassay diagrammed in Fig. 1.

Abnormal body patterns reminiscent of chironomid double cephalons and double abdomens can be generated in *Drosophila* eggs by leakage and transplantation of cytoplasm (44). To produce the double abdomen pattern in *Drosophila* eggs, it is necessary to remove anterior cytoplasm and transplant posterior cytoplasm to the anterior pole region. Either operation by itself causes mostly reduction or deletion of head structures, with an expansion of thorax and abdomen toward the anterior. Some of the operated embryos show supernumerary telson structures, such as anal plates and sensory cones, at the anterior end. However, no mirror-image duplications of the abdominal segments are visible in the larval cuticle. Similar results are obtained by anterior uv irradiation of *Drosophila* eggs (26). This comparison suggests that in chironomids posterior determinants are uv-resistant, present in both anterior and posterior egg halves, and repressed anteriorly by uv-sensitive anterior determinants. This was shown most conclusively when uv-irradiated isolated anterior fragments of *Smittia* embryos produced abdomens instead of heads (W. Ritter and K. Sander, unpublished; reviewed in Ref. 25). If posterior determinants in *Drosophila* were similarly present but repressed in the anterior half of *Drosophila* eggs, the repressor would appear to be uv-resistant. Alternatively, posterior determinants might be absent from the anterior half of *Drosophila* eggs (see below).

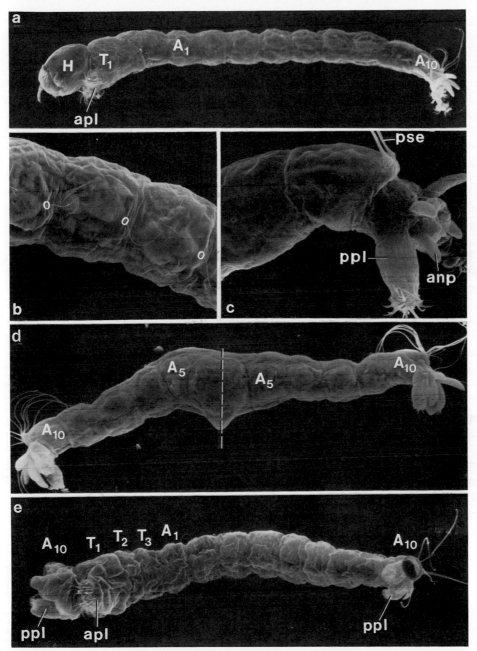

FIGURE 2. *Chironomus* larvae showing normal (*a–c*), symmetrical double abdomen (*d*), and asymmetrical double abdomen (*e*) patterns. H = head, T_1 to T_3 = thoracic segments, A_1 to A_{10} = abdominal segments, *anp* = anal papilla, *apl* = anterior proleg, *ppl* = posterior proleg, *pse* = posterior seta. Origins of lateral setae (circled in part *b*) indicate posterior margins of segments. Hatched line in *d* indicates plane of symmetry and polarity reversal in symmetrical double abdomen. Specimen in *e* shows the normal series of segments T_1 through A_{10} and, attached to the anterior face of T_1, the terminal abdominal structures of A_{10} with reversed polarity.

The analysis of embryonic axis determination has been advanced enormously by the study of maternal-effect mutations in *Drosophila*. Such mutations affect maternal gene products deposited in the egg cytoplasm during oogenesis. If these maternal products are critically needed before the onset of embryonic transcription, deficiencies cannot be corrected by the introduction of a paternal wild-type allele. Throughout this chapter, we conveniently refer to offspring from maternal-effect mutant females as mutant embryos. For instance, we call offspring from *bicoid* (*bcd*) females "*bcd* embryos." Also for convenience, we often refer to wild-type alleles of genes or their products by their mutant names, such as "*bcd* protein." Only where we see a potential ambiguity do we add the "plus" symbol to indicate the wild-type allele, as in "*bcd*+ activity." Nüsslein-Volhard et al. (44) have classified 18 maternal-effect genes that specify the anteroposterior body pattern into three groups that affect the (i) anterior, (ii) posterior, or (iii) terminal regions of the embryo. On the basis of allele frequencies, it is estimated that the majority of such genes have now been identified.

The genes affecting the posterior part of the body pattern have often been epitomized by the *oskar* (*osk*) mutant, which shows deletions of abdominal segments while head, thorax, and telson are normal (29). In addition, *osk* embryos lack polar granules and do not form pole cells. Both defects can be rescued by transplanting posterior cytoplasm (0 to 10 percent egg length) from wild-type donors. Transplantation to the posterior pole of the recipient can rescue pole-cell formation, while the abdominal defect is rescued most effectively by transplantation into the prospective abdominal region (20 to 50 percent egg length). Similar properties are shown by six other mutants, *staufen* (*stau*), *tudor* (*tud*), *vasa* (*vas*), *valois* (*vls*), *nanos* (*nos*), and *pumilio* (*pum*), except that *nos* and *pum* do not affect pole-cell formation (Ref. 30; 57; R. Lehmann, personal communication).

The key abdominal determinant seems to be the *nos* product. The functions of *osk, stau, tud, vas,* and *vls* appear to be in localizing and stabilizing both the source of abdominal determinant activity and the germ-cell determinants in the posterior pole plasm. This view is supported by the following observations. Nurse cell cytoplasm from most of these mutants, with the exception of *nos*, rescues the abdominal phenotype of itself and the others; nurse cell cytoplasm from *nos* nurse cells does not rescue the embryos of *nos* or other mutants. In contrast, posterior egg cytoplasm from a given mutant, with the exception of *pum*, does not rescue the abdominal phenotype of itself or other mutant embryos. Only *pum* posterior plasm shows the same localized activity as wild-type posterior cytoplasm, i.e., it will rescue *pum* and other mutant embryos if injected into the abdominal region (Ref. 30; R. Lehmann, personal communication). Thus, the *pum* product seems to be required specifically for the transport of the abdominal determinant activity from its source, the posterior pole plasm, to its site of action, the abdominal region. Since the role of the *osk* gene was analyzed before *nos*, many subsequent investigations have used *osk* as the prototype of genes affecting posterior determinants. Many of the *osk*+-dependent activities revealed in these studies are likely to be caused more directly by the *nos*+ product, which presumably fails to become localized or is rendered inactive in *osk* embryos.

Among the genes affecting the anterior body pattern in *Drosophila* embryos, the key gene appears to be *bicoid (bcd)*. Offspring from *bcd* mothers resemble the embryos obtained after leakage of anterior cytoplasm from wild-type eggs. Strong mutant alleles lack head and thorax completely and show an extended abdomen joined at the anterior end with a duplication of the telson, i.e., the structures beyond the eighth abdominal segment. Transplanting cytoplasm from wild-type donors can rescue *bcd* embryos. Again, the rescuing activity is localized, in this case between 85 and 100 percent egg length (0 percent = posterior pole). Moreover, transplantation of anterior wild-type cytoplasm to the middle of *bcd* embryos induces heterotopic formation of anterior structures accompanied by local polarity reversal (17). Two other maternal-effect mutants, *exuperantia (exu)* and *swallow (sww)*, affect the anterior body pattern by interfering with the localization of *bcd* mRNA (see below). Mutations at the three *bicaudal* loci (*bic, Bic-C* and *Bic-D*) each produce the same wide range of anterior deletions and posterior duplications. Symmetrical bicaudal phenotypes consist of a mirror-image duplication of the one to five most posterior abdominal segments, except that pole cells are formed only posteriorly (43). These phenotypes differ from the *bcd* phenotype by duplications of abdominal segments in addition to the telson.

The terminal structures of the *Drosophila* larva are affected by *torso (tor)*, *trunk (trk)*, *torsolike (tsl)*, *fs(1) polehole (fs(1)ph)*, and *fs(1) Nasrat (fs(1)N)*. Each of these mutations delete the labrum and acron anteriorly, and the posterior midgut, Malpighian tubules, telson, eighth abdominal segment, and part of the seventh abdominal segment, posteriorly. The pole plasm is present in these embryos, and functional pole cells are formed. Since the phenotypes of both *bic* and *bcd* show an anterior duplication of the telson, the anterior terminal region appears to become transformed into its posterior counterpart in both of these mutants.

Relationship between Abdominal and Germ-Cell Determinants

The pleiotropic effects on pole-cell formation and abdominal segmentation observed in *osk, stau, tud, vas*, and *vls* raise the question whether any gene products act exclusively as either germ-cell or abdominal determinants. Alternatively, a morphogen gradient with a maximum at the posterior pole could determine pole formation, while abdominal segments would be specified at lower morphogen concentrations toward the egg equator. Strong mutant phenotypes of *pum* and *nos* show abdominal defects but have normal polar granules and form functional pole cells, indicating that these gene products are necessary for abdomen, but not for germ-cell, development (30, 44). Conversely, mutants in other loci lack pole cells or fail to form gametes, but are otherwise normal. For instance, *gs(1)N26* eggs show normal polar granules and a normal body pattern, but fail to form pole cells. The movement of cleavage nuclei into the posterior pole region is the critically disturbed function which needs to be on time for pole-cell formation but not for abdomen devel-

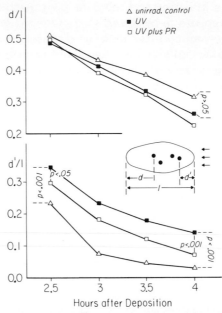

FIGURE 3. Effects of posterior irradiation on nuclear migration in *Smittia* embryos. Ultraviolet irradiation was applied between 55 and 60 min after egg deposition, and photoreactivating light (PR) was applied between 2 and 3 h later. *Upper panel*: Relative distance (*d/l*) of anteriormost nucleus from anterior pole. *Lower panel*: relative distance (*d'/l*) of posteriormost nucleus from posterior pole. In uv-irradiated embryos (■), the migration of the posteriormost nucleus is significantly delayed or curtailed as compared to the unirradiated controls (△). The effect of uv is mitigated to a significant extent by photoreactivating light, which repairs uv damage to RNA (□). Each point is based on the analysis of approximately 18 whole mounts obtained in two independent experiments. Probabilities (*p*) are based on *t* test for differences between means. From Ref. 71.

opment (45). This nuclear movement, at least in *Smittia*, is controlled by uv-sensitive components in the posterior pole plasm (Fig.3). Pole-cell and abdomen formation are also separated in bicaudal phenotypes of *Drosophila* and in double abdomens of chironomids, which show pole cells in the posterior abdomen but not in its anterior counterpart (26, 43). We conclude that pole-cell and abdomen development require separate, and probably localized, cytoplasmic determinants. The pleiotropic effects of *osk, stau, tud, vas,* and *vls* may reflect common mechanisms for localizing or stabilizing abdominal and germ-cell determinants.

Determinants of Dorsoventral Polarity

Unlike the anteroposterior axis, the dorsoventral axis cannot be reprogrammed easily by uv irradiation or other experimental manipulations. Consequently, most of the progress in identifying dorsoventral determinants has come from genetic analysis rather than from experimental embryology. Twelve maternal-effect mutations which specifically disturb dorsoventral polarity in the *Drosophila* embryo have been identified (2, 3, 58). Since these mutations have no major effect on anteroposterior polarity, it can be inferred that different cytoplasmic determinants participate in specifying each axis.

The dorsoventral mutations can be subdivided into ventralizers and dorsalizers according to the shifts in cell fate caused by each mutation. For instance, cells are generally switched to dorsal programs of development in *dorsal* (*dl*) embryos (i.e., offspring from homozygous *dl* females). In the null phenotype, cells that would normally form mesoderm and the neurogenic ectoderm (ventral neurogenic region) develop as dorsal epidermis. Homozygous *cactus* (*cac*) embryos, in contrast, are ventralized; their neurogenic ectoderm is expanded at the expense of dorsal epidermis (D. Godt, unpublished; reviewed in Ref. 2; 58). Two genes, *easter* (*ea*) and *Toll* (*Tl*), are unusual because mutations at these loci can give rise to either ventralized or dorsalized embryos. Recessive *Tl* (Tl^-) and *ea* alleles are dorsalizing; dominant *Tl* (Tl^D) and *ea* alleles are ventralizing. Mutations in the other eight dorsoventral maternal-effect genes are dorsalizing.

Several of the gene products (*dl*, *Tl*) involved in dorsoventral patterning are distributed evenly throughout the egg as inactive precursors. They become activated locally to establish the polarity of the embryo. The most dramatic evidence for this conclusion comes from experiments in which Tl^- embryos are injected with wild-type cytoplasm or Tl^+ mRNA. The normal range of dorsoventral structures is restored by this treatment, but the ventral pattern elements always develop near the site of injection. Thus, the normal dorsoventral polarity is locally inverted by injecting Tl^+-containing fractions into the original dorsal side of Tl^- embryos, which can still be identified in relation to morphological features of the chorion. Moreover, cytoplasm from the dorsal side of wild-type donors (again, as defined by chorion morphology) is just as effective as total egg cytoplasm in inducing ventral structures in Tl^- recipients. In agreement with these results, in situ hybridizations show a uniform dorsoventral distribution of maternal *Tl* mRNA at the preblastoderm and blastoderm stages (20). Taken together, these results show that: (i) inactive Tl^+ mRNA or protein is found throughout the egg, (ii) Tl^+ mRNA or protein is activated on the prospective ventral side in normal development, and (iii) activation occurs at the site of injection under conditions of the rescue experiment (2, 3).

Similar rescue experiments have been used to define the spatial distribution of other maternal gene products affecting dorsoventral polarity. For example, *dl* rescuing activity is first distributed uniformly and becomes enriched on the ventral side at the preblastoderm stage (54). Consistent with this, in situ hybridization and immunostaining experiments show that the *dl* mRNA is uniformly distributed throughout the embryo while the *dl* protein is most concentrated on the ventral side (63). Even before pole-cell formation *pelle* (*pll*) activity is enriched ventrally (41). In contrast to rescued Tl^- embryos, ventral structures always form on the original ventral side and not near the site of injection. The ability to orient the dorsoventral axis is a property so far only observed for the Tl^+ product. No differences in the spatial distribution of rescuing activity are found for *snake* (*snk*), *tube*, or *ea*. However, *ea* mutations can be either ventralizing or dorsalizing, suggesting that *ea* may be spatially regulated like *Tl* (3).

The hierarchy of gene products affecting the dorsoventral axis has been deduced from an analysis of temperature-sensitive alleles and double mutants. The best candidates for the "downstream" loci ultimately responsible for establishing dorsoventral polarity are *Tl, dl,* and possibly *cac. Tl* has a temperature-sensitive period extending from about 45 to 150 min postfertilization (25°C), and the dominant alleles are partially epistatic to *ea, gastrulation defective, pipe, nudel,* and *snk*. All of these properties are consistent with the idea that *Tl* acts late to determine the dorsoventral axis (3). Since *dl* alleles are in turn epistatic to Tl^D mutations, one must postulate a later requirement for *dl*. Given that *cac* is the only maternal product specifically required for dorsal structures, it may also have a direct effect on cell fate.

Based on these genetic and rescue experiments, Anderson and Nüsslein-Volhard (3) propose that the net effect of the maternal gene products on dorsoventral polarity is to generate a morphogen gradient. This model provides an elegant, though purely formal, interpretation of the weak mutant phenotypes, which exhibit losses of extreme dorsoventral structures and proportionate shifts of the remaining structures on the blastoderm fate map. Certain other developmental phenomena are also explained most simply by a maternal gradient. For instance, in the neurogenic ectoderm, the tendency of cells to differentiate as neuroblasts increases progressively from dorsal to ventral (67). Immunostaining of preblastoderm and blastoderm embryos shows that the *dl* protein is distributed in a ventral to dorsal gradient in peripheral nuclei, with the high point being found ventrally (63). The critical test of any putative morphogen gradient is to demonstrate that different concentrations of the candidate molecule have different regulatory effects on embryonic nuclei. In the case of *dl*, it should be possible to test this by (i) determining whether the phenotypes of *dl* mutations correlate with changes in the *dl* gradient and (ii) seeing whether the genes regulated by *dl* are controlled in a concentration-dependent fashion by the *dl* protein.

Molecular Nature of Localized Cytoplasmic Determinants

In *Drosophila*, the availability of mutants has made it possible to clone and characterize several of the genes involved in cytoplasmic localization. In species without established genetics, such as *Chironomus* and *Smittia*, individual gene products may be identified by means of bioassays or by low stringency hybridization using probes from *Drosophila*.

The bicoid Gene in Drosophila

Given the key role of the *bcd* gene in determining the anteroposterior polarity in *Drosophila* eggs, one expects to find in the *bcd* protein some of the characteristics of other gene regulatory proteins. The embryonic *bcd* protein appears in western blots as two bands with apparent molecular weights of 55,000 and 57,000. The two variants may result from a minor difference in mRNA splicing and/or protein phosphorylation ˙(4, 11). The *bcd* protein contains a homeodomain, which, in its presumptive DNA-binding region, differs from the *Antennapedia*- and *Ultrabithorax*-like homeodomains. The presence of additional repetitive elements, including a *paired* repeat (16), and a putative RNA binding domain (47a), suggests that *bcd* may be part of a gene regulatory network. In situ hybridizations show a rather sharp localization of *bcd* RNA near the anterior pole of oocytes, eggs, and early embryos (4). Transplantation experiments suggest that the localization of *bcd* RNA may be linked with its translational control (see below). Immunologically detectable quantities of *bcd* protein are first observed after egg deposition but independently of fertilization. Immunostaining of *bcd* protein in whole mounts of preblastoderm and blastoderm stages shows a gradient profile with a maximum near the anterior pole. This profile is relevant to the question whether gap genes may become activated by different concentrations of *bcd* protein (see below). The stain intensity along the periphery of the embryo, when plotted versus egg length (0 percent = posterior pole), shows a shoulder of maximum intensity between 100 and 90 percent followed by an exponential decrease until a background level of around 30 percent is reached. Most of the stain is concentrated in the nuclei as expected for a homeodomain-containing protein (11).

Both *bcd* RNA and protein interact with other maternal gene products in early embryos. The anterior localization of *bcd* mRNA depends on the activities of *exu* and *sww* (see below). Conversely, *bcd* protein appears to shape a gradient of *caudal* (*cad*), another localized gene product in *Drosophila* embryos. Maternal *cad* transcripts are distributed uniformly during the first 12 nuclear cycles. A gradient of *cad* RNA with a maximum at the posterior pole is formed thereafter, presumably by selective RNA degradation. Prior to the appearance of the RNA gradient, a distinct gradient of *cad* protein is built up beginning at cell cycle 10. Selective translation of maternal *cad* mRNA in the posterior region could establish the protein gradient and at the same time stabilize the mRNA where it is translated (32). The translation of *cad* mRNA might be inhibited anteriorly by *bcd*[+] protein because in *bcd* embryos the concentration of maternal *cad*[+] protein is uniformly high (38). The function of the maternal *cad* gradient is not obvious since embryos derived from *cad* mothers can be rescued almost completely by a paternally derived *cad*[+] gene.

Mutual inhibition between the gene products of *bcd* and *nos* are indicated by the following observations. Products of *nos* seem to inhibit *bcd* products because the ectopic formation of cephalic structures after transplantation of anterior cytoplasm, as well as the longevity of *bcd* mRNA in the posterior halves of *exu* and *sww* embryos, are enhanced in *osk* and *vas* mutant backgrounds (4,

18). Conversely, *bcd* products appear to suppress *nos* products because the expression domain of the gap gene *Krüppel* (see below) is shifted—not merely expanded—anteriorly in a *bcd* mutant background (19).

Anterior Determinants in Chironomids

The regional suppression of the synthesis of a particular protein by anterior determinants has also been shown in *Smittia*. The synthesis of a "posterior indicator protein" (designated PI_1, molecular weight of approximately 55,000 and isoelectric point of approximately 5.5) is limited to the posterior half (excluding pole cells) in normal embryos before cellular blastoderm formation. As is the case with *cad* in *Drosophila*, the PI_1 protein in *Smittia* seems to be translated first from maternal mRNA and then from embryonic transcripts. Inactivation of anterior determinants by uv irradiation is followed by both double abdomen formation and embryonic PI_1 synthesis in both posterior and anterior fragments. Photoreactivating treatment, which repairs uv damage to egg RNA, restores both normal development and the restriction of PI_1 synthesis to the posterior half of the embryo (26).

The RNA showing anterior determinant activity in *Chironomus* eggs has been analyzed using the *Chironomus* rescue bioassay (Fig. 1). In eggs and early embryos, the activity is associated with poly(A)$^+$ RNA, and with an RNA size class estimated between 250 and 500 nucleotides. After blastoderm formation, the activity is no longer associated with RNA but is still present in cytoplasm. Heterospecific transplantations indicate that anterior determinant activity is also present, and localized, in *Smittia* and *Drosophila* egg cytoplasm although the activity decreases with phylogenetic distance between donor and recipient (13). *Smittia* and *Chironomus* represent two subfamilies within the family Chironomidae and the suborder Nematocera, which is considered to be more primitive than the suborder Brachycera, to which *Drosophila* belongs.

Localization of Anterior and Posterior Determinants in Drosophila as Compared to Chironomids

The unusually small size (250 to 500 nucleotides) of anterior determinants in *Chironomus* eggs contrasts with the size of corresponding maternal RNAs in *Drosophila* eggs, including *bcd* (2.6 kilobases) and *hunchback* (*hb*, 3.1 kilobases). The discrepancy adds to the apparent difference in the localization of posterior determinants discussed above. In chironomids, posterior determinants are present in both egg halves. They are repressed, but not degraded, anteriorly by localized determinants including the short poly(A)$^+$ RNA identified in the rescue bioassay. In *Drosophila*, anterior and posterior determinant activities inhibit each other when they are mixed by transplantation or as a result of disturbed localization. Normally, these activities seem to be well segregated to their respective pole regions. Both *bcd* RNA and protein are localized anteriorly as described above. The *nos* activity seems to be localized as well since *nos* embryos can only be rescued by posterior pole plasm from wild-type donors (R. Lehmann, personal communication). The localization of *nos* product will leave any homologue to the small anterior determinant in *Chironomus* without apparent function in *Drosophila*. According to this hypothesis, the heterospecific activity of anterior cytoplasm from *Drosophila* eggs would be vestigial or would be caused by cross-activity of other gene products.

The role of the *bicaudal* gene products is not easy to understand. While the original *bic* mutant has been characterized as a hypomorph (43), dominant mutations in two other loci (*Bic-D* and *Bic-C*) produce the same wide range of phenotypes (40). The *bicaudal* products cannot be just auxiliary to *bcd* because strong *bcd* phenotypes do not show the replacement of anterior with posterior segments that characterizes the bicaudal phenotype. The latter must result from the presence of posterior determinants in the anterior egg half of *bicaudal* embryos. This is shown most clearly by rescue experiments using *Bic-D* donors and *osk* recipients. In *Bic-D* donors, *osk*-rescuing activity is present in both anterior and posterior pole plasm, but not in midegg cytoplasm, whereas in wild-type donors the activity is limited to the posterior pole region. Moreover, the epistasis of *osk* over *Bic-D* indicates that *osk*$^+$-dependent activity is required for the development of the bicaudal phenotype (29). The results suggest that the functions of the *bicaudal* loci prevent the establishment of a second source of *osk*$^+$-dependent activity, presumably of *nos*$^+$ product, near the anterior pole. However, this second source is not a complete duplication of the posterior pole plasm since pole cells form only posteriorly in *bicaudal* embryos. Once the *bicaudal* products from *Drosophila* and the small anterior determinant from *Chironomus* are cloned, it will be interesting to see whether homologues exist in the other species and how their functions compare.

Dorsoventral Determinants in Drosophila

The molecular analysis of *dl, Tl,* and *snk* has provided some hints as to how the spatial regulation of dorsoventral determinants is achieved and how they mediate patterning. The *snk* gene encodes a 430 amino acid protein with strong homology to serine proteases. This homology includes conservation of the catalytic amino acid residues and of cysteines involved in disulfide bonding. Many serine proteases are synthesized as zymogens and are activated by specific cleavage at the peptide bond preceding the sequence Ile-Val-Gly-Gly. Since the *snk* protein also contains this sequence, it is probably regulated in the same way. These features suggest that a spatially regulated proteolytic cascade is involved in establishing dorsoventral polarity (10).

Tl encodes a 5.3 kilobase poly(A)$^+$ RNA that can direct the synthesis of a membrane protein (22). Two repetitive sequences of potential biologic significance are found in the extracytoplasmic domain of the *Tl* protein. There are 17 Asn-X-Ser/Thr tripeptide sequences that could be N-linked glycosylation sites. In addition, there are 15 repeats of a leucine-rich sequence.

Structure-function relations in other known proteins implicate such leucine-rich repeats in membrane anchoring or in mediating the binding of proteins as specific ligands. On the basis of these structural motifs, Anderson and her associates have pointed out several possibilities for the function of the *Tl* protein (22). One possibility is that the *Tl* protein might function as a receptor involved in signal transduction. Given the potential membrane binding properties of the leucine-rich repeats, it is also conceivable that the maternal *Tl* protein is involved directly in forming the ventral furrow during gastrulation. Consistent with this idea, there is some evidence that zygotically expressed *Tl* protein may act as a cell-adhesion molecule during late embryogenesis (20).

The amino terminus of the *dl* protein shares extensive homology with proteins encoded by the *rel* proto-oncogenes (62). With either v-*rel* or c-*rel*, the degree of conservation over a 295 amino acid stretch is about 50 percent. If one includes conservative amino acid changes, then the degree of similarity increases to 80 percent. The *dl* protein is found mainly or exclusively in the peripheral somatic nuclei of preblastoderm and blastoderm embryos and shows a graded decrease in concentration from the ventral to dorsal midline (63). Taken together, these observations raise the possibility that *dl* may encode a gene-regulatory protein. Since *dl* mRNA is evenly distributed during oogenesis and early embryogenesis, the dorsoventral differences in *dl* activity must be caused by translational or posttranslational control mechanisms. Whatever their nature, these mechanisms appear to be activated as a consequence of fertilization (63). This is in marked contrast to the gradients of *bcd* and *cad* protein, which can be generated in unfertilized eggs.

On the basis of these observations, several authors have proposed the following working hypothesis for dorsoventral patterning in *Drosophila* (10, 22, 62). An unknown asymmetric cue may initiate a proteolytic activation cascade that leads to the synthesis or activation of *Tl* protein on the ventral side. Active *Tl* protein may then transduce a signal that regulates *dl* activity, which in turn would exert some regulatory effect in the nuclei of blastoderm cells. This hypothesis is admittedly speculative, but it does provide a useful guide to further experimentation.

Early Effects of Localized Cytoplasmic Determinants

The effects of localized cytoplasmic determinants have been analyzed, for the most part, in morphological terms. However, these processes are preceded by less overt reactions which provide clues to some of the intermediate links in the causal chain of events. We will first consider some early effects of posterior pole plasm on nuclear behavior, which presumably do not involve transcription of the embryonic genome, and then review some effects of localized determinants on patterns of early embryonic gene expression.

Effects on Nuclear Migration and Chromosome Elimination

In *Drosophila* embryos during the ninth interphase, a few migrating nuclei appear in the posterior periplasm, creating cytoplasmic bulges called pole buds. When the remainder of the migrating nuclei appear elsewhere in the periplasm during the 10th interphase and enter mitosis, the pole buds divide simultaneously and the daughter cells are pinched off as pole cells (14). The timely arrival of nuclei in the posterior periplasm is critical to pole-cell formation. Delaying the arrival of nuclei by temporary ligation causes a time-dependent decrease in the percentage of embryos forming pole cells. Such embryos can be rescued by transplanting posterior plasm from younger donors (45). Thus, the posterior pole plasm seems to lose its ability for instructive interactions with nuclei about 30 min after their normal arrival time in the posterior pole plasm. The nature of this interaction is unknown.

In *Smittia* embryos, the first pole cell buds off much earlier when one of the first four cleavage nuclei arrives at the posterior pole, which is about seven nuclear cycles prior to the arrival of the somatic nuclei in the periplasm. The first pole cell divides twice, giving rise to four pole cells, which then stop dividing until they enter the fifth mitosis in the larval gonad. The precocious migration of the first pole-cell nucleus depends on a localized activity in the posterior pole plasm. This is indicated by the delay of the migration after uv irradiation of the posterior pole (Fig. 3). The effect is observed only after irradiation of a target area that includes the oosome, which corresponds to the polar granules in *Drosophila*. Control irradiations of areas just anterior to the oosome area do not inhibit the migration of the prospective pole-cell nucleus. The effective targets of the uv light appear to be cytoplasmic RNP particles as indicated by the photoreversibility of the uv effect and by the action spectrum for pole-cell inhibition (Fig. 3; see also Ref. 26). How RNP particles located near the oosome might serve to attract a single nucleus that is located about 50 μm away remains to be investigated. Again, the timing of this peculiar nuclear migration is critical. If delayed by one mitotic cycle or longer, pole cells may still be formed but are not preserved and gametes do not develop. The lack of viability of these pole cells presumably results from chromosome elimination during mitosis, which normally occurs only in somatic nuclei but not in pole cells at this stage (71). Such protection of dividing pole cells from chromosome elimination under the influence of posterior pole plasm has been observed previously in other insects (9). In summary, the posterior pole plasm may control several events in early development, including the timing of nuclear migration and chromosomal modifications.

Dorsoventral differences in cell shape, mitotic properties, and morphogenetic behavior also manifest themselves during the blastoderm, gastrula, and germ-band stages. At least some of these processes depend, directly or indirectly, on the maternal gene products involved in dorsoventral polarity (3). The causal role of zygotic dorsoventral genes in these processes is not yet known.

Effects on Embryonic Gene Expression

The formation of a segmented germ band in *Drosophila* is preceded by the expression of certain embryonic genes in very distinct patterns (1). The first zygotically active genes include the "gap genes," *hunchback* (*hb*), *Krüppel* (*Kr*), and *knirps* (*kni*) (Fig. 4). Strong mutant phenotypes show large deletions ("gaps") in the larval segment pattern, which are generally associated with cell death after gastrulation. The *Kr* gene is transcribed as early as the 12th nuclear cycle when the embryo is still in a plasmodial state. By stage 13, *Kr* RNA is detected in a domain extending from the posterior prothoracic compartment to the anterior first abdominal compartment (parasegments 4 to 6 in Fig. 4). *Kr* protein, visualized by antibody staining, covers a similar domain in cellular blastoderm embryos. Mutations affecting posterior determinants including *osk* cause a posterior extension of the *Kr* expression domain, whereas *bcd* mutations generally produce an anterior shift and enlargement of the *Kr* domain (19).

In addition to being controlled by localized maternal gene products, the gap genes also delimit each other in their expression. In situ hybridizations with a labeled *Kr* DNA probe reveal an anterior expansion of the *Kr* domain in *hb* embryos as well as a posterior expansion in *kni* embryos (23). Interactions between gap genes are also reflected in the fact that the segment deletions in gap mutant phenotypes are larger than the expression domains of the corresponding genes on the blastoderm fate map. In accordance with the cross-regulation among the gap genes, and with their control over other segmentation genes, both the *Kr* and the *hb* proteins show homology to transcription factor IIIA in the zinc finger domain, a putative nucleic acid–binding structure (66). In fact a point mutation causing a single amino acid substitution (Cys to Ser) in the putative metal-binding region of the *Kr* protein eliminates the *Kr* function (48). A different type of transcription regulator belonging to the steroid-thyroid receptor superfamily is encoded by the *kni* gene (42).

Another class of genes expressed in anteroposterior patterns during preblastoderm and blastoderm stages are the "pair-rule genes" including *fushi tarazu* (*ftz*). After in situ hybridization or immunolabeling, *ftz* RNA and protein show a distinct pattern of seven transverse stripes (Fig. 4). This pattern is again affected by maternal gene products. In *bcd* embryos, for example, the *ftz* stripes are reduced in number, shifted anteriorly, and broadened (39). The *ftz* expression pattern also depends on the activity of gap genes and some of the other pair-rule genes. Thus, the maternal effects on pair-rule gene expression may be caused indirectly by disruption in the activity of other early embryonic genes.

Maternal gene products that define dorsoventral polarity control a set of embryonic genes expressed in dorsoventral patterns at the preblastoderm and blastoderm stages and during gastrulation. Mutations in these embryonic genes can also be classified as ventralizers and dorsalizers (2, 3). Their phenotypes are attributed to changes in cell fate since there is no overt cell death. Dorsalizing mutations of embryonic genes, such as *twist* (*twi*) and *snail* (*sna*), block the formation of the ventral furrow and the subsequent development of

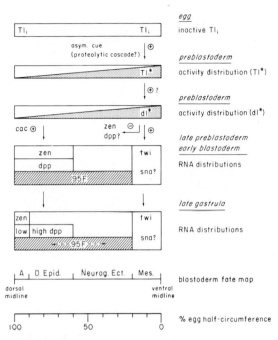

FIGURE 4 Competitive gap gene activation model for *Drosophila* embryos. In eggs and early embryos, activities dependent on two maternally expressed genes, *bicoid* (*bcd*) and *nanos* (*nos*), are localized near the anterior and posterior pole, respectively. During nuclear migration, *bcd* protein forms a gradient with a maximum near the anterior pole. At the same time, a *nos*⁺-dependent signal spreads from the posterior pole towards the anterior; its propagation depends on the activity of another maternally expressed gene, *pumilio* (*pum*). At preblastoderm (stages 12 and 13), the embryonic gap genes *hunchback* (*hb*), *Krüppel* (*Kr*), and *knirps* (*kni*) are expressed in certain cellular domains. It is proposed that *hb* is activated by the *bcd* protein and *kni* by the *nos*⁺-dependent signal. *Kr* is thought to be expressed spontaneously, except that it is repressed by the *bcd* and *nos* products and by *hb* and *kni* (indicated by double-headed arrows in diagram). The *hb* and *kni* domains are also delimited by two further domains of embryonic gene expression (*atll* and *ptll*) that become established near the egg poles. The activation of the gene(s) expressed in both *atll* and *ptll* depends on the products of maternally expressed genes including *torso* (*tor*). The activation of the *atll* genes depends in addition on the *bcd*⁺ product. Interactions between genes expressed in adjacent domains, such as *hb* and *Kr*, or *kni* and *ptll*, are thought to entrain the activity of pair-rule and other embryonic genes, for example, *evenskipped* (*eve*) and *fushi tarazu* (*ftz*).

mesodermal derivatives and internal organs. In situ hybridizations show that *twi* RNA is expressed in a narrow longitudinal strip of cells centered on the ventral midline at the cellular blastoderm stage and in the ventral furrow and presumptive mesoderm during gastrulation (Fig. 5). Antibody staining at the cellular blastoderm stage reveals a somewhat larger expression domain for

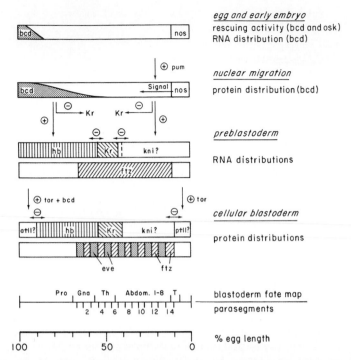

FIGURE 5 Maternal gene products and embryonic genes involved in establishing dorsoventral polarity in *Drosophila*. The dorsoventral axis as shown extends from the ventral to dorsal midline, corresponding to half the embryonic circumference in a cross-section through the thorax or abdomen, and maps onto the embryonic circumference only until the blastoderm stage. Coordinates defining the domains of zygotic gene expression and the boundaries of amnioserosa (A), dorsal epidermis (D. Epid.), neurogenic ectoderm (Neurog. Ecto.) and mesoderm (Mes.) are approximate. The spatial distributions labeled *Tl** and *dl** indicate approximate levels of activity as inferred from cytoplasmic transplantation experiments but are not meant to imply linear gradients. The patterns of embryonic gene expression summarize the RNA distributions as determined by in situ hybridizations at late preblastoderm–early cellular blastoderm (stage 14 a) and late gastrula stages.

After fertilization, an unknown asymmetric cue is thought to initiate a process leading to the enhancement of *Tl** and *dl** activity on the prospective ventral side. The maternal signal specifying "ventral" activates the *twi* gene in presumptive mesoderm and represses *zen* on the ventral side. The distribution and specific functions of *cac* are not known, but it is required, directly or indirectly, for the transcription of *zen* dorsally. Interactions between zygotic dorsoventral genes have not been demonstrated experimentally but may occur. (The exact dorsoventral distribution of the 95F gene is not shown for the late gastrula stage, but it is still excluded from the mesoderm. The expression profile of *sim* is not shown.)

the *twist* protein. The protein is still restricted to a narrow ventral band but extends dorsally around both the anterior and posterior poles to label ectodermal and endodermal primordia as well as the presumptive mesoderm (68). Ventralizing mutations map to five loci, *decapentaplegic (dpp)*, *zerknüllt (zen)*, *twisted gastrulation, tolloid*, and *shrew*. Null homozygotes of dpp^{Hin-} resemble Tl^D ventralized embryos. The other four mutants have less severe phenotypes, i.e., deletion of the amnioserosa and, in some cases, of a part of dorsal epidermis (2, 3). Transcripts of *zen* and *dpp* are found in a broad longitudinal strip of blastoderm cells centered on the dorsal midline and encompassing about 40 percent of the dorsoventral circumference in wild-type embryos (49, 60). In the anterior and/or posterior pole regions, the transcripts are expressed ventrally as well. At gastrulation, *zen* expression disappears from the pole regions and is restricted dorsally to the presumptive amnioserosa and possibly the optic lobe (about 10 percent of the dorsoventral circumference); *dpp* RNA becomes enriched in the presumptive dorsal epidermis and decreases in the presumptive amnioserosa during gastrulation (Fig. 5).

Three zygotic genes [single-minded *(sim)*, pointed *(pnt)*, and rhomboid *(rho)*] are thought to function specifically or primarily in the formation of neural and epidermal derivatives in the neurogenic ectoderm (35, 69). Defects in neural and epidermal pattern elements are evident in mutant embryos, but it is not known what happens to the affected cells. The *sim* gene is first transcribed at the cellular blastoderm stage in two narrow longitudinal stripes in the neurogenic ectoderm (69). The expression of *sim*, *pnt*, and *rho* probably depends on maternally provided dorsoventral cues, but this has not been demonstrated directly.

The embryonic expression of *twi* and *zen* is profoundly affected by maternal-effect mutations. In the dorsalized offspring of *pll, ea, Tl*, or *dl* females, embryonic *twi* expression is abolished (68), while *zen* expression is expanded so that *zen* protein is found in all blastoderm nuclei. However, the reduction of *zen* protein concentration near the poles occurs normally during cellularization, as it does in wild-type embryos. Ventralizing mutations (Tl^D and cac^-) abolish *zen* expression in the midbody regions but do not eliminate normal *zen* expression near the poles. The expression of *dpp* may be regulated in the same way as *zen*. Formally, these results can be accommodated by a model in which maternal determinants regulate the activity of a transcriptional repressor of *zen* and possibly *dpp* (50). The transcriptional control of *zen*, *twi*, and other zygotic dorsoventral genes may be mediated, at least in part, by the *dl* protein since it is present in embryonic nuclei. The polar expression of *zen* (and possibly *dpp*) is regulated separately and requires the activity of two maternal-effect anteroposterior patterning genes, *tor* and *trk*. The polar activity shown by *dpp* and *zen* is not a general property of all early dorsoventral zygotic genes. The *g2* locus of the *gooseberry-zipper* region, for instance, is transcribed in a similar pattern to *zen* dorsally but is not expressed at the anterior and posterior poles (8). Likewise, the integrin PS2α gene is expressed ventrally in the presumptive mesoderm but never in the polar regions of the cellular blastoderm embryo (5). The functions, if any, of these two genes during cellularization and gastrulation are not known.

The maternal control of zygotic dorsoventral gene activity could reflect the activation or repression of different genes at different threshold concentrations of a single morphogen gradient. Alternatively, the different regions of embryonic gene expression could also be established by a combination of zygotic gene interactions and the direct regulatory effects of one or more localized maternal gene products. In either case, the *dl* protein is likely to play an important role, given its nuclear location. Regulatory interactions between the zygotic dorsoventral genes are certainly plausible. The patterns of expression of *zen* and *dpp* overlap extensively at first and then gradually separate into distinct domains (49, 60). Likewise, the dorsoventral domains of *twi* (and perhaps *sna*) and a *Notch*-related gene (95F) are, for the most part, mutually exclusive (27, 68; Fig. 5). The *zen* and *sna* proteins contain DNA-binding motifs (homeobox and zinc fingers, respectively), and *dpp* encodes a putative growth factor (7, 46, 49). Hence, they may be involved directly in gene regulation or cell-cell signaling.

Maternal Morphogen Gradients versus Competitive Gene Activation

During gastrulation and germ-band formation, a distinct pattern of embryonic segments emerges from a seemingly uniform layer of blastoderm cells. Clonal analysis in *Drosophila* indicates that the fates of blastoderm cells are limited to one segment or compartment thereof. Moreover, transplantation experiments reveal that cells are stably determined within a segment (59). The data reviewed so far show that localized cytoplasmic determinants play an important role in these processes. However, the amount of detailed information encoded in these determinants is still debatable.

Gradient Model

A basic concept in developmental biology is the specification of body patterns under the control of morphogen gradients. In classical terms, the idea is that qualitatively different pattern elements, such as the segments of an insect, are evoked by quantitative differences in some scalar parameter, such as the local concentration of a diffusible signal. Meinhardt (36) proposed that the anteroposterior pattern of insect embryos is controlled by an overall gradient shaped by autocatalysis and diffusion. The morphogen may be synthesized anywhere in the embryo but, because of some symmetry-breaking event, the concentration peak becomes stabilized at the posterior pole. (This feature of the model is necessary for its ability to explain the formation of double abdomen patterns.) Several predictions derived from this model were tested experimentally and were at variance with the observed results (24).

A modern interpretation of the gradient concept is that certain key genes become activated at morphogen concentrations between different threshold levels. In a recent extension of his model, Meinhardt (37) suggested that the posterior-anterior gradient in *Drosophila* embryos specifies two "marginal" regions at the egg poles and four "cardinal" regions in between. The cardinal regions, each 3.5 segments long, are considered to be the expression domains of the three gap genes mentioned above (*hb, Kr,* and *kni*) plus another gene, *giant* (*gt*), although the pattern disturbances observed in *gt* mutants fall within the expression areas of the three gap genes. The borders between the cardinal regions are thought to organize the expression patterns of pair-rule genes, which in turn induce the expression of segment polarity genes.

Recent results on the localization of *bcd* gene products in *Drosophila* embryos have been interpreted along the lines of a gradient model, with the *bcd* protein serving as the morphogen and the gap genes as the presumed targets. To determine the relationship between *bcd* protein levels and cell fate, Driever and Nüsslein-Volhard (12) compared these parameters in offspring from females with one, two, three, or four copies of the *bcd*[+] gene. Increases or decreases in embryonic *bcd* protein levels were associated with posterior or anterior shifts of cell fates, respectively. However, these shifts were less dramatic than the changes in *bcd* protein concentration, and all four genetic conditions gave rise to normal hatching larvae. Yet the authors propose, referring to the Meinhardt's (37) model, that different concentration ranges of the *bcd* protein activate different gap genes. Indeed the zygotic expression of the *hb* gene is positively controlled by *bcd* product via a regulatory region that is upstream of the *hb* promoter (55, 65). However, there is little evidence for the activation of other gap genes (see Fig. 4) by higher or lower concentrations of *bcd* protein. In the anterior pole region adjacent to the *hb* domain, i.e., between approximately 90 and 100 percent egg length, the concentration of *bcd* protein does not increase significantly over its highest level in the *hb* domain (Fig. 6C in Ref. 11). Moreover, proper activation of embryonic genes in this area (*atll* domain in Fig. 4 of this chapter) requires a combination of *bcd* and other maternal gene products. This is best seen in *tor* embryos which show normal *bcd* protein levels (12) but still lack both acron and labrum. The *Kr* gene, which is expressed in the domain posterior to the *hb* domain, is repressed rather than activated by the *bcd* product (see below). The embryonic genes expressed in the *kni* and *ptll* domains do not seem to be affected at all by the *bcd* product in its wild-type distribution, which becomes very shallow in the posterior egg half.

Competitive Gap Gene Activation Model

It seems to us that viewing the gap genes as being controlled by a morphogen gradient implies singling out one localized cytoplasmic determinant where three or more of them play equally important roles. The gradient model also tends to impose a uniform type of control on all gap genes where in fact different types of control are observed. As an alternate model, which we call the "competitive gap gene activation model," we propose that gap genes are activated and inhibited by several maternal gene products and that the gap gene expression domains are further delineated by interactions among these embryonic genes. For simplicity, we consider only the major expression domains of the gap genes *hb, Kr,* and *kni,* which abut and cover the range from about 10 to 90 percent on the blastoderm fate map (Fig. 4). We also assume that two additional domains of embryonic gene expression become established near the anterior pole (90 to 100 percent) and the posterior pole (approximately 0 to 10 percent). The blastoderm cells in these areas form the terminal segments plus anterior and posterior midgut. The *tailless* (*tll*) gene seems to be expressed in both domains since *tll* mutant embryos lack both anterior and posterior terminalia (64). To distinguish the anterior from the posterior pole region, we refer to them loosely as the anterior tailless (*atll*) and posterior tailless (*ptll*) domains, respectively. We assume that both domains express *tll* but that they differ in embryonic gene activity by one or more additional genes. These may include *cad* or *Kr,* which are transcribed in part of the posterior but not in the anterior *tll* domain (23, 38).

We propose that the *hb* and *kni* genes are activated by the maternal gene products of *bcd* and *nos,* respectively, the absence of which causes similar mutant phenotypes. The intervening gap gene domain (of *Kr*) does not seem to be under positive control of a cytoplasmic determinant since no maternal-effect mutant with a phenotype similar to *Kr* has been found. Rather, the *Kr* gene is negatively controlled by the gap genes expressed in the abutting domains, i.e., *hb* and *kni,* and by other activities depending on the maternal gene products of *bcd* and *nos.* As a result, the *Kr* gene becomes activated where both *bcd* and *nos* are absent. Since the maternal *bcd* and *nos* products inhibit each other (44), their simultaneous presence at low concentrations also leads to *Kr* activation. In addition, we propose that maternal gene products of *tor, trk, tsl, fs(1)ph,* and *fs(1)N* activate embryonic genes in the *atll* and *ptll* regions. For convenience, we refer to the products of these genes collectively as the "*tor* products." We suggest that the *bcd* and *tor* products together activate the gene(s) characteristic of *atll,* whereas the *tor* products on their own activate the gene(s) characteristic of *ptll.* In embryos with reduced amounts of *bcd* product, as in the offspring of mothers with weak *bcd* alleles, neither *atll* nor *ptll* genes become activated properly in the anterior pole region.

The competitive gap gene activation model differs from the corresponding part of Meinhardt's (37) model. Instead of the four cardinal genes and two marginal genes, our model works only with three gap genes (*hb, Kr,* and *kni*) and two terminal embryonic genes (*tll* and one unknown), which together cover the entire ectodermal embryo in terms of phenotypic defects, and presumably, expression domains. For the rest of this section, we conveniently use the term "gap genes" for *hb, Kr,* and *kni,* as well as for *tll* and the unknown gene which distinguishes the *atll* from the *ptll* domain. We wish to point out that these genes, instead of being specified by an overall concentration gradient of one diffusible morphogen, are controlled in different ways by several maternal gene products. In particular, *Kr* seems to be under negative, rather than positive, control of maternal gene products. Also, the *tor* products seem to have functions beyond the activation of *tll,* since *tll* affects somewhat smaller terminal domains than *tor* (64). In agreement with other parts of Meinhardt's model, we suggest that the borders between the gap gene domains serve to entrain the overlapping patterns of different pair-rule genes and their anteroposterior polarity.

This model is simplified in several respects. (i) The *nos* and *tor* gene products represent classes of maternal-effect mutants with similar phenotypes. (ii) The gradients of maternal *hb* and *cad* products have been neglected, because their absence is almost completely rescuable by the embryonic expression of these genes. (iii) The expression domains of *hb* and *Kr* are stage dependent and more complex than shown. (iv) Under experimental or mutant conditions, the maternal products of *bcd* and *nos* may be present in the same egg region, inhibiting each other, and allowing *Kr* to become expressed.

The competitive gap gene activation model provides a straightforward explanation for the striking similarities between the phenotypes of *bcd* and *hb* as well as *nos* and *kni* (1). Our model also incorporates the cross-regulatory interactions between the gap genes, which result in a posterior expansion of the *hb* expression domain in *Kr* embryos and corresponding expansions of the *Kr* expression domain in *hb* and *kni* embryos (23). In addition, our model explains similar expansions and shifts of the *Kr* domain in *bcd* and *osk* embryos as well as the generation of a large central *Kr* domain in *bcd/osk* double mutants, whereas *tor* and *trk* mutants show virtually normal *Kr* domains (19). Similar interactions seem to occur between the *tll* domains and the abutting *hb* and *kni* regions. In both *tor* and *tll* embryos, the intervening body regions are expanded, and the expression domains of pair-rule genes at the blastoderm stage are shifted correspondingly (34, 57, 64). In the following paragraphs, we interpret several mutant phenotypes and experimentally induced pattern abnormalities in terms of the competitive gap gene activation model. These interpretations can be tested directly to the extent that the abnormal phenotypes can be generated reproducibly and molecular probes for the gap gene products become available.

Symmetrical double abdomen and bicaudal phenotypes can be interpreted as resulting from the gap gene activation pattern *ptll-kni-ptll*. In chironomids, this pattern may result from derepression of resident posterior determinants in the anterior egg half, as discussed above. The opposite polarities of the two sets of posterior abdominal segments could be entrained by the opposite polarities of the borders between the *kni* domain and the two *ptll* domains. In *Drosophila bicaudal* mutants, the *ptll-kni-ptll* pattern may be generated by a faulty distribution of *nos*[+] activity and its inhibition of *bcd*[+] activity. The hypothetical *ptll-kni-ptll* pattern would also explain the bicaudal phenotypes observed in strong homozygous *sww/bcd* double mutants (18). The lack of *bcd* product would prevent the activation of *atll* and *hb*, and *Kr* might be inhibited because the *nos* product is not sufficiently restricted to the posterior pole region. The phenotype of *bcd/osk* double mutants, which shows no polarity or overt segmentation, is ascribed to a gap gene activation pattern of *ptll-Kr-ptll*. The abuttal of these normally disparate domains may lead to a failure of normal pair-rule gene activation. However, injection of wild-type posterior pole plasm into the middle region of *bcd/osk* recipients should again cause the gap gene activation pattern *ptll-kni-ptll*, and the resulting embryos should show the segmented bicaudal phenotype, with the telsons near the egg poles rather than at the site of injection. This result is in fact observed (44).

Asymmetrical double abdomens in *Chironomus*, as well as strong *bcd* phenotypes in *Drosophila*, can be rationalized as resulting from a *ptll-Kr-kni-ptll* gap gene activation pattern. In *Drosophila*, lack of *bcd* in the presence of *tor* product causes (i) activation of *ptll* instead of *atll*, (ii) failure to activate *hb*, and (iii) a corresponding expansion of *Kr* and *kni*. In *Chironomus*, the activation of *Kr* may result from a marginal inactivation of anterior determinants, causing a "stalemate" between these and their antagonistic posterior determinants. This view is supported by the observations that asymmetrical double abdomens are generally rare (less than 2 percent of all double abdomens), except after uv irradiation following the ninth mitosis when there may be insufficient time for either antagonist to prevail (M. Laurel and K. Kalthoff, unpublished). The postulated abuttal of *ptll* and *Kr* domains may explain the lack of proper segmentation often observed around the plane of polarity reversal in asymmetrical double abdomens of *Chironomus* (47) and the lack of thoracic segments in strong *bcd* phenotypes (18). Weak *bcd* phenotypes, which show mainly head defects, are ascribed to an *(atll-hb-)Kr-kni-ptll* gap gene activation pattern, where the *atll* genes and *hb* are activated incompletely because of a *bcd* protein level that is too low to generate a normal pattern but too high to allow *ptll* gene activation.

The formation of double cephalon patterns in chironomids and the *dicephalic (dic)* phenotype in *Drosophila* (31) are explained as a result of the gap gene activation patterns, *atll-hb-atll* or *atll-hb-Kr-hb-atll*, which in turn can be ascribed to the presence of anterior determinants, or *bcd*$^+$ product, in both anterior and posterior egg halves (see below). The dominance of *bcd*$^+$ activity may be facilitated by faulty localization of its posterior counterpart, presumably *nos*$^+$ activity, in *Drosophila*. Since *exu* mutants show an almost even distribution of *bcd*$^+$ protein (12), the same explanation applies to *exu/osk* and *exu/vas* double mutants which show similar phenotypes. The defects in cephalic structures, which are often observed in these as well as in *dic* phenotypes, are thought to result from incomplete activation of *atll* genes and of *hb* by a marginal concentration of *bcd* product.

Gaps in the segmentation pattern have been generated in many insect species by ligation before the blastoderm stage (51, 52). Generally, the size of the gap, i.e., the number of missing segments, decreases with increasing age of the embryo at the time of ligation. The character of the missing segments corresponds to the position of the ligature. Schubiger and Newman (56) noted that *Drosophila* embryos ligated at preblastoderm stages resembled the phenotypes of gap mutants. In terms of our model, we suggest that the ligations cause a buildup of *bcd* and/or *nos* product, with the result that the *hb* and/or *kni* domains, respectively, become enlarged, and that the *Kr* domain becomes reduced or absent. Such enlargements or reductions may occur in increments of several segments, as some authors observed even before gap genes became widely known (51, 70).

Origin and Maintenance of Cytoplasmic Localization

Some of the cytoplasmic localizations in insect eggs are initiated during oogenesis. The localization of germ-cell determinants in *Drosophila* oocytes at stages 13 and 14 has been demonstrated directly by heterotopic transplantation (Chap. 11, this volume). The anteroposterior egg polarity is clearly related to the arrangement of oocytes and nurse cells in ovarian follicles (6). Although first established in the ovary, the localization of anterior determinants is also modified later during embryonic development. With respect to dorsoventral polarity, there is no evidence so far from rescue experiments or in situ hybridizations for clearly localized and long-lived cytoplasmic determinants in oocytes or eggs. Rather, as discussed above it appears that subtle and transient cues become amplified during early embryogenesis.

Cytoplasmic Streaming in Ovarian Follicles

The ovarian follicle of *Drosophila* consists of one oocyte, an adjacent cluster of 15 nurse cells, and an envelope of follicle cells. The oocyte and its sister nurse cells originate from one germ-line cell (cystoblast) by four mitotic divisions with incomplete cytokinesis. As a result, the oocyte remains connected to the nurse cells by cytoplasmic bridges that are wide enough to transfer streams of cytoplasm containing mitochondria and other organelles. Virtually all egg RNA is synthesized in the polyploid nurse cells and is transferred to the oocyte via the cytoplasmic bridges. The transferred cytoplasm enters near the anterior pole since the nurse cells are adjacent anteriorly to the oocyte. Laterally and posteriorly, the oocyte is surrounded by follicle cells. These are somatic cells involved in vitellogenin uptake and chorion synthesis. The follicular structure of chironomids and other insects with polytrophic oogenesis is similar, except that the number of nurse cells varies between species.

Cytoplasmic streaming in *Drosophila* ovarian follicles between stages 10 and 14 has been analyzed by time-lapse cinematography, histology, and inhibitor applications (Ref. 21; Chap. 13, this volume). During stages 10B to 12, nurse cell cytoplasm flows into the oocyte ("nurse cell streaming"), where it becomes rapidly mixed with ooplasm as a result of cytoplasmic streaming within the oocyte ("oocyte streaming"). Microtubule inhibitors interfere with oocyte streaming but not with nurse cell streaming. Under these conditions, the inflowing cytoplasm becomes stratified in the oocyte. At the same time, the oocyte nucleus, which is normally held in place anteriorly by a dense network of microtubules, becomes displaced toward the posterior pole. These observations are relevant to the problem of cytoplasmic localization. First, time-dependent stratification of inflowing cytoplasm alone cannot explain cytoplasmic localization because the cytoplasm is normally mixed by oocyte streaming. Second, the behavior of the nucleus shows that even large organelles are held in place, despite cytoplasmic streaming, by cytoskeletal components. Together, the observations support the view that cytoplasmic components might be retained and concentrated locally by receptors associated with cytoskeletal components that function as selective anchors. Cytoskeleton-based mechanisms for localizing mRNAs may also be utilized in vertebrate eggs.

Localization of Anterior and Posterior Determinants in Drosophila Oocytes

Using a labeled cDNA probe, Frigerio et al. (16) and Berleth et al. (4) have observed striking patterns of *bcd* RNA distribution in sectioned ovaries, eggs, and embryos. In ovaries, the hybridization signal is found over the nurse cell cytoplasm, and most strongly in a thin crescent of anterior oocyte periplasm adjacent to the nurse cells. These data provide a straightforward explanation for the dicephalic phenotype. Homozygous *dic* females show some abnormal ovarian follicles with clusters of nurse cells adjacent anteriorly and posteriorly to the oocyte. Eggs from such mothers give rise to embryos with mirror-image duplications of cephalic, thoracic, and anterior abdominal structures in the absence of posterior abdomen (6, 15, 31). Together, the observations indicate that maternally expressed *bcd* mRNA is transcribed in nurse cells, transferred to the oocyte via cytoplasmic bridges, and selectively retained near the anterior oocyte pole by anchoring components which would not necessarily have to be localized themselves. The anchorage of *bcd* mRNA depends on a segment of about 625 bases in the 3' untranslated region which is capable of forming extensive secondary structure (33).

The localization of *bcd* depends on the activity of two other maternal gene products, *sww* and *exu*. These mutant phenotypes share with weak *bcd* alleles the lack of anterior head structures. In contrast to weak *bcd* alleles, removal of anterior cytoplasm does not exacerbate the *sww* or *exu* phenotype. Conversely, anterior cytoplasm from *sww* or *exu* donors has little rescuing activity when transplanted to *bcd* recipients (18). Moreover, in situ hybridization of *bcd* probes to *sww* or *exu* oocytes and embryos show almost even distributions of *bcd* RNA (4, 61). These observations indicate that the products of exu^+ and sww^+ are directly or indirectly required for "anchoring" bcd^+ mRNA near its point of entry into the oocyte. The entire phenotypic effect of *exu* seems to be mediated by *bcd*, whereas *sww* causes additional disturbances. The *sww* protein may therefore be involved in several cytoskeletal functions while the *exu* protein may serve solely to link the *bcd* mRNA to the *sww* protein (4, 18).

The timing and mechanisms of localization appear to be different for anterior versus posterior determinants. The abdominal phenotype of *osk* can be rescued with cytoplasm from wild-type oocytes and nurse cells at stage 10. The rescuing activity, i.e., presumably the nos^+ product, is present in excess quantity since 5 to 10 embryos can be rescued with cytoplasm from a single nurse cell cluster. No corresponding rescuing activity for *bcd* embryos can be obtained from wild-type nurse cells or immature oocytes. In fact, *bcd* recipients of such transplantations develop the symmetrical bicaudal phenotype just as they do after transplantation of posterior egg cytoplasm (53). The lack of *bcd* activity does not result from inhibition by posterior determinants because *bcd* rescue was also not obtained with nurse cell cytoplasm from *nos* mutants lacking posterior determinant activity (R. Lehmann, personal communication). These results must be interpreted in the context that *bcd* RNA is present and localized in oocytes as early as stage 8, whereas *bcd* protein is not detectable by western blotting or immunohistology until after egg deposition (4, 11). The accumulation of *bcd* protein in embryonic nuclei and the presence of a homeodomain strongly suggest that the protein and not the mRNA itself acts as an anterior determinant. However, translation of *bcd* mRNA seems blocked until egg deposition. Upon cytoplasmic transfer from oocytes or nurse cells to embryos, *bcd* mRNA may also not be translated effectively. Some rescuing activity is transplantable only with cytoplasm from stage 14 oocytes (53). It seems that translatability of *bcd* mRNA depends on some maturation process and/or proper anchorage of *bcd* mRNA in the egg. The excess quantity of posterior determinants suggests that they may become degraded during late oogenesis, except in the posterior pole plasm where they are stabilized along with germ-cell determinants. This would be in accord with the pleiotropic effects of *osk, stau, tud, vas,* and *vls* described above.

Changing Localization of Anterior Determinants during Early Embryogenesis

The localization of *bcd* mRNA in *Drosophila* embryos, as revealed by in situ hybridization, changes during early embryonic development. In cleavage stage embryos, most of the signal is concentrated in a cone-shaped area behind the anterior pole. In preblastoderm embryos, most of the signal becomes relocalized to the periplasm between 75 and 100 percent egg length (4, 16). A similar relocalization of anterior determinants in *Smittia* embryos has been inferred from the stage-dependent response to irradiation with a uv microbeam (26). Between egg deposition and nuclear migration, double abdomen patterns are produced with high yields by microbeam irradiation of a small target area (diameter 20 μm) behind the anterior pole. This area contains a yolk-free cytoplasmic cone located behind the anterior pole of newly deposited eggs (73). With the onset of nuclear migration, irradiation of this area becomes less efficient while irradiation of adjacent areas yields more double abdomens. During the same interval, the double abdomen yield observed after uv irradiation of the entire anterior surface also increases dramatically. The same changes have been observed in eggs of *Chironomus samoensis* (28). Together, the data indicate that in chironomid embryos a major fraction of anterior determinants shifts from a sequestered position behind the anterior pole to a wider distribution over the anterior periplasm.

A maternal-effect mutant of *Chironomus samoensis, spontaneous double abdomen (sda),* produces double abdomen embryos spontaneously, albeit with variable penetrance. Embryos from the *sda* strain are more responsive than wild-type embryos to uv induction of double abdomens. The uv irradiation seems to exacerbate a lack of anterior determinants in the periplasm resulting from an incomplete shift during nuclear migration. In addition to spontaneous double abdomens, the *sda* strain occasionally produces spontaneous double cephalons. Cytologically, embryos from the *sda* strain show signs of disturbed egg architecture, including yolk extrusions and detached cells. Therefore, the unstable anteroposterior polarity in *sda* embryos may be ascribed to alterations in cytoskeletal components involved in anchoring anterior determinants and/or segregating them into anterior blastoderm cells (28). This situation is reminiscent of the anchorage of *bcd* mRNA being dependent on *exu* and *sww* products (which seem to be directly or indirectly associated with the cytoskeleton in *Drosophila* oocytes). Failure in either the anchoring step during oogenesis in *Drosophila* or the translocation step during nuclear migration in chironomids produces the same phenotypes that are observed when anterior determinants are diminished or lacking. The identification of the anchoring mechanisms and their possible relation to the activation of anterior determinants is one of the challenges for the future.

Acknowledgments

We are grateful to Kathryn Anderson, Herwig Gutzeit, Herbert Jäckle, Paul Krieg, Ruth Lehmann, Judith Lengyel, Christiane Nüsslein-Volhard, Klaus Sander, Edwin Stephenson, and Ruth Steward, who provided helpful comments on the manuscript and/or shared data prior to publication. Our work is supported by the National Science Foundation (KK) and the Damon Runyon–Walter Winchell Cancer Research Fund Fellowship DRG-935 (MR).

General References

ANDERSON, K., and NUSSLEIN-VOLHARD, C.: 1986. Dorsal-group genes of *Drosophila*. In: *Gametogenesis and the Early Embryo*. (J. Gall, ed.), pp. 177–194, Liss, New York.

NUSSLEIN-VOLHARD, C., FROHNHOFER, H. G., and LEHMANN, R.: 1987. Determination of anteroposterior polarity in *Drosophila*. *Science* **238**:1675–1681.

SANDER, K.: 1984. Embryonic pattern formation in insects: Basic concepts and their experimental foundations. In: *Pattern Formation*. (G. M. Malacinski, and S. V. Bryant, eds.), pp. 245–268, Macmillan, New York.

References

1. Akam, M. 1987. The molecular basis for metameric pattern in the *Drosophila* embryo. *Development (Camb.)* **101**:1–22.

2. Anderson, K. 1987. Dorsal-ventral embryonic pattern genes of *Drosophila*. *Trends Genet.* **3**:91–97.

3. Anderson, K., and Nüsslein-Volhard, C. 1986. Dorsal-group genes of *Drosophila*. In: *Gametogenesis and the Early Embryo*. (J. Gall, ed.), pp. 177–194, Liss, New York.

4. Berleth, T., Burri, M., Thoma, G., Bopp, D., Richtenstein, S., Frigerio, G., Noll, M.,

and Nüsslein-Volhard, C. 1988. The role of localization of *bicoid* RNA in organizing the anterior pattern of the *Drosophila* embryo. *EMBO J.* **7**:1749–1756.

5. Bogaert, T., Brown, N., and Wilcox, M. 1987. The *Drosophila* PS2 antigen is an invertebrate integrin that, like the fibronectin receptor, becomes localized to muscle attachments. *Cell* **51**:929–940.

6. Bohrmann, J., and Sander, K. 1987. Aberrant oogenesis in the patterning mutant *dicephalic* of *Drosophila melanogaster*: time-lapse recordings and volumetry in vitro. *Roux's Arch. Dev. Biol.* **196**:279–285.

7. Boulay, J., Dennefeld, C., and Alberga, A. 1987. The *Drosophila* developmental gene *snail* encodes a protein with nucleic acid binding fingers. *Nature (Lond.)* **330**: 395–398.

8. Côte, S., Preiss, A., Haller, J., Schuh, R., Kienlin, A., Seifert, E., and Jäckle, H. 1987. The *gooseberry-zipper* region of *Drosophila*: five genes encode different spatially restricted transcripts in the embryo. *EMBO J.* **6**:2793–2801.

9. Counce, S. J. 1973. Causal analysis of insect embryogenesis. In: *Developmental Systems: Insects*, vol. II. (S. J. Counce, and C. H. Waddington, eds.), pp. 1–156, Academic Press, New York.

10. DeLotto, R., and Spierer, P. 1986. A gene required for the specification of dorsal-ventral pattern in *Drosophila* appears to encode a serine protease. *Nature (Lond.)* **323**:688–692.

11. Driever, W., and Nüsslein-Volhard, C. 1988. A gradient of *bicoid* protein in *Drosophila* embryos. *Cell* **54**:83–93.

12. Driever, W., and Nüsslein-Volhard, C. 1988. The *bicoid* protein determines position in the *Drosophila* embryo in a position-dependent manner. *Cell* **54**:95–104.

13. Elbetieha, A., and Kalthoff, K. 1988. Anterior determinants in embryos of *Chironomus samoensis*: Characterization by rescue bioassay. *Development (Camb.)* (in press).

14. Foe, V. E., and Alberts, B. M. 1983. Studies of nuclear and cytoplasmic behavior during the five mitotic cycles that precede gastrulation in *Drosophila* embryogenesis. *J. Cell Sci.* **61**:31–70.

15. Frey, A., Sander, K., and Gutzeit, H. 1984. The spatial arrangement of germ line cells in ovarian follicles of the mutant *dicephalic* in *Drosophila melanogaster*. *Roux's Arch. Dev. Biol.* **193**:388–393.

16. Frigerio, G., Burri, M., Bopp, D., Baumgartner, S., and Noll, M. 1986. Structure of the segmentation gene *paired* and the *Drosophila* PRD gene set as part of a gene network. *Cell* **47**:735–746.

17. Frohnhöfer, H. G., and Nüsslein-Volhard, C. 1986. Organization of anterior pattern in the *Drosophila* embryo by the maternal gene *bicoid*. *Nature (Lond.)* **324**: 120–125.

18. Frohnhöfer, H. G., and Nüsslein-Volhard, C. 1987. Maternal genes required for the anterior localization of *bicoid* activity in the embryo of *Drosophila*. *Genes & Dev.* **1**:880–890.

19. Gaul, U., and Jäckle, H. 1987. Pole region–dependent repression of the *Drosophila* gap gene *Krüppel* by maternal gene products. *Cell* **51**:549–555.

20. Gertulla, S., Jin, Y., and Anderson, K. 1988. Zygotic expression and activity of the *Drosophila Toll* gene, a gene required maternally for embryonic dorsal-ventral pattern formation. *Genetics.* **119**:123–133.

21. Gutzeit, H. 1986. The role of microtubules in the differentiation of ovarian follicles during vitellogenesis in *Drosophila*. *Roux's Arch. Dev. Biol.* **195**:173–181.

22. Hashimoto, C., Hudson, K., and Anderson, K. 1988. The *Toll* gene of *Drosophila*, required for dorsal-ventral embryonic polarity, appears to encode a transmembrane protein. *Cell* **52**:269–279.

23. Jäckle, H., Tautz, D., Schuh, R., Seifert, E., and Lehmann, R. 1986. Cross-regulatory interactions among the gap genes of *Drosophila*. *Nature (Lond.)* **324**: 668–670.

24. Kalthoff, K. 1978. Pattern formation in early insect embryogenesis—data calling for modification of a recent model. *J. Cell Sci.* **29**:1–15.

25. Kalthoff, K. 1979. Analysis of a morphogenetic determinant in an insect embryo. In: *Determinants of Spatial Organization.* (S. Subtelny, and I. Konigsberg, eds.), pp. 97–126, Academic Press, New York.

26. Kalthoff, K. 1983. Cytoplasmic determinants in Dipteran eggs. In: *Time, Space, and Pattern in Embryonic Development.* (W. R. Jeffery, and R. A. Raff, eds.), pp. 313–348, Liss, New York.

27. Knust, E., Dietrich, U., Tepass, U., Bremer, K., Weigel, D., Vässin, H., and Campos-Ortega, J. 1987. EGF homologous sequences encoded in the genome of *Drosophila melanogaster*, and their relation to neurogenic genes. *EMBO J.* **6**:761–766.

28. Kuhn, K. L., Percy, J., Laurel, M., and Kalthoff, K. 1987. Instability of the anteroposterior axis in *spontaneous double abdomen (sda)*, a genetic variant of *Chironomus samoensis* (Diptera, Chironomidae). *Development (Camb.)* **101**: 591–603.

29. Lehmann, R., and Nüsslein-Volhard, C. 1986. Abdominal segmentation, pole cell formation, and embryonic polarity require the localized activity of *oskar*, a maternal gene in *Drosophila*. *Cell* **47**:141–152.

30. Lehmann, R., and Nüsslein-Volhard, C. 1987. Involvement of the *pumilio* gene in the transport of an abdominal signal in the *Drosophila* embryo. *Nature (Lond.)* **329**:167–170.

31. Lohs-Schardin, M. 1982. *Dicephalic*: A *Drosophila* mutant affecting polarity in follicle organization and embryonic patterning. *Wilhelm Roux's Arch. Dev. Biol.* **191**: 28–36.

32. Macdonald, P. M., and Struhl, G. 1986. A molecular gradient in early *Drosophila* embryos and its role in specifying the body pattern. *Nature (Lond.)* **324**:537–545.

33. Macdonald, P. M., and Struhl, G. 1988. *Cis*-acting sequences responsible for anterior localization of *bicoid* mRNA in *Drosophila* embryos. *Nature (Lond.)* **336**:595–598.

34. Mahoney, P. A., and Lengyel, J. A. 1987. The zygotic segmentation mutant *tailless* alters the blastoderm fate map of the *Drosophila* embryo. *Dev. Biol.* **122**:464–470.

35. Mayer, U., and Nüsslein-Volhard, C. 1988. A group of genes required for pattern formation in the ventral ectoderm of the *Drosophila* embryo. *Genes & Dev.* **2**:1496–1511.

36. Meinhardt, H. 1977. A model of pattern formation in insect embryogenesis. *J. Cell Sci.* **23**:117–139.

37. Meinhardt, H. 1986. Hierarchical inductions of cell states: a model for segmentation in *Drosophila*. *J. Cell Sci. (Suppl.)* **4**:357–381.

38. Mlodzik, M., and Gehring, W. J. 1987. Expression of the *caudal* gene in the germ line of *Drosophila*: Formation of an RNA and protein gradient during early embryogenesis. *Cell* **48**:465–478.

39. Mlodzik, M., De Montrion, C. M., Hiromi, Y., Krause, H. M., and Gehring, W. J. 1987. The influence on the blastoderm fate map of maternal-effect genes that affect the antero-posterior pattern in *Drosophila*. *Genes & Dev.* **1**:603–614.

40. Mohler, J., and Wieschaus, E. F. 1986. Dominant maternal-effect mutations of *Drosophila melanogaster* causing the production of double abdomen embryos. *Genetics* **112**:803–822.

41. Müller-Holtkamp, F., Knipple, D., Seifert, E., and Jäckle, H. 1985. An early role of maternal mRNA in establishing the dorsoventral pattern in *pelle* mutant *Drosophila* embryos. *Dev. Biol.* **110**:238–246.

42. Nauber, U., Pankratz, M. J., Kienlin, A., Seifert, E., Klemm, U., and Jäckle, H. 1988. Abdominal segmentation of the *Drosophila* embryo requires a hormone receptor-like protein encoded by the gap gene *knirps*. *Nature (Lond.)* **336**:489–492.

43. Nüsslein-Volhard, C. 1977. Genetic analysis of pattern formation in the *Drosophila melanogaster* embryo. Characterization of the maternal effect mutant *bicaudal*. *Wilhelm Roux's Arch. Dev. Biol.* **183**:249–268.

44. Nüsslein-Volhard, C., Frohnhöfer, H. G., and Lehmann, R. 1987. Determination of anteroposterior polarity in *Drosophila*. *Science* **238**:1675–1681.

45. Okada, M. 1986. Cytoplasmic function segregating germ line in *Drosophila* embryogenesis. *Zool. Sci. (Tokyo)* **3**:573–583.

46. Padgett, R., St. Johnston, R., and Gelbart, W. 1987. A transcript from a *Drosophila* pattern gene predicts a protein homologous to the transforming growth factor-beta family. *Nature (Lond.)* **325**:81–84.

47. Percy, J., Kuhn, K. L., and Kalthoff, K. 1986. Scanning electron microscopic analysis of spontaneous and UV-induced abnormal segment patterns in *Chironomus samoensis* (Diptera, Chironomidae). *Wilhelm Roux's Arch. Dev. Biol.* **195**:92–102.

47a. Rebagliati, M. 1989. An RNA recognition motif in the *bicoid* protein. *Cell*, in press.

48. Redemann, N., Gaul, U., and Jäckle, H. 1988. Disruption of a putative Cys-zinc interaction eliminates the biological activity of the *Krüppel* finger protein. *Nature (Lond.)* **332**:90–92.

49. Rushlow, C., Doyle, H., Hoey, T., and Levine, M. 1987. Molecular characterization of the *zerknüllt* region of the Antennapedia gene complex in *Drosophila*. *Genes & Dev.* **1**:1268–1279.

50. Rushlow, C., Frasch, M., Doyle, H., and Levine, M. 1987. Maternal regulation of *zerknüllt*: a homoeobox gene controlling differentiation of dorsal tissues in *Drosophila*. *Nature (Lond.)* **330**:583–586.

51. Sander, K. 1976. Specification of the basic body pattern in insect embryogenesis. *Adv. Insect Physiol.* **12**:125–238.

52. Sander, K. 1984. Embryonic pattern formation in insects: Basic concepts and their experimental foundations. In: *Pattern Formation.* (G. M. Malacinski, and S. V. Bryant, eds.), pp. 245–268, Macmillan, New York.

53. Sander, K., and Lehmann, R. 1988. *Drosophila* nurse cells produce a posterior signal required for embryonic segmentation and polarity. *Nature (Lond.)* **335**:68–70.

54. Santamaria, P., and Nüsslein-Volhard, C. 1983. Partial rescue of *dorsal*, a maternal effect mutation affecting the dorso-ventral pattern of the *Drosophila* embryo, by the injection of wild-type cytoplasm. *EMBO J.* 2:1695–1699.

55. Schröder, C., Tautz, D., Seifert, E., and Jäckle, H. 1988. Differential regulation of the two transcripts from the *Drosophila* gap segmentation gene *hunchback*. *EMBO J.* 7:2881–2887.

56. Schubiger, G., and Newman, S. 1982. Determination in *Drosophila* embryos. *Am. Zool.* 22:47–55.

57. Schüpbach, T., and Wieschaus, E. 1986. Maternal-effect mutations altering the anterior-posterior pattern of the *Drosophila* embryo. *Roux's Arch. Dev. Biol.* 195: 302–317.

58. Schüpbach, T., and Wieschaus, E. 1989. Female sterile mutations on the second chromosome of *Drosophila melanogaster*. I. Maternal effect mutations. *Genetics* 121:101–117.

59. Simcox, A. A., and Sang, J. H. 1983. When does determination occur in *Drosophila* embryos? *Dev. Biol.* 97:212–221.

60. St. Johnston, R., and Gelbart, W. 1987. *Decapentaplegic* transcripts are localized along the dorsal-ventral axis of the *Drosophila* embryo. *EMBO J.* 6:2785–2791.

61. Stephenson, E. C., Chao, Y.-C., and Fackenthal, J. D. 1988. Molecular analysis of the *swallow* gene of *Drosophila melanogaster*. *Genes & Dev.* 2:1655–1665.

62. Steward, R. 1987. *Dorsal*, an embryonic polarity gene in *Drosophila*, is homologous to the vertebrate proto-oncogene, *c-rel*. *Science* 238:692–694.

63. Steward, R., Zusman, S., Huang, L., and Schedl, P. 1988. The *dorsal* protein is distributed in a gradient in early *Drosophila* embryos. *Cell* 55:487–495.

64. Strecker, T. R., Kongsuwan, K., Lengyel, J. A., and Merriam, J. R. 1986. The zyotic mutant *tailless* affects the anterior and posterior ectodermal regions of the *Drosophila* embryo. *Dev. Biol.* 113, 64–76.

65. Tautz, D. 1988. Regulation of the *Drosophila* segmentation gene *hunchback* by two maternal morphogenetic centres. *Nature (Lond.)* 332:281–284.

66. Tautz, D., Lehmann, R., Schnürch, H., Schuh, R., Seifert, E., Kienlin, A., Jones, K., and Jäckle, H. 1987. Finger protein of novel structure encoded by *hunchback*, a second member of the gap class of *Drosophila* segmentation genes. *Nature (Lond.)* 327:383–389.

67. Technau, G. 1987. A single cell approach to problems of cell lineage and commitment during embryogenesis of *Drosophila melanogaster*. *Development (Camb.)* 100: 1–12.

68. Thisse, B., Stoetzel, C., Gorostiza-Thisse, C., and Perrin-Schmitt, F. 1988. Sequence of the *twist* gene and nuclear localization of its protein in endomesodermal cells of early *Drosophila* embryos. *EMBO J.* 7:2175–2183.

69. Thomas, J. B., Crews, S. T., and Goodman, C. S. 1988. Molecular genetics of the *single-minded* locus: A gene involved in the development of the *Drosophila* nervous system. *Cell* 52:133–141.

70. Vogel, O. 1978. Pattern formation in the egg of the leafhopper *Euscelis plebejus* fall. (Homoptera): Developmental capacities of fragments isolated from the polar egg regions. *Dev. Biol.* 67:357–370.

71. Von Brunn, A., and Kalthoff, K. 1983. Photoreversible inhibition by ultraviolet light of germ line development in *Smittia* sp. *Dev. Biol.* 100:426–439.

72. Yajima, H. 1983. Production of longitudinal double formation by centrifugation or by UV-partial irradiation in the *Chironomus samoensis* egg. *Entomol. Gen.* 8:171–191.

72a. Yisraeli, J., and Melton, D. 1988. The maternal mRNA Vgl is correctly localized following injection into *Xenopus* oocytes. *Nature (Lond.)* 336:592–595.

73. Zissler, D., and Sander, K. 1982. The cytoplasmic architecture of the insect egg cell. In: *Insect Ultrastructure*, vol. 1. (R. C. King, and H. Akai, eds.), pp. 189–221, Plenum Press, New York.

Questions for Discussion with the Editors

1. *Do you expect that the insect morphogen gradient models will be general and apply to other systems such as vertebrates (e.g., mammals)?*

First, we still need to be convinced that morphogen gradient models are adequate descriptions of the situation in dipteran embryos. The minimal requirement for a true morphogen would be that at least two genes are activated exclusively by this morphogen in a concentration-dependent fashion. Currently, the best candidate for such a morphogen in insect embryos is the *bicoid* (*bcd*) protein. It is involved in activating the *hunchback* (*hb*) gene and in the control of another one or more genes whose embryonic expression distinguishes the anterior from the posterior *tailless* domain (*atll* and *ptll* in our Fig. 4). In wild-type embryos, the *atll* genes are activated where the *bcd* gradient has a shoulder of maximum concentration. However, in offspring from females with only one *bcd*$^+$ gene, the *atll*-specific genes seem to be almost normally controlled although the *bcd* protein concentration is lowered to a level where the *hb* domain becomes established in the wild type. The salient feature is that activation of *atll*-specific genes near the anterior pole requires the *bcd* and the *torso* (*tor*) products whereas *bcd* by itself activates *hb*. This situation is better described by a combinatorial model than by a gradient model. Along the dorsoventral axis, *dl* is a candidate for a morphogen gradient but direct tests of its role as a morphogen are required. Second, even if new data would prove gradients to be active in insect embryos, this is unlikely to be general. Insect embryos are special in that cells remain connected by wide cytoplasmic bridges until gastrulation, when cell potencies have become restricted to half a segment. Until that time, the incipient cells communicate by macromolecules which diffuse through cytoplasm and accumulate in the target nuclei. In most other developing systems communication between cells is presumably limited by the size of gap junctions and the signaling capabilities of cell surface proteins and second messengers.

2. *You allude to the "anchoring mechanism" which localizes morphogens. Please speculate about the molecular features such a mechanism might be expected to exhibit.*

In the simplest case, a bivalent protein could bind to a component of the cytoskeleton (actin, tubulin, or intermediate filaments) and to a unique protein or RNA sequence in the *bcd* mRNP. This would be analogous to known cross-linking proteins like ankyrin, which binds tightly to spectrin, a cytoskeletal protein, and to band 3, a membrane protein, through two separate domains. However, a localization mechanism would not necessarily have to involve the cytoskeleton. Another possibility is that the cytoplasmic domain of a membrane protein could act as a "hook" or receptor for anteriorly localized mRNPs. This would resemble the mechanism for targeting of mRNAs to the endoplasmic reticulum (ER), which involves the binding of an mRNP–ribosome–signal receptor particle complex to a docking protein protruding from the ER membrane. This mechanism has the interesting property that correct targeting to the ER is required for translation of the mRNA. The nurse cell–oocyte border is a morphologically distinct region and could contain a specialized anchoring system or microenvironment. However, a localized anchoring system of this sort is not strictly required. The topography of the nurse cell–oocyte complex per se would be sufficient to ensure correct localization at the anterior. It is harder to imagine how posterior determinants are localized. Conceivably, posterior determinants might be distributed uniformly at first and then degraded anteriorly or translocated to the posterior pole. The latter possibility would be similar to the way mRNAs are apparently localized to the vegetal pole of amphibian oocytes.

CHAPTER **15**

Cytoplasmic Organization and Cell Lineage in Early Mammalian Development

Martin H. Johnson

It is convenient and common practice to distinguish two types of mechanism for generating cell diversity during development. Some cells respond to extrinsic signals such as may be provided by inductive interactions or relative position in a developmental field. The presence (or strength) of such a signal determines whether or not any given cell in a homogeneous, responding population differentiates in a particular way. Thus, the spatial distribution of the signal dictates the spatial pattern of cell diversification. Alternatively, rather than setting up a pattern of diversification in a *population* of cells, some developmental strategies involve first setting up a spatial pattern within a *single* cell and then distributing elements of this pattern differentially at cell division to generate a "mosaic" embryo in which cells with different endowments have different fates. Such a strategy is observed more commonly, but not exclusively, in the early stages of the development of metazoans, or in protozoans, when cells tend to be larger. The distinction between these two types of mechanism is useful but may be artificial. As we understand more about the ways in which asymmetry is set up in an individual cell, it becomes clear that interactions of a quasi-inductive type often appear to dictate the pattern of asymmetry generated. Similarly, as the nature of the cellular reorganizations involved in setting up asymmetry within a single cell are understood, they appear to resemble in many ways the responses of cells to inductive stimuli.

In the early development of the mammal both intercellular signaling and differential cleavage appear to be important in generating cell diversity. In

this chapter, the roles that each of these types of processes occupy are reviewed. A more detailed consideration of the precise subcellular and molecular mechanisms involved was recently reviewed elsewhere (12).

The Early Development of the Mouse

Most work on mechanisms in early mammalian development has been carried
out on the mouse. The principal early developmental features, illustrated
schematically in Fig. 1, are as follows.

(i) In the first stage, the oocyte arrested at the second meiotic division, is
fertilized. (ii) The spindle rotates, meiosis is reactivated, and the second polar
body is extruded; a fertilization cone forms at the site of sperm entry. The
oolemma overlying the spindle and the sperm head lacks microvilli. (iii)
Pronuclei form 4 to 6 h after sperm entry and migrate centrally 12 to 18 h
after sperm entry. (iv) Pronuclear membranes break down at 18 to 21 h, and
the chromosomes mingle on the first mitotic spindle at first cleavage. (v) The
2-cell stage lasts about 18 h, during which activation of the embryonic genome
occurs. (vi) The 4-cell stage lasts about 12 h. Note that cleavage of the two

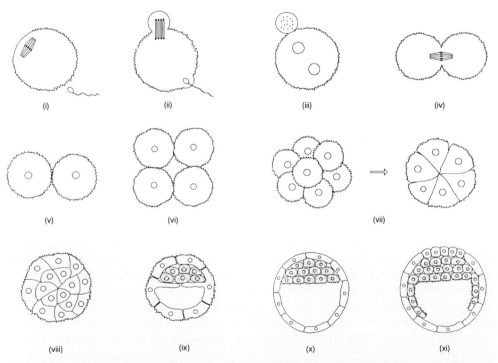

FIGURE 1. Schematic summary of the preimplantation development of the mouse.
Throughout most of these stages the embryo is surrounded by the zona pellucida which is
not shown here. (From Ref. 12).

2-cell blastomeres is not synchronous. (vii) During the 12 h of the 8-cell stage, compaction occurs, which involves close cell-cell apposition and the polarization of cells radially. (viii) The 12 h of the 16-cell morula stage mark the first time that inner and outer cell populations are detectable, the inner cells being apolar and the outer cells polar. (ix) Late in the 32-cell stage, a blastocoel forms, so that by the 64-cell stage (3.5 days into development) a clear blastocoelic cavity and two distinct cell populations with different fates are present. The inner population constitutes the inner cell mass (shaded) which will form all the embryonic and extraembryonic tissues except for the trophoblast of the placental chorion. The outer population consists of trophectoderm cells which will form the definitive trophoblast of the placenta. Trophectodermal processes separate the inner cell mass from the blastocoel. (x) By 4.5 days the trophectodermal processes have withdrawn and the inner cell mass has differentiated into a primitive endodermal epithelial layer (fronting the blastocoel) enclosing primitive ectodermal cells. The zona pellucida is shed at this stage, and the blastocyst commences implantation. (xi) The endoderm migrates over the inner surface of the trophectoderm, which itself consists of two subpopulations. They are a proliferating group of polar trophectoderm cells over the inner cell mass and mural trophectoderm cells forming the walls of the blastocoel, which do not proliferate but undergo endoreduplication of their DNA to give polytene chromosomes in the primary giant cells.

Two points require attention. First, the pace of early development is relatively slow, a feature that is almost certainly imposed on the conceptus by the dual role of the maternal uterus. The endometrium must be transformed from an environment congenial for spermatozoal transport and maturation to one in which implantation can occur. This process takes several days. Second, morphogenetic and differentiative activity within the preimplantation conceptus is concerned exclusively with the segregation of extraembryonic ("preplacental") tissues from those that will give rise to the embryo itself. Both the trophectoderm and the primary endoderm, and all their descendant cells, will contribute exclusively to the nourishment of first the embryo and then the fetus by tapping maternal resources at and after implantation. The mammalian conceptus is a "yolk gatherer" rather than a "yolk carrier" (26). In the preimplantation conceptus, the progenitor cells of the embryo are restricted to the primary ectoderm, which also serves as the source of all extraembryonic mesoderm, the amnion, and the allantois. It is not until 6.5 to 7.5 days of mouse development that the allocation of cells specifically and exclusively to the embryo has occurred; in consequence, up to this time identical twinning can and does occur (2). Technically, therefore, an embryo cannot be identified until 6.5 to 7.5 days; prior to this time the total product of fertilization is properly termed the conceptus, preembryo, or proembryo. However, conventionally, if improperly, the term "embryo" is adopted to describe first the total product of fertilization and then that part of the product that will give rise to the fetus. For the sake of convention and brevity, the term embryo is so used in this chapter.

A set of general rules governing the emergence of mural and polar trophectoderm as well as primitive endoderm and ectoderm was established by the late-1970s. Since that time, the more detailed exposition of underlying cellular and molecular mechanisms has been confined largely to the primary divergence of the inner cell mass and the nascent trophectoderm, which, therefore, constitutes the major topic of this chapter.

The First Differentiative Divergence: Inner Cell Mass and Trophectoderm

Lineage Analysis in the Early Embryo

The early mouse embryo is remarkably plastic. The pioneering experiments of Tarkowski, Mintz, Graham, and their colleagues revealed that two or more embryos could be aggregated together, and a single embryo could be split into two or more parts or its constituent cells rearranged; nonetheless, development was often remarkably unimpaired. Moreover, if some of the constituent cells of a cleaving embryo were made identifiable by marking them prior to their rearrangement, it was found that their developmental fate depended on the position to which they had been relocated. From these experiments, two general conclusions emerged.

1. Cell diversification in early murine development was considered to arise not by a process of differential cleavage of a mosaic egg and embryo, but by a process of positional recognition deriving from interaction between erstwhile identical and equivalent cells.
2. This cell interaction involved cells recognizing whether they were on the outside or the inside of the embryo and so developing into trophectoderm or inner cell mass (ICM), respectively.

Subsequent work has shown that these interpretations were not entirely correct, being arrived at as a result of the erroneous, although commonly held, belief that mosaic and determinate development were one and the same. The observation of developmental plasticity does not preclude a role for differential cleavage (6, 40).

Careful analysis of embryos between the 2- and 64-cell stages reveals that the earliest time at which two distinguishable cell populations are present is the 16-cell stage. Prior to this stage, a part of the cell membrane of each blastomere is exposed on the surface of the embryo (underlying the zona pellucida), but thereafter some blastomeres are totally surrounded by others. The cells in these internal and external populations differ not only in their relative position but also in their phenotypes. If the exposed blastomeres in intact 16- to 64-cell embryos are marked and the embryos are then disaggregated to single cells, the marked outer cells are found to be polar in phenotype, each having a clear apicobasolateral axis. In contrast, each inside cell is organized symmetrically and is nonpolar. The fact that these phenotypic differences, which can also be observed in situ in sectioned or whole-mount embryos, survive disaggregation shows that they are not simply transient adaptations to cell packing or position but rather have a more enduring basis. Indeed, when polar and nonpolar blastomeres isolated in this way are reaggregated together under appropriate conditions, they reconstitute an embryo in which the phenotypically distinct cells relocate to their original position. Thus, the positional and phenotypic differences seem to be mutually reinforcing. By the 64-cell stage, the inside nonpolar cells constitute the nascent ICM and the outside polar cells constitute the primary trophectoderm (25).

What is the origin of the two cell populations that are first observed at the 16-cell stage? Careful examination of embryos at the 8-cell stage and both during and immediately after division to the 16-cell stage has revealed that these two subpopulations are distinguishable from the moment of their formation. They do not arise, therefore, as a result of recognizing their relative positions immediately after their formation. Rather, during the ten or so hours of the preceding 8-cell stage, each constituent blastomere undergoes a radical reorganization of its cytoskeleton, cytoplasm, cytocortex, and surface to transform from an essentially symmetrical to a highly asymmetrical or polarized phenotype. Elements of this asymmetry, once established, are then stably maintained both after the isolation of 8-cell blastomeres from the embryo and throughout their division to yield two 16-cell blastomeres (Fig. 2). Thus, the two subpopulations of cells observed at this latter stage arise by a process of differential cleavage—the inside apolar cells deriving from the basolateral region of polarized 8-cells and the outside polar cells inheriting apical regions (28). However, not all polarized 8-cell blastomeres in every embryo divide

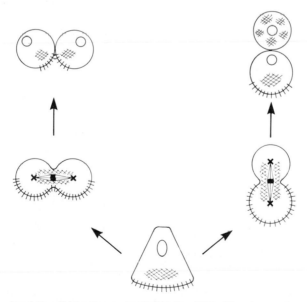

FIGURE 2. Polarized 8-cell blastomeres divide to yield either two polar cells or one polar and one nonpolar cell depending on the plane of cleavage.

differentiatively; some divide along, rather than across, the long axis of their polarity and thereby generate two polar progeny. These observations are summarized schematically in Fig. 2.

It is clear that the ratio of inside cells to outside cells in any 16-cell embryo will be determined by the ratio of differentiative to nondifferentiative (conservative) cleavage divisions of the eight blastomeres in the parent embryo. Recent use of lineage markers, which label single cells or defined groups of cells in situ, has revealed that the number of cells allocated to the inside (ICM) lineage by the process of differentiative division varies among embryos, being occasionally as low as zero or as great as eight, but usually in between, giving an average of about five inside cells for a large population of 16-cell embryos (11, 35). A second allocation of cells to the inside (ICM) lineage occurs at the division from the 16- to the 32-cell stage, and does so in an exactly analogous manner to the first allocation. Thus, polar 16-cell blastomeres can divide in either a differentiative or a conservative manner, and only in the former case are inside cells generated. Interestingly, it is found that, for any given embryo, the number of polar 16-cells dividing differentiatively is related inversely to the number of 8-cell blastomeres that previously divided differentiatively. Thus, a 16-cell embryo with a lot of inside cells tends to generate rather few during its transition to a 32-cell embryo compared with a 16-cell embryo with a low number of inside cells. There is evidence to suggest that when the ratio of inside to outside cells is high, the outside cells are deformed tangentially by the large internal cell cluster so that they are constrained to divide conservatively (11). Once the two allocations of cells to the inside have been achieved by the 32-cell stage, further significant allocations are not a feature of normal development and two separate lineages have effectively been established (9).

Mechanisms Underlying Spatial Reordering in the Early Embryo

It is clear from the foregoing analysis that the establishment of a polar phenotype and its retention during differential cleavage are central elements in the generation of the ICM and trophectoderm. How is a stable polar phenotype established and why does it first appear at the 8-cell stage? The evidence bearing on these questions has been considered in detail elsewhere (12, 27), and the principal conclusions are summarized here together with reference to more recently published work.

The axis of the spatial reordering that occurs in each blastomere at the 8-cell stage appears to be directed by cell contact patterns. The long axis of polarity develops orthogonal to the sum of asymmetric cell contacts made by each cell, ensuring that the apical polar region of each of the cells faces outward (Fig. 3). Since this apical region is relatively nonadhesive compared with the basolateral membrane, the embryo that results maximizes its intraembryonic contacts and minimizes adhesion to other embryos. It is in effect a closed protoepithelial cyst (10, 12). The positional signal that is trans-

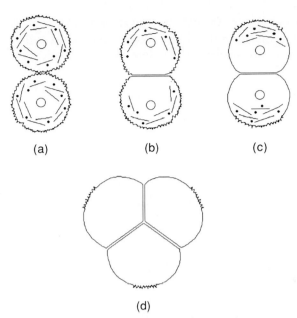

(a)   (b)   (c)

(d)

FIGURE 3. (*a*) A pair of newly formed 8-cell blastomeres. During the 8-cell stage cells polarize and flatten on each other. The axis of polarity develops normal to the point(s) of contact between cells. (*b*) Initially flattening and the polarization of the cytoskeleton and cytoplasmic contents are most evident, but (*c*) later in the cell cycle the cell surface is also visibly polarized as evidenced by the distribution of microvilli. (*d*) When the intercellular contacts are more extensive, the size of the polar region of microvilli on each cell is smaller. From Ref. 36.

mitted between blastomeres to determine the axis of polarization involves (not necessarily exclusively) the calcium-dependent cell adhesion glycoprotein uvomorulin (also known as E-cadherin or L-CAM), and neutralization of its activity leads to abnormalities of polarization. The response to the uvomorulin-mediated signal is first observable in a reorganization of the cytoskeleton that is adjacent to the points of contact with other cells (Fig. 3). This reorganization appears to be a secondary consequence of a local change in the cytocortex of the cell, involving a reduction in stabilizing (or possibly cross-linking and/or nucleating) activity such that microtubules and microfilaments disassemble locally (23). This local cytocortical change then spreads progressively from the contact point(s) between cells for a limited distance around the perimeter of the cell. If there is relatively little contact between cells, then a relatively large portion of the cytocortex remains unaffected, whereas extensive contacts leave a correspondingly smaller region uninfluenced (Fig. 3; Ref. 36). The region of the cytocortex that escapes this contact-mediated cytoskeletal destabilizing activity early in the 8-cell stage becomes associated at some point later in the same cell cycle with a cytoskeletal *stabilizing* activity. The end result is a heterogeneous cytocortex, divided into two zones. The apical zone that is remote from contact points stabilizes cytoskeletal elements, and, in this region of the cell surface, microvilli remain, being largely absent basolaterally. The apical zone is also associated with a subcortical concentration of microfilaments, microtubules, pericentriolar material, and many associated organelles, in contrast to the basolateral region which is characterized by the relative absence of these cytoplasmic features. This transformation, which has been achieved by the end of the 8-cell stage, provides the basis for the differential cleavage which then may follow. Throughout that cleavage, the cytocortex retains clear evidence of its heterogeneity of organization. Indeed, mitosis can be arrested artificially for many hours yet the cortical heterogeneity persists (28).

After division is completed, any interphase 16-cell daughter blastomeres that inherit all or part of the apical polar region of stabilizing cytocortex have a polar organization imposed on their cytoplasmic and surface properties, while those blastomeres that do not inherit any of the polar cytocortex remain nonpolar. In this way, cell diversity is achieved. The polar cytocortical domain resembles a "determinant" to the extent that any cell receiving part of it is evidently constrained to be polar, and, therefore, must include trophectodermal cells among its descendents (25).

The precise nature of the changes that give heterogeneity to the cytocortex is under study, and presumably they will involve cytoskeletal associated proteins such as spectrin (39) or Microtubule Associated Proteins (MAPS) (33). Moreover, since the whole process of cytocortical polarization can occur in the absence of protein synthesis (31), the changes must involve the turnover, modification, or altered activity of preexisting proteins. If the whole complex transformation from a symmetrical to a highly asymmetrical cell occurs using posttranslational mechanisms, does this mean that the changes observed are prefigured in some as yet unrecognized asymmetrical organization that can be traced back through cleavage to the oocyte? The answer appears to be no. There is no evidence to support the idea (and some evidence against it) that prior to the 8-cell stage each blastomere is in any permanent sense "regionalized." Some regional differences may be observed, but these appear to be transient responses to local conditions or unrelated to the generation of ICM and trophectoderm. Rather it seems that early in the 8-cell stage a change occurs in each blastomere that permits the appropriate polarizing response to uvomorulin-mediated contact. All the required structural elements are in place and "ready to go"; indeed many of them are present from the early 4-cell stage (31). The nature of the maturational change that permits this polarizing response to contact is unknown, but there is clear evidence that it does not depend on a mechanism that involves the simple counting of rounds of cytokinesis, DNA replication, or reduction in the nucleocytoplasmic ratio (38).

The experiments summarized above have led to the conclusion that the differential cleavage of asymmetrical cells is an essential component of the process of cell diversification leading to ICM and trophectoderm formation. Moreover, the cytocortex has been identified as the likely locus of the relatively stable primary asymmetry. This asymmetry then provides the focus for second order reorganizations of surface and cytoplasm that may ultimately prove to include asymmetries of mRNA distribution. Thus, the mouse may not differ so fundamentally from other types of organism and embryo in which a cytocortical localising action appears to be operating [e.g., *Stentor* (7); *Paramecium* (34); *Fucus* (24); *Xenopus* (19, 20); see Chap. 2, this volume, for a discussion of such examples]. However, the mouse embryo does appear to part company from many other early embryos that use a similar mechanism in that the earliest signs of cell asymmetry do not appear until cleavage is underway and several blastomeres are present. Presumably this late setting up of a mosaic pattern is simply a reflection of the relatively slow pace of early development in the mouse alluded to earlier. Moreover, other types of embryos that may show a degree of cytoplasmic asymmetry prior to fertilization, do, nonetheless, show elaboration of more complex asymmetries after fertilization and during cleavage, the organization of which often seems to reflect patterns of cell interaction, analogous to the situation seen in the mouse (e.g., Refs. 5, 14, and 20).

Order of Division and Cell Allocation

Since we now understand the cellular dynamics by which ICM and trophectoderm emerge, it is possible to explain in more detail subtle influences on cell allocation, such as the effect of the order of division of cells within an embryo. It was noted by Graham and his colleagues (29) and confirmed by others that cells within the developing mouse embryo do not divide synchronously and that the earlier dividing cells tend to contribute disproportionately more of their descendents to the ICM than do the later dividing cells. Given that ICM cells are now known to be derived by the differentiative division of polarized 8- and 16-cell blastomeres, it seemed reasonable to suppose that early dividing cells might undergo more differentiative divisions than later dividing cells. This has been shown to be so (16). It appears that those cells within an embryo that divide first from the 4- to the 8-cell stage are also the first to enter the process of polarization and the maximization of basolateral intercellular contact that accompanies it. The establishment of more extensive contacts reduces the area of exposed cortex that will ultimately form the stabilizing cytocortical pole (Fig. 3). The lower the ratio of polar to basolateral cytocortex, the greater the chance of a differentiated division (36).

Developmental Plasticity and Differential Cleavage

How can the original conclusions drawn about the nature of developmental processes in the mouse (see the preceding section "Lineage Analysis in the Early Embryo") be squared with the newer observations and conclusions reported in the later parts of the last section? The basis for the belief that the differential cleavage of asymmetrically organized cells could not underlie the divergence of the ICM and trophectoderm lineages was the evident plasticity of early mouse development. A reexamination of this plasticity in the light of our current understanding reveals that there is no conflict since:

1. Any rearrangement of the relative positions of blastomeres up to this stage should not disturb development since stable cell asymmetries are not set up until the 8-cell stage.

2. Rearrangements of blastomeres after this stage can be followed by cell sorting and reorientation so that the cells reassume their appropriate relative positions. This sorting out occurs because the apical polar membrane is relatively nonadhesive compared to the basolateral membrane and the whole membrane of the apolar cells. Thus, polar cells tend to enclose apolar cells, their apical zones facing outward. The embryos are able to adjust to these cellular rearrangements and development is not necessarily impaired.

3. A drastic disturbance of the post 8-cell embryo, e.g., the complete removal of the entire population of either the polar or the apolar cells from a 16-cell embryo, will result in a second line of regulation that also gives plasticity to development. Thus, a population consisting entirely of apolar cells forms a compact aggregate in which the outer cells are exposed to asymmetrical contacts. They then proceed to polarize, just as early 8-cell blastomeres did, and so generate an outer layer of polar cells. Thus, an embryo with both cell types is reconstituted. Conversely, an aggregate consisting exclusively of polar cells tends to generate a large number of apolar cells during the next mitotic division (see above), again restoring balance to the population.

These types of regulatory responses to the loss of one cell type or another at the 16-cell stage appear to operate in situ in a small population of embryos in which repair may have been required (9). None of these regulatory responses is incompatible with the differential distribution of cellular elements at division that is implicit in the mosaic model.

Does Cytoplasmic Organization Have a Role to Play in the Subsequent Events of Early Mouse Development?

The formation of the definitive ICM and trophectoderm marks the end of the first differentiative event of mouse development. The next events then focus on differentiation within each of these two tissues. The polar trophectoderm cells overlying the ICM proliferate to yield ectoplacental cone tissue and extraembryonic ectoderm, while the remaining mural trophectoderm undergoes polyploidization and forms the primary giant cells (Fig. 1). Within the ICM, an epithelial-like primitive endoderm cell layer forms adjacent to the blastocoelic cavity and the remaining group of primitive ectodermal cells is sandwiched between it and the overlying polar trophectoderm (Fig. 1). It is from these ectoderm cells that all of the embryo proper, together with the extraembryonic mesoderm, the amnion, and probably the allantois will develop, while the primitive endoderm contributes largely or exclusively to the yolk sac (2, 27, 30, 37). While the general nature of these differentiative events was described several years ago, the underlying mechanisms that determine how these various cell populations emerge remains relatively unexplored.

Origins of Primitive Endoderm

While the primitive endoderm cells occupy the region of the ICM adjacent to the blastocoel, it is not clear whether, and if so why, they arise there. Is there a causal relationship between their position and what they become? If so, what feature of their position is important and how is it recognized? Alternatively, is there a distinct subpopulation of cells within the ICM that is destined to generate endoderm and that locates to the blastocoelic surface? If so, where does this subpopulation come from and what properties cause it to form the endoderm lineage?

Experiments have been performed that analyze the contribution that the descendents of individual ICM cells that were injected into blastocysts make to the tissues of the postimplantation embryo. These experiments have found that up until about 4.5 days of development an individual cell may contribute to endoderm and/or ectoderm lineages, but thereafter only one or other lineage is colonized (17). While we have learned from the results of this experiment the approximate time of commitment to a restricted lineage, we still do not know anything about the mechanisms that operate prior to commitment. This is true because, as now appears to be the case for the proto-ICM and prototrophectoderm cells, a mixed population of apolar protoectoderm and asymmetrically organized protoendoderm cells could give the observed result as long as any single protoendoderm cell was able to divide differentiatively prior to its commitment. While early evidence suggested that ICM cells on the blastocoelic surface of the ICM were the ones that formed endoderm, examination of ICMs early in the process of endoderm delamination has suggested that the protoendodermal cells may be found distributed much more widely around and within the ICM and may become restricted to the blastocoelic surface only later (3, 18). These results suggest that simple asymmetrical exposure of progenitor cells (analogous to that observed at the 8-cell stage and described above) cannot provide the signal to form endoderm. Support for this view also comes from the observation that the trophectodermal processes, which normally cover the entire blastocoelic surface of the ICM during blastocyst expansion (Fig. 1; Ref. 13), can remain present until *after* primitive endoderm has begun to form (3). Indeed, it is possible that the withdrawal of these processes is a consequence rather than a cause of endoderm delamination.

The behavior of embryonic carcinoma cells induced to undergo endodermal delamination in vitro has also been taken as evidence that cell position is an important determinant of endodermal differentiation, but the exact cell type in situ to which these cell lines correspond is not entirely clear. For example, whereas embryonic carcinoma cells are prevented from forming endoderm by exposure to soluble laminin, suggesting that this extracellular matrix molecule might form part of the positional signaling mechanism (22), the same does not appear to be true for the ICM. Neither soluble laminin nor antibodies to it impair endoderm formation (Ref. 3 and unpublished data). Thus, decisive evidence that favors positional recognition as a necessary or exclusive component of the process of endoderm formation is lacking at present.

In contrast, there is recent evidence to suggest that the cells of the developing ICM are heterogeneous with respect to their origin and properties. Thus, ICM cells that form at the first (8 to 16 cell) allocation tend to have few, if any, cytokeratin filaments, whereas those ICM cells that form at the second (16 to 32 cell) allocation do have cytokeratin filaments and also tend to be positioned more superficially within the ICM (4). Since primitive endoderm cells also form superficially and assemble cytokeratin filaments early in their development, it is possible that they are largely or exclusively derived from this second population of cells. If so, the influence of cytoplasmic organization on the generation of cell diversity may linger further into development than was hitherto thought. Clearly, it will be important to resolve the question of the relative roles of cell interaction and cellular history in the process of primitive endoderm delamination. It was observed that retinoic acid can stimulate some embryonic carcinoma cell lines to delaminate into endoderm (1) and that both retinoic acid and its receptor appear to form part of a naturally occurring morphogenetic signaling system (32, 42). This observation raises the possibility that it is the acquisition of a functional retinoic acid receptor system that determines whether cells become primitive endoderm. Acquisition of such a system could be influenced by the previous history and origins of the ICM cells.

Origins of Trophectodermal Heterogeneity

The experiments of Gardner (17) and his colleagues established that the proliferation of trophectoderm in the polar region was due to an underlying stimulus from the ICM. The nature of that stimulus has not been established. It seems likely that *any* trophectoderm cells are competent to respond to the stimulus since introduction of a second ICM (or the transplantation of the ICM) to any region of mural trophectoderm around the blastocyst wall, if performed early enough, results in a second (or new) center of trophectodermal proliferation. The very nature of these experiments means that they cannot prove that *all* trophectodermal cells are capable of proliferation. Indeed, it has been suggested that trophectodermal cells that are limited to forming primary giant cells might exist and that they are situated at the site of the nascent blastocoelic cavity and can mark the position of the abembryonic pole [that is, the region of mural trophectoderm directly opposite the ICM which is the earliest site of primary giant cell formation (8, 41)]. Recently, experiments by Garbutt and her colleagues (15) have shown that the site of both the nascent blastocoel and abembryonic pole formation is in fact influenced by the division of order of cells at earlier stages. Thus, the nascent blastocoel tends to be associated with the trophectodermal descendents of early-dividing cleavage stage blastomeres. This observation could be reasonably explained if the capacity to form blastocoelic fluid was acquired during the sixth cell cycle and that the first cells to reach this cycle (or some point in it) therefore marked the earliest site of fluid accumulation. The important points for this chapter are that (1) the distinction between early-dividing and late-dividing blastomeres is established at the 2-cell stage and (2) the progeny of each 2-cell blastomere then retain their relative difference in division order to the blastocyst stage (21). Thus, there is a persisting influence of 2-cell cytoplasmic properties and organization on the morphogenesis of the blastocyst and the fate (if not the potential) of trophectoderm cells at this stage.

Conclusions

It is common sense to suppose that the cytoplasmic endowment inherited by a cell will provide a major limitation on, and determinant of, that cell's properties, developmental fate, and potential. It is also evident from numerous developmental examples that cells are responsive to environmental signals. A balanced view requires us to take account of each of these influences in arriving at an understanding of the underlying cellular and subcellular mechanisms at work in shaping the details of the developing embryo. Larger cells with a higher inertial component will tend to inherit cytoplasmic features with longer biologic half-lives and exert a stronger influence on cell behavior, an influence likely to be particularly marked when combined with short cell cycles. However, the setting up and expression of these cytoplasmic features often appear to involve prior or concurrent interactions with extrinsic signals, which assume an increasingly dominant role later in development. It is our contention that the mouse early embryo is not exceptional in the mechanisms that it uses to generate complexity, combining the influence of a relatively stable cytoplasmic organization with the processes of cell interaction.

Acknowledgments

I would like to thank Tom Fleming for his advice during the preparation of this manuscript. The work described in this chapter was supported by grants to the author and his colleagues from the Medical Research Council and the Cancer Research Campaign.

General References

ROSSANT, J., and PEDERSEN, R. A.: 1986. *Experimental Approaches to Mammalian Embryonic Development.* Cambridge University Press, New York.

PEDERSEN, R. A.: 1987. Early Mammalian Embryogenesis. In: *The Physiology of Reproduction*, vol. 1. (E. Knobil, and J. D. Neill, eds.), pp. 187–230, Raven Press, New York.

References

1. Alonso, A., Weber, T., and Jorcano, J. L. 1987. Cloning and characterisation of keratin D, a murine endodermal cytoskeletal protein induced during in vitro differentiation of F9 teratocarcinoma cells. *Roux's Arch. Dev. Biol.* **196**:16–21.

2. Beddington, R. 1986. Analysis of tissue fate and prospective potency in the egg cylinder. In: *Experimental Approaches to Mammalian Embryonic Development.* (J. Rossant and R. A. Pedersen, eds.), pp. 121–150, Cambridge University Press, New York.

3. Chisholm, J. C. 1986. Cell Diversification in the Mouse Early Embryo. Dissertation for Doctorate of Philosophy, University of Cambridge.

4. Chisholm, J. C., and Houliston, E. 1987. Cytokeratin filament assembly in the preimplantation mouse embryo. *Development (Camb.)* **101**:565–582.

5. Conklin, E. G. 1905. Mosaic development in Ascidian eggs. *J. Exp. Zool.* **2**:146–223.

6. Davidson, E. 1986. *Gene Activity in Early Development,* 3d ed. Academic Press, New York.

7. De Terra, N. 1985. Cytoskeletal discontinuities in the cell body cortex initiate basal body assembly and oral development in the ciliate *Stentor. J. Embryol. Exp. Morphol.* **87**:249–257.

8. Denker, H. W. 1983. Cell lineage, determination and differentiation in earliest developmental stages in mammals. *Bibl. Anat.* **24**:22–58.

9. Dyce, J., George, M. A., Goodall, H., and Fleming, T. P. 1987. Do cells belonging to trophectoderm and inner cell mass in the mouse blastocyst maintain discrete lineages? *Development (Camb.)* **100**:685–698.

10. Fleming, T. P. 1986. Endocytosis and epithelial biogenesis in the mouse early embryo. *Bioessays* **4**:105–109.

11. Fleming, T. P. 1987. A quantitative analysis of cell allocation to trophectoderm and inner cell mass in the mouse blastocyst. *Dev. Biol.* **119**:520–531.

12. Fleming, T. P., and Johnson, M. H. 1988. From egg to epithelium. *Annu. Rev. Cell Biol.* **4**:459–485.

13. Fleming, T. P., Warren, P. D., Chisholm, J. C., and Johnson, M. H. 1984. Trophectodermal processes regulate the expression of totipotency within the inner cell mass of the mouse expanding blastocyst. *J. Embryol. Exp. Morphol.* **84**:63–90.

14. Freeman, G. 1983. The role of egg organisation in the generation of cleavage patterns. In: *Time, Space and Pattern in Embryonic Development.* (W. R. Jeffery, and R. A. Raff, eds.), pp. 171–196, Liss, New York.

15. Garbutt, C. L., Chisholm, J. C., and Johnson, M. H. 1987. The establishment of the embryonic:abembryonic axis in the mouse embryo. *Development (Camb.)* **100**:125–134.

16. Garbutt, C. L., Johnson, M. H., and George, M. A. 1987. When and how does division order influence cell allocation to the inner cell mass of the mouse blastocyst? *Development (Camb.)* **100**:325–332.

17. Gardner, R. L. 1983. Origin and differentiation of the extraembryonic tissues in the mouse. *Int. Rev. Exp. Pathol.* **24**:63–133.

18. Gardner, R. L. 1985. Regeneration of endoderm from primitive ectoderm in the mouse embryo: fact or artefact? *J. Embryol. Exp. Morphol.* **88**:303–326.

19. Gerhart, J. C., Vincent, J.-P., Scharf, S. R., Black, S. D., Gimlich, R. L., and Danilchik, M. 1984. Localisation and induction in early development of *Xenopus*. *Philos. Trans. R. Soc. Lond. B Biol. Sci.* **307**:319–330.

20. Gerhart, J., Danilchik, M., Roberts, J., Rowning, B., and Vincent, J.-P. 1986. Primary and secondary polarity of the amphibian oocyte and egg. In: *Gametogenesis and the Early Embryo*. (J. G. Gall, ed.), pp. 305–319, Liss, New York.

21. Graham, C. F., and Deussen, Z. A. 1978. Features of cell lineage in preimplantation mouse development. *J. Embryol. Exp. Morphol.* **48**:53–72.

22. Grover, A., Andrews, G., and Adamson, E. D. 1983. Role of laminin in epithelium formation by F9 aggregates. *J. Cell Biol.* **97**:137–144.

23. Houliston, E., Pickering, S. J., and Maro, B. 1987. Redistribution of microtubules and pericentriolar material during the development of polarity in mouse blastomeres. *J. Cell Biol.* **104**:1299–1308.

24. Jaffe, L. F. 1968. Localization in the developing *Fucus* egg and the general role of localizing currents. *Adv. Morphol.* **7**:295–328.

25. Johnson, M. H., Chisholm, J. C., Fleming, T. P., and Houliston, E. 1986. A role for cytoplasmic determinants in the development of the mouse early embryo. *J. Embryol. Exp. Morphol.* **97**(Suppl):97–121.

26. Johnson, M. H., and Everitt, B. J. 1988. *Essential Reproduction*, 3d ed. Blackwell Scientific, Oxford.

27. Johnson, M. H., and Maro, B. 1986. Time and space in the mouse early embryo: a cell biological approach to cell diversification. In: *Experimental Approaches to Mammalian Embryonic Development*, (J. Rossant, and R. A. Pedersen, eds.), pp. 35–66, Cambridge University Press, New York.

28. Johnson, M. H., Pickering, S. J., Dhiman, A., Radcliffe, G. S., and Maro, B. 1988. Cytocortical organisation during natural and prolonged mitosis of mouse 8-cell blastomeres. *Development (Camb.)* **102**:143–158.

29. Kelly, S. J., Mulnard, J. G., and Graham, C. F. 1978. Cell division and cell allocation in early mouse development. *J. Embryol. Exp. Morphol.* **48**:37–51.

30. Lawson, K. A., and Pedersen, R. A. 1987. Cell fate, morphogenetic movement and population kinetics of embryonic endoderm at the time of germ layer formation in the mouse. *Development (Camb.)* **101**:627–652.

31. Levy, J. B., Johnson, M. H., Goodall, H., and Maro, B. 1986. Control of the timing of compaction: a major developmental transition in mouse early development. *J. Embryol. Exp. Morphol.* **95**:213–237.

32. Maden, M., and Summerbell, D. 1986. Retinoic acid-binding protein in the chick limb bud: identification at developmental stages and binding affinities of various retinoids. *J. Embryol. Exp. Morphol.* **97**:239–250.

33. Maro, B., Houliston, E., and Paintrand, M. 1988. Purification of meiotic spindles and cytoplasmic asters from mouse oocytes. *Dev. Biol.* **129**:275–282.

34. Milhausen, M., and Agabian, N. 1983. *Caulobacter* flagellin mRNA segregates asymmetrically at cell division. *Nature (Lond.)* **302**:630–632.

35. Pedersen, R. A., Wu, K., and Balakier, H. 1986. Origin of the inner cell mass in mouse embryos: cell lineage analysis by microinjection. *Dev. Biol.* **117**:581–595.

36. Pickering, S. J., Maro, B., Johnson, M. H., and Skepper, J. N. 1988. The influence of cell contact on the cleavage of mouse 8-cell blastomeres. *Development (Camb.)* **103**:353–363.

37. Rossant, J. 1986. Development of extraembryonic lineages in the mouse embryo. In: *Experimental Approaches to Mammalian Embryonic Development.* (J. Rossant, and R. A. Pedersen, eds.), pp. 97–120, Cambridge University Press, New York.

38. Smith, R. K. W., and Johnson, M. H. 1985. DNA replication and compaction in the cleaving embryo of the mouse. *J. Embryol. Exp. Morphol.* **89**:133–148.

39. Sobel, J. S., and Alliegro, M. A. 1985. Changes in the distribution of a spectrin-like protein during development of the preimplantation mouse embryo. *J. Cell Biol.* **100**:333–336.

40. Stent, B. 1985. The role of cell lineage in development. *Philos. Trans. R. Soc. Lond. B Biol. Sci.* **312**:3–20.

41. Surani, M. A. H., and Barton, S. C. 1984. Spatial distribution of blastomeres is dependent on cell division order and interactions in mouse morulae. *Dev. Biol.* **102**: 335–343.

42. Thaller, C., and Eichele, G. 1987. Identification and spatial distribution of retinoids in the developing chick limb bud. *Nature (Lond.)* **327**:625–628.

Questions for Discussion with the Editor

1. *What mechanisms might bridge the temporal and spatial gaps between cytoplasmic organization systems (e.g., polarized cytoskeleton) and regional gene expression patterns in the mammalian embryo?*

It has been known for several years now that by the time the blastocyst has formed, its two phenotypically distinct cell populations (ICM and trophectoderm) have different biosynthetic profiles of "tissue specific" proteins [J. VanBlerkom, S. C. Barton, and M. H. Johnson. 1976. *Nature (Lond.)* **259**:319–321] that seem to reflect at least in part different mRNA activity profiles (M. H. Johnson. 1979. *J. Embryol. Exp. Morphol.* **53**:335–344; Duprey et al. 1985. *Proc. Natl. Acad. Sci. U.S.A.* **82**: 8353–8539). Some of these distinguishing marker proteins have been identified as part of the intermediate filament system of the embryo (P. Brulet et al. 1980. *Proc. Natl. Acad. Sci. U.S.A.* **77**:4113–4117) and are first synthesized at earlier morula stages (A. H. Handyside and M. H. Johnson. 1978. *J. Embryol. Exp. Morphol.* **44**: 191–199). It seems that initially all cells in the embryo synthesize and assemble intermediate filament proteins, but that as the two lineages diverge, there is up-regulation of synthesis and assembly in the trophectodermal lineage and down-regulation in the ICM lineage (4). It is possible that continuing interactions between the two cell types modulate assembly, thereby influencing the levels of free monomer that then can regulate synthesis. Differential gene expression may be product-driven. The absence of a wider range of suitable molecular markers makes it difficult to know how general such a mechanism might be. However, it may be important that the earliest and most evident differences between the two emergent lineages are found in their cytoskeletal organization, since this could secondarily affect the stability, localization, and/or activity of other mRNAs that do not code for cytoskeletal proteins.

2. *Please expand upon your statement that "the mouse may not differ so fundamentally from other organisms in which a cytoskeletal localizing activity appears to be operating (e.g., Stentor, Paramecium, etc.)."*

Both (1) the systematic spatial localization of developmental information within a cell and, if it occurs, (2) the regulated orientation of division plane in relation to the axis of that localization must depend upon a spatial organization that is relatively stable with respect to both (1) the length of the cell cycles involved and (2) the changes in cell organization that accompany cell activities such as movement or cleavage. The cytocortex, and particularly its role in organizing the cytoskeletal system and thereby the cytoplasm, seems to provide in a number of systems such a stable organization. Thus, in *Caenorhabditis elegans* a defined area of the anterior cytocortex of blastomere Po organizes not only the anterior-posterior axis of the cell but also the regular orientation of the cleavage plane, and does so via effects on the cytoskeletal system of the cell (A. A. Hyman, 1988. Establishment of division axes in the early embryonic divisions of *C. elegans*. Doctoral Dissertation, University of Cambridge). In *Xenopus* and *Rana* the dorsoventral axis of the embryo is specified by a rotation of the cytoplasmic contents of the egg within the cytocortex. This rotation is oriented by, and may be driven by, parallel arrays of microtubules associated with the vegetal cytocortex (R. Elinson, and R. Rowning. 1988. *Dev. Biol.* **128**: 185–197).

The Extracellular Matrix as an Information System

CHAPTER **16**

Extracellular Matrix Control of Cell Migration during Amphibian Gastrulation

Kurt E. Johnson, Norio Nakatsuji
and Jean-Claude Boucaut

Introduction

The Nature of the Problem

DURING THEIR EARLY DEVELOPMENT, all metazoan organisms undergo a period of development where changes in cell shape, cell number, and cell-cell associations produce fundamental rearrangements in the morphology of the embryo. These changes lead to the generation of new shapes by a process known as *morphogenesis*. In many cases, certain designated cell groups move actively about in the embryo. At present, the molecular mechanisms involved in this designation and subsequent morphogenetic selectivity are poorly understood. Typically, morphogenetic cell movements are controlled in a temporal and spatial pattern that is precisely repeated in each embryo. Cells move along specific pathways within the embryo, moving from one location to another along a *particular* pathway. It is difficult enough to understand how cells move from one location to another inside the embryo, but it is even more mysterious why they choose one particular pathway for this locomotion from among the large number of pathways theoretically available to them. By way of analogy, imagine that you were planning to travel from Washington to Paris to Tokyo. There are any number of itineraries that you could plan, based on the time available, your funds, your preference for mode of travel, and your strength and stamina. A great number of factors would go into deciding to fly to Paris and then Tokyo by way of the Concorde rather than taking a boat to Europe, riding a bicycle to Paris, taking a train across Europe and Asia, and finally swimming to Japan. Similarly, morphogenetic cell movements occur along a limited number of paths as cells start in one location and end up in another at a later time in development. We examine some of the factors that we feel are crucial for guiding these morphogenetic cell movements along specific pathways during amphibian gastrulation.

Anatomy of Gastrulation

The eggs of amphibians are on the order of 1 to 3 mm in diameter. They are asymmetrical at ovulation, having a flattened and heavily pigmented animal pole and a more rounded and less pigmented vegetal pole. The center of gravity of the egg lies below the geometric center, so that after fertilization they rest in water with their pigmented sides up, facing the warming rays of the sun. Within a few hours after fertilization, eggs begin to cleave rapidly. Usually the first two cleavage planes pass from the animal pole to the vegetal pole, dividing the egg into two and then four equal blastomeres (Fig. 1a to c). The third cleavage furrow passes perpendicular to the first two but closer to the animal pole, creating four smaller animal pole blastomeres and four larger vegetal pole blastomeres (Fig. 1d). As additional cleavages ensue quickly, several thousand cells are generated without an increase in the volume of the embryo (Fig. 1e to i). During formation of the *blastula*, a hollow ball of cells, small intercellular gaps between intraembryonic blastomeres coalesce to create an eccentric cavity called the *blastocoel*, which is located above the vegetal blastomeres and below the animal blastomeres. The animal blastomeres make up an epithelial sheet several cell layers thick. This epithelium constitutes the roof of the blastocoel. The vegetal blastomeres make up a yolky mass of cells which forms the floor of the blastocoel. Between the small animal blastomeres and the large vegetal blastomeres, one finds an equatorial group of intermediate-size cells located in the *marginal zone*.

At the beginning of gastrulation, *bottle cells* form near the blastopore. Their apical surfaces become extremely contracted as a result of the action of microfilament bundles (1). This initial step in gastrulation leads to the formation of an invagination site in the marginal zone (Fig. 2a). The blastopore first forms as a pigmented depression in the marginal zone and rapidly grows into a small curved slit (Fig. 2b). As the blastopore grows into a crescent-shaped structure, more and more bottle cells become recruited into the blastopore (Fig. 2c). In addition, cells above the dorsal lip of the blastopore begin to migrate toward and then over the dorsal lip in a movement known as *involution*. A broad sheet of cells, representing the primordium of the chordamesoderm and located above the dorsal lip of the blastopore, converges on the blastopore and moves toward it. Simultaneously, this group of cells becomes rearranged from a short, broad group into a long, narrow group. This movement has been called *convergent extension* by Keller (21), and it thought to be the main engine producing the driving force for gastrulation. Convergent extension ap-

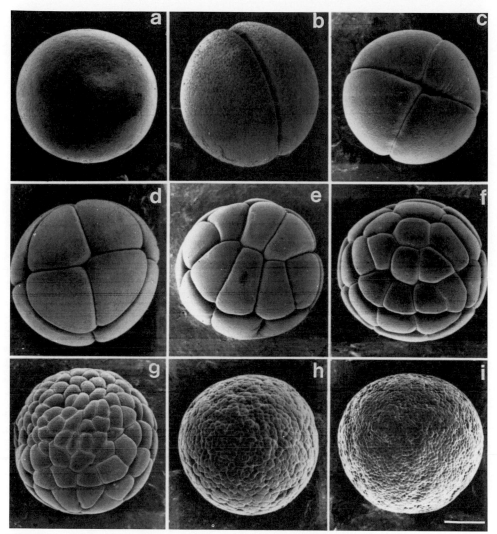

FIGURE 1. Early development of *Pleurodeles waltlii* embryos. (*a*) Fertilized, uncleaved egg, (*b*) 2-cell stage, (*c*) 4-cell stage, (*d*) 8-cell stage, (*e*) 16-cell stage, (*f*) 32-cell stage, (*g*) early blastula, (*h*) midblastula, and (*i*) late blastula. Bar, 400 μm.

pears to be an autonomous movement because it occurs spontaneously in isolated tissue fragments in vitro (21–23). The main engine of gastrulation has been identified. Now we need to investigate what factors *guide* the moving system once it has been set in motion.

As chordamesodermal cells involute, their vacated space on the surface of the embryo is occupied by spreading presumptive ectodermal cells. The spreading of the ectoderm is an extensive movement of a sheet of cells known as *epiboly*, an event which is closely coordinated with convergent extension (Fig. 2*d* to *g*). Eventually, all of the presumptive mesoderm leaves the surface

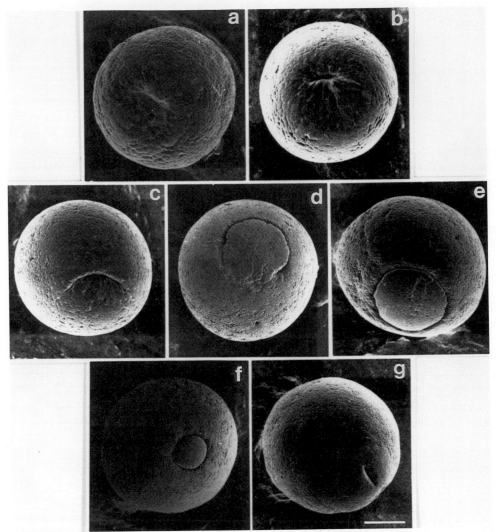

FIGURE 2. Gastrulation of *Pleurodeles waltlii* embryos from the vegetal aspect. (*a*) Blastopore is a pigmented area just beginning to invaginate. (*b*) Blastopore deepens as invagination continues. (*c*) Blastopore grows into a crescent. (*d*) Blastopore is nearly a complete circle. (*e*) Large circular blastopore and large yolk plug. (*f*) Small circular blastopore and small yolk plug. (*g*) Yolk plug disappears from surface and blastopore closes. Bar, 400 μm.

of the embryo and is replaced by the epibolic spreading of presumptive ectoderm. The blastopore continues to grow from a crescent into a circle, and then the diameter of the circular blastopore shrinks by constriction (Fig. 2*d* to *g*). This is due to both epiboly and the movement of presumptive endoderm inside the embryo. By the end of gastrulation, the presumptive endoderm is drawn completely inside the embryo (Fig. 2g). At this point, the presumptive ectoderm has completely surrounded the embryo.

During gastrulation, a new cavity, called the *archenteron*, grows from the blastopore as the invagination site at the blastopore deepens and expands toward the animal pole inside the embryo. The archenteron represents the precursor of the lumen of the gut tube. As the archenteron expands, it grows at the expense of the shrinking blastocoel. By the end of gastrulation, the blastocoel is nearly completely obliterated.

The leading edge of the archenteron consists of groups of cells that adhere to and spread across the inner surface of the roof of the blastocoel. The cells at the leading edge of the growing archenteron form broad lamellipodial attachments on the basal surfaces of cells making up the roof of the blastocoel. It seems that these cells attach preferentially to this inner surface and are stimulated to move away from the blastopore by several factors which we discuss later.

During this complex series of morphogenetic movements, the blastula is converted into a triploblastic embryo with an ectoderm, a mesoderm and notochord, and an endoderm. This process of gastrulation brings the chordamesodermal inductors of the central nervous system from the outer surface of the embryo to inside and underneath the ectodermal tissue that is to be induced during neurulation, which, in turn, leads to the formation of the central nervous system. By the end of gastrulation, a cranial and caudal end of the embryo have been established; i.e., the cranial direction is most distant from the yolk plug, and the caudal direction is toward the yolk plug. The notochord is formed during gastrulation and lies along the future body axis. It also defines an axis of bilateral symmetry and allows a left and right side to be identified. Soon after gastrulation has been completed, *neurulation* converts a *neural plate* first into a *neural groove* and then into a *neural tube*. The neural plate folds up on itself, forms a tubular structure, and then sinks beneath the surface ectoderm. Once neurulation and the morphogenesis of the intraembryonic mesoderm and endoderm have been completed, the basic vertebrate body plan has been established. One important question has recently arisen: Do different classes of amphibians show significant differences in the detailed mechanics of gastrulation?

Recent work has revealed that there are significant differences in mesodermal cell formation in different amphibian species. Smith and Malacinski (40) confirmed Keller's (19, 20) earlier observations in the anuran *Xenopus laevis* to show that mesodermal cells arise from deep cells in the dorsal marginal zone (DMZ). In addition, Smith and Malacinski (40) found that mesodermal cells arise in the superficial DMZ in the urodele *Ambystoma mexicanum*. Furthermore, Lundmark (26) has provided evidence that in *A. mexicanum* mesodermal cell migration may play a significant role in convergent extension. Recently, Shi et al. (39) have investigated the capacity for extension of the DMZ in the urodele *Pleurodeles waltlii* gastrulae. They showed that intercalation plays a role in convergent extension but also showed that when rotated 90 or 180°, grafted DMZ explants still involuted normally and extended in accord with the appropriate animal pole–blastopore axis of host embryos. Furthermore, if the roof of the blastocoel was removed at the blastula stage, i.e., when mesodermal cells had not yet undergone migration on the blastocoel roof, involution and extension of the DMZ remained extremely limited. These results suggest that mesodermal cell migration may play an important role in convergent extension in *Pleurodeles*. In contrast, Keller (21) has shown that in the anuran *Xenopus laevis*, DMZs rotated 90° and then grafted into hosts failed to involute; they, instead, showed extension in the appropriate direction with respect to the graft animal pole–blastopore axis rather than with respect to the same axis in the host. Clearly then, there are significant differences in the detailed mechanisms of gastrulation in anuran and urodele embryos.

Morphogenetic cell movements during embryogenesis occur inside the three-dimensional cell mass, thus preventing experimental manipulations or even simple observations of cell movement except in a few fortunate cases (see Ref. 43 for review). Amphibian embryos are not transparent because of the large number of yolk platelets and pigment granules. Thus, it is necessary to study amphibian gastrula cell behavior in vitro or to develop methods for disrupting gastrulation in vivo. Kubota and Durston (24) cut open gastrulae to observe cell migration. Alternative ways to study cell migration and its control are to (1) isolate embryonic cells and study their behavior in well-controlled culture conditions, (2) inhibit cell locomotion in vivo with specific probes (4–6, 10), or (3) utilize formation of interspecific, arrested hybrid embryos (13). An inherent difficulty of these approaches is to determine what represents normal cell behavior inside embryos and what is artifact due to the artificial culture conditions or nonspecific effects of probes and hybridization.

Morphological studies of fixed samples (28, 31) have revealed that migrating mesodermal cells in gastrulae have rounded cell bodies and extend lamellipodia and filopodia mostly in the direction of their migration, i.e., toward the animal pole (31). Filopodia usually extend from the margin of a lamellipodium. The rate of the cell migration in *Xenopus laevis* embryos was estimated from histological sections made from serially fixed samples (28). The mesoderm cells as a group seemed to migrate at a rate of about 6 μm/min.

Extracellular Fibrils Provide a Contact Guidance System

Contact Inhibition of Mesodermal Cells Occurs in Vitro

After developing cell culture conditions adequate to study mesodermal cell migration in vitro, Johnson (15) and Nakatsuji and Johnson (33) showed that mesodermal cells exhibit contact inhibition of movement. When lamellipodia or filopodia of one cell came into contact with lamellipodia or filopodia from another cell, the cellular protrusions showed typical contact paralysis. In culture, under appropriate conditions, gastrula mesodermal cells usually became bipolar and make lamellipodia and filopodia at opposing ends of the cell body (see Fig. 7a and c). In these cells, considerable amounts of energy and time were consumed by the competition between opposite ends of the cell body. On the other hand, scanning electron microscopy (SEM) showed that the migrating mesoderm cells in gastrulae usually have lamellipodia only on the leading edge (toward the animal pole) (31). This observation led us to the conclusion that there must be additional factors that direct the mesoderm cells to produce locomotory organelles only toward the animal pole and, thus, guide the cell in that direction. One such factor is probably contact inhibition of movement, which would cause dispersion of the migratory cells from the crowded blastopore region toward the less-crowded animal pole region. Another factor was the discovery of a network of anastomosing extracellular matrix (ECM) fibrils on the inner surface of the roof of the blastocoel.

Ubiquitous Nature of Fibrillar ECM in Amphibian Gastrulae

Nakatsuji et al. (31) and Boucaut and Darribère (2, 3) discovered a dense network of extracellular fibrils lining the basal surface of epithelial cells that make up the roof of the blastocoel. These fibrils are sparse prior to gastrulation, appear at the beginning of gastrulation, and continue to accumulate during gastrulation (Fig. 3). They also found that these fibrils were significantly aligned parallel to the animal pole–blastopore axis. These fibrils have been observed in all eight species of amphibian embryos examined to date, which are the urodeles *Ambystoma maculatum, A. mexicanum, Pleurodeles waltlii,* and *Cynops phyrrhoghaster* and the anurans *Xenopus laevis, Rana pipiens, Rana sylvatica,* and *Bufo bufo* (2, 3, 16, 29, 34–38). It is also known that these fibrils contain fibronectin (FN) (2–8, 18, 25, 38) (Fig. 4) and laminin (9, 32). FN synthesis occurs at low levels prior to gastrulation but increases during gastrulation. FN-containing fibrils accumulate preferentially on the inner surface of the roof of the blastocoel (2, 3, 25) presumably due to the differential distribution of a 140×10^3 dalton cell surface receptor for FN (FNR) (10) (Fig. 5). Could we obtain solid experimental evidence showing that this ECM played a role of guiding cell locomotion in vitro and in vivo?

Conditioning of Culture Substrata with Extracellular Fibrils Transferred from the Inner Surface of the Ectoderm Layer

Migrating mesodermal cells in amphibian gastrulae use an anastomosing network of extracellular matrix fibrils on the inner surface of the ectoderm layer as their substratum for cell migration (31, 34, 37) (Fig. 4). Boucaut et al. (4) have shown that when the animal cap was inverted 180° so that the surface facing the perivitelline space now faced into the blastocoel, grafted explants do not provide a suitable substratum for mesodermal cell migration because extracellular matrix fibrils were not available. Filopodia of migrating mesodermal cells often show close association with the fibrils, suggesting that the fibrils serve as a preferential adhesion site for filopodia and lamellipodia. Furthermore, the presence of the statistically significant alignment of such

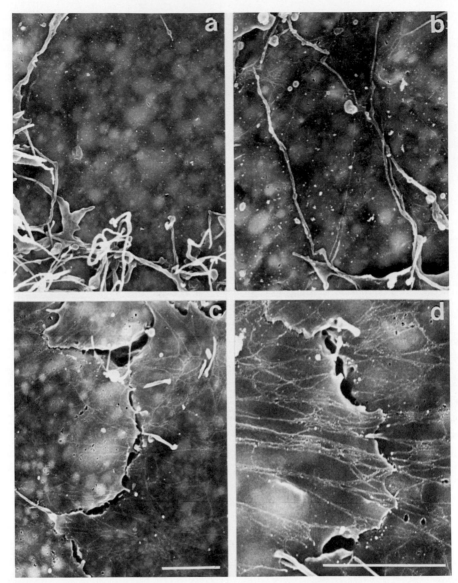

FIGURE 3. Scanning electron micrographs of the inner surface of the ectodermal layer of *Ambystoma maculatum* embryos. Bars, 5 μm. (*a*) Early blastula, (*b*) late blastula, (*c*) and (*d*) early gastrula. (*a–c*) All the same magnification. From N. Nakatsuji and K. E. Johnson. 1983. *J. Cell Sci.* **59**:61–70. Used with permission of Company of Biologists Ltd.

fibrillar network along the blastopore–animal pole axis (31), which is probably produced by the stretching of the ectodermal layer along this axis during gastrulation movements (36), indicates an interesting possibility for an actual role in vivo for a long postulated contact guidance system (44) by an aligned fibrillar network (29, 31).

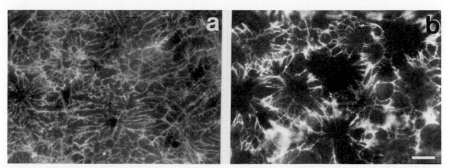

FIGURE 4. In situ immunofluorescent distribution of FN-related polypeptides on whole-mount specimens of presumptive ectodermal cells in (a) *Pleurodeles waltlii* early gastrulae and (b) *Rana pipiens* early gastrulae. Bar, 10 μm.

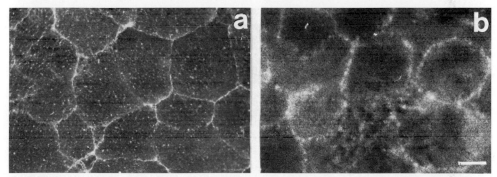

FIGURE 5. In situ immunofluorescent distribution of FNR-related polypeptides on whole-mount specimens of presumptive ectodermal cells in (a) *Pleurodeles waltlii* early gastrulae and (b) *Rana pipiens* early gastrulae. Bar, 5 μm. Part *a* from T. Darribère, K. M. Yamada, K. E. Johnson, and J.-C. Boucaut. 1988. *Dev. Biol.* **126**: 182–194. Used with permission of Academic Press.

To test this hypothesis, Nakatsuji and Johnson (34) transferred the extracellular fibrillar network to a coverslip surface (Fig. 6). A rectangular piece of the ectodermal layer was dissected from the dorsal part of early gastrulae of *Ambystoma maculatum* and explanted to a plastic coverslip with its inner surface facing down on the coverslip surface. After 5 h of culture, an outline of the explant and an arrow marking the direction toward the animal pole of the original embryo were scratched on the coverslip. After the explant was mechanically removed from the coverslip, dissociated mesoderm cells isolated from gastrulae were seeded on the conditioned surface. Cell movement was recorded by time-lapse cinemicrograpy, followed by fixation of the coverslip for SEM.

Mesodermal cells adhered rapidly to such conditioned surfaces. They extended lamellipodia and filopodia and migrated actively on the conditioned surface. Mesodermal cells did not attach to the coverslip surface outside the

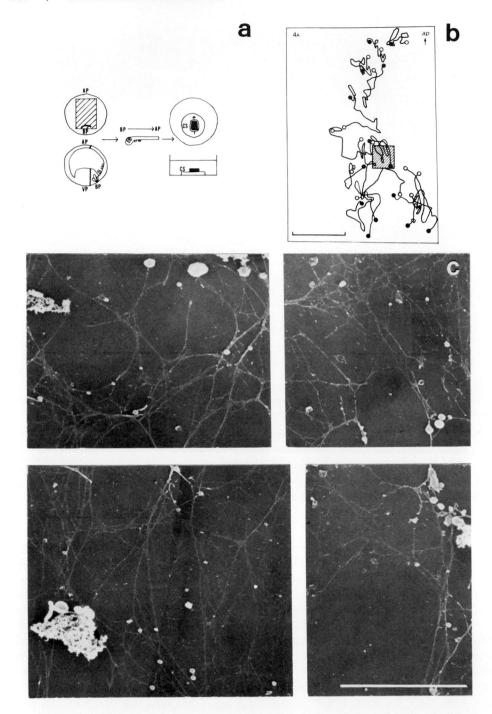

conditioned area. *Ambystoma* mesodermal cells produced large lamellipodia and moved persistently in one direction for long periods. Scanning electron microscopy revealed that ECM fibrils were transferred onto the coverslip surface from the inner surface of the ectodermal layer. Filopodia of the attached mesodermal cells adhered preferentially to these fibrils. Mesodermal cells from *Xenopus* gastrulae also moved on surfaces conditioned by *Ambystoma* ectodermal tissue.

Marking the approximate center of the cell bodies in time-lapse films gave cell trails. These cell trails appeared somewhat aligned along the blastopore–animal pole axis of the conditioning ectodermal explant (after the explant had been removed from the coverslip). Nakatsuji and Johnson (34) developed a computer program in order to analyze alignment of cell trails and fibrillar networks on the coverslip surfaces. Curved or zigzag lines were divided into short segments. Displacement (dy and dx) along the y axis (usually adjusted to the blastopore–animal pole axis) and along the perpendicular x axis of each segment was measured. The computer added the absolute values of each displacement ($\Sigma| dy |$, $\Sigma| dx |$). Their ratio ($\Sigma| dy |/\Sigma| dx |$) was then calculated and designated r. If there were no alignment, r would be equal to 1. If there were an alignment along the y axis, r would be greater than 1. In the opposite case, r would be between 0 and 1. To obtain a more symmetrical parameter, suitable for statistical manipulation, we took $R = \log 2r$. In the case of random cell movement, R would be 0. An alignment along the y axis and x axis would give positive and negative values, respectively, for R.

By using this computer program, Nakatsuji and Johnson (34) analyzed apparently random movement of the *Xenopus* mesodermal cells on a homogeneous surface coated with collagen and serum (33). They obtained an R value of -0.03 ± 0.42 ($n = 58$), supporting the prediction that the mean value should be equal or close to 0. On the other hand, cell trails on the conditioned surface gave an R value of $+0.22 \pm 0.66$ ($n = 65$) for *Ambystoma* cells and $+0.25 \pm 0.45$ ($n = 63$) for *Xenopus* cells. In both cases, the shift of the mean to the positive value was statistically significant at a level of $P < 0.01$.

Analysis of the fibrillar network observed by SEM revealed that orientation of the cell trails coincided with alignment of the fibrils in the same area. For example, where cell trails appeared unoriented, the fibrillar network was not significantly aligned. On the other hand, where several cell trails were

FIGURE 6. Experiment illustrating conditioning of substrata with fibrils deposited by a conditioning explant. (*a*) Diagram illustrating experimental setup. AP, animal pole, BP, blastopore; CS coverslip; VP, vegetal pole. (*b*) Cell trails drawn from time-lapse film. Solid circles, starting points; open circles, finishing points after 1 h. Cell trails are aligned along the animal pole–blastopore axis (vertical axis) ($R = + 0.46$), bar, 100 μm. (*c*) A *scanning electron micrograph* of the crosshatched square area. The R-value for fibrils here is $+0.41$, bar, 10 μm. Data shown in part *a* are from: N. Nakatsuji, 1984. *Amer. Zool.* 24:615–627. Used with permission of American Society of Zoologists. Data shown in parts *b* and *c* are from N. Nakatsuji and K. E. Johnson, 1983. *J. Cell Sci.* 59:43–60. Used with permission of Company of Biologists Ltd.

oriented strongly along the y axis, the fibrillar network was strongly aligned along the same axis (Fig. 6).

Contact Guidance by an Artificial Alignment of Fibrils Produced by Mechanical Tension

We now knew that an alignment of the extracellular fibrillar network on the inner surface of the ectoderm layer guided migrating mesodermal cells during gastrulation. We hypothesized that this fibrillar alignment was produced by stretching of the ectoderm along the blastopore–animal pole axis during convergent extension and epiboly. To test this hypothesis, we created artificial mechanical tension on the conditioning ectodermal layer. Figure 7 describes a trick used by Nakatsuji and Johnson (36) to apply mechanical tension continuously to the ectodermal layer during conditioning. A rectangular piece of the ectodermal layer was dissected from an early gastrula so that the long dimension of the rectangle was parallel to the animal pole–blastopore axis of the embryo. This explant was draped across a slender strip of a coverslip so that the long axis of the explant was parallel to the long axis of the coverslip. This was done to create mechanical tension forces *perpendicular* to the normal tension axis in vivo. Both ends of the ectodermal explant were free to droop in the medium so that their own weight constantly applied tension to the central part of the explant. If the coverslip strip was set horizontally, both edges of the coverslip held the explant tightly, thus preventing transfer of the tension toward the central part of the explant. Therefore, the strip was tilted 30° to loosen the lower edge of the explant. In this way, constant tension was applied to the central part of the explant during the conditioning.

After 5 h of culture, the explant was removed from the coverslip surface. Dissociated mesodermal cells were then seeded on the conditioned surface. Scanning electron microscopy revealed that the transferred fibrillar network was indeed realigned along the axis of mechanical tension. Analysis of cell movement showed that cell trails were also strongly aligned along the axis of tension force. When the ectoderm explant from early gastrulae (stage 10) was put on the coverslip in an orientation so that its original blastopore–animal pole axis becomes perpendicular to the axis of tension (as shown in Fig. 6), the R value of the cell trails with y axis adjusted to the tension axis was $+0.29 \pm 0.73$ ($n = 55$). The mechanical tension and resulting stretching of the ectoderm layer erased the original alignment along the blastopore–animal pole axis and produced a stronger alignment along the tension axis. When the ectoderm layer from stage 10 gastrulae with weak alignment, as described before, was used for the conditioning, the R value was $+0.36 \pm 0.78$ ($n = 40$). As a control, when the ectoderm layer from stage 10– gastrulae was put on a horizontally set coverslip strip, the R value was -0.01 ± 0.90 ($n = 52$), showing no oriented cell movement.

Mechanical tension and stress can be produced frequently during morphogenetic movements in the embryo. The idea that such tension or stress produces alignment in the extracellular matrix and causes the contact guid-

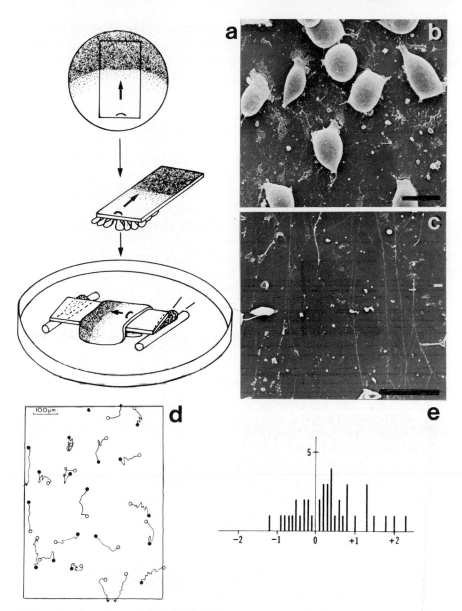

FIGURE 7. Experiment illustrating fibril alignment as a result of mechanical tension artificially applied to a conditioning explant. (*a*) Diagram illustrating experimental set up. (*b*) Scanning electron micrographs of cells and (*c*) of aligned fibrils. The axis of tension in both SEMs is the vertical axis of the photograph. (*d*) Cell trails drawn from time-lapse film. Open circles, starting points; solid circles, finishing points after 1 h. Cell trails are aligned along the axis of tension (vertical axis). The overall R value calculated for all cell trails here is +0.47. (*e*) Histogram showing distribution of cell trails according to their R values. The distribution is clearly shifted to the right (+0.36 ± 0.78, n = 40). The shift is statistically significant (P < .005). From N. Nakatsuji and K. E. Johnson, 1984. *Nature* **307**:453–455. Used with permission of Macmillan Journals Ltd.

ance or other kinds of orientation effects on the embryonic cells probably has wide application during embryogenesis. One example is that an alignment of the extracellular collagen gel caused by tissue masses can give orientation to the migration and growth of cells and tissues (41).

Conditioning Experiments Using Arrested Hybrid Embryos

Most hybrid embryos between various *Rana* species stop their development at the early gastrula stage and show almost no mesodermal cell migration (13, 27). In these embryos, the extracellular fibrils are absent or very much reduced in number on the inner surface of the ectoderm layer (11, 18, 37). Using embryos hybrid between *Rana* species, Nakatsuji and Johnson (37) did conditioning experiments with various combinations of the ectodermal layer and mesodermal cells. As expected, normal *Rana pipiens* mesodermal cells attached to and actively migrated on surfaces conditioned by the ectodermal layer from normal embryos. Seeded mesodermal cells moved at a mean rate of 4.1 μm/min. The computer analysis of the cell trails, however, revealed very weak alignment along the blastopore–animal pole axis of the ectoderm layer ($R = 0.12 \pm 0.72$, $n = 84$). Such very weak alignment compared to the *Ambystoma* embryo may relate to the fact that the mesodermal cells in anuran embryos migrate as a packed mass of many cells. In contrast, mesodermal cells in urodele embryos migrate as individual cells, sometimes completely separated from other cells and thus, perhaps needing more guidance for the oriented migration.

The ectodermal layer from arrested hybrid embryos deficient in fibrils in vivo had almost no conditioning effect on adhesion by mesodermal cells of either hybrid embryos or normal embryos. Another interesting finding was that if the mesodermal cells from arrested hybrid embryos were seeded on surfaces conditioned by normal ectodermal layer, they adhered to the surface and moved at moderate rates. For example, mesodermal cells from the hybrid, *R. pipiens* females × *R. catesbeiana* males moved at a mean rate of 1.8 μm/min on surfaces conditioned by ectodermal layer from *R. pipiens* normal embryos. This result suggests that the migratory deficiency in hybrid embryos is partially rescued by deposited extracellular fibrils from normal embryos.

Adhesion and Movement by the Mesoderm Cells on Substrata Coated with Fibronectin or Laminin

Culture dish surfaces coated with fibronectin or laminin were found to be good substrata for attachment and migration by *Xenopus* and *Pleurodeles* gastrula mesodermal cells (Fig. 8) (10, 30). For example, about 80 percent of *Xenopus* mesodermal cells attached to FN-coated substrata within 1 h and moved actively at a mean rate of 2.8 μm/min. Johnson (17) also observed adhesion and migration of mesodermal cells attached to Sepharose beads coupled with FN.

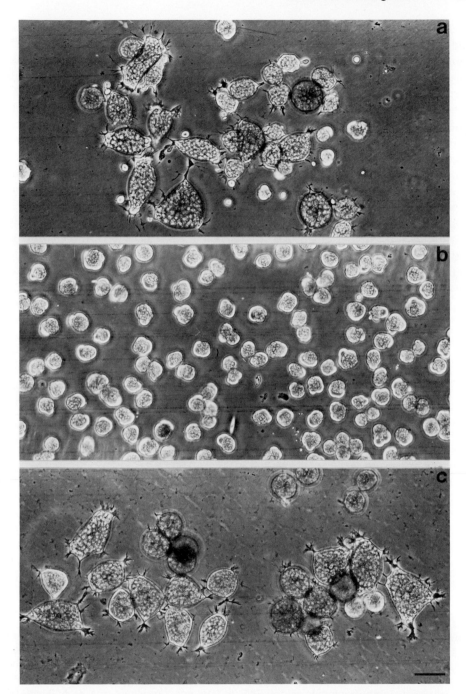

FIGURE 8. Phase contrast light micrographs showing *Xenopus* gastrula mesoderm cells attached to (*a*) fibronectin-coated or (*c*) laminin-coated substrata. (*b*) Gastrula ectoderm cells show circus movement but do not attach to the fibronectin- or laminin-coated surface. Scale bar, 50 μm.

Coating with laminin isolated from mouse EHS sarcoma also provided good substrata for *Xenopus* cell adhesion and movement (Fig. 8c). The strength of the adhesion, however, was weaker compared to the fibronectin-coated surface, and the mean rate of movement was lower (1.8 μm/min). Furthermore, Darribère et al. (10) showed that Fab' fragments of anti-FNR caused detachment of *Pleurodeles* cells previously attached to FN-coated substrata (Fig. 9). Nakatsuji (30) also showed that type IV collagen and heparan sulfate supported neither adhesion nor movement of *Xenopus* gastrula mesodermal cells. Another important finding was that a heterogeneous distribution of fibronectin or laminin molecules in vitro caused guiding effects on the adhesion and migration of the mesodermal cells. For example, Fig. 10 shows an accumulation of *Bufo bufo japonicus* gastrula mesodermal cells inside an area coated with fibronectin. The cells clearly preferred the coated over the uncoated areas in the culture dish. Furthermore, under certain conditions, laminin molecules make fibrillar aggregates in vitro. When a network of such fibrils was aligned accidentally, mesodermal cells elongated and moved along the axis of fibril alignment, illustrating a clear case of contact guidance in vitro. These observations show that it is possible to guide the mesodermal cells by heterogeneous distributions of ECM in vivo and in vitro.

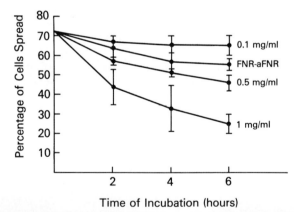

FIGURE 9. Dose-response curves for detachment of ectomesodermal cells by Fab' fragments of anti-FNR. Cells were obtained from early gastrula stage *Pleurodeles waltlii* embryos (stage 8b). They were cultured for 90 min on FN-coated substrates. They were then incubated in medium containing 1000 μg/mL Fab' fragments of anti-FNR absorbed with FNR (control) or Fab' fragments of anti-FNR at final concentrations of 100, 500, or 1000 μg/mL. After 2, 4, and 6 h of culture, attached and spread cells were counted. These results are expressed as the percentages of total cells that remained attached and spread. Each point is the average of four determinations. Error bars represent standard deviation. From: T. Darribère, K. M. Yamada, K. E. Johnson, and J.-C. Boucaut. 1988. *Dev. Biol.* **126**: 182–194. Used with permission of Academic Press.

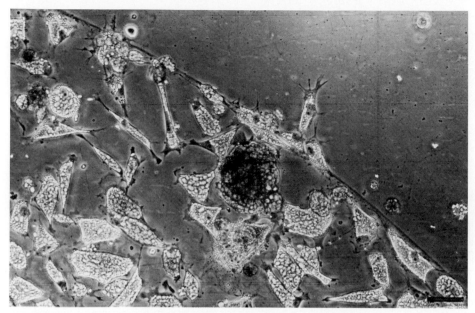

FIGURE 10. Dissociated *Bufo* gastrula mesoderm cells accumulate inside the fibronectin-coated area (lower-left part) but rarely cross the border between the coated and noncoated surfaces. Scale bar, 100 μm.

Studies with Probes to Disrupt Function of ECM Molecules in Vivo

Studies with Probes to Disrupt Cell-Fibronectin Interaction

If these extracellular fibrils serve as a contact guidance system for migrating mesodermal cells, one would expect that probes that disrupt cell-fibronectin interaction would have a profound effect on cell migration. Boucaut et al. (4, 5) have provided evidence that this prediction was correct. First, when living embryos were injected with Fab′ fragments of anti-FN IgG at the early gastrula stage, gastrulation was blocked (Fig. 11). The antibodies interacted with FN-containing fibrils in vivo in such a way as to prevent cell adhesion to the fibrils. Similar injections at the late gastrula stage had no noticeable effect on neurulation (4). Second, when early gastrula were probed with synthetic peptides representing the cell-binding domain of FN, again gastrulation was blocked completely. A peptide representing the collagen-binding domain of FN had no effect on gastrulation (5, 6). We interpreted these results in a similar way, i.e., the peptides blocked the cell surface receptor for FN, thus preventing cells from adhering to FN-fibrils. Third, when Darribère et al. (10) injected Fab′ fragments of anti-FNR IgG into living early gastrulae, gastrulation was once again completely inhibited (10) (Fig. 12). In this case, we interpreted these results to mean that the antibody blocked the FNR, thus once

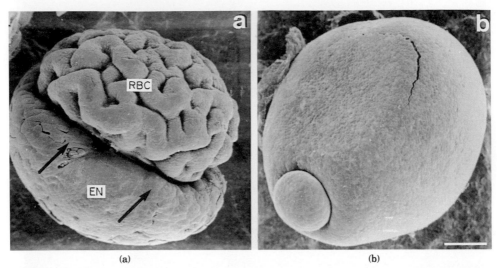

(a)   (b)

FIGURE 11. Scanning electron micrograph of the in vivo effect of Fab' fragments of anti-FN during gastrulation. *Pleurodeles waltlii* embryos were microinjected at the early gastrula stage (stage 8b) and observed 24 h later. (*a*) Injection of Fab' fragments of anti-FN. This result is typical of the more severely arrested embryos. A complete inhibition of gastrulation was observed. Note the highly convoluted roof of the blastocoel (RBC), circular blastopore (arrow), and smooth exposed endodermal mass (EN). (*b*) Control embryo injected with Fab' preimmune IgG. The embryo undergoes normal gastrulation. An early neural plate has formed in this control embryo indicating that primary embryonic induction has taken place. Bar. 100 μm. From J.-C. Boucaut, T. Darribère, H. Boulekbache, and J. P. Thiery. 1984. *Nature* 307:364–367. Used with permission of Macmillan Journals, Ltd.

again preventing the interaction between migrating mesodermal cells and their FN-rich ECM.

All three probes prohibit the interaction between migrating mesodermal cells and the FN-rich ECM on the inner surface of the roof of the blastocoel, albeit by different mechanisms. The anti-FN antibodies coat FN-fibrils (8), preventing access by migrating mesodermal cell receptor. The anti-FNR antibodies bind to cell surface FNR and thereby prevent cell adhesion to FN-containing fibrils. The synthetic peptides bind to cell surface FNR and thus prohibit cells from adhering to the FN-rich ECM. Each probe has the net effect of preventing the crucial interaction between migrating mesodermal cells and ECM deposited on the inner surface of the roof of the blastocoel.

Scanning Electron Microscopy of Probed Embryos

Boucaut et al. (4–6) and Darribère et al. (10) performed extensive SEM observations on embryos injected with Fab' anti-FN, Fab' anti-FNR, and synthetic peptides to the cell-binding domain of FN. In all three different experimental

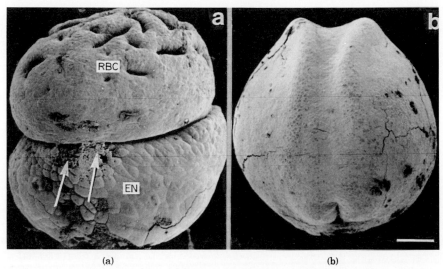

(a)   (b)

FIGURE 12. Scanning electron micrograph of the in vivo effect of Fab′ fragments of anti-FNR during gastrulation. *Pleurodeles waltlii* embryos were microinjected at the early gastrula stage (stage 8b) and observed 24 h later. (a) Injection of Fab′ fragments of anti-FNR. This result is typical of the more severely arrested embryos. A complete inhibition of gastrulation was observed. Note the highly convoluted roof of the blastocoel (RBC), circular blastopore (arrow), and smooth exposed endodermal mass (EN). (*b*) Control embryo injected with Fab′ preimmune IgG. The embryo undergoes normal gastrulation. An early neural plate has formed in this control embryo indicating that primary embryonic induction has taken place. Bars, 100 μm. From T. Darribère, K. M. Yamada, K. E. Johnson, and J.-C. Boucaut. 1988. *Dev. Biol.* **126**: 182–194. Used with permission of Academic Press.

conditions, there were striking similarities between the probed *Pleurodeles waltlii* embryos (4–6, 10). Injected embryos formed a blastopore in an appropriate location. The blastopore formed a conspicuous circular constriction in the equatorial region of the blocked embryo which divided the embryo into two hemispheres. The animal hemisphere became extensively folded with convolutions and deep furrows. In contrast, the vegetal hemisphere remained smooth.

When probed embryos were fractured and examined in the SEM, they showed a large blastocoel that was not collapsed. The outer surface of the roof of the blastocoel was extensively convoluted. In contrast, the inner surface was remarkably smooth (Fig. 13). Migrating mesodermal cells formed a ring-like collection in the marginal zone but they failed to migrate across the inner surface of the roof of the blastocoel presumably because they were unable to gain an appropriate foothold there. Migrating mesodermal cells could be seen to project small filopodia and lamellipodia toward the FN-fibril strewn substratum, but examination of stereopairs of images showed that the mesodermal cells were unable to attach these peripheral protrusions to the extracellular matrix.

FIGURE 13. Scanning electron micrograph of the in vivo effect of the cell-binding peptide of fibronectin during gastrulation. *Pleurodeles waltlii* embryos were microinjected at the early gastrula stage (stage 8b) and observed 24 h later. Embryos were fractured along the axis of bilateral symmetry. The migration of marginal zone cells is clearly inhibited, and they come to rest on the floor of the blastocoel cavity (arrow). A substantial blastocoel is apparent in the arrested gastrula. Bar, 100 μm. From J.-C. Boucaut, T. Darribère, D. L. Shi, H. Boulekbache, K. M. Yamada, and J. P. Thiery. 1985. *J. Embryol. Exp. Morph.* **89**: (Suppl.) 211–227. Used with permission of the Company of Biologists Ltd.

Interpretation of Results

Our results demonstrating an arrest of gastrulation in antibody-injected embryos also point to the important role in gastrulation of mesodermal cell migration in *Pleurodeles waltlii* (4–6, 10). In the most extensively arrested of these probed embryos, a circular blastopore formed, the endodermal mass remained on the outer surface of the embryo, and the roof of the blastocoel became extensively convoluted. Arrested embryos had a large blastocoel. Keller et al. (23) have claimed that these probed embryos have an ensemble of abnormalities that are difficult to associate solely with failure of mesodermal cell migration. This assertion may well be true if the results were obtained with *Xenopus laevis* embryos. The four major morphogenetic movements of gastrulation are, "...epiboly of the animal cap, bottle cell formation, migration of cells on the gastrula wall, and convergent extension" (22, 26). Since mesodermal cells are found in the superficial layers of urodeles, unlike *Xenopus laevis*, perhaps failure of mesodermal cell migration *can* explain the abnormalities seen in our probed embryos. We propose that an abortive type of ectodermal spreading occurs in these embryos, albeit not normal epiboly. Perhaps ectodermal spreading by intercalation continued unabated in probed embryos, but mesodermal cell migration failed because appropriate interaction between mesodermal cells and the ECM coating the inner surface of the roof of the blastocoel failed to occur. Under these circumstances, one might expect convolutions to appear in the roof of the blastocoel. After all, if a tissue layer spread actively (by whatever mechanism but most probably by cellular intercalation) but lacked an appropriate *space* to spread into, then perhaps it would fold up on itself instead. Since mesodermal cells cannot migrate out of the periblastoporal region subsequently occupied by ectodermal cells, then, perhaps epiboly does not occur normally but instead results in convolution of the roof of the blastocoel.

A complete circular blastopore forms in anti-FN (4, 6) and anti-FNR probed *Pleurodeles waltlii* embryos (10). This result may indicate that stages of blastopore formation in *Pleurodeles waltlii* are independent of FN function. We do know that bottle-cell formation is dependent on some paternal gene function because it rarely occurs normally in arrested hybrid embryos, although it does occur in one arrested hybrid known to show paternal gene expression (12–14, 27).

The endodermal mass remains exposed on the outer vegetal surface of the embryo because there is no epibolic spreading of ectodermal cells to engulf it. We feel that all of these morphological traits of anti-FNR probed embryos can be attributed to a failure of mesodermal cell migration. In unpublished experiments, we have probed *Xenopus laevis* embryos with Fab' fragments of anti-FNR. We found that these embryos showed complete exogastrulation. Ectodermal cells sat as an isolated cap at one end of a strikingly elongated chordamesodermal and endodermal mass. It appears that convergent extension in *Xenopus laevis* is a powerful force that is independent of FN function. Nevertheless, even in *Xenopus laevis*, mesodermal cell interaction with the FN-rich extracellular matrix on the inner surface of the roof of the blastocoel appears to be required for normal gastrulation.

Conspicuous convolution of the roof of the blastocoel occurs in three interspecific arrested hybrid embryos: *Rana pipiens* female × *Rana sylvatica* male (14), *Rana pipiens* female × *Rana temporaria* male, and *Rana pipiens* female × *Rana esculenta* male. It does not occur in other arrested hybrids, (e.g., *Rana pipiens* female × *Rana catesbeiana* male or *Rana pipiens* female × *Rana clamitans* male) (13). These observations, along with recent studies showing defects in FN-function in arrested hybrid embryos (18) suggest that arrested hybrid embryos share similarities with *Pleurodeles waltlii* gastrulae probed with molecules capable of disrupting cell-FN interaction. In the future, we plan to study FN function in normal and hybrid embryos hoping to gain more sophisticated insights into the role of FN in amphibian gastrulation.

Conclusions and Future Directions

Where We Stand in Our Model of the Role of the ECM in Controlling Amphibian Gastrulation

In this presentation, we have described evidence to support a complex new model of amphibian gastrulation. Cellular rearrangements such as intercalation and exchange of neighbors (22) seem to provide the driving forces for gastrular morphogenetic cell movements. Once these forces are generated, an oriented network of fibronectin- and laminin-containing extracellular fibrils appears to serve as a contact guidance system for directing mesodermal cell migration. If convergent extension provides the motive force as a main engine, perhaps oriented fibrillar extracellular matrix provides direction to the morphogenetic cell movements of gastrulation. These fibrils are especially prominent on the inner surface of the roof of the blastocoel, a location particularly enriched in receptors for these ECM macromolecules. Morphogenetic cell movements of the roof of the blastocoel probably produce tensions in vivo which serve to orient the ECM into an aligned fibrillar network. In addition, migrating mesodermal cells have the ability to recognize this ECM by virtue of their own receptors. Migrating mesodermal cells, in conjunction with contact inhibition of locomotion caused by cells following them through the blastopore, show a net displacement away from the blastopore, across the inner surface of the roof of the blastocoel, toward the animal pole.

Future Directions

We are on the threshold of a new era in research on the control of morphogenetic cell movements during gastrulation. Now that we have a much clearer understanding of the role of FN and other ECM components in gastrulation, we can begin to ask questions concerning the control of FN synthesis and distribution during gastrulation. In the future, we hope to learn more about the genetic control of the spatial and temporal regulation of FN synthesis and the factors that control the appearance of the FNR. Recombinant DNA technology has allowed cloning of the gene for integrin, a 140×10^3 dalton integral membrane glycoprotein which serves as a receptor linkage between FN and intracellular actin (42). Perhaps by creating appropriate probes, we can learn about the developmental control of gene expression for the synthesis of extracellular matrix components and their cell surface receptors.

Ultimately, we feel that genetically programmed synthetic events lead to the production of new extracellular matrix components. The expression of appropriate cell surface receptors may both control the intraembryonic distribution of extracellular matrix components and also enable cells to respond specifically to changes in their pericellular environment. Finally, one set of morphogenetic movements such as epiboly may impose important epigenetic organizational order on the extracellular matrix, allowing specific contact guidance of other moving cells. Together, these genetic and epigenetic events direct morphogenetic cell movements.

Acknowledgments

The work presented in this paper was supported by NSF Grant DCB 8400256 to Kurt E. Johnson. J.-C. Boucaut is grateful to Drs. J. P. Thiery and K. M. Yamada for their continual help and encouragement. The work presented in this paper was supported by grants from CNRS (UA 1135), Ministère de l'Education (ARU), ARC and by Université P. M. Curie (France).

General References

Boucaut, J.-C., Darribere, T., Shi, D. L., Boulebache, H., Yamada, K. M., and Thiery, J. P.: 1985. Evidence for the role of fibronectin in amphibian gastrulation. *J. Embryol. Exp. Morphol.* **89**(Suppl.):211–227.

Keller, R. E.: 1986. The cellular basis of amphibian gastrulation. In: *Developmental Biology. A Comprehensive Synthesis*, vol. 2. *The Cellular Basis of Morphogenesis* (L. W. Browder, ed.), pp. 241–327, Plenum, New York.

References

1. Baker, P. C. 1965. Fine structure and morphogenic movements in the gastrula of the treefrog, *Hyla regilla. J. Cell Biol.* **24**:95–116.

2. Boucaut, J.-C., and Darribère, T. 1983. Presence of fibronectin during early embryogenesis in the amphibian *Pleurodeles waltlii. Cell Differ.* **12**:77–83.

3. Boucaut, J.-C., and Darribère, T. 1983. Fibronectin in early amphibian embryos: migrating mesodermal cells are in contact with a fibronectin-rich fibrillar matrix established prior to gastrulation. *Cell Tissue Res.* **234**:135–145.

4. Boucaut, J.-C., Darribère, T., Boulekbache, H., and Thiery, J. P. 1984. Prevention of gastrulation but not neurulation by antibody to fibronectin in amphibian embryos. *Nature (Lond.)* **307**:364–367.

5. Boucaut, J.-C., Darribère, T., Poole, T. J., Aoyama, H., Yamada, K. M., and Thiery, J. P. 1984. Biologically active synthetic peptides as probes of embryonic development: A competitive peptide inhibitor of fibronectin function inhibits gastrulation in amphibian embryos and neural crest cell migration in avian embryos. *J. Cell Biol.* **99**:1822–1830.

6. Boucaut, J.-C., Darribère, T., Shi, D. L., Boulekbache, H., Yamada, K. M., and Thiery, J. P. 1985. Evidence for the role of fibronectin in amphibian gastrulation. *J. Embryol. Exp. Morphol.* **89**(Suppl.):211–227.

7. Darribère, T., Boucher, D., Lacroix, J.-C., and Boucaut, J.-C. 1984. Fibronectin synthesis during oogenesis and early development of the amphibian *Pleurodeles waltlii. Cell Differ.* **14**:171–177.

8. Darribère, T., Boulekbache, H., Shi, D. L., and Boucaut, J.-C. 1985. Immunoelectron microscopic study of fibronectin in gastrulating amphibian embryos. *Cell Tissue Res.* **239**:75–80.

9. Darribère, T., Riou, J.-F., Shi, D. L., Delarue, M., and Boucaut, J.-C. 1986. Synthesis and distribution of laminin-related polypeptides in early amphibian embryos. *Cell Tissue Res.* **246**:45–51.

10. Darribère, T., Yamada, K. M., Johnson, K. E., and Boucaut, J.-C. 1988. The 140 kD fibronectin receptor complex is required for mesodermal cell adhesion during gastrulation in the amphibian *Pleurodeles waltlii. Dev. Biol.* **126**:182–194.

11. Delarue, M., Darribère, T., Aimer, C., and Boucaut, J.-C. 1985. Bufonid nucleocytoplasmic hybrids arrested at the early gastrula stage lack a fibronectin-

containing fibrillar extracellular matrix. *Wilhelm Roux's Arch. Dev. Biol.* **194**: 275–280.

12. Gregg, J. R. 1957. Morphogenesis and metabolism of gastrula-arrested embryos in the hybrid *Rana pipiens* female × *Rana sylvatica* male. In: *The Beginnings of Embryonic Development* (A. Tyler, A. C. von Borstel, and C. Metz, eds.), pp. 231–261, American Association for the Advancement of Science, Washington.

13. Johnson, K. E. 1970. The role of changes in cell contact behavior in amphibian gastrulation. *J. Exp. Zool.* **175**:391–428.

14. Johnson, K. E. 1971. A biochemical and cytological investigation of differentiation in the interspecific hybrid amphibian embryo *Rana pipiens* female × *Rana sylvatica* male. *J. Exp. Zool.* **177**:191–206.

15. Johnson, K. E. 1976. Ruffling and locomotion in *Rana pipiens* gastrula cells. *Exp. Cell Res.* **101**:71–77.

16. Johnson, K. E. 1984. Glycoconjugate synthesis during gastrulation in *Xenopus laevis*. *Am. Zool.* **24**:605–624.

17. Johnson, K. E. 1985. Frog gastrula cells adhere to fibronectin-Sepharose beads. In: *Molecular Determinants of Animal Form* (G. M. Edelman, ed.), pp. 271–292, Liss, New York.

18. Johnson, K. E., Boucaut, J.-C., Darribère, T., and Riou, J.-F. 1987. Fibronectin in normal and gastrula arrested hybrid frog embryos. *Anat. Rec.* **218**:68A.

19. Keller, R. E. 1975. Vital dye mapping of the gastrula and neurula of *Xenopus laevis*. I. Prospective areas and morphogenetic movements in the superficial layer. *Dev. Biol.* **42**:222–241.

20. Keller, R. E. 1976. Vital dye mapping of the gastrula and neurula of *Xenopus laevis*. II. Prospective areas and morphogenetic movements in the deep region. *Dev. Biol.* **51**:118–137.

21. Keller, R. E. 1984. The cellular basis of gastrulation in *Xenopus laevis*: Postinvolutional convergence and extension. *Am. Zool.* **25**:589–602.

22. Keller, R. E. 1986. The cellular basis of amphibian gastrulation. In: *Developmental Biology. A Comprehensive Synthesis*, vol. 2. The Cellular Basis of Morphogenesis (L. W. Browder, ed.), pp. 241–327, Plenum, New York.

23. Keller, R. E., Danilchik, M., Gimlich, R., and Shin, J. 1985. Convergent extension by cell intercalation during gastrulation in *Xenopus laevis*. In: *Molecular Determinants of Animal Form* (G. M. Edelman, ed.), pp. 111–141, Liss, New York.

24. Kubota, H. Y., and Durston, A. J. 1978. Cinematographical study of cell migration in the opened gastrula of *Ambystoma mexicanum*. *J. Embryol. Exp. Morphol.* **44**: 71–80.

25. Lee, G., Hynes, R., and Kirschner, M. 1984. Temporal and spatial regulation of fibronectin in early *Xenopus* development. Cell **36**:729–740.

26. Lundmark, C. 1986. Role of bilateral zones of ingressing superficial cells during gastrulation of *Ambystoma mexicanum*. *J. Embryol. Exp. Morphol.* **97**:47–62.

27. Moore, J. A. 1955. Abnormal combinations of nuclear and cytoplasmic systems in frogs and toads. *Adv. Genet.* **7**:139–182.

28. Nakatsuji, N. 1975. Studies on the gastrulation of amphibian embryos: Cell movement during gastrulation in *Xenopus laevis* embryos. *Wilhelm Roux's Arch. Dev. Biol.* **178**:1–14.

29. Nakatsuji, N. 1984. Cell locomotion and contact guidance in amphibian gastrulation. *Am. Zool.* **24**:615–627.

30. Nakatsuji, N. 1986. Presumptive mesodermal cells from *Xenopus laevis* gastrulae attach to and migrate on substrata coated with fibronectin or laminin. *J. Cell Sci.* **86**:109–118.

31. Nakatsuji, N., Gould, A., and Johnson, K. E. 1982. Movement and guidance of migrating mesodermal cells in *Ambystoma maculatum* gastrulae. *J. Cell Sci.* **56**:207–222.

32. Nakatsuji, N., Hashimoto, K., and Hayashi, M. 1985. Laminin fibrils in newt gastrulae visualized by immunofluorescent staining. *Dev. Growth & Differ.* **27**:639–643.

33. Nakatsuji, N., and Johnson, K. E. 1982. Cell locomotion *in vitro* by *Xenopus laevis* gastrula mesodermal cells. *Cell Motil.* **2**:149–161.

34. Nakatsuji, N., and Johnson, K. E. 1983. Conditioning of a culture substratum by the ectodermal layer promotes attachment and oriented locomotion by amphibian gastrula mesodermal cells. *J. Cell Sci.* **59**:43–60.

35. Nakatsuji, N., and Johnson, K. E. 1983. Comparative study of extracellular fibrils on the ectodermal layer in gastrulae of five amphibian species. *J. Cell Sci.* **59**:61–70.

36. Nakatsuji, N., and Johnson, K. E. 1984. Experimental manipulation of a contact guidance system in amphibian gastrulation by mechanical tension. *Nature (Lond.)* **307**:453–455.

37. Nakatsuji, N., and Johnson, K. E. 1984. Substratum conditioning experiments using normal and hybrid frog embryos. *J. Cell Sci.* **68**:49–67.

38. Nakatsuji, N., Smolira, M. A., and Wylie, C. C. 1985. Fibronectin visualized by scanning electron microscopy immunocytochemistry on the substratum for cell migration in *Xenopus laevis* gastrula. *Dev. Biol.* **107**:264–268.

39. Shi, D. L., Delarue, M., Darribère, T., Riou, J.-F., and Boucaut, J.-C. 1987. Experimental analysis of the extension of the dorsal marginal zone in *Pleurodeles waltlii* gastrulae. *Development (Camb.)* **100**:147–161.

40. Smith, J. C., and Malacinski, G. M. 1983. The origin of the mesoderm in the anuran, *Xenopus laevis*, and a urodele, *Ambystoma mexicanum*. *Dev. Biol.* **98**:250–254.

41. Stopak, D., Wessels, N. K., and Harris, A. K. 1985. Morphogenetic rearrangement of injected collagen in developing chick limb buds. *Proc. Nat. Acad. Sci. U.S.A.* **82**:2804–2808.

42. Tamkun, J. W., DeSimone, D. W., Fonda, D., Patel, R. S., Buck, C., Horwitz, A. F., and Hynes, R. O. 1986. Structure of integrin, a glycoprotein involved in the transmembrane linkage between fibronectin and actin. *Cell* **46**:271–282.

43. Trinkaus, J. P. 1984. *Cells into Organs: Forces That Shape the Embryo*, 2d ed. Prentice-Hall, Englewood Cliffs.

44. Weiss, P. 1945. Experiments on cell and axon orientation *in vitro*: The role of colloidal exudates in tissue organization. *J. Exp. Zool.* **100**:353–386.

Questions for Discussion with the Editor

1. *Please speculate on how the guidance of gastrular cell movements in amphibia might differ from guidance in avian embryos and mammalian embryos.*

   Much less is known about the guidance of cells in avian and mammalian embryos at the gastrula stage. Fibronectin and a 140×10^3 dalton glycoprotein complex which serves as a receptor for fibronectin and laminin are abundant in the avian gastrula stage embryo and are especially enriched on the basal surface of the epiblast. We presume that similar mechanisms are operating in avian embryos. The lateral emigration of invaginated mesodermal cells away from the primitive streak probably also follows oriented networks of FN-rich extracellular matrix. This would be simple enough to test experimentally, first by careful direct observation in the SEM or with fluorescent microscopy. It would also be possible to use chick epiblasts to condition artificial substrata with deposited extracellular matrix and then seed migrating mesodermal cells from gastrulae onto such substrata. We also imagine that similar mechanisms are involved in gastrulation in mammalian embryos.

2. *What do you expect will provide the gateway to the study of the molecular biology of gastrulation? For example, how can we identify the genes which direct blastopore cells to initiate migration?*

Once we identify molecules involved in some crucial way in gastrulation, such as fibronectin or integrin, we will be in a position to do some very exciting experiments bearing on the genetic control of morphogenesis. For example, the gene for avian integrin has been cloned. Soon we will be able to use techniques of molecular biology to create radioactive probes capable of detecting integrin gene expression by in situ hybridization techniques. By combining histological sections with probes for gene function, we would be able to learn more about the regional expression of genes within the whole embryo. Amphibian embryos would be ideal for such experiments because they are large, thus, allowing for reasonable spatial resolution. They are also easily obtained in large numbers and all at the same developmental stage. As far as the genes involved in bottle-cell formation are concerned, first we will need to identify the proteins that are crucial for the process of bottle-cell formation. Perhaps microfilament proteins such as actin are important for initiating bottle-cell formation. Regulation of actin gene expression might cause certain cells to become bottle cells. Once bottle cells are formed, presumably changes in cell surface proteins are involved in initiating cell migration. In interspecific arrested hybrids, one cross shows moderate but incomplete invagination at the blastopore, another cross shows only the slightest indication of bottle-cell formation but no invagination, and a third cross shows evidence for neither bottle-cell formation nor invagination. Such variants in the extent of the initial stages of gastrulation may prove useful for identifying proteins involved in the initiation of bottle-cell formation and the subsequent events of invagination. Specifically, if new protein synthesis is necessary for initiation of bottle-cell formation and migration, one could isolate the blastopore region of normal and arrested hybrid embryos, incubate them in labeled protein precursors, and then search for differences in the pattern of protein synthesis by two-dimensional sodium dodecyl sulfate–polyacrylamide gel electrophoresis and autoradiography. Significant differences in the protein synthesis pattern between normal and hybrid embryos might point to an important role for particular proteins in bottle-cell formation. The next step would then be to raise antibodies to spots from the gels and then use these antibodies in an attempt to disrupt bottle-cell behavior in vivo and in vitro.

CHAPTER 17

Role of the Extracellular Environment in Neural Crest Migration

Marianne Bronner-Fraser

Introduction

THE EXTRACELLULAR MATRIX (ECM) is composed of large molecules produced by cells and secreted into the extracellular spaces. Identified components of the extracellular matrix include the glycoproteins fibronectin, laminin, and collagens; the proteoglycans heparan and chondroitin sulfate proteoglycan; and the glycosaminoglycan hyaluronan. Many ECM molecules have multiple binding regions which are recognized by various cell types as well as by other ECM molecules. Because of this property, these molecules have a tendency to aggregate into a stable, organized "matrix" which can serve both as a structural scaffolding for tissues and as a substrate for cell migration.

During development the extracellular matrix is dynamic in both space and time. Changes in the composition of the ECM correlate with important morphogenetic events, leading to the proposal that many of these molecules may play an important role in embryonic development. For example, extracellular matrix molecules are thought to be involved in aspects of gastrulation (Refs. 2 and 3; see Chap. 17, this volume), gland morphogenesis (15), primordial germ-cell migration (20), and neural crest cell migration (3, 6, 7, 11).

This chapter summarizes some experiments that illustrate the functional importance of interactions between the cell surface and the extracellular environment in morphogenesis. This is not intended as a comprehensive review but rather as a synopsis of recent findings which address the role of cell-matrix interactions in neural crest development. Neural crest (NC) cells first appear during neurulation. In most species, they migrate away from the neural tube shortly after neural tube closure. However, some neural crest cells, such as those in the cranial region of the mouse, begin migrating prior to fusion of the neural folds. After leaving the neural tube, NC cells enter a primarily cell-free space that is filled with extracellular matrix (ECM) molecules including fibronectin, laminin, tenascin-cytotactin, collagens, heparan sulfate proteoglycans, and numerous other molecules.

The neural crest has several unique properties which make it an ideal system for studying the mechanisms involved in cell migration and differentiation. First, these cells migrate extensively along characteristic pathways during embryogenesis. Second, neural crest cells give rise to diverse and numerous derivatives, including pigment cells, adrenal chromaffin cells, and the ganglia of the peripheral nervous system. Third, neural crest cells are accessible to surgical, immunological, and biochemical manipulations during both initial and certain later stages in their development.

After their initial emigration from the neural tube or neural folds, neural crest cells intermingle with other cell types, permitting interactions with ECM or cell surface molecules on these cells. In the trunk region of avian embryos, neural crest cells follow two primary migratory pathways: a dorsolateral route underneath the skin and a ventral route through the somite (Fig. 1). In addition, a few cells migrate in the space between adjacent somites. Within the somites, neural crest cells preferentially migrate through the rostral half of each sclerotome but are absent from the caudal sclerotome (8, 38). Motor axons exhibit a similar pattern of distribution (23). Furthermore, the neural crest–derived dorsal root and sympathetic ganglia form in alignment with the rostral portion of each somite (26). Rotation of the somites in the rostrocaudal plane results in an inversion in the pattern of neural crest (43, 44) and motor axon outgrowth (23), suggesting that the somites play a critical role in the patterning of neural crest cells and in the development of the peripheral nervous system. These findings highlight the importance of the tissue environment in the patterning of NC cells during their migration and subsequent differentiation.

The pathways followed by neural crest cells appear to be characteristic of the axial level through which they migrate (Fig. 2; reviewed in Ref. 27). Cranial neural crest cells, unlike their counterparts in the trunk, do not encounter somitic tissue, but rather migrate underneath the ectoderm and through the cranial mesenchyme. In the vagal region, neural crest cells migrate laterally between the ectoderm and somite and eventually invade the gut to form enteric ganglia. Thus, neural crest cells from different axial levels migrate along distinct pathways, localize in different regions of the embryo, and differentiate into divergent cell types.

Approaches for Studying the Role of the ECM in Developing Systems

In establishing a possible role for an extracellular matrix molecule in the developing embryo, a useful strategy is to determine:

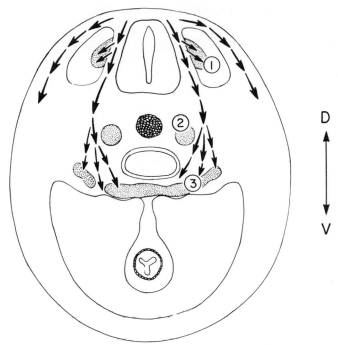

FIGURE 1. Schematic diagram illustrating the migratory pathways and derivatives of trunk neural crest cells. Cells following the ventral pathway localize in three main areas: (1) the dorsal root ganglia, (2) the sympathetic ganglia, and (3) the adrenomedullary cells. Neural crest cells following the dorsal pathway migrate under the ectoderm and become skin melanocytes (From Ref. 4).

1. The ability of cells to interact with the isolated ECM molecule in vitro
2. The spatiotemporal distribution of the molecule
3. The effects of functional inactivation of the molecule in vivo. Some of the approaches currently used to examine cell-matrix interactions during development are described below.

Analysis of Cell Behavior in Vitro

Neural crest cells can be grown in vitro by explanting neural tubes prior to neural crest migration. Neural crest cells emigrate from the dorsal side of the explanted neural tube and form a monolayer on the substrate (12). In a typical tissue culture experiment, the neural crest cells are grown on substrates composed of purified ECM molecules and their migratory and/or differentiated responses are monitored. Purified ECM molecules can be applied as a thin layer onto the tissue culture plate (two-dimensional substrate) or as a thicker layer often incorporated into a collagen type I gel (three-dimensional substrate). In addition to using defined components, neural crest cells can be grown on

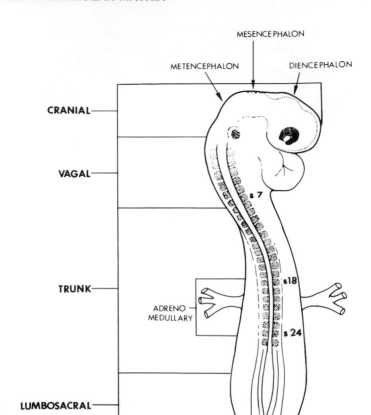

FIGURE 2. Schematic diagram illustrating regions along the neural axis which differ in their range of neural crest derivatives and neural crest migratory pathways. The cranial neural crest emerges from levels of the neural tube above the otic vesicles. The vagal neural crest arises from the neural tube between somitic levels 1 and 7. The trunk neural crest emerges from axial levels between somites 8 and 28 along with those neural crest cells that contribute to the adrenal gland arising from somitic levels 18–24. The lumbosacral neural crest emerges from axial levels beyond somite 28 (From Ref. 5).

"native" or less well-defined matrices synthesized by other cell types in culture. Complex matrices can also be extracted from tumors which overproduce ECM molecules, such as the EHS sarcoma, which has abundant laminin, nidogen, and collagen type IV. Antibodies or other reagents that competitively inhibit specific matrix components can be added to these cultures and the resulting perturbations in cell migration and differentiation examined.

Tissue culture has the distinct advantage that experimental conditions can be controlled in large part by the experimenter, making it possible to utilize defined conditions in comparison to those encountered in the embryo. This permits detailed examination of interactions between cells and specific extracellular matrix molecules. Because cells can be grown on transparent substrates, it is possible to examine their migratory response directly or by time-lapse cinematography. These experiments utilize a comparatively simple system in which the substrates, cell types, and composition of the medium can be relatively defined.

Possible shortcomings of these studies are:

1. The cells may behave differently in culture than in the embryo since they are isolated from neighboring cells and tissues.
2. The ECM molecules may adsorb onto the two- or three-dimensional substrates in a manner which does not correspond to their conformation in situ.
3. The finding that a molecule may serve as a good substrate in vitro does not necessarily reflect a similar function in vivo.

Although the results of such experiments do not definitively address the cells' behavior in situ, they can be informative about the *potential* interactions available to that cell type.

Analysis of Molecular Distribution in Vivo

An important first step in establishing the potential role of a molecule in a complicated developmental event such as neural crest migration is to examine the distribution of that molecule. The advantage of this approach is that it is technically easy and many of the necessary probes are readily available. Furthermore, when the molecular distribution correlates spatially and temporally with the event of interest, then that molecule is a reasonable candidate for being involved in the interaction. Immunocytochemical techniques using specific antibodies which are available for many ECM components are most commonly used to define molecular distribution. For glycosaminoglycans, their localization can be determined by histochemical stains such as Alcian blue or ruthenium red. Although these stains recognize a class of molecules rather than a single molecular component, it is often possible to distinguish the predominantly stained molecules by using specific enzymes to remove a single component and comparing the resultant distribution pattern to a nonenzymatically treated section. Recently, specific binding proteins for hyaluronan have become available, making it possible to use a selective cytochemical stain.

One must be careful not to overinterpret the results of such correlational studies. The presence of an ECM molecule along a migratory pathway cannot be taken as proof that the molecule is involved in cell migration. For example, the molecule may exist along the pathway but may not directly interact with the cells of interest. In addition, some ECM molecules may have common regions recognized by the same antibody, suggesting that the observed immunoreactivity may actually reflect the distribution of a family of molecules rather than a specific ECM molecule. Conversely, the absence of immunoreactivity for a certain ECM component does not prove that the molecule is absent from that embryonic region. Several possible explanations can account for the lack of immunoreactivity. First, the molecule may not be present. Alternatively, some antibodies may recognize only one form of a molecule and not others which are present within the embryo. This is particularly problematic for proteoglycans which may have a common core protein but different side chains or vice versa. Third, the molecule may be present but masked or destroyed by the fixation used for tissue preparation. Thus, the results of correlational studies should be interpreted with caution. They represent only a first step in establishing a role for an ECM molecule and should be used in conjunction with more direct experimental evidence.

In Vivo Perturbation Experiments

One approach for testing the involvement of cell-matrix interactions in situ is to introduce into the embryo function-perturbing antibodies or peptides that block interactions of cell surface receptors with extracellular matrix molecules. These substances can be injected onto pathways followed by migrating cells. For example, in the study of neural crest cell migration, antibodies are microinjected lateral to the neural tube in chick embryos during the initial stages of neural crest cell migration in the cranial, trunk, or other regions of the embryo. Similarly, perturbation experiments can be performed using a peptide containing the sequence Arg-Gly-Asp (RGD) which competitively inhibits cell adhesion to fibronectin and other adhesive molecules (36, 40). If the embryos develop abnormally after introduction of these reagents which disrupt cell-matrix interactions, the results suggest that the blocked interactions are important for normal development.

The advantage of a perturbation approach is that it is possible to test the function of the molecule in its native state, in contrast to tissue culture studies and cytochemical approaches. In vivo perturbation experiments have the disadvantage of being prone to many potential artifacts. Therefore, great care must be taken to perform proper control experiments. Antibody molecules which are multivalent, for example, can result in cell agglutination without blocking a cell-matrix interaction. An important control experiment that eliminates this possibility is to use antibodies that bind to the same cells, and ideally the same molecules, but do not block function. It is best to use monovalent fragments of the antibody molecule since they have only one binding site and cannot, therefore, cause cell aggregation. Synthetic peptides, which are small molecules, circumvent the possible agglutination problem incurred with antibodies. However, these reagents also suffer from a variety of pitfalls. First, since the peptides tend to have only a few amino acids, the sequence is often shared by more than one protein. For example, in the case of the RGD peptide, this sequence is found in fibronectin, vitronectin, collagens, fibrinogen, and numerous other molecules and may affect the binding of all of them to both common and distinct receptors (40). This potentially reduces the specificity of the reagent. Second, very high concentrations of peptide are typically required to achieve an effect. This introduces the problem of possible cytotoxicity to the cells. The appropriate control reagents for these experiments are peptides synthesized in exactly the same way as the experimental reagent but having a single amino acid substitution.

The final point that must be kept in mind with respect to in vivo perturbation experiments is that even if removal of one molecular interaction blocks a developmental event, that does not preclude the involvement of other molecules in that event. Morphogenesis is likely to consist of a cascade of interwoven events, blockage of any one of which will influence the process as a whole. Perturbation experiments cannot inform the investigator about the number of interactions or whether the molecules under examination play a primary or a tertiary role. Similarly, failure to find an effect does not necessary mean that the molecule is not involved since there may be redundant mechanisms acting on a single cell type (48). Despite these caveats, in vivo perturbations can be highly informative and offer insights into molecules involved in cell migration.

Distribution of Cell Surface and ECM Molecules along Neural Crest Pathways

It has been proposed that several extracellular matrix molecules are important in aspects of neural crest migration. These molecules include fibronectin, laminin, and cytotactin-tenascin. Since neural crest cells can attach to numerous ECM molecules, they must possess cell surface receptors which can interact with the ECM. A few of these receptors have been characterized. For example, neural crest cells at all levels of the neural axis possess integrin receptors (24). The integrins are a family of heterodimers which are ligands for several extracellular matrix molecules. In avians, the antibodies CSAT (21) and JG22 (18) recognize the $beta_1$ subunit of the alpha-beta integrin heterodimer. This subunit is thought to be common to a family of receptors for fibronectin, laminin, type IV collagen, and perhaps other molecules. In addition to the integrins, neural crest cells possess other cell surface molecules. For example, neural crest cells have a surface molecule(s) recognized by the HNK-1 antibody, which is directed against a carbohydrate epitope present on a family of adhesion molecules (25) as well as some unidentified molecules. Still other potential receptors remain undefined. Cell surface receptors for extracellular matrix molecules serve as important links between the ECM and the cell's motile response to the matrix. The integrin receptor, for instance, is a transmembrane protein with a cytoplasmic domain which binds to talin (22) and is, therefore, a direct link between the cell surface and the cytoskeleton. Thus, cell surface receptors may act as signal transducers between the extracellular matrix and the cell's motile apparatus.

In addition to cell-substrate adhesion, neural crest–derived neurons possess cell-cell adhesion molecules such as the neural cell adhesion molecule (N-CAM), which has been proposed to be important for proper formation of the nervous system. As a first step in testing their role in vivo, numerous investigators have examined the distribution of these molecules in the embryo. All of them show an interesting distribution which correlates with some aspects of neural crest cell migration and/or differentiation. The pattern of immunoreactivity and correlation with neural crest cell distribution in specific regions of the embryo is described below for some cell surface and ECM molecules. For each axial level, neural crest cells follow pathways which possess differentially distributed ECM molecules.

Cranial Neural Crest

In the mesencephalon, premigratory neural crest cells appear rounded in morphology and reside within an extracellular matrix–rich region of the dorsal neural tube. They are morphologically distinct from the columnar epithelial cells of the neural tube. Upon leaving the neural tube, mesencephalic neural crest cells migrate underneath the ectoderm as a crescent-shaped population of cells within the cranial mesenchyme. With time, the cells progress further ventrally, where they enter the branchial arches and eventually populate much of the face. These cells give rise to glia and some neurons of the cranial ganglia, cartilage, and muscle in the face.

Premigratory neural crest cells, neural tube cells, and all other cells in the early embryo have lows levels of immunocytochemically detectable integrin receptors on their surfaces (Fig. 3). The levels of integrin-immunoreactivity increase on the neural crest cell surface with time. In contrast, the levels of receptor in the mesenchyme do not appear to increase. This suggests that levels of integrin are developmentally regulated on the neural crest cell surface.

The cranial mesenchyme through which neural crest cells move is rich is extracellular matrix molecules. We have examined the distribution along cranial neural crest pathways of fibronectin, of a heparan sulfate proteoglycan (HSPG), and of laminin which is thought to be present in a complex with HSPG. The distributions of fibronectin and laminin are illustrated in Fig. 3. All three molecules exhibit similar patterns of immunoreactivity: prominent immunoreactivity was noted around the neural tube, under the ectoderm, and within the cranial mesenchyme in fibrillar form as part of a loose matrix. Intriguingly, all of these molecules are found in a dense matrix surrounding and interdigitating with premigratory neural crest cells. Thus, the distribution pattern of fibronectin, laminin, and HSPG is consistent with the hypothesis that they play a role in either the initiation of neural crest migration or in later migration through the cranial mesenchyme.

We have recently examined the distribution along cranial neural crest migratory pathways of tenascin, a glycoprotein which is related or identical to myotendinous antigen (13), cytotactin, (19), hexabrachion (16, 50), and the glioma mesenychymal extracellular matrix protein (1). This molecule appears to be present on the surface of cells, in a loose interstitial matrix, and in a basement membrane–type matrix. Like fibronectin and laminin, tenascin immunoreactivity is prominent around premigratory neural crest cells. However, it is not detectable in the cranial mesenchyme prior to neural crest mi-

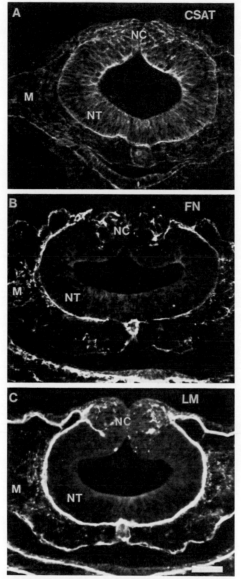

FIGURE 3. Transverse sections in the head just prior to cranial neural crest migration in stage 10 embryos (*A*) Integrin (CSAT) immunoreactivity was observed surrounding most cells in the embryo, including the neural tube (NT) and the premigratory neural crest (NC) cells (X312). (*B*) Fibronectin (FN) was observed on the basal surface of the neural tube, ectoderm, notochord, and endoderm, within the cranial mesenchyme (M), and in the area surrounding the premigratory neural crest cells (×312). (*C*) Laminin (LM) immunoreactivity was observed on the basal surface of the neural tube, ectoderm, endoderm, notochord, within the cranial mesenchyme and surrounding the premigratory neural crest cells (X312). From Ref. 24.

gration and, after the onset of migration, appears to colocalize with neural crest cells (Fig. 4). Along the neural tube, tenascin appears to be graded in a dorsoventral fashion such that the staining is higher on the dorsal neural tube and progressively decreases ventrally. Thus, there are some interesting similarities and differences between the distribution of tenascin, fibronectin, and laminin along cranial neural crest pathways. Their differential distributions may reflect functional differences between these molecules.

Trunk Neural Crest

The description of trunk neural crest pathways has been somewhat controversial in the past because of the difficulty in distinguishing these cells from the tissues through which they migrate. The availability of antibodies that recognize migrating neural crest cells (49, 51), however, has made it possible to describe accurately the distribution of neural crest cells in this region. NC cells migrate from the dorsal side of the trunk neural tube and embark upon two predominant pathways, either underneath the ectoderm or along the neural tube and into the sclerotome. Within the sclerotome, the distribution of neural crest cells is discontinuous such that the NC cells are only present in the *rostral* half of each somite and are absent from the *caudal* half (8, 38). This periodicity might result from the caudal somite being a nonpermissive substrate for neural crest cells (see Chap. 5, this volume). During their migration, neural crest cells possess distinct integrin immunoreactivity which is higher than that found on sclerotomal cells (Fig. 5). Once the neural crest cells have migrated ventrally past the level of the somite, their distribution becomes continuous in the rostrocaudal plane. However, there exists a NC cell–free space around the notochord; it has been suggested that this tissue inhibits neural crest migration (32).

We have examined the distribution of fibronectin, laminin, and HSPG in the trunk to see if these molecules are good candidates for playing a role in neural crest migration. All three molecules demonstrate a similar distribution pattern and are located around the basal laminae of the neural tube, notochord, and ectoderm (cf., Refs. 33, 46). They are also distributed in fibrillar form in the sclerotome. However, no apparent differences were noted between the distribution of fibronectin, laminin (Fig. 4), and HSPG in the rostral versus the caudal half of the sclerotome. Thus, none of these molecules has the selective distribution necessary for being a sole "guiding" molecule in trunk neural crest migration. Although one cannot rule out the possibility that these molecules serve as permissive substrates for cell migration, the data suggest that molecules other than fibronectin, laminin, and HSPG must be responsible for the selective distribution of neural crest cells within the rostral sclerotome. It is possible, for example, that another molecule which is distributed selectively in one-half of the somite alters the cell's binding to fibronectin and/or laminin. This type of modulatory role has been proposed for cytotactin-tenascin (29, 45), which is preferentially distributed in the rostral half of each somite. It has been suggested that laminin, reported to be more abundant on the rostral

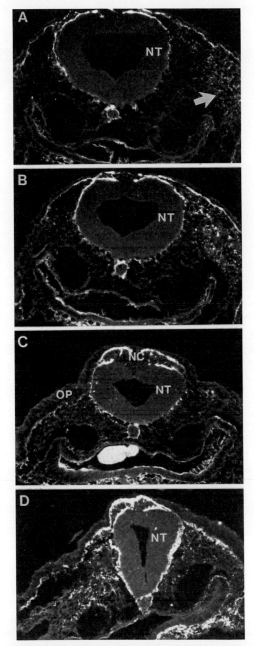

FIGURE 4. Fluorescence photomicrographs of transverse sections illustrating the distribution of tenascin in the head of an eleven somite embryo prepared by the freeze-substitution method. (*A*) and (*B*) Sections through the mesencephalon. Tenascin immunoreactibity was present in a dorsoventral gradient along the neural tube (NT), under the ectoderm, around the notochord, and in the cranial mesenchyme (indicated by arrows) in the vicinity of neural crest cells. (*C*) A section through the rhombencephalon with abundant tenascin immunoreactivity surrounding the premigratory neural crest (NC) cells. Some immunoreactivity was also noted underneath the otic placode (OP). (*D*) A section through the posterior rhombencephalon with abundant staining around the neural tube, notochord, underneath the ectoderm, and within the mesenchyme. (×168) From Ref. 10.

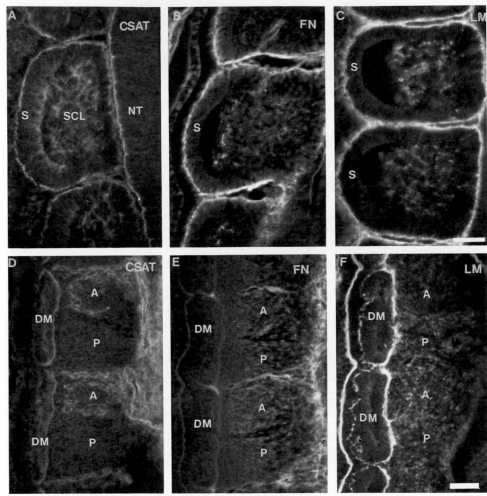

FIGURE 5. Longitudinal sections through the trunk at the onset of neural crest migration (*A–C*; stage 15 embryos) and during active neural crest migration (*D–F*; stage 17 embryos). (*A*) Integrin immunoreactivity (CSAT) outlined the somite (S) and neural tube (NT), and was uniform within the sclerotome (SCL) (×335). (*B*) Fibronectin (FN) and(*C*) laminin (LM) immunoreactivities were present around the somite and within the sclerotome. (×360; ×375). (*D*) During active neural crest migration, integrin was observed around the dermamyotome (DM) and surrounding neural crest cells within the anterior (A) half of the sclerotome. In contrast, the posterior (P) half of the sclerotome had little integrin immunoreactivity (×250). (*E*) Fibronectin and (*F*) laminin reactivities were observed around the dermamyotome and were within both anterior and posterior halves of the sclerotome (×250). From Ref. 24.

basal lamina than on the caudal basal lamina of the dermamyotome (28), is responsible for the pattern of neural crest migration through the rostral half-sclerotome. However, we find that neural crest cells migrate through the rostral half of the sclerotome even after rotation of the neural tube 180° about its dorsoventral axis. In these cases, neural crest cells emerge adjacent to the ventral sclerotome and do not contact the basal lamina of the dermamyotome during their initial migration (Ref. 44; see section "Role of the Tissue Environment" later in this chapter). This suggests that the sclerotome, not the dermamyotome, is responsible for the rostrocaudal patterning.

It has been proposed that fibronectin concentrations are high during the active phase of neural crest migration and become progressively diminished with time (44). This dynamic change in fibronectin distribution was taken as evidence for a guiding role for this molecule in NC migration. However, in a more detailed analysis correlating the distribution of fibronectin with the distribution of neural crest cell identified by the HNK-1 antibody, it appears that the fibronectin immunoreactivity in the sclerotome where NCs migrate is highest *prior* to NC migration and is comparatively low during migration. After gangliogenesis, the fibronectin distribution again appears to increase in the sclerotome (24). The distributions of laminin and HSPG also appear to change with time in a similar manner to that of fibronectin.

One molecule that appears to be distributed selectively in the rostral half of each somite is cytotactin-tenascin (29, 45). We have examined the distribution of tenascin and have confirmed that tenascin immunoreactivity is observed in the rostral sclerotome at the time of advanced neural crest migration. However, ablation of premigratory neural crest cells results in absence of tenascin immunoreactivity in the rostral sclerotome (C. D. Stern et al., in preparation), suggesting that tenascin immunoreactivity may be on neural crest cells or induced by the presence of neural crest cells. Furthermore, during early migratory stages, as the first NC cells are entering the sclerotome, tenascin appears to be *uniformly* distributed in the interstitial ECM of the sclerotome. At later stages, the tenascin immunoreactivity in the sclerotome seems to disappear except in the immediate vicinity of neural crest cells (Fig. 6). Thus, tenascin appears to be an unlikely molecule for initially guiding neural crest cells through the rostral sclerotome.

The neural cell adhesion molecule has been proposed to play an important role in the dispersion and aggregation of neural crest cells (47). N-CAM is present on the surface of premigratory neural crest cells within the neural tube but is lost from the NC cell surface during early migratory stages. N-CAM then reappears on cells in the neural crest–derived ganglia such as the dorsal root and sympathetic ganglia. Because of this developmentally regulated change in N-CAM immunoreactivity, it has been proposed that this molecule plays an important role in aggregation of neural crest–derived ganglia. We have examined in detail the correlation between the time of formation of the dorsal root and sympathetic ganglia and the expression of neuronal traits such as neurofilament and N-CAM immunoreactivity. Since all neurons are known to express neurofilament proteins, this serves as a convenient assay for the state of neuronal differentiation. We found in the case of the sympathetic

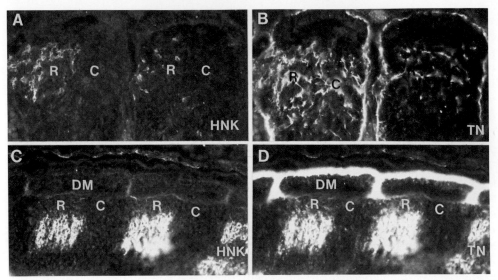

FIGURE 6. Fluorescent micrographs illustrating longitudinal sections through chick embryos that have been double-labeled with the HNK-1 antibody and an antibody against tenascin (TN). (*A* and *B*) During the early stages of neural crest migration, neural crest cells are in the rostral (R) half of each sclerotome, but tenascin is uniformly distributed in an interstitial matrix in both rostral and caudal (C) halves of the sclerotomes (×230). (*C* and *D*) At advanced stages of migration, HNK-1 and tenascin immunoreactivity colocalize in the rostral half of each sclerotome. There is also prominent tenascin immunoreactivity around the dermamyotome (DM) (×115).

ganglia that ganglionic condensation, neurofilament expression, and N-CAM expression all occurred simultaneously. However, for the dorsal root ganglia (DRG), the ganglia condensed and expressed neurofilament immunoreactivity several stages before the onset of N-CAM expression (Fig. 7; Ref. 26). These results demonstrate that the appearance of N-CAM immunoreactivity does not always correlate with the time of gangliogenesis. Therefore, N-CAM may not be the causative agent for the formation of all ganglia and may in some cases be a consequence of gangliogenesis. This finding highlights the importance of a detailed examination of the distribution of molecules proposed to be involved in developmental events as a first test of their potential role in cell adhesive interactions.

Functional Tests of Cell-ECM Interactions by Antibody Perturbation

Perturbations experiments represent one approach for examining the importance of cell surface–ECM interactions in developing systems. Antibodies or other reagents can be used to disrupt one or more of the adhesive interactions and the subsequent developmental effects are examined. This approach has

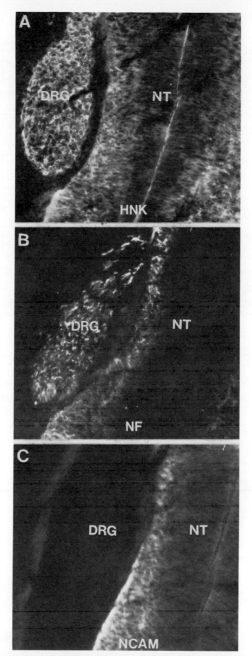

FIGURE 7. Fluorescent photomicrographs of adjacent cryostat sections cut in the transverse plane through the trunk of a stage 21 chick embryo at the level of the 15th somite. Sections were stained with (A) HNK-1 antibody, (B) anti-neurofilament, and (C) anti-N-CAM. (A) The cells of the forming dorsal root ganglia (DRG) appear to have condensed, as assayed by HNK-1 immunoreactivity (×210). (B) A majority of the cells in the DRG express neurofilament, indicating that they are differentiated neurons (×210); (C) No apparent N-CAM immunoreactivity was detectable within the DRG, though N-CAM was seen within the neural tube (NT) and myotome (×210). (From Ref. 26).

been used to demonstrate the functional importance of N-CAM in retinal axon guidance and map formation (17, 41) and of fibronectin or its receptor during gastrulation (2, 3) and cranial neural crest migration (3, 6, 7). A number of monoclonal antibodies directed against cell surface and extracellular matrix molecules have been used in perturbation experiments in order to study the possible involvement of these antigens in neural crest migration. These reagents were used to perturb cell-matrix interactions both in vitro and in situ. Using an in vitro assay, one can determine whether the antibody under examination functionally disrupts interactions between cells and extracellular matrix molecules.

In Vitro Analysis of Neural Crest Cell Migration

Neural crest cells migrate avidly on substrates containing fibronectin (39), laminin (31), collagen (35), as well as other molecules. On fibronectin substrates, neural crest cell migration and attachment is progressively enhanced with increasing fibronectin concentration. In the case of laminin, however, higher concentrations of substrate can actually inhibit cell migration (34).

The in vitro assay represents a means by which the effects of the antibody can be tested under relatively defined conditions. For the case of neural crest cells grown on fibronectin substrates, antibodies against fibronectin can block their migration (39). Antibodies against the integrin receptor disrupt neural crest cell adhesion to fibronectin and laminin substrates (6, 7). The effects are rapid, causing many cells to detach from the dish or aggregate with other cells within 15 min, and are readily reversible upon removal of the antibody. These results suggest that binding and dissociation of integrin receptors and matrix molecules occur rapidly. Interestingly, not all neural crest cells are affected by the antibodies. Thus, the antibody may recognize receptors present on only a subset of neural crest cells or only expressed at certain times during the cell cycle.

HNK-1 is another cell surface epitope which is present on migrating neural crest cells, neural crest–derived neurons, and some other cells. Addition of HNK-1 antibody to neural crest cells in tissue culture caused detachment and aggregation of the cells grown on laminin substrates (9). The effects of HNK-1 antibody were slow, with changes in cellular morphology becoming apparent only after several hours. The effect was reversible, with neural crest cells returning to normal morphology after removal of the antibody. In contrast, HNK-1 antibody had no effects on neural crest cells grown on fibronectin. This finding suggests that the HNK-1 epitope may be present on a molecule involved in binding to laminin.

In Vivo Perturbation Experiments

Antibodies were introduced along cranial neural crest pathways by microinjection lateral to the mesencephalic neural tube. Embryos ranging from the neural-fold stage to the nine somite stage were used for injection. Embryos with greater than ten somites at the time of injection had no detectable abnormalities, suggesting that they were sensitive to the injected antibodies for only a limited time during their development. After injection, antibody molecules diffused freely on the injected side of the embryo but were barely detectable on the uninjected side, as if they did not readily cross the midline. This made it possible to use the uninjected side as an internal control. The injected antibody remained detectable for approximately one day, during which time the concentration gradually decreased.

An Antibody Against the Integrin Receptor

By microinjecting antibodies that bind and functionally inactivate the β_1 subunit of the integrin receptor into the mesencephalic mesenchyme, we sought to test the role of the integrin receptor in cranial neural crest cell migration in situ. Antibody-injected embryos were fixed and examined during the period of neural crest migration. Integrin antibodies caused major defects including reduced numbers of neural crest cells on the injected side, neural crest cells within the lumen of the neural tube (Fig. 8), ectopic neural crest cells external to the neural tube, and neural tube anomalies. The observed results are similar to those obtained with synthetic peptides containing the fibronectin cell-binding sequence (3) or with antibodies against fibronectin (37). In contrast to the antibodies which block cell-ECM interactions, several control monoclonal antibodies which bind to integrins but do not block cell-matrix interactions had no detectable affect on cranial neural crest or neural tube development. These findings support the notion that antibody-induced

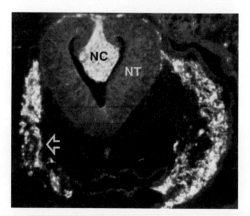

FIGURE 8. Fluorescence photomicrographs of transverse sections showing the effects of the CSAT antibody on cranial neural crest migration in an embryo fixed 18 h after injection. This embryo had an aggregate of neural crest (NC) cells protruding into the lumen of the neural tube (NT) and a 58 percent reduction in the neural crest cell volume on the injected side (indicated by arrow) relative to the control side (× 155). From Ref. 8.

perturbations in cranial morphogenesis result from functional block of the integrin receptor and suggest that the receptor complex is important in the normal development of the cranial neural crest and neural tube.

PERTURBATIONS AFTER INJECTION OF THE HNK-1 ANTIBODY

In order to examine if the surface molecules bearing this epitope are involved in some aspects of neural crest migration, the antibody was injected onto cranial neural crest pathways and the subsequent effects on neural crest cells were assessed. Injection of HNK-1 resulted in neural crest and neural tube anomalies including ectopic neural crest cells external to and within the neural tube. In contrast to experiments in which the integrin antibodies were microinjected into the mesencephalon, the most common defect was ectopic aggregates of neural crest cells outside of the neural tube; no reduction in neural crest cell number was detected. However, both integrin and HNK-1 antibodies caused accumulation of neural crest cells within the neural tube of some embryos and neural tube deformities. Simultaneous injection of integrin and HNK-1 antibodies resulted in additive effects, suggesting that the two antibodies work on different sites. We are currently in the process of identifying the HNK-1 antigen(s) on neural crest cells to gain insight into the function of this molecule(s).

One molecule which possesses an HNK-1 epitope and is present on the surface of cranial neural crest cells in tenascin. Antibodies against tenascin-cytotactin also perturb cranial neural crest migration, yielding abnormalities which are very similar to those obtained with HNK-1 antibodies (10). Tenascin may, therefore, represent the molecule which accounts for the observed HNK-1 mediated affects. Alternatively, tenascin may represent a distinct molecule that is important for normal neural crest development.

AN ANTIBODY AGAINST A LAMININ–HEPARAN SULFATE PROTEOGLYCAN COMPLEX

When the results of experiments using integrin antibodies or a synthetic decapeptide that competitively inhibits cell binding to fibronectin are taken together, they suggest an important function for fibronectin in cranial neural crest migration. However, these reagents are not entirely specific. For example, the cell-binding sequence of fibronectin is also present in numerous other extracellular matrix molecules. Likewise, integrin antibodies block neural crest cell adhesion to both laminin and fibronectin. In addition to using antibodies that interfere with cell binding to fibronectin, we have used an antibody which functionally perturbs cell adhesion to laminin as a first attempt to distinguish between the respective roles of these matrix molecules.

Laminin is thought to occur in a complex with HSPG in its native state. The inhibitor of neurite outgrowth (INO) antibody recognizes and functionally blocks cell interactions with this laminin-HSPG complex (14, 30). We have injected INO antibody along neural crest pathways in the mesencephalon in order to examine the possible role of laminin in cranial neural crest migration

(11). One day after injection, the embryos had severe abnormalities in cranial neural crest migration including ectopic neural crest cells external to the neural tube, neural crest cells within the lumen of the neural tube, and neural tube deformities. In contrast, embryos injected with antibodies against laminin or HSPG were unaffected. These results indicate that functional blockage of a laminin-HSPG complex perturbs cranial neural crest migration, providing the first evidence that laminin and/or HSPG is involved in aspects of neural crest migration in vivo.

The defects observed after INO antibody injection were somewhat different than those observed after injection of integrin or fibronectin antibodies. One possible explanation for these differences is that laminin and fibronectin may be critical for different subpopulations of neural crest cells or may be important at different phases of cranial neural crest migration. The finding that both the laminin-HSPG complex and fibronectin may play roles in neural crest development highlights the fact that multiple interactions may be important during complicated morphogenetic events.

Regional Differences between Neural Crest Cell Populations

Cranial and trunk neural crest cells follow distinct migratory pathways and give rise to a different range of derivatives. This suggests that some inherent differences may exist between these two populations. Microinjection of integrin antibodies (6), FN synthetic peptides (M. Bronner-Fraser, unpublished observation), or INO antibodies (11) have caused no apparent defects in trunk neural crest migration in situ. This is in contrast to the profound defects observed after injection of the same antibodies at the same or lower concentrations into the cranial region. Several possible explanations could account for the differences between cranial and trunk neural crest cells. First, it is possible that both regions use similar migratory mechanisms but that microinjection of antibody is less efficient in the trunk region. If this is true, higher concentrations or alternative methods of antibody delivery will be required. Alternatively, inherently different strategies of migration may be used by cranial and trunk neural crest cells. The extracellular matrix may be quite different along the pathways followed by cranial and trunk neural crest cells. In support of this idea, integrin antibodies profoundly affect trunk neural crest cells in vitro on fibronectin or laminin substrates even though they are ineffective in vivo. Experiments to discriminate between these possible cranial and trunk differences are currently in progress.

Role of the Tissue Environment

The peripheral nervous system is segmented along the neural axis in a manner which correlates with the segmental pattern of the somites. For each somite, there is one dorsal ganglion, one sympathetic ganglion, and one ventral root. What controls this precise segmentation? Two possible explanations are that the information resides within the nervous system itself (i.e., the developing neural tube) or that the information is present within the environment through which cells and neurites move. These issues can be resolved by studying the role of tissue interactions during establishment of the peripheral nervous system, i.e., during neural crest migration and motor axon outgrowth.

It is now clear that both neural crest cells and axons emanating from the ventral neural tube preferentially move through the rostral half of each somite (8, 23, 38). The similarities between the distribution of neural crest cells and motor axons suggest that both cell types may be subject to similar guiding cues. To examine whether the segmental information was present in the somites themselves or within the neural tube, Keynes and Stern (23) rotated either the neural tube or the segmental plate along the rostrocaudal axis prior to somitogenesis and examined the pattern of motor axon outgrowth. They found that neural tube rotation did not alter the axonal pattern, whereas rotation of the somites resulted in an inverted pattern of axon outgrowth. We have performed similar experiments and examined the effects of somite rotation on neural crest cell patterning (44). The orientation of the somites was inverted 180° about the rostrocaudal axis by surgically rotating the segmental plate prior to somitogenesis. Morphologically normal epithelial somites differentiated within a few hours after the operation, followed by their transformation into distinct dermamyotomes and sclerotomes. In contrast to the unoperated regions of the embryo where neural crest cells were observed in the rostral half of the sclerotome, neural crest cells in the grafted region of the embryos were always observed in the *caudal* half of the sclerotome, i.e., the half that would have been *rostral* in its original orientation (Fig. 9A). The results of the segmental plate rotations suggest that neural crest cells are guided in their migration by rostrocaudal differences that are inherent to the somites.

The nature of the cues which govern the movement of neural crest cells and axons through the rostral sclerotome is the subject of much interest. The molecular composition of the rostral and caudal somite may be responsible for this segmental arrangement (see Chap. 5, this volume). Several molecular differences between the rostral and caudal somites have been discerned. For example, Stern et al. (42) have reported that peanut agglutinin preferentially binds to the caudal half of each sclerotome. Similarly, cytotactin-tenascin have been reported to be selectively distributed in the rostral half of each sclerotome (29, 45). A possible explanation for the selective distribution of neural crest migration in the rostral sclerotome is that inhibitory molecules may be present in the caudal sclerotome and/or attractive-permissive molecules may be present in the rostral sclerotome.

Neural crest cells also demonstrate a characteristic directionality in their migration in the dorsoventral plane. NC cells originate from the dorsal neural tube. The majority of the cells then migrate ventrally, with many cells localizing between the neural tube and the sclerotome to form the dorsal root ganglia or around the dorsal aorta to form sympathetic ganglia and

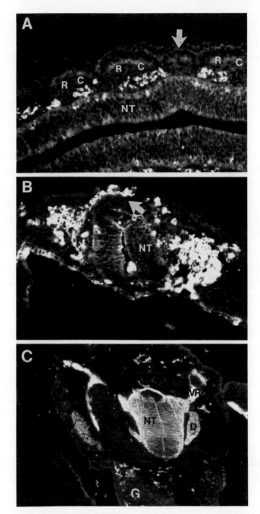

FIGURE 9. Sections stained with the HNK-1 antibody to recognize chick neural crest cells. (A) Longitudinal section through an embryo in which the segmental plate was rotated 180° about the rostrocaudal axis at stage 12. Caudal to the graft border (indicated by arrow), neural crest cells were observed in the rostral (R) half of each sclerotome. At the level of the graft, neural crest cells were seen within the caudal (C) half of each sclerotome (×220). (B) An embryo in which the neural tube was rotated 180° in the dorsoventral plane and the notochord was removed. Neural crest cells migrated up into the sclerotome and were also observed above the dorsal neural tube subjacent to the ectoderm (indicated by arrow) (×238). (C) A stage 25 embryo in which the neural tube was rotated 180° at stage 12 in the dorsoventral plane in the absence of a notochord. The neural tube (NT) forms normally but with inverted dorsoventral polarity. The dorsal root ganglia (D) form normally relative to the inverted neural tube. The ventral roots (VR) project from the dorsal portion of the neural tube and project to the limb; G = gut (×100). From Ref. 44.

adrenomedullary cells. In order to examine whether neural crest cells have any inherent directionality in their migration, the neural tube and notochord were removed from the host embryo and turned upside down in the dorsoventral plane. After dorsoventral inversion, neural crest cells emerged from the portion of the neural tube which was originally dorsal but had now been grafted to a ventral orientation. The neural crest cells were able to migrate up into the rostral sclerotome (but not the caudal sclerotome). In the presence of a dorsal notochord, neural crest cells were absent from an approximately 50-μm radius surrounding the notochord. This is consistent with the idea that the notochord may produce a diffusible substance that inhibits neural crest migration (32). However, when the notochord was removed, neural crest cells filled in all the permissive spaces, including the dorsal midline subjacent to the ectoderm (Fig. 9B). This indicates that neural crest cells have no inherent directionality in their migration but are equally capable of migrating dorsally as migrating ventrally.

In order to examine the role that neural tube orientation plays in gangliogenesis, embryos with dorsoventrally inverted neural tubes were allowed to survive until after the time of dorsal root and sympathetic ganglion condensation. In embryos with inverted neural tubes, aggregates of HNK-1 immunoreactive cells with the morphological appearance of dorsal root and sympathetic ganglia formed normally relative to the orientation of the grafted neural tube but with inverted dorsoventral polarity relative to the host embryo. The dorsal root ganglia were teardrop-shaped, with opposite polarity to that found in unoperated embryos (Fig. 9C). The sympathetic-like ganglia branched off from the ventral roots, but were located in the dorsal-most region of the embryo. The motor axons also appeared normal relative to the neural tube and seemed to project appropriately into the wing. These results suggest that the dorsoventral information that governs the patterning of dorsal root and sympathetic ganglia and the initial outgrowth of the ventral root predominantly resides within the neural tube. Analogously, the somite rotation experiments suggest that much of the rostrocaudal polarity resides within the somites. Thus, the tissue environment provides specific polarity cues in both the rostrocaudal and dorsoventral plane which influence the pattern of neural crest migration and gangliogenesis.

Conclusion

The experiments discussed in this chapter illustrate the important role of interactions between cells and the extracellular environment in neural crest migration. A variety of approaches can be used to study cell-matrix interactions in the neural crest system. First, cells can be grown in tissue culture on defined substrates. This technique allows for examination of cell behavior under relatively defined conditions but suffers from the drawback that the in vitro conditions may not accurately mimic the embryonic environment. Second, the molecular distribution of potentially interesting molecules can be examined in situ. This approach helps define candidate molecules that have the correct spatiotemporal distribution to be involved in cell-matrix interactions but is not informative about their function. Third, in vivo perturbation experiments, when properly controlled, can test the function of particular cell-matrix interactions. This technique cannot, however, distinguish whether the molecule under examination plays a primary or tertiary role in a complicated cascade of events. In any of these experimental paradigms, a negative finding does not necessarily suggest that the molecules under examination play no role in neural crest migration. Since all of these approaches have their respective strengths and weaknesses, a combinatorial scheme utilizing all of these methods is most valuable for analyzing interactions between neural crest cells and their environment. Furthermore, experimental manipulations in which various regions of the embryo are rotated into new configurations can be highly informative about the tissues which may influence the pattern of cell migration. This type of experiment, therefore, can help define the tissues and axial components which may provide important guidance information.

Using combinations of these approaches, several, but by no means all, of the molecules involved in cell migration have been identified. Because multiple ECM molecules have been shown to influence neural crest cells, it is clear that a scheme in which one molecule is responsible for one event cannot explain complicated morphogenetic events like neural crest migration. Rather, numerous molecules are likely to be involved either simultaneously or sequentially. In addition, redundant mechanisms (see Chap. 7, this volume) may exist in order to assure that development occurs properly, and this may complicate analysis.

Because the tissue and extracellular matrix environments appear very different in the head and trunk, it is possible that different strategies may be important for the migration of cranial versus trunk neural crest cells. In the cranial region, we have established that integrin receptors, laminin, and, most recently, tenascin are involved in some aspects of neural crest migration. In the trunk region, it is clear that regional differences between the rostral and caudal somites are important for neural crest and motor axon guidance. However, no role has currently been established for the molecules which have differential rostrocaudal distributions. Other potentially important molecules have yet to be identified. Defining the differences in molecular composition between cranial and trunk neural crest pathways may help explain the regionally specific choice of migratory routes followed by neural crest cells at various axial levels. Such analyses may distinguish inherent differences between neural crest cell populations in the head and trunk from differences in the environment through which these cells migrate.

Acknowledgments

I thank Dr. Scott Fraser for helpful comments on the manuscript and Georgia Guillory for excellent technical assistance. Parts of this work were supported by USPHS Grant HD-15527 and by Basic Research Grant 1-896 from the March of Dimes. The author is a Sloan Foundation Research Fellow.

General References

NEWGREEN, D. F., and ERICKSON, C. A.: 1986. The migration of neural crest cells. *Int. Rev. Cytol.* **103**:89–145.

LEDOUARIN, N. M.: 1982. *The Neural Crest.* Cambridge University Press, New York.

References

1. Bourdon, M. A., Wikstrand, C. J., Furthmayr, H., Matthews, T. J., and Bigner, D. D. 1983. Human glioma mesenchymal extracellular matrix antigen defined by a monoclonal antibody. *Cancer Res.* **43**:2796–2805.

2. Boucaut, J. C., Darribere, T., Boulekbache, H., and Thiery, J. P. 1984. Prevention of gastrulation but not neurulation by antibodies to fibronectin in amphibian embryos. *Nature (Lond.)* **307**:364–367.

3. Boucaut, J. C., Darribere, T., Poole, T. J., Aoyama, H., Yamada, K. M., and Thiery, J. P. 1984. Biologically active synthetic peptides as probes of embryonic development: a competitive peptide inhibitor of fibronectin function inhibits gastrulation in amphibian embryos and neural crest migration in avian embryos. *J. Cell Biol.* **99**:1822–1830.

4. Bronner, M. E., and Cohen, A. M. 1979. Migratory patterns of cloned neural crest melanocytes injected into host chicken embryos. *Proc. Natl. Acad. Sci. U.S.A.* **76**: 1843–1847.

5. Bronner-Fraser, M. 1980. The neural crest: what can it tell us about cell migration and determination? *Curr. Top. Dev. Biol.* **15**:1–25.

6. Bronner-Fraser, M. 1985. Alterations in neural crest migration by a monoclonal antibody that affects cell adhesion. *J. Cell Biol.* **101**:610–617.

7. Bronner-Fraser, M. 1986. An antibody to a receptor for fibronectin and laminin perturbs cranial neural crest development in vivo. *Dev. Biol.* **117**:528–536.

8. Bronner-Fraser, M. 1986. Analysis of the early stages of trunk neural crest migration in avian embryos using monoclonal antibody HNK-1. *Dev. Biol.* **115**:44–55.

9. Bronner-Fraser, M. 1987. Perturbations of cranial neural crest migration by the HNK-1 antibody. *Dev. Biol.* **123**:321–331.

10. Bronner-Fraser, M. 1988. Distribution and function of tenascin during cranial neural crest development in the chick. *J. Neurosci. Res.* **21**:135–147.

11. Bronner-Fraser, M., and Lallier, T. 1988. A monoclonal antibody against a laminin-heparan sulfate proteoglycan complex perturbs cranial neural crest migration *in vivo*. *J. Cell Biol.* **106**:1321–1330.

12. Cohen, A. M., and Konigsberg, I. 1975. A clonal approach to the problem of neural crest determination. *Dev. Biol.* **46**:262–280.

13. Chiquet, M., and Fambrough, D. M. 1984. Chick myotendinous antigen. I. A monoclonal antibody as a marker for tendon and muscle morphogenesis. *J. Cell Biol.* **98**:1937–1946.

14. Chiu, A. Y., Matthew, W. D., and Patterson, P. H. 1986. A monoclonal antibody

that blocks the activity of a neurite regeneration-promoting factor: studies on the binding site and its localization in vivo. *J. Cell Biol.* **103**:1383–1398.

15. David, G., and Bernfield, M. 1981. Type I collagen reduces the degradation of basal lamina proteoglycan by mammary epithelial cells. *J. Cell Biol.* **91**:281–286.

16. Erickson, H. P., and Taylor, H. C. 1987. Hexabrachion proteins in embryonic chicken tissues and human tumors. *J. Cell Biol.* **105**:1387–1394.

17. Fraser, S., Murray, B. A., Chuong, C.-M., and Edelman, G. M. 1984. Alteration of the retinotectal map in *Xenopus* by antibodies to neural cell adhesion molecules. *Proc. Natl. Acad. Sci. U.S.A.* **81**:4222–4226.

18. Greve, J. M., and Gottlieb, D. I. 1982. Monoclonal antibodies which alter the morphology of cultured chick myogenic cells. *J. Cell Biochem.* **18**:221–229.

19. Grumet, M., Hoffman, S., Crossin, K. L., and Edelman, G. M. 1985. Cytotactin, an extracellular matrix protein of neural and non-neural tissues that mediates glia-neuron interactions. *Proc. Natl. Acad. Sci. U.S.A.* **82**:8075–8079.

20. Heaseman, J., Hynes, R. O., Swan, A. P., Thomas, V., and Wylie, C. C. 1981. Primordial germ cells of *Xenopus* embryos: the role of fibronectin during migration. *Cell* **27**:437–447.

21. Horwitz, A. F., Duggan, K., Greggs, R., Decker, C., and Buck, C. 1985. The CSAT antigen has properties of a receptor for laminin and fibronectin. *J. Cell Biol.* **101**:2134–2144.

22. Horwitz, A. F., Duggan, K., Buck, C., Beckerle, M. C., and Burridge, K. 1986. Interaction of plasma membrane fibronectin receptor with talin, a transmembrane linkage. *Nature (Lond.)* **320**:531–533.

23. Keynes, R. J., and Stern, C. D. 1984. Segmentation in the vertebrate nervous system. *Nature (Lond.)* **310**:786–789.

24. Krotoski, D., Domingo, C., and Bronner-Fraser, M. 1986. Distribution of a putative cell surface receptor for fibronectin and laminin in the avian embryo. *J. Cell Biol.* **103**:1061–1072.

25. Kruse, J., Mailhammer, R., Wenecke, H., Faissner, A., Sommer, I., Goridis, C., and Schachner, M. 1984. Neural cell adhesion molecules and myelin-associated glycoprotein share a common carbohydrate moiety recognized by monoclonal antibodies L2 and HNK-1. *Nature (Lond.)* **311**:153–155.

26. Lallier, T., and Bronner-Fraser, M. 1988. A spatial and temporal analysis of dorsal root and sympathetic ganglion formation in the avian embryo. *Dev. Biol.* **127**:99–112.

27. LeDouarin, N. M. 1982. *The Neural Crest.* Cambridge University Press, New York.

28. Loring, J. F., and Erickson, C. A. 1987. Neural crest cell migratory pathways in the trunk of the chick embryo. *Dev. Biol.* **121**:220–236.

29. Mackie, E. J., Tucker, R. P., Halfter, W., Chiquet-Ehrismann, R., and Epperlein, H. H. 1988. The distribution of tenascin coincides with pathways of neural crest cell migration. *Development (Camb.)* **102**:237–250.

30. Matthew, W. D., and Patterson, P. H. 1983. The production of a monoclonal antibody that blocks the action of a neurite outgrowth–promoting factor. *Cold Spring Harbor Symp. Quant. Biol.* **48**:625–631.

31. Newgreen, D. F. 1984. Spreading of explants of embryonic chick mesenchyme and epithelia on fibronectin and laminin. *Cell Tissue Res.* **236**:265–277.

32. Newgreen, D. F., Scheel, M., Kastner, V. 1986. Morphogenesis of sclerotome and

neural crest in avian embryos: in vivo and in vitro studies on the role of notochordal extracellular matrix. *Cell Tissue Res.* **244**:299–313.

33. Newgreen, D. F., and Thiery, J. P. 1980. Fibronectin in early avian embryos: synthesis and distribution along the migration pathways of neural crest cells. *Cell Tissue Res.* **211**:269–291.

34. Perris, R., and Johansson, S. 1987. Amphibian neural crest cell migration on purified extracellular matrix components: a chondroitin sulfate proteoglycan inhibits locomotion on fibronectin substrates. *J. Cell Biol.* **105**:2511–2522.

35. Perris, R., Paulsson, M., and Bronner-Fraser, M. 1989. Mechanisms of neural crest cell migration on isolated extracellular matrix molecules: receptor specificity and role of cell surface heparan sulfate proteoglycans. *Dev. Biol.* (in press).

36. Pierschbacher, M. D., and Ruoslathi, E. 1984. The cell attachment activity of fibronectin can be duplicated by small synthetic fragments of the molecule. *Nature (Lond.)* **309**:30–33.

37. Poole, T. J., and Thiery, J. P. 1986. Antibodies and synthetic peptides that block cell-fibronectin adhesion arrest neural crest migration in vivo. In: *Progress in Developmental Biology* (H. Slavkin, ed.), pp. 235–238, Liss, New York.

38. Rickmann, M., Fawcett, J. W., and Keynes, R. J. 1985. The migration of neural crest cells and the growth of motor axons through the rostral half of the chick somite. *J. Exp. Morphol. Embryol.* **90**:437.

39. Rovasio, R. A., Delouvee, A., Yamada, K. M., Timpl, R., and Thiery, J. P. 1983. Neural crest cell migration: Requirements for exogenous fibronectin and high cell density. *J. Cell Biol.* **96**:462–473.

40. Ruoslahti, E., and Pierschbacher, M. D. 1987. New perspectives in cell adhesion: RGD and integrins. *Science* **238**:491–497.

41. Silver, J., and Rutishauser, U. 1984. Guidance of optic axons in vivo by a preformed adhesive pathway on neuroepithelial endfeet. *Dev. Biol.* **106**:485–499.

42. Stern, C. D., Sisodiya, S. M., and Keynes, R. J. 1986. Interactions between neurites and somite cells: inhibition and stimulation of nerve growth in the chick embryo. *J. Embryol. Exp. Morphol.* **91**:209–226.

43. Stern, C. D., and Keynes, R. J. 1987. Interactions between somite cells: the formation and maintenance of segment boundaries in the chick embryo. *Development (Camb.)* **99**:261–272.

44. Stern, C. D., and Bronner-Fraser, M. 1990. Role of somite and neural tube orientation in chick neural crest migration and gangliogenesis. (in prep.).

45. Tan, S.-S., Crossin, K. L., Hoffman, S., and Edelman, G. M. 1987. Asymmetric expression in somites of cytotactin and its proteoglycan ligand is correlated with neural crest cell distribution. *Proc. Natl. Acad. Sci. U.S.A.* **84**:7977–7981.

46. Thiery, J. P., Duband, J. L., and Delouvee, A. 1982. Pathways and mechanism of avian trunk neural crest migration and localization. *Dev. Biol.* **93**:324–343.

47. Thiery, J. P., Duband, J. L., Rutishauser, U., and Edelman, G. M. 1982. Cell adhesion molecules in early chick embryogenesis. *Proc. Natl. Acad. Sci. U.S.A.* **79**:6737–6741.

48. Tomaselli, K. J., Reichardt, L. F., and Bixby, J. L. 1986. Distinct molecular interactions mediate neuronal process outgrowth on nonneuronal cell surfaces and extracellular matrices. *J. Cell Biol.* **103**:2659–2672.

49. Tucker, G. C., Aoyama, H., Lipinski, M., Tursz, T., and Thiery, J. P. 1984. Identical reactivity of monoclonal antibodies HNK-1 and NC-1: conservation in verte-

brates on cells derived from the neural primordium and on some leukocytes. *Cell Differ.* **14**:223–230.

50. Vaughan, L., Huber, S., Chiquet, M., and Winterhalter, K. H. 1987. A major, six-armed glycoprotein from embryonic cartilage. *EMBO J.* **6**:349–353.

51. Vincent, M., and Thiery, J. P. 1984. A cell surface marker for neural crest and placodal cells: Further evolution in the peripheral and central nervous system. *Dev. Biol.* **103**:468–481.

Questions for Discussion with the Editor

1. *Do neural crest cells change their phenotype (i.e., gene expression pattern) as they read the ECM they migrate through? That is, do you suppose the ECM, in addition to guiding migration, might play a subtle role in regulating neural crest cell differentiation?*

The possible role of the extracellular matrix in neural crest cell differentiation is a subject of great interest. Both fibronectin and complex matrices containing laminin plus other molecules have been shown to enhance catecholaminergic expression by avian neural crest cells in culture, suggesting that some matrix molecules can at least influence levels of gene expression. The most compelling evidence that the extracellular matrix and/or associated molecules can *direct* neural crest cell differentiation comes from a recent paper (Perris, 1988. *Science* **241**:86–89) in which axolotl neural crest cells were grown on microcarrier filters conditioned with matrix from embryonic regions corresponding to the site of pigment cell differentiation (subepidermal matrix) or the site where dorsal root ganglia form (preganglionic matrix). Those neural crest cells grown on subepidermal matrix differentiated into pigment cells. In contrast, identical cultures grown on preganglionic matrix gave rise to neurons. This suggests that some molecule(s) associated with the matrix was able to induce differentiation of or select for a specific neural crest cell phenotype.

2. *Is there any prospect on the horizon for experiments that would use developmental genetics to gain insight into the way in which different neural crest cell populations might be preprogrammed for specific migration patterns?*

In the avian embryo, which is the best studied neural crest cell model, the prospects for developmental genetics in the near future are poor. Few avian mutants are available, with the possible exception of pigment pattern mutants, and little is known about avian genetics. A few mutants are available in other species. For example, the white axolotl has a defect in the subepidermal extracellular matrix which is nonpermissive for neural crest migration at the appropriate developmental stage. Axolotl mutants offer the greatest promise for studying factors regulating neural crest migration since there are accumulating data regarding neural crest migratory pathways and associated extracellular matrix molecules in this species. In mice, there are some mutants, such as *splotch*, which appear to affect the neural crest and other structures. These may prove useful in defining factors important for murine neural crest development. As more becomes known about neural crest migratory pathways in mouse embryos, interpretation of the mutant phenotype will be facilitated. Furthermore, it may be possible to utilize transgenic mice to study these issues.

Despite the current lack of developmental genetics, there is accumulating evidence that many neural crest cells are not preprogrammed to specific migration patterns. In this review, I discuss experiments in which the neural tube was inverted about its dorsoventral axis. After rotation, some neural crest cells migrate upward. However, neural crest derivatives form normally relative to the inverted neural tube despite the fact that they followed abnormal migratory pathways. Other experiments from this laboratory suggest that many premigratory neural crest cells are multipotent and differentiate according to cues acquired during or after their migration. Thus, it seems that many aspects of neural crest migration and differentiation occur in response to the environment rather than being preprogrammed within the neural crest. This does not, however, preclude the possibility that some neural crest cells may be preprogrammed for specific migration and differentiation patterns.

Cell Surface Organization Systems

CHAPTER 18

Cell-Substratum Adhesion: Mechanism and Regulation

Aaron W. Crawford and Mary C. Beckerle

Introduction

CELLS ARE CAPABLE OF establishing stable associations with other cells as well as with extracellular substrata. The ability of cells to adhere to extracellular ligands in a specific and regulated manner is essential to a number of biologic processes, including phagocytosis, lymphocyte recirculation, unidirectional killing by T-cells, platelet-mediated clot retraction, and controlled morphogenetic movements during embryogenesis. The specificity of cell adhesion events was illustrated quite dramatically in the early experiments of Wilson in which cells from two species of marine sponges were mixed; the individual cells aggregated only with cells of the same species to regenerate two distinct sponges (78). Later Holtfreter demonstrated that dissociated embryonic cells interacted specifically with other cells derived from the same parent tissue to give rise to what he called "histotypic aggregates" (36). These experiments pointed out that cells could recognize determinants on the surfaces of other cells and could exhibit nonrandom adhesions. It is this ability of cells to recognize cues in the extracellular environment that enables neural crest cells to migrate long distances over specific pathways during nervous system morphogenesis and that contributes significantly to important biologic activities such as lymphocyte homing and wound healing.

In addition to being ligand-specific, cell adhesion must be regulated. There are now a number of cases in which highly regulated cell adhesion has been documented. The blood platelet provides a striking example of a cell in which adhesion events are exquisitely controlled. When circulating in the blood, unactivated platelets are completely nonadhesive; they do not interact with each other or with fibrinogen, the precursor of the extracellular clot matrix material. However, upon exposure to a torn blood vessel, the platelets undergo a series of dramatic changes, including a major alteration in morphology and increase in adhesive properties. Once activated the platelets are able to bind fibrinogen as well as other platelets to form an impenetrable barrier which prevents the flow of blood beyond the vessel wall. Unregulated platelet adhesion would have devastating consequences for the organism: the supply of unactivated platelets would be depleted, and aggregates of platelets within the circulation would drastically increase the occurrence of ischemia resulting from blood vessel blockage. This extreme example of the requirement for regulated adhesion is described here for purposes of illustration; however, numerous other situations exist in which cells exhibit regulated adhesion. Some of these are discussed later.

Fibroblast Adhesion in Culture

The general question we want to consider in this chapter is how do cells adhere? That is, how do cells establish specific and regulated associations with extracellular ligands? As described above, cells can adhere to ligands present on other cells or in the extracellular matrix (see Chap. 17, this volume). For the sake of simplicity, we consider only the mechanism of cell-substratum adhesion here. There are many similarities between cell-cell and cell-substratum adhesion, although some of the molecular details are distinct.

Much of what has been learned about the mechanism of cell-substratum adhesion has come from studies on isolated fibroblasts in vitro. Fibroblasts in culture adhere via specialized regions of the plasma membrane called adhesion plaques or focal contacts. It is possible to visualize these sites of cell-substratum adhesion directly by use of a specialized light microscopic technique called interference reflection microscopy (27, 44). By this approach the adhesion plaques, areas where the cell membrane is separated from the substratum by only 10 to 15 nm, appear black (Fig. 1). Surrounding these black patches are diffuse areas that appear dark gray by interference reflection microscopy; in these regions, referred to as "close contacts," the cell membrane is typically separated from the substratum by about 15 to 30 nm. Interference reflection microscopy has provided a valuable tool for studying a cell's association with its substratum. Since this method can be used with living cells, it has been possible to document the presence of adhesion plaques on unfixed material. Moreover, from this approach it is apparent that cells in culture associate with extracellular substrata at discrete regions of the plasma membrane rather than along their entire ventral surfaces.

If one views adhesion plaques at higher resolution in the electron microscope, as was done several years ago by Singer (Fig. 2), it appears that extracellular matrix materials interact with the plasma membrane specifically at the region of the electron-dense adhesion plaque (68). Moreover, the extracellular matrix filaments, in particular those composed of fibronectin, appear to be structurally continuous with the bundles of actin filaments that impinge upon the cytoplasmic face of the plasma membrane. In some thin sec-

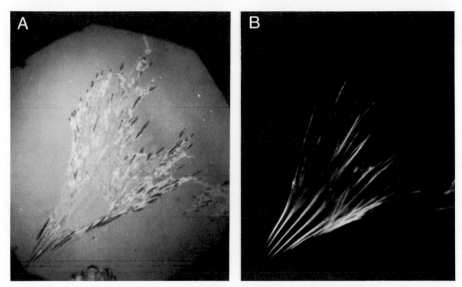

FIGURE 1. Stress fibers terminate at adhesion plaques. (*A*) A chicken embryo fibroblast viewed by interference reflection microscopy reveals the regions of close membrane substratum contact. The adhesion plaques appear black by this optical method. (*B*) The cell is stained with rhodamine-phalloidin in order to visualize actin filaments. There is a striking correlation between the distribution of the adhesion plaques and the termini of the actin filament bundles.

tions, for example the one shown in Fig. 2*B*, it almost appears that there is a collinear relationship between the extracellular fibronectin fibers and the cytoplasmic actin filament bundles, the two filament sets appearing to intermingle at the adhesion plaque. From such images it was postulated that a structural connection between the extracellular matrix and the cytoskeleton is established at the adhesion plaque. Singer termed this region of apparent transmembrane connection the "fibronexus."

Several experimental results suggest that the cell's ability to organize cytoplasmic actin and to adhere are mutually interdependent. For example, if a well-spread, adhesive cell is exposed to cytochalasin B, an agent that causes the disassembly of filamentous actin within cells, the treated cells will rapidly exhibit a loss of organized stress fibers; concomitant with the loss of filamentous actin, the cells become detached from the substratum and exhibit a loss of fibronectin from the cell surface (2). Likewise, if one blocks the ability of cells to adhere to a substratum, for example, by blocking receptors for extracellular ligands with antibodies (50), one finds that in addition to the expected loss of cell adhesion, the cells also fail to exhibit stress fiber formation.

One of the most dramatic demonstrations of the apparent connection between the extracellular matrix and the cytoskeleton comes from studies on virus-transformed cells. Transformed cells typically exhibit reduced substratum adhesion as well as a disorganized actin cytoskeleton. Exposure of trans-

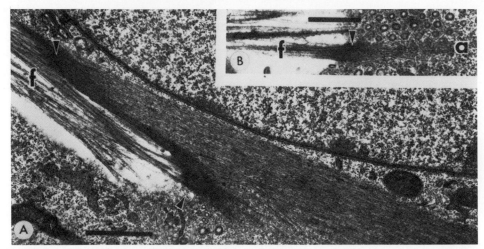

FIGURE 2. The fibronexus. An electron micrograph of the transmembrane connection between the extracellular matrix protein fibronectin (Fn) and a bundle of actin filaments is beautifully illustrated in this electron micrograph by Singer. (*A*) Fibronectin associates with the cell membrane at the region where the bundle of actin filaments (stress fiber) terminates at the electron-dense adhesion plaque (arrowheads). (*B*) The collinear relationship between the extracellular fibronectin filaments and the intracellular actin filaments is seen clearly in the insert. (Reprinted with permission from I. I. Singer. 1979. *Cell* **16**:675–685. Copyright © 1979, Cell Press.)

formed cells to the extracellular matrix protein fibronectin restores cell-substratum adhesion and induces the reorganization of actin filaments into stress fibers (3, 80). It is quite striking that the addition of an extracellular protein can so drastically affect the intracellular organization of actin filaments. Taken together, we believe that these observations indicate that a transmembrane connection between the extracellular matrix and the actin cytoskeleton occurs at adhesion plaques. In the next section we consider what is known about the molecular nature of this transmembrane linkage.

Structure and Composition of the Adhesion Plaque

On the extracellular face of the plasma membrane, a number of extracellular matrix components, including fibronectin, vitronectin, and heparan sulfate proteoglycans, have been reported to accumulate at the adhesion plaque (for review see Ref. 19). Of these, one of the most extensively studied extracellular matrix molecules is fibronectin. Early workers identified fibronectin as an extracellular glycoprotein that was dramatically reduced in abundance in transformed cells (39). Because of this characteristic, for a number of years the protein was referred to as the LETS (large extracellular transformation-sensitive) protein. Fibronectin is composed of two very similar, but not necessarily identical, polypeptide chains of 230,000 to 250,000 daltons. The individual α and β chains both contain a series of functional domains specialized for interacting with a variety of ligands including other extracellular molecules such as heparin and gelatin (collagen) as well as the cell surface (41). The biologic role of fibronectin is complex. It stimulates cell adhesion as well as cell migration (see Chap. 17, this volume) and is thought to play a significant role in a variety of cell-mediated processes, including wound healing, gastrulation (see Chap. 16, this volume), hemostasis, and malignant transformation.

Fibronectin has been shown to be associated with adhesion plaques when cells are maintained under stationary growth conditions (69). Recently the region of the fibronectin molecule that enables it to associate with cells was defined (57, 65). Although fibronectin is a very large protein, the essential component of the cell-binding domain of fibronectin was shown to be a tetrapeptide with the sequence Arg-Gly-Asp-Ser (RGDS in the single letter amino acid code). When attached to a coverslip, this tetrapeptide promotes the attachment and spreading of cells on the substratum (70). In addition, if cells are exposed to synthetic peptides containing RGDS sequences prior to plating, cell attachment to fibronectin is prohibited (57).

Given the ability of fibronectin to associate specifically with cell surfaces, it was anticipated that cells would possess receptors for fibronectin. A number of approaches have now been successfully utilized to identify such cell-surface receptors. Some investigators have identified cell-surface proteins that bind fibronectin by passing detergent-treated cell extracts over an affinity column containing a 120×10^3 dalton chymotryptic fragment of fibronectin that retained cell-binding but not heparin- or collagen-binding activity. Bound proteins were eluted from the column with a hexapeptide containing the specific cell attachment sequence of fibronectin, RGDS. By this approach a transmembrane glycoprotein complex that migrates as a single 140×10^3 dalton polypeptide on reduced sodium dodecyl sulfate (SDS) gels was identified as a specific receptor for fibronectin on human osteosarcoma cells (59). When incorporated into liposomes, this glycoprotein complex mediated the attachment of the liposomes to a fibronectin-coated substratum.

A related avian fibronectin receptor was identified independently by two groups using an immunological approach. Neff and co-workers (50) prepared monoclonal antibodies against myoblast membranes and screened the hybridoma supernatants for their ability to inhibit or disrupt myoblast adhesion. One monoclonal antibody, referred to as CSAT, dramatically affected myoblast adhesion. The CSAT monoclonal antibody immunoprecipitates a glycoprotein complex with component polypeptides migrating at 160×10^3 to 180×10^3 daltons on reduced SDS-polyacrylamide gels. Greve and Gottlieb (33) similarly identified two monoclonal antibodies, referred to as JG9 and JG22, that bind to the surfaces of myotubes and cause them to detach from the substratum. JG9, JG22, and CSAT monoclonal antibodies all appear to recognize the same glycoprotein complex composed of integral membrane proteins that migrate differently on reduced and nonreduced SDS-polyacrylamide gels. The glycoprotein complexes isolated by their affinity to either CSAT or JG22 monoclonal antibodies bind fibronectin (1, 38); the CSAT antigen has also been shown to interact with laminin and vitronectin (17, 38).

Extracellular matrix receptors from a variety of species and cell types have recently been characterized. Many of these, including the avian and mammalian fibronectin receptors described above, are now known to be members of a large family of transmembrane receptor complexes referred to as "integrins" (40). Integrins typically consist of noncovalently linked α and β subunits which exhibit extensive inter- and intrachain disulfide bonding. Each subunit has a small cytoplasmic domain, a transmembrane domain, and a large extracellular domain. The β-chains contain conserved cysteine repeat regions that participate in extensive intrachain disulfide bonding. The α-chains are larger than the β-chains and frequently consist of two disulfide-linked subunits.

Integrin extracellular matrix receptors have now been localized in adhesion plaques in a number of cell types (25, 28). Given the evidence that a connection between the extracellular matrix and the cytoskeleton occurs at adhesion plaques, it was a distinct possibility that the transmembrane integrin surface receptors were involved in mediating these associations.

Prior to the characterization of the fibronectin receptor and demonstration that such receptors are localized at sites of cell-substratum adhesion, a number of cytoplasmic adhesion plaque components were identified. One of these is a 225×10^3 dalton polypeptide called "talin" (20), a name derived from the Greek word for "ankle." Indirect immunofluorescence experiments have shown that talin is localized at adhesion plaques underlying the plasma membrane at fibronectin-rich regions and in ruffling edges of cells. By use of an approach called "equilibrium gel filtration," a direct interaction between talin and avian integrin has been demonstrated (37). Equilibrium gel filtration experiments enable detection of interactions that are characterized by rapid exchange. In this case talin and radiolabeled integrin were passed over a small gel filtration column that was preequilibrated with talin. This experimental design ensures that the integrin will have access to potential binding sites on talin during the entire course of column elution. In such equilibrium gel filtration experiments, the elution profile of the labeled receptor is determined in the presence and absence of talin and the two profiles are compared. Interaction between the receptor (integrin) and ligand (talin) is detected as a shift in the elution profile of the receptor toward the void volume, indicating that the receptor has migrated through the column as a larger complex in the presence of talin (Fig. 3). Other proteins, including other adhesion plaque components, do not induce a shift in the elution profile of integrin. However, a synthetic peptide corresponding to the putative cytoplasmic domain of the integrin β-chain inhibits the association between talin and integrin (17). Such inhibition studies provide strong evidence that the interaction between talin and integrin detected by equilibrium gel filtration is specific and could be relevant in living cells. The association between talin and integrin detected in vitro is not a high affinity interaction. It is not clear what this means for the cell. It is possible that low affinity interactions may be preferable to high affinity ones when the associations are necessarily dynamic, such as occur at sites of cell-substratum adhesion. In addition, there may be other protein-

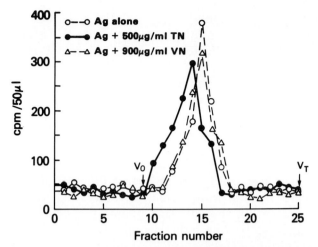

FIGURE 3. Talin binds integrin. Equilibrium gel filtration was used to detect the association of avian integrin (Ag) with talin (TN). Radioactively labeled integrin is passed over a column pre-equilibrated with talin (●—●). The elution profile shifts toward the void volume relative to the profile obtained when the integrin is gel-filtered without talin (○—○). This shift reflects the fact that integrin elutes as a larger complex in the presence of talin, indicating that the two proteins interact. Another adhesion plaque protein, vinculin (VN) has no effect on the elution profile of integrin (△—△). The dissociation constant for the binding of talin to integrin was calculated to be approximately 0.7 μM. (Reprinted with permission from A. Horwitz, K. Duggan, C. Buck, M. C. Beckerle, and K. Burridge. 1986. *Nature (Lond.)* **320**: 531–533. Copyright © 1986, Macmillan Journals Limited.)

protein interactions, as yet uncharacterized, that contribute to the overall stability of the transmembrane linkage that occurs at adhesion plaques.

Talin has also been shown to interact with another cytoplasmic component of adhesion plaques (22), a 130×10^3 dalton polypeptide called vinculin (21, 31). Vinculin was originally postulated to be the protein that linked actin to the plasma membrane at adhesion plaques. Early on, it was reported that vinculin associated directly with actin. However, the reported interaction between vinculin and actin is now thought to be due to a collection of low molecular weight contaminants in the vinculin preparation. Antisera raised against these low molecular weight vinculin-contaminating proteins stain adhesion plaques and recognize 150×10^3 and 200×10^3 dalton proteins in addition to the low molecular weight polypeptide used as the immunogen (77). It has been suggested that the low molecular weight proteins are generated from the 150×10^3 and 200×10^3 dalton proteins (now referred to as "tensin") by proteolysis. The specific role of tensin in adhesion plaques remains to be determined; however, there is recent evidence that the 150×10^3 dalton protein is a vinculin-binding protein (76).

Although vinculin interacts with talin with reasonable affinity in vitro, vinculin has been shown to associate with adhesion plaques in the apparent absence of talin in experiments using isolated adherent membranes (6). Therefore, it appears that vinculin's association with the adhesion plaque, although possibly relying on talin to some extent, does not depend exclusively on the presence of talin. The mechanism of talin-independent association of vinculin with the adhesion plaque is as yet unknown.

Over time the molecular complexity of the adhesion plaque has become more and more evident. We have recently identified yet another novel component of adhesion plaques by analysis of a nonimmune rabbit serum that stains focal contacts in indirect immunofluorescence experiments (7). Analysis of this nonimmune serum revealed that the antibody responsible for the adhesion plaque staining specifically recognizes an 82×10^3 dalton protein. This ubiquitous 82×10^3 dalton protein has an interesting distribution in cells (Fig. 4); it is found within the adhesion plaque proper as well as along the actin filament bundles near where they terminate at the adhesion plaque. The subcellular distribution of the 82×10^3 dalton protein places it in a position where it may function as a link between actin filaments and some component of the adhesion plaque. Interestingly, the 82×10^3 dalton protein is present at low levels in cells, i.e., substoichiometric with respect to any of the other adhesion plaque proteins discussed thus far. (By way of speculation, perhaps the 82×10^3 dalton protein is an actin filament capping protein or serves some regulatory function at the adhesion plaque.) We are currently developing procedures for purification of this new adhesion plaque component so that we can learn more about its role at sites of cell-substratum adhesion.

Several other proteins have been localized in adhesion plaques, including a 200×10^3 dalton cytoplasmic protein (48) as well as the membrane proteins herpes simplex virus glycoprotein D (52), the FC-1 antigen (53), and the 30B6 antigen (60), an integrin-like protein. The specific functions of these adhesion plaque proteins remain to be determined.

As described above, actin filaments associate with the plasma membrane at adhesion plaques. These actin filaments are associated with a number of actin-binding proteins including fimbrin, filamin, myosin, and α-actinin. These proteins may participate in organizing actin filaments into the linear arrays found in the stress fibers or in generating motive or contractile forces along the length of the filaments. Although some of these actin-binding proteins are found in adhesion plaques as well as along the length of the actin filaments, none appear to be significantly involved in the linkage of actin to the membrane at the adhesion plaque. α-Actinin has been shown to interact with vinculin in gel overlays (72) as well as in solution; however, the interaction does not appear to involve a very high affinity. Moreover, α-actinin and vinculin are not consistently colocalized in cells, consequently the significance of their association in vitro is not totally clear. Actin itself has been reported to interact to some degree with integrin molecules from the platelet (54), so a direct interaction between actin and integrin may account, in part, for the transmembrane connection between the extracellular matrix and the cytoskeleton that occurs at adhesion plaques.

The state of our current understanding of adhesion plaque composition and organization is summarized in Table 1 and Fig. 5. It should be noted at

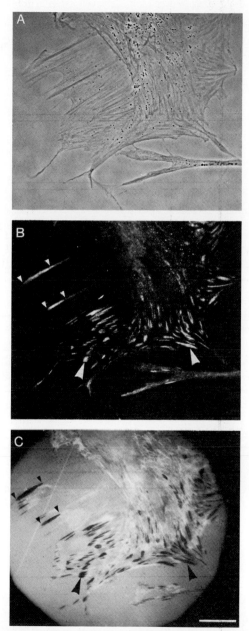

FIGURE 4. Distribution of the 82×10^3 dalton protein in a chicken embryo fibroblast. (A) Phase contrast microscopy and (B) indirect immunofluorescence microscopy using antibodies directed against the 82×10^3 dalton protein. (C) Interference reflection microscopy. The 82×10^3 dalton protein is found in adhesion plaques (note larger arrowheads), but in some cases extends beyond the domain of the adhesion plaque as defined by interference reflection microscopy (note small arrowheads bracketing adhesion plaques). (Reprinted by permission from M. C. Beckerle. 1986. *J. Cell Biol.* **103**:1679–1687, by copyright permission of the Rockefeller University Press.)

TABLE 1 Proteins Localized at Adhesion Plaques

Extracellular	Transmembrane	Cytoplasmic
Structural Components		
Fibronectin (69)*	Integrin (25, 28)	Actin
Vitronectin (5)	30B6 (60)	α-Actinin (73)
Heparan sulfate	FC-1 (53)	Fimbrin (15)
proteoglycan (79)	HSV glycoprotein D	Vinculin (21, 31)
	(52)	82×10^3 dalton (7)
		200×10^3 dalton (48)
		Talin (20)
		Tensin (77)
Regulatory Components		
120×10^3 and	None known	Calcium-dependent
150×10^3 dalton		protease II (8)
proteases (24, 26)		pp60^{src} (61)
Plasminogen activa-		p120$^{gag\text{-}abl}$ (62)
tor (urokinase) (58)		p90$^{gag\text{-}yes}$ (32) tyrosine
		p80$^{gag\text{-}yes}$ (32) kinases
		82×10^3 dalton (?) (7)

*Numbers in parentheses represent references.

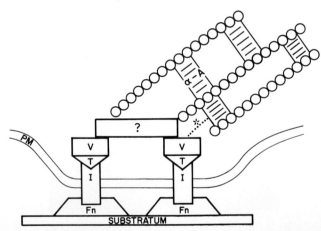

FIGURE 5. This diagram shows a simplified outline of the protein-protein interactions thought to occur at regions of cell-substratum adhesion. At these sites a transmembrane connection between actin filaments and the extracellular matrix is established. For example, fibronectin (Fn) associates with the cell via its receptor, integrin (I). Integrin binds the cytoplasmic protein talin (T) which also interacts with vinculin (V). The mechanism by which this protein complex is associated with the actin filaments is unknown (?); however there is some evidence that vinculin may interact directly (*) with the actin-binding protein α-actinin (α-A).

this point, that this outline of the molecular interactions that occur at sites of cell-substratum adhesion is incomplete. Many other proteins that reside in adhesion plaques have been identified but were not included in the summary diagram since the nature of their associations at the adhesion plaque has not yet been defined. Research in the area of cell adhesion is presently very intensive, and new components of adhesion plaques are constantly being identified. Moreover, the complexity of the interactions that occur at these specialized and dynamic regions of the plasma membrane is becoming more and more evident. Only a few years ago a single protein was postulated to link actin to the membrane at sites of adhesion. At present an order of magnitude more components have been identified. Still, the number of proteins at adhesion plaques is not infinite, and in fact, since many of the already identified proteins have been shown to associate with other known proteins, we suspect that many of the key structural players have now been recognized. One of the major challenges at present is to define the mechanism by which the associations at the adhesion plaque are regulated by cells.

Regulation of Adhesion Plaque Assembly and Disassembly

The formation of adhesion plaques has been extensively characterized in an interference reflection microscopy study by Izzard and Lochner (45). These investigators studied spreading cells by time-lapse cinematography. In early stages of cell spreading, lamellipodia or microspikes that extend upward from the cell surface are generated. When these "touch down," they establish close contacts with the substratum. These close contacts appear to stimulate localized protrusive movement of cytoplasm into the leading lamella causing it to advance and initiate the formation of an adhesion plaque. Formation of a focal contact occurs rapidly, often in less than a second. The short time required for development of an adhesion plaque coupled with the fact that linear bundles of microfilaments are present at the site of presumptive adhesion plaque development prompted these investigators to suggest that adhesion plaques are generated via the association of preformed bundles of actin filaments with the substratum-attached membrane.

An alternative model for adhesion plaque formation has been offered by Geiger and coworkers. They have suggested that the association of membrane receptors with their extracellular ligands induces receptor aggregation. Upon clustering, the receptors could become immobilized and recruit cytosolic components into the adhesion plaque. According to this model, the aggregate of proteins at the plasma membrane serves as the nucleation site for microfilament assembly (4). There is some evidence in support of each of these models; however, the precise pathway governing the assembly of adhesion plaques has yet to be resolved. It would be very interesting to examine whether, for example, fibronectin binding to fibroblasts stimulates actin polymerization in cells, a result that might be expected if Geiger's model is correct.

Once assembled, how are adhesion plaques turned over by cells? In a study on cell movement in culture, Chen (23) showed that elongating leading lamellae generate tension in the cell that ultimately induces the severing of the cell's posterior foothold and a forward surging of cytoplasm. The release of posterior attachments to the substratum results literally from a rupturing of the cell at sites of adhesion which leaves fragments of the adhesion plaque and associated cytoplasm attached to the substratum in its original location. Intuitively, this means of severing an association with the substratum seems harsh on the cell, and it probably only occurs under conditions in which the cell has an unusually high affinity for an extracellular ligand, such as might be expected to occur under artificial conditions in culture. In contrast, there are now many documented examples of the cell's ability to disassemble adhesion plaques directly. For example, Izzard and Lochner (45) observed that as new focal contacts are established near the cell margin, those that remain behind had a tendency to be lost. This direct observation illustrates that cells have the ability to disassemble adhesion plaques as well as initiate their formation. Likewise when cells in culture enter mitosis, they round up, losing their organized adhesion plaques. By immunofluorescence, the components of adhesion plaques appear to be uniformly distributed in the cytoplasm of mitotic cells (our unpublished observations). Adhesion plaques are also disassembled by some cells in response to tumor-promoting phorbol esters (66) or certain growth factors such as platelet-derived growth factor (34).

Thus it has been clear for some time that adhesion plaques are dynamic. But it has been very difficult to determine the mechanism(s) by which these specialized regions of the plasma membrane are assembled and disassembled. Certainly in some cases, a cell may permanently give up its adhesive properties, for example, during the transition from a bone-marrow restricted adhesive reticulocyte to a circulating, nonadhesive mature erythrocyte. Alternatively, a migrating cell may be developing adhesive contacts with a substratum in one region while relinquishing these associations in another area. Two key questions that persist at present are (1) how do cells modulate their adhesions and (2) how does an individual cell coordinately control the development and loss of adhesions according to their spatial location?

One mechanism by which a cell might modulate its adhesion to the substratum is by regulated selective proteolysis of adhesion plaque components. We have recently determined that the calcium-dependent protease type II (CDP-II) resides in adhesion plaques (Fig. 6) (8). The adhesion plaque component, talin, is an excellent substrate for CDP-II and, in fact, appears to be dramatically more susceptible to calcium-dependent proteolysis than other adhesion plaque components such as vinculin and α-actinin (Fig. 7). CDP-II cleaves talin into two fragments, generating polypeptide fragments of 190×10^3 to 200×10^3 daltons and 46×10^3 daltons. It is easy to imagine that proteolytic cleavage of talin could dramatically affect the organization of proteins in the adhesion plaque and thus alter the stability of the substratum contacts. There is evidence that cleavage of talin occurs in vivo. Most strikingly, talin is cleaved by the calcium-dependent protease in living platelets during platelet aggregation and clot retraction (9, 30). The proteolysis of talin

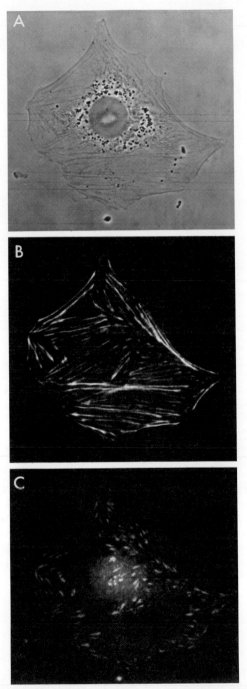

FIGURE 6. The calcium-dependent protease is a component of adhesion plaques. (A) Phase contrast microscopy. (B) Actin filament distribution as revealed by rhodamine-phalloidin. (C) Distribution of the calcium-dependent protease type II. The protease is localized in adhesion plaques.

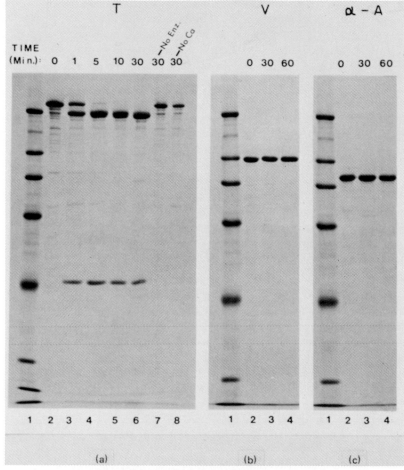

FIGURE 7. Susceptibility of adhesion plaque proteins to calcium-dependent proteolysis. The proteolytic cleavage of talin (T), vinculin (V) and α-actinin (α-A) by CDP-II. (*panel a*) CDP-II rapidly cleaves talin, whereas (*panel b*) vinculin and (*panel c*) α-actinin are resistant to proteolysis. Cleavage of talin is completely dependent on the presence of exogenous protease (panel A, lane 7) and calcium (panel A, lane 8). (Reprinted with permission from M. C. Beckerle, K. Burridge, G. N. DeMartino, and D. E. Croall. 1987. *Cell* 51:567–577. Copyright © 1987, Cell Press.)

in vivo in platelets suggests that the cleavage has physiological significance. Given the timing of talin cleavage in platelets, it appears that talin cleavage may be involved in the regulation of platelet adhesion to other cells or to fibrin. Current research is under way to determine the mechanism of CDP-II activation and how proteolysis of talin is related to the structural integrity of the adhesion plaque in living cells.

Extracellular proteases may also be used by cells to regulate adhesion to a substratum. In particular, transformed cells have been shown to express degradative enzymes that dissolve extracellular matrices (24, 26). For example, Rous sarcoma virus–transformed cells produce proteases that preferentially degrade fibronectin (26). These proteases have apparent molecular weights of 120,000 and 150,000 and are active at sites of cell-substratum adhesion (24). Recently the plasminogen activator, urokinase, has also been reported to be localized at focal contacts (58). Plasminogen activator is a protease that cleaves plasminogen to generate the active protease, plasmin. Plasmin has been implicated in extracellular matrix degradation by transformed and normal cells. It is possible that cells employ a mechanism such as degradation of the extracellular matrix to affect cell-substratum adhesion and invade connective tissue, an event which is a prerequisite to the processes of embryo implantation, tissue remodeling and repair, and tumor metastasis.

A posttranslational modification that has received much attention recently as a possible mechanism for influencing adhesion plaque stability is phosphorylation. Talin, vinculin, and integrin have all been shown to be substrates for the tyrosine kinase oncogene product encoded by the transforming virus Rous sarcoma virus (RSV). In response to transformation by this virus, cells lose their adhesions to the substratum and develop a rounded morphology. Transformed cells exhibit a dramatic redistribution of adhesion plaque components, typically into small adhesive structures called "rosettes" that are concentrated under the nucleus. By indirect immunofluorescence, the transforming tyrosine kinase of Rous sarcoma virus, pp60^{src}, has been localized to adhesion plaques (61). Because of the tyrosine kinase activity of pp60^{src}, it has been suggested that phosphorylation of adhesion plaque components may contribute to the development of the rounded transformed phenotype. Numerous attempts have been made to correlate the level of phosphorylation of adhesion plaque components with the degree of cell-substratum adhesion in transformed and normal cells. Vinculin was found to contain considerably higher levels of phosphotyrosine in transformed cells, approaching eight times the amount present in control cells (67). It was subsequently revealed that phosphorylation of the tyrosine residues of talin also increases in transformed cells (29, 55). Partial transformation mutants of RSV (63) have been used to determine the correlation between the presence of pp60^{src} in the adhesion plaque with the concomitant development of the transformed phenotype and the phosphorylation of talin and vinculin. No correlation between the level of phosphotyrosine in vinculin (46, 51) or talin (29) and the manifestation of a transformed phenotype has been found. These results suggest that the tyrosine phosphorylation of talin and vinculin in transformed cells does not significantly affect cell adhesion.

More recently, however, it was discovered that the β-chain of integrin is also phosphorylated in vivo in virally transformed cells (35). One of the putative tyrosine phosphorylation sites occurs at residue 788 in the integrin β-chain, a position which falls in the region of the cytoplasmic domain of the protein (71). This phosphorylation site is in the region of the β-chain that is thought to interact with talin. Interestingly, integrin isolated from RSV-transformed cells fails to bind talin in equilibrium gel filtration assays (17), raising the possibility that phosphorylation of the β-chain prohibits this association. This is a very provocative finding; however, it has not yet been proven whether it is the tyrosine phosphorylation of the β-chain that specifically prohibits its interaction with talin.

Protein kinase C, a lipid- and calcium-dependent kinase involved in signal transduction, has also been shown to phosphorylate adhesion plaque components. Both vinculin (75) and talin (10, 47) are substrates for protein kinase C in vitro, and vinculin has also been shown to be phosphorylated by protein kinase C in vivo (74). Tumor-promoting phorbol esters that stimulate protein kinase C activity also induce phosphorylation of talin in vivo, but it has not been determined whether this increased phosphorylation is due to the direct action of protein kinase C (10). Interestingly, recent experiments have demonstrated that in lymphocytes, the redistribution of talin to colocalize with immunoglobulin-capped integrin is stimulated by treatment of the cells with tumor-promoting phorbol esters (18); this observation suggests that protein kinase C–mediated phosphorylation may regulate the association of talin and integrin. If true, this could provide an important control for adhesion plaque assembly.

Phosphorylation is a rapidly reversible protein modification that could be used to regulate dynamic cell-substratum adhesions. However, in some situations longer-term effects on cell adhesion might be required. In this case, transcriptional or translational control of the absolute concentrations of adhesion plaque components could be utilized to control the extent of cell adhesion. Already there is some precedence for this approach to regulating cell adhesion. For example, the level of fibronectin receptors is regulated during erythropoiesis. Fibronectin receptors are present on the cell surfaces of reticulocytes that are retained in the extracellular matrix–rich bone marrow. However, the abundance of extracellular matrix receptors declines as the reticulocytes mature, and by the time they become circulating erythrocytes in the blood, surface fibronectin receptors are no longer present (56). The modulation of the level of fibronectin receptors during erythrocyte differentiation provides a mechanism by which the immature reticulocytes are selectively retained in the bone marrow and mature erythrocytes lacking extracellular matrix receptors are released into the circulation.

Studies using 3T3-L1-mouse preadipocytes, chicken embryo fibroblasts, or rat fibroblasts have revealed that synthesis of receptors for extracellular matrix molecules increases in response to transforming growth factor-beta (TGF-β) (43). Increased synthesis of extracellular matrix receptors predictably increases the adhesion of the TGF-β treated cells to extracellular matrix components. TGF-β appears to stimulate increased production of fibronectin receptors by two separate mechanisms. Exposure of cells to TGF-β results in elevated levels of mRNA coding for the surface extracellular matrix receptor, and, in addition, increases the rate of conversion of the receptor β-subunit precursor to the mature form (43). TGF-β also stimulates production of fibronectin, various forms of collagen, and matrix proteoglycans in many cell types (42), providing yet another mechanism that encourages cell adhesion. These cellular responses to TGF-β appear to have significant biologic relevance.

TGF-β stored in platelets is likely to be an agonist in wound healing, being secreted upon platelet activation to promote adhesion of fibroblasts in the area of the wound and the subsequent production of connective tissue components to rebuild the subendothelial layer.

Interestingly, cells are also capable of responding directly to the nature of their environment by regulating the expression of adhesion plaque components. For example, cells plated on an adhesive substratum exhibit increased transcription and translation of vinculin mRNA, whereas on a nonadhesive substratum vinculin expression is reduced (11).

Thus it appears that cells are able to regulate their associations with extracellular matrices by a variety of mechanisms. The complexity of cell adhesion modulation is only beginning to be elucidated. In the next few years this will surely be an area of intensive investigation. Already it appears that cells can modulate their adhesive properties by regulating extracellular, transmembrane, and cytoplasmic components of adhesion plaques in ways that result in either long-term effects or potentially transient effects on the association of the cell with the extracellular environment.

Role of Adhesion in Directed Cell Migration

In order for a cell to migrate, it must be able to regulate its adhesion. Migrating cells establish new adhesions at their leading edges and relinquish older associations with the substratum in order to move forward. Although motile cells in culture do not exhibit well-developed adhesion plaques, we suspect that the components of adhesion plaques, though less highly organized and prominent in moving cells, do participate in the adhesive events prerequisite to cell migration. Adhesion plaques are, in some senses, artifacts of tissue culture. Nevertheless, we believe they exhibit the same sets of protein-protein interactions, albeit elaborated, that occur at any site of cell-substratum adhesion. Support for this belief comes from the immunocytochemical localization of adhesion plaque components in tissues. Adhesion plaque proteins are typically found in association with plasma membrane at regions where a cell contacts the basal lamina or at areas of the cell membrane specialized for actin-membrane interactions, such as the dense plaques of smooth muscle.

It is clear that some of the known components of adhesion plaques are involved in cell migration. For example, neural crest migration is enhanced by fibronectin but not by other extracellular matrix components such as laminin (64). Anti-fibronectin antibodies added to the medium cause reduced cell adhesion and prevent migration. Thus adhesion to fibronectin appears to stimulate cell migration in vitro. This now appears to be the case in vivo as well. Microinjection of anti-fibronectin antibodies into embryos prior to or at the onset of gastrulation prevents the invagination of presumptive mesodermal cells (Ref. 13; also see Chap. 16, this volume). Furthermore, if a section of the roof of the blastocoel is inverted, mesodermal cell migration is inhibited in this area; because fibronectin is present on the internal but not the external surface of the blastocoelic roof, it is believed that the lack of a fibronectin-rich matrix in the area of the inverted roof causes a reduction in cell migration in that area (13).

Fibronectin receptors have also been implicated in cell migration since microinjection of antibodies against avian integrin disrupts neural crest cell migration in developing chicken embryos (Ref. 16; also see Chap. 17, this volume). Moreover, microinjection of the cell-binding sequence of fibronectin, RGDS, prevents gastrulation in amphibian (14) and *Drosophila* embryos (49) and blocks neural crest cell migration in avian embryos (14). This tetrapeptide is known to interact with fibronectin receptors but may also bind other receptors for RGD-containing adhesion molecules.

Finally, there is genetic evidence for a role of integrins in embryogenesis. The *Drosophila* embryonic lethal mutation, lethal myospheroid, is now known to be caused by a defect in a gene encoding an integrin β-chain (47a). Embryos carrying this mutation develop normally until about 14 h of embryogenesis at which time the muscles fail to attach effectively to the body wall; when the unanchored muscles begin to contract, the force generated disrupts the internal organization of the embryo, resulting in a lethal phenotype. Interestingly, the *Drosophila* embryonic position-specific antigens also appear to be members of the integrin family of cell adhesion molecules (12).

Perspectives

Adhesion plaques provide a model system for studying the mechanism by which cells establish and regulate their associations with the extracellular matrix. With the identification and characterization of proteins localized in cells at sites of substratum adhesion, it has been possible to build a conceptual framework for how the transmembrane connections between the extracellular environment and the cytoskeleton are developed by cells. The regulation of some of the components of adhesion plaques is also now being elucidated; advances in the area of regulation of adhesion are enabling us to visualize on a molecular level how adhesion plaques might be assembled and disassembled. Much work, of course, remains to be done in this research area. A major challenge for the future will be to begin to address more of these questions in living cells and organisms in order to define the relationship between our biochemical observations and the dynamics of adhesive interactions in vivo.

Acknowledgments

We are grateful to Maurine Vaughan for preparation of this manuscript. Our work on cell-substratum adhesion is supported by grants from the National Science Foundation (DCB 8602131) and the American Heart Association.

General References

EDELMAN, G. M., and THIERY, J. P., EDS.: 1985. *The Cell in Contact: Adhesions and Junctions as Morphogenetic Determinants*. John Wiley, New York.

BURRIDGE, K.: 1986. Substrate adhesions in normal and transformed fibroblasts: organization and regulation of cytoskeletal, membrane and extracellular matrix components at focal contacts. *Cancer Rev.* 4:18–78.

References

1. Akiyama, S. K., Yamada, S. S., and Yamada, K. M. 1986. Characterization of a 140 kd avian cell surface antigen as a fibronectin-binding molecule. *J. Cell Biol.* 102:442–448.

2. Ali, I. U., and Hynes, R. O. 1977. Effects of cytochalasin B and colchicine on attachment of a major surface protein of fibroblasts. *Biochim. Biophys. Acta* 471:16–24.

3. Ali, I. U., Mautner, V., Lanza, R. P., and Hynes, R. O. 1977. Restoration of normal morphology, adhesion and cytoskeleton in transformed cells by addition of a transformation-sensitive surface protein. *Cell* 11:115–126.

4. Avnur, Z., and Geiger, B. 1981. Substrate-attached membranes of cultured cells. Isolation and characterization of ventral membranes and the associated cytoskeleton. *J. Mol. Biol.* 153:361–379.

5. Baetscher, M., Pamplin, D. W., and Bloch, R. J. 1986. Vitronectin at sites of cell-substrate contact in cultures of rat myotubes. *J. Cell Biol.* 103:369–378.

6. Ball, E. H., Freitag, C., and Gurofsky, S. 1986. Vinculin interaction with permeabilized cells: disruption and reconstitution of a binding site. *J. Cell Biol.* 103:641–648.

7. Beckerle, M. C. 1986. Identification of a new protein localized at sites of cell-substrate adhesion. *J. Cell Biol.* 103:1679–1687.

8. Beckerle, M. C., Burridge, K., DeMartino, G. N., and Croall, D. E. 1987. Colocalization of calcium-dependent protease II and one of its substrates at sites of cell adhesion. *Cell* 51:569–577.

9. Beckerle, M. C., O'Halloran, T., and Burridge, K. 1986. Demonstration of a relationship between talin and P235, a major substrate of the calcium-dependent protease in platelets. *J. Cell. Biochem.* 30:259–270.

10. Beckerle, M. C., O'Halloran, T., Earp, S., and Burridge, K. 1985. Evidence for functional similarities between talin and platelet P235. *J. Cell Biol.* 101:411a.

11. Bendori, R., Salomon, D., and Geiger, B. 1987. Contact-dependent regulation of vinculin expression in cultured fibroblasts: a study with vinculin-specific cDNA probes. *EMBO J.* 6:2897–2905.

12. Bogaert, T., Brown, N., and Wilcox, M. 1987. The *Drosophila* PS2 antigen is an invertebrate integrin that, like the fibronectin receptor, becomes localized to muscle attachments. *Cell* 51:929–940.

13. Boucaut, J. C., Darribère, T., Boulekbache, H., and Thiery, J. P. 1984. Prevention of gastrulation but not neurulation by antibodies to fibronectin in amphibian embryos. *Nature (Lond.)* 307:364–367.

14. Boucaut, J. C., Darribère, T., Poole, T. J., Aoyama, H., Yamada, K. M., and Thiery, J. P. 1984. Biologically active synthetic peptides as probes of embryonic development: A competitive peptide inhibitor of fibronectin function inhibits gastrulation in amphibian embryos and neural crest cell migration in avian embryos. *J. Cell Biol.* **99**:1822–1830.

15. Bretscher, A., and Weber, K. 1980. Fimbrin, a new microfilament-associated protein present in microvilli and other cell surface structures. *J. Cell Biol.* **86**:335–340.

16. Bronner-Fraser, M. 1985. Alteration in neural crest migration by a monoclonal antibody that affects cell adhesion. *J. Cell Biol.* **101**:610–617.

17. Buck, C. A., and Horwitz, A. F. 1987. Integrin, a transmembrane glycoprotein complex mediating cell-substratum adhesion. *J. Cell Sci. Suppl.* **8**:231–250.

18. Burn, P., Kupfer, A., and Singer, S. J. 1988. Dynamic membrane-cytoskeletal interactions: specific association of integrin and talin arises *in vivo* after phorbol ester treatment of peripheral blood lymphocytes. *Proc. Natl. Acad. Sci. U.S.A.* **85**: 497–501.

19. Burridge, K. 1986. Substrate adhesions in normal and transformed fibroblasts: organization and regulation of cytoskeletal, membrane and extracellular matrix components at focal contacts. *Cancer Rev.* **4**:18–78.

20. Burridge, K., and Connell, L. 1983. A new protein of adhesion plaques and ruffling membranes. *J. Cell Biol.* **97**:359–367.

21. Burridge, K., and Feramisco, J. R. 1980. Microinjection and localization of a 130k protein in living fibroblasts: a relationship to actin and fibronectin. *Cell* **19**:587–595.

22. Burridge, K., and Mangeat, P. 1984. An interaction between vinculin and talin. *Nature (Lond.)* **308**:744–746.

23. Chen, W.-T. 1981. Mechanism of retraction of the trailing edge during fibroblast movement. *J. Cell Biol.* **90**:187–200.

24. Chen, J.-M., and Chen, W.-T. 1987. Fibronectin-degrading proteases from the membranes of transformed cells. *Cell* **48**:193–203.

25. Chen, W.-T., Greve, J. M., Gottlieb, D. I., and Singer, S. J. 1985. Immunocytochemical localization of 140 kD adhesion molecules in cultured chicken fibroblasts and in chicken smooth muscle and intestinal epithelial tissues. *J. Histochem. Cytochem.* **33**:576–586.

26. Chen, W.-T., Olden, K., Bernard, B. A., and Chu, F. 1984. Expression of transformation-associated protease(s) that degrade fibronectin at cell contact sites. *J. Cell Biol.* **98**:1546–1555.

27. Curtis, A. S. G. 1964. The mechanism of adhesion of cells to glass. A study by interference reflection microscopy. *J. Cell Biol.* **20**:199–215.

28. Damsky, C. H., Knudsen, K. A., Bradley, D., Buck, C. A., and Horwitz, A. F. 1985. Distribution of the cell substratum attachment (CSAT) antigen on myogenic and fibroblastic cells in culture. *J. Cell Biol.* **100**:1528–1539.

29. DeClue, J. E., and Martin, G. S. 1987. Phosphorylation of talin at tyrosine in Rous sarcoma virus–transformed cells. *Mol. Cell Biol.* **7**:371–378.

30. Fox, J. E. B., Goll, D. E., Reynolds, C. C., and Phillips, D. R. 1985. Identification of two proteins (actin-binding protein and P235) that are hydrolyzed by endogenous Ca^{2+}-dependent protease during platelet aggregation. *J. Biol. Chem.* **260**:1060–1066.

31. Geiger, B. 1979. A 130K protein from chicken gizzard: its localization at the termini of microfilament bundles in cultured chicken cells. *Cell* 18:193–205.

32. Gentry, L. E., and Rohrschneider, L. R. 1984. Common features of the *yes* and *src* gene products defined by peptide-specific antibodies. *J. Virol.* 51(2):539–546.

33. Greve, J. M., and Gottlieb, D. I. 1982. Monoclonal antibodies which alter the morphology of cultured chick myogenic cells. *J. Cell. Biochem.* 18:221–229.

34. Herman, B., and Pledger, W. J. 1985. Platelet-derived growth factor–induced alterations in vinculin and actin distribution in BALB/c-3T3 cells. *J. Cell Biol.* 100: 1031–1040.

35. Hirst, R., Horwitz, A., Buck, C., and Rohrschneider, L. 1986. Phosphorylation of the fibronectin receptor complex in cells transformed by oncogenes that encode tyrosine kinases. *Proc. Natl. Acad. Sci. U.S.A.* 83:6470–6474.

36. Holtfreter, J. 1948. Significance of the cell membrane in embryonic processes. *Ann. N. Y. Acad. Sci.* 49:709–760.

37. Horwitz, A., Duggan, K., Buck, C., Beckerle, M. C., and Burridge, K. 1986. Interactions of plasma membrane fibronectin receptor with talin—a transmembrane linkage. *Nature (Lond.)* 320:531–533.

38. Horwitz, A., Duggan, K., Greggs, R., Decker, C., and Buck, C. 1985. The CSAT antigen has properties of a receptor for laminin and fibronectin. *J. Cell Biol.* 101: 2134–2144.

39. Hynes, R. O. 1973. Alteration of cell-surface proteins by viral transformation and by proteolysis. *Proc. Natl. Acad. Sci. U.S.A.* 70:3170–3174.

40. Hynes, R. O. 1987. Integrins: a family of cell surface receptors. *Cell* 48:549–554.

41. Hynes, R. O., and Yamada, K. M. 1982. Fibronectins: multifunctional modular glycoproteins. *J. Cell Biol.* 95:369–377.

42. Ignotz, R. A., and Massagué, J. 1986. Transforming growth factor-β stimulates the expression of fibronectin and collagen and their incorporation into the extracellular matrix. *J. Biol. Chem.* 261(9):4337–4345.

43. Ignotz, R. A., and Massagué, J. 1987. Cell adhesion protein receptors as targets for transforming growth factor-β action. *Cell* 51:189–197.

44. Izzard, C. S., and Lochner, L. R. 1976. Cell-to-substrate contacts in living fibroblasts: an interference reflexion study with an evaluation of the technique. *J. Cell Sci.* 21:129–159.

45. Izzard, C. S., and Lochner, L. R. 1980. Formation of cell-to-substrate contacts during fibroblast motility: an interference-reflexion study. *J. Cell Sci.* 42:81–116.

46. Kellie, S., Patel, B., Mitchell, A., Critchley, D. R., and Wigglesworth, N. M. 1986. Comparison of the relative importance of tyrosine-specific vinculin phosphorylation and the loss of surface-associated fibronectin in the morphology of cells transformed by Rous sarcoma virus. *J. Cell Sci.* 82:129–142.

47. Litchfield, D. W., and Ball, E. M. 1986. Phosphorylation of the cytoskeletal protein talin by protein kinase C. *Biochem. Biophys. Res. Commun.* 134:1276–1283.

47a. MacKrell, A. J., Blumberg, B., Haynes, S. R., and Fessler, J. H. 1988. The lethal myospheroid gene of *Drosophila* encodes a membrane protein homologous to vertebrate integrin β- subunits. *Proc. Natl. Acad. Sci. U.S.A.* 85:2633–2638.

48. Maher, P., and Singer, S. J. 1983. A 200-kd protein isolated from the fascia adherens membrane domains of chicken cardiac muscle cells is detected immunologically in fibroblast focal adhesions. *Cell Motil.* 3:419–429.

49. Naidet, C., Sémériva, M., Yamada, K. M., and Thiery, J. P. 1987. Peptides containing the cell-attachment recognition signal Arg-Gly-Asp prevent gastrulation in *Drosophila* embryos. *Nature (Lond.)* **325**:348–350.

50. Neff, N. T., Lowrey, C., Decker, C., Tovar, A., and Damsky, C. 1982. A monoclonal antibody detaches embryonic skeletal muscle from extracellular matrices. *J. Cell Biol.* **95**:654–666.

51. Nigg, E. A., Sefton, B. M., Singer, J. J., and Vogt, P. K. 1986. Cytoskeletal organization, vinculin-phosphorylation, and fibronectin expression in transformed fibroblasts with different cell morphologies. *Virology* **151**:50–65.

52. Norrild, B., Virtanen, I., Lehto, V. P., and Pedersen, B. 1983. Accumulation of Herpes Simplex virus Type I Glycoprotein D in adhesion areas of infected cells. *J. Gen. Virol.* **64**:2499–2503.

53. Oesch, B., and Birchmeier, W. 1982. New surface component of fibroblast's focal contacts identified by monoclonal antibody. *Cell* **31**:671–679.

54. Painter, R. G., Prodouz, K. N., and Gaarde, W. 1985. Isolation of a subpopulation of glycoprotein IIb-IIIa from platelet membranes that is bound to membrane actin. *J. Cell Biol.* **100**:652–657.

55. Pasquale, E. B., Maher, P. A., and Singer, S. J. 1986. Talin is phosphorylated on tyrosine in chicken embryo fibroblasts transformed by Rous sarcoma virus. *Proc. Natl. Acad. Sci. U.S.A.* **83**:5507–5511.

56. Patel, V. P., and Lodish, H. F. 1986. The fibronectin receptor on mammalian erythroid precursor cells: characterization and developmental regulation. *J. Cell Biol.* **102**:449–456.

57. Pierschbacher, M. D., and Ruoslahti, E. 1984. Cell attachment activity of fibronectin can be duplicated by small synthetic fragments of the molecule. *Nature (Lond.)* **309**:30–33.

58. Pöllanen, J., Hedman, K., Nielsen, L. S., Dano, K., and Vaheri, A. 1988. Ultrastructural localization of plasma membrane-associated urokinase-type plasminogen activator at focal contacts. *J. Cell Biol.* **106**:87–95.

59. Pytela, R., Pierschbacher, M. D., and Ruoslahti, E. 1985. Identification and isolation of a 140 kd cell surface glycoprotein with properties expected of a fibronectin receptor. *Cell* **40**:191–198.

60. Rogalski, A. A., and Singer, S. J. 1985. An integral glycoprotein associated with the membrane attachment sites of actin microfilaments. *J. Cell Biol.* **101**:785–801.

61. Rohrschneider, L. R. 1980. Adhesion plaques of Rous sarcoma virus–transformed cells contain the *src* gene product. *Proc. Natl. Acad. Sci. U.S.A.* **77**:3514–3518.

62. Rohrschneider, L. R., and Najita, L. M. 1984. Detection of the *v-abl* gene product at cell-substratum contact sites in Abelson murine leukemia virus-transformed fibroblasts. *J. Virol.* **51**:547–552.

63. Rohrschneider, L., and Reynolds, S. 1985. Regulation of cellular morphology by the Rous sarcoma virus *src* gene: analysis of fusiform mutants. *Mol. Cell. Biol.* **5**: 3097–3107.

64. Rovasio, R. A., Delouvee, A., Yamada, K. M., Timpl, R., and Thiery, J. P. 1983. Neural crest cell migration: requirements for exogenous fibronectin and high cell density. *J. Cell Biol.* **96**:462–473.

65. Ruoslahti, E., and Pierschbacher, M. D. 1986. Arg-gly-asp: a versatile cell recognition signal. *Cell* **44**:517–518.

66. Schliwa, M., Nakamura, T., Porter, K. R., and Euteneur, V. 1984. A tumor

promotor induces rapid and coordinated reorganization of actin and vinculin in cultured cells. *J. Cell Biol.* **99**:1045–1059.

67. Sefton, B. M., Hunter, T., Ball, E. H., and Singer, S. J. 1981. Vinculin: a cytoskeletal target of the transforming protein of Rous sarcoma virus. *Cell* **24**:165–174.

68. Singer, I. I. 1979. The fibronexus: a transmembrane association of fibronectin-containing fibers and bundles of 5 nm microfilaments in hamster and human fibroblasts. *Cell* **16**:675–685.

69. Singer, I. I. 1982. Association of fibronectin and vinculin with focal contacts and stress fibers in stationary hamster fibroblasts. *J. Cell Biol.* **92**:398–408.

70. Singer, I. I., Kawka, D. W., Scott, S., Mumford, R. A., and Lark, M. W. 1987. The fibronectin cell attachment sequence arg-gly-asp-ser promotes focal contact formation during early fibroblast attachment and spreading. *J. Cell Biol.* **104**:573–584.

71. Tamkun, J. W., DeSimone, D. W., Fonda, D., Patel, R. S., and Buck, C. 1986. Structure of integrin, a glycoprotein involved in the transmembrane linkage between fibronectin and actin. *Cell* **46**:271–282.

72. Wachsstock, D. H., Wilkins, J. A., and Lin, S. 1987. Specific interaction of vinculin with α-actinin. *Biochem. Biophys. Res. Commun.* **146**:554–560.

73. Wehland, J., Osborn, M., and Weber, K. 1979. Cell-to-substratum contacts in living cells: A direct correlation between interference-reflexion and indirect-immunofluorescence microscopy using antibodies against actin and alpha-actinin. *J. Cell Sci.* **37**:257–273.

74. Werth, D. K., and Pastan, I. 1984. Vinculin phosphorylation in response to calcium and phorbol esters in intact cells. *J. Biol. Chem.* **259**:5264–5270.

75. Werth, D. K., Niedel, J. E., and Pastan, I. 1983. Vinculin, a cytoskeletal substrate for protein kinase C. *J. Biol. Chem.* **258**:11423–11426.

76. Wilkins, J. A., Risinger, M. A., Coffey, E., and Lin, S. 1987. Purification of a vinculin binding protein from smooth muscle. *J. Cell Biol.* **104**:130a.

77. Wilkins, J. A., Risinger, M. A., and Lin, S. 1986. Studies on proteins that co-purify with smooth muscle vinculin: identification of immunologically related species in focal adhesions of nonmuscle and Z-lines of muscle cells. *J. Cell Biol.* **103**:1483–1494.

78. Wilson, H. V. 1907. On some phenomena of coalescence and regeneration in sponges. *J. Exp. Zool.* **5**:245–258.

79. Woods, A., Höök, M., Kjellén, L., Smith, C. G., and Rees, D. A. 1984. Relationship of heparan sulfate proteoglycans to the cytoskeleton and extracellular matrix of cultured fibroblasts. *J. Cell Biol.* **99**:1743–1753.

80. Yamada, K. M., Yamada, S. S., and Pastan, I. 1976. Cell surface protein partially restores morphology adhesiveness and contact inhibition of movement to transformed fibroblasts. *Proc. Natl. Acad. Sci. U.S.A.* **73**:1217–1221.

Questions for Discussion with the Editor

1. *Adhesion plaque formation sounds very complex, as you note in the legend of Fig. 5. How many genes do you suppose are directly involved in assembling and/or disassembling an adhesion plaque?*

It is not surprising that adhesion plaques are complex in their composition and regulation since these elaborations of the plasma membrane perform so many different functions. They serve as points of attachment of the cytoskeleton to the plasma membrane. They link a region of the plasma membrane to an extracellular ligand, and they are probably regions where certain types of signal transduction events occur. However, despite the number and complexity of adhesion plaque functions, we do not envision an endless list of adhesion plaque constituents and controlling elements. The fact that a number of the known components of adhesion plaques have been shown to interact with each other suggests that we are already seeing a large part of the picture.

2. *Do you expect that developing systems as diverse as marine sponges and mammalian embryos will employ cell adhesion mechanisms which are similar at the molecular level?*

Some of the adhesion molecules that are utilized by mammalian cells have been highly conserved through evolution. For example, proteins related to members of the integrin family of extracellular matrix receptors have been identified by immunological criteria in vertebrates, invertebrates, and even fungi (E. E. Marcantonio and R. O. Hynes. 1988. *J. Cell Biol.* **106**:1765–1772). Marine sponges may have a simpler repertoire of cell surface receptors than do mammals; however, it would not be surprising to us if they utilized a subset of the strategies available to more complex organisms.

The remarkable conservation of integrin sequences throughout evolution also raises the possibility that integrin-binding proteins may likewise exhibit a wide phylogenetic distribution. In particular, since the work of Marcantonio and Hynes has shown that the C-terminal cytoplasmic domain of integrin β-chain is both conserved and widely distributed, we suspect that talin, which interacts with this domain on integrin, may likewise be present in many vertebrate and invertebrate species.

CHAPTER **19**

Roles of Cell Adhesion Molecules: Reinnervation of Skeletal Muscle as a Case Study

Joshua R. Sanes and Jonathan Covault

MOST CELLS CONTACT other cells and extracellular matrices as they differentiate. The influences of these contacts on cellular functions are particularly evident during the development and regeneration of the nervous system, and the search for molecules involved in such contact-mediated phenomena is now a major preoccupation of neurobiologists. Central to the field is the assumption that the fundamentally important interactions of cells with other surfaces are adhesive in nature. Accordingly, relevant molecules are sought by assays of cell attachment, and the molecules found are called cell adhesion or substratum adhesion molecules. It is our aim here to consider some biologic roles of these adhesion molecules. However, because the area is vast (in fact, nearly a discipline unto itself), we have taken a very selective approach. We begin by making some general points about adhesion molecules; then we consider how adhesive interactions might figure in one well-studied process, reinnervation of skeletal muscle; and finally we discuss ways to assess the importance of particular adhesion molecules in reinnervation.

Neural Adhesion Molecules

During the last decade, more than a dozen neural cell adhesion molecules have been identified. Table 1 summarizes key features of the better described of these molecules. They are all glycoproteins, and they are all associated with

429

TABLE 1 Neural Cell Adhesion Molecules

Molecule	Mol. wt., × 10³	Expressed by	Key features	Putative functions
NCAM	120, 140, and 180 are most prominent, varies between cells	Many embryonic cell types, adult neurons, and denervated muscle	Homophilic adhesion. May interact with heparan sulfate proteoglycans. Developmentally regulated forms produced by alternate exon usage. Contains variable amounts of an unusual polysialic acid which can modulate adhesion. Present on both neuron cell body and processes. Expression in adult skeletal muscle is regulated by nerve-induced activity	Neuron-neuron, neuron-glia, neurite-muscle and glia-glia cell adhesion. Histogenesis and axon fasciculation
L1/NgCAM/NILE*	135 and 200 are most prominent, varies between cells	Postmitotic neurons. Some Schwann cells	Particularly prevalent in axon tracts. High levels on axons, low on neuron cell bodies. Induced by nerve growth factor in sensory ganglia neurons	Axon fasciculation, neurite outgrowth, and cerebellar granule cell migration
MAG	100	Oligodendroglia, myelinating Schwann cells	Restricted to the periaxonal membrane of myelinating glial cells	Axon-myelin sheath adhesion, maintenance of 12–15 nm periaxonal spacing
PO	29	Myelinating Schwann cells	Transmembrane glycoprotein, makes up 50% of PNS† myelin protein. Expression corresponds with the development of PNS myelin	Adhesion between external leaflets of Schwann cell membrane in compact myelin
N-Cadherin	130	Many embryonic cell types, adult neurons	Calcium-dependent adhesion. Similar or identical to the adherens junction cell adhesion molecule, A-CAM.	Neuron-neuron, and neurite-muscle adhesion. Histogenesis

TABLE 1 Neural Cell Adhesion Molecules (*Continued*)

Molecule	Mol. wt., × 10^3	Expressed by	Key features	Putative functions
Integrin	125–130 and 155 heterodimer	Many cell types including neurons	Family of membrane-bound matrix receptors, including a laminin-collagen receptor	Participates in cell binding to laminin and fibronectin
Fibronectin	Dimer of similar 220 subunits	Serum-hepatocytes. Cellular—many cell types especially fibroblasts and PNS glia	Secreted glycoprotein, functionally distinct forms produced by differential RNA splicing. Multiple functional domains bind to ECM and cell surface receptors	Neural crest cell migration, and axon outgrowth in the PNS
Laminin	900 complex: A, 400; B1, 210; and B2, 200	Many cell types. Fibroblasts, Schwann cells, and CNS glia of frog and fish. Optic tract of other vertebrates during development	Secreted glycoprotein common to all basal laminae. Multiple functional domains for binding other ECM molecules and cell surface receptors	Axon outgrowth in PNS, and some areas of CNS
J1/tenascin/cytotactin*	Major forms of 200 and 220	Variety of cell types, especially glial cells and fibroblasts	Secreted glycoprotein associated with surface of CNS glial cells and some ECM binds to chondroitin sulfate proteoglycan and collagen fibrils. Related to myotendinous antigen and hexabrachion	Neuron-glia adhesion, and cerebellar granule cell migration
AMOG	50	Cerebellar astrocytes and oligodendrocytes	Cell surface glycoprotein	Neuron-glial adhesion, cerebellar granule cell migration
Astrotactin	100	Cerebellar granule cell neurons	Cell surface glycoprotein. Markedly reduced levels on Weaver granule cells	Cerebellar granule cell, and Bergmann glia cell adhesion-recognition
HNK-1/L2*	Carbohydrate epitope	Subset of neurons and glial cells	Sulfated glucuronic acid attached to several adhesion molecules—NCAM, L1/NgCAM, MAG, and J1/cytotactin	Antibodies to the epitope can inhibit neuron-glia adhesion

*Different names given by groups that discovered these molecules independently. †Abbreviations used: CNS, central nervous system; ECM, extracellular matrix; PNS, peripheral nervous system. Note that only some sites of expression are listed.

cell surfaces in some way. In other respects, however, they make up a diverse group. Rather than plow our way through the list, we prefer to enumerate some features common to sets of adhesion molecules. Recent reviews provide further details and ample documentation of the assertions we make here (5, 11, 33, 34).

1. *The term "adhesion molecule" refers to an assay and not necessarily to a biologic function.* Adhesion molecules are given their name because they mediate adhesion of cells to each other, to other cells, or to substrata in short-term cell-binding assays. NCAM, the first neural cell adhesion molecule to be characterized (whence its name) was isolated by an immunological approach: polyspecific antibodies were prepared that inhibited the aggregation of retinal cells, and NCAM was purified on the basis of its ability to neutralize the inhibitory effect of the antibodies. Several other adhesion molecules have subsequently been isolated using this approach. In other cases, molecules isolated using assays unrelated to adhesion have been found to promote adhesion when coated on substrata or to elicit antibodies that block adhesion. In all these cases, the underlying assumption is that important cell-cell or cell-substratum interactions are adhesive in nature, and a growing body of evidence testifies to the validity—or at least to the utility—of this assumption. It does not follow, however, that the biologic roles of adhesion molecules in vivo are limited to adhesion, or even that adhesion per se is of paramount importance. For example, binding of ligands to transmembrane adhesion molecules may lead to rearrangement of cell surface components or to generation of intracellular signals which, in turn, trigger processes of cellular differentiation. In short, while some adhesion molecules may function to glue cells together in appropriate ways, others may, instead or in addition, play more complex and subtle roles in neural development.

2. *Many adhesion molecules are members of gene families.* As cDNAs encoding adhesion molecules are cloned and sequenced, it is becoming apparent that many adhesion molecules can be grouped into families based on conserved features of their primary structure. One family includes NCAM, L1/NgCAM (neuron-glia CAM; actually a misnomer), MAG (myelin-associated glycoprotein), and PO all of which contain similar, repeating domains which are thought to form a series of disulfide-bonded loops (Fig. 1). The amino acid sequences and positions of cysteine residues within these repeats show significant homology to repeating units in immunoglobulins, suggesting that the adhesion molecules are members of an "immunoglobulin superfamily" (46). A second family, the cadherins, includes the neural cell adhesion molecule N-cadherin as well as E-cadherin (also called L-CAM or uvomorulin) and P-cadherin, which are present in various nonneural tissues. These molecules, named after the source from which each was initially isolated (neural, epithelial, and placental tissue, respectively), are expressed in an overlapping, tissue-specific manner and appear to be the principal adhesive components of cellular adherent junctions (42). Interestingly, adhesion mediated by members of these two gene families shows a striking functional difference, being essentially calcium-independent for the immunoglobulin

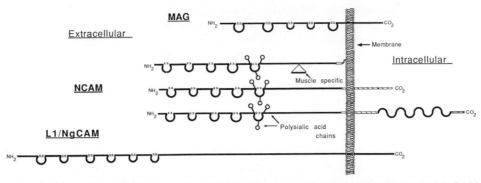

FIGURE 1. Schematic diagram of the primary structure for MAG, NCAM, and L1/NgCAM, members of the immunoglobulin-related adhesion molecule family. All three contain homologous repeating extracellular domains. The four polypeptide forms shown for NCAM result from translation of differently spliced RNAs. Myofiber NCAM mRNA contains an additional 108 nucleotides between exons 12 and 13; this insert is indicated by a triangle.

homologues and strictly calcium-requiring for the cadherins. A third adhesion system consists of a family of receptors for extracellular matrix molecules called integrins or cytoadhesins. Each integrin is a heterodimer of α and β subunits; at least 10 α's and 3 β's have been isolated (30). Finally, the extracellular matrix proteins laminin and collagen, to which cells adhere, are each encoded by a family of homologous genes (17, 27).

3. *Individual adhesion molecule genes can produce several distinct molecular species.* Here again, NCAM provides a good example. First, alternative splicing generates at least four different mRNAs from the NCAM gene (9, 10), each of which encodes a different polypeptide (Fig. 1). The longest is a transmembrane protein with a long cytoplasmic segment that can interact with cytoskeletal elements, the middle one is a transmembrane protein with a shorter cytoplasmic domain, and the shortest one has no transmembrane segment but is anchored to the membrane by a glycolipid-bearing tail. A variant of the shortest one bears an additional 38 amino acids encoded by an exon which appears to be expressed only in muscle. Second, NCAM molecules are variably glycosylated in developmentally regulated and cell type–specific patterns. Individual molecules vary both in their content of sialic acid and in their decoration by an unusual, sulfated-glucuronic acid–containing polysaccharide recognized by the L2 and HNK-1 monoclonal antibodies. Together, these variations in polypeptide and carbohydrate structure generate a bewildering array of NCAM molecules from one gene (33). The functional significance of these variations is just beginning to be explored.

4. *Both peptide and carbohydrate components of adhesion molecules can contribute to their adhesive properties.* Studies on the mechanism of action of adhesion molecules has rightly focused on protein-protein interactions. One triumph of this approach has been the discovery of short amino acid sequences in fibronectin and other extracellular matrix molecules that are recognized by their receptors, the integrins (30). On the other hand, it has long been hypoth-

esized that some adhesive interactions are a result of the recognition of carbohydrate moieties by proteins. The best evidence for such sugar-mediated adhesion currently comes from nonneural systems (e.g., sperm-egg recognition; Ref. 43), but similar mechanisms are likely to operate in the nervous system as well (discussed in Ref. 33). One example is the glucuronic acid–containing moiety L2/HNK-1 mentioned above; there is now reason to believe that this sugar itself is involved in the adhesive process (22). Conversely, carbohydrates can inhibit protein-protein adhesion. For example, NCAM that is highly sialylated (bearing long chains of an unusual α-2,8-polysialyl linkage) binds less well to other NCAM molecules than does NCAM that is desialylated. Furthermore, desialylation of NCAM on living cells can enhance the effectiveness of even non-NCAM-dependent adhesion; presumably the sialic acid shields both NCAM molecules and other, neighboring molecules from contact with partners on other cells (32). The developmental variations in the polysialic acid content of NCAM mentioned above are therefore likely to have important functional consequences.

   5. *Individual cell-cell interactions can involve multiple adhesion molecules.* An early example here was an experiment showing that aggregation of a clonal population of neuroblastoma cells was inhibited better by antibodies to NCAM plus antibodies to L1 than by either antibody alone (29). More recently, several studies have shown that neurite outgrowth in both neuron-glia and neuron-myotube cocultures can involve multiple adhesion molecules (1, 2, 37). For example, in neuron-myotube cocultures, neurite outgrowth was markedly inhibited by a combination of antibodies to NCAM, N-cadherin, and integrin, whereas each antibody individually was only marginally effective (2). This multiplicity of adhesive mechanisms raises the intriguing possibility of a combinational basis for some aspects of specific connectivity in the nervous system. However, it also poses a vexatious technical problem: inhibition of cell adhesion by a specific antibody cannot be taken to mean that the antigen in question is the only one involved and lack of inhibition by an antibody cannot be interpreted as meaning that the antigen in question is not involved.

   6. *The distinction between cell-cell and cell-substratum adhesion molecules is getting blurry.* A few years ago, it was defensible to distinguish between "cell adhesion molecules" and "substratum adhesion molecules," with the substratum taken to be the in vitro homologue of the extracellular matrix (see Chap. 18, this volume). More recent studies, however, have revealed a disquieting ability for individual molecules to straddle the fence. Thus, NCAM and L1, both transmembrane proteins and prototypic cell adhesion molecules, are each found associated with extracellular matrices in some circumstances (26). In the immune system one of the integrins, the family of receptors for extracellular matrix molecules, has now been shown to have an NCAM-related protein as its ligand (38). Thus, while it still makes good biologic sense to distinguish cellular attachment to other cells from adhesion to extracellular matrices, some individual adhesion molecules may participate in both processes.

Adhesive Interactions during Reinnervation of Muscle

Because of its relative simplicity and the wealth of information available about it, the skeletal neuromuscular junction (NMJ) provides a useful system in which to study neurobiological processes involving adhesive interactions. While both embryonic synaptogenesis and reinnervation of adult muscle following nerve injury have been studied in detail, we focus here on the latter.

When the motor nerve supplying a muscle is cut or crushed, axonal segments distal to the injury degenerate and are phagocytized, leaving the muscle denervated. Unlike central axons, whose regenerative capacity is limited (at least in mammals), the motor axons sprout profusely near the point of injury and extend processes that reenter and reinnervate the muscle. The new neuromuscular junctions that form resemble the original ones structurally and functionally. At least five phenomena that occur during reinnervation seem likely to involve adhesion molecules; immunochemical studies have provided clues as to what some of these molecules might be.

Axons Regenerate along Denervated Nerve Trunks

While nerve injury leads to degeneration of the distal segments of axons, other elements of the distal nerve stump survive. Unless they are deliberately displaced, regenerating axons generally grow back to and into the muscle through these surviving nerve trunks. Long ago, Ramón y Cajal recognized that the attraction of nerve trunks for axons included both tropic and adhesive components (28). On the one hand, axons cross gaps of several millimeters to enter the distal stumps of nerve trunks (Fig. 2a), suggesting the existence of long-range signals—presumably soluble, chemotactic molecules—to which axons respond. On the other hand, having reached the nerve trunk, axons continue to grow, and they do so in a very specific pattern: they grow within the tubes of basal lamina (BL) that once ensheathed axon–Schwann cell units (Fig. 2b). More recently, electron microscopy has shown that growth cones of regenerating axons adhere closely to the inner surface of the Schwann cell BL and that axons grow along Schwann cell BL even after the Schwann cells themselves have been killed by freezing (19). In addition, regenerating axons contact Schwann cell surfaces within the tubes and dissociated neurons extend neurites along Schwann cell surfaces in vitro. Thus, adhesive interactions of axons with Schwann cell membranes and BL both seem to be important determinants of axonal regeneration.

What are the molecules that promote neurite extension on the membranes and BL of Schwann cells? The first point to make is that until recently axonal growth was viewed as being guided largely by mechanical factors, with axons growing along paths of least resistance or mechanical inhomogeneities in their substrata. Instrumental in overturning this view (which was called stereotropism, contact guidance, or tactile adhesion) were the experiments of

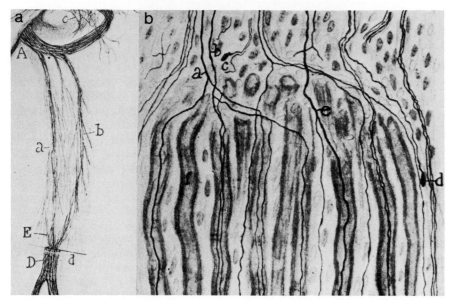

FIGURE 2. Ramón y Cajal's drawings of axons regenerating after transection of the sciatic nerve. (a) Axons (A) cross a wide gap to enter the surviving distal stump (D). (b) Within the distal stump, axons grow preferentially within Schwann cell sheaths. (From Ref. 28.)

Letourneau (25). He cultured neurons on patterned substrata to which cells adhered differentially and found that neurites extended preferentially along the more adhesive stripes. The conclusion he drew—that adhesive interactions promote neurite extension—has been a powerful guide to subsequent work, although, not surprisingly, some exceptions to a strict correlation between adhesive strength and neurite-promoting ability are now coming to light (16).

When tested separately in vitro, many components of basal laminae are capable of binding neurons and/or promoting axonal growth; these components include fibronectin, laminin, heparan sulfate proteoglycans, and collagen. However, most attention has focused on laminin, a major structural glycoprotein component of most BLs. In vitro, laminin is the most potent promoter of neurite outgrowth of the isolated matrix components that have been tested, and laminin has been shown to account for most of the neurite-promoting activity secreted by a variety of nonneural cells, including Schwann cells. Interestingly, laminin frequently appears to be complexed with a heparan sulfate proteoglycan, and even though pure laminin promotes neurite extension well, the neurite-binding site exposed in situ may involve epitopes on both components of the complex.

As for the axonal molecules that recognize laminin, the best candidate at present is a member of the integrin family of receptors for extracellular matrix molecules; antibodies to the integrin β_1 subunit markedly inhibit neurite extension on laminin. However, a variety of nonintegrin membrane-associated laminin binding proteins have been isolated from neural and nonneural cells (reviewed in Ref. 34), and it remains possible that integrin is only part of the story.

For molecules that promote neurite extension along Schwann cell membranes, one looks to the cell adhesion molecules. Immunohistochemical studies of normal and regenerating nerve have shown that both NCAM and L1 are present on regenerating axons and on "denervated" Schwann cells, implicating these two molecules. Indeed, two recent studies (1, 37) on embryonic cells in vitro show that neurite extension on Schwann cell is markedly inhibited by antibodies to L1/NgCAM and to N-cadherin; anti-NCAM has a less pronounced (37) or no (1) effect, while antibodies to integrin inhibit outgrowth only when combined with anti-L1/NgCAM or anti-N-cadherin. Determination of which of these molecules are active during regeneration in vivo awaits use of bioassays of the sort described below.

Axons Reinnervate Original Synaptic Sites

When regenerating axons reach the surface of denervated muscle fibers, they preferentially form new synapses at original synaptic sites (Fig. 3a). While new or "ectopic" synaptic sites are formed under special circumstances (see below), the degree of precision is, in most situations, remarkable. For example, when motor axons reinnervate frog muscles following nerve crush, over 95 percent of the nerve-muscle contacts formed are at original sites, even though these sites occupy less than 0.1 percent of the muscle fiber surface. Delivery of axons to original sites by nerve trunks could account for some of this precision, but it cannot account for it all. First, having left the nerve trunks, the axons retrace their original routes on the muscle fiber surface for hundreds of microns. Second, some reinnervating axons form "sprouts" that extend beyond original synaptic sites; these sprouts, which grow outside of nerve sheaths, frequently reach and innervate additional original sites (Fig. 3a). Thus, it appears that there are factors concentrated at original sites that axons can recognize.

Some information about the nature of these factors was provided by the experiment sketched in Fig. 3b. Each muscle fiber is coated by a layer of BL, and this BL survives mechanical injuries severe enough to induce death and degeneration of the muscle fibers. By denervating and damaging muscles and then using x-ray irradiation to prevent regeneration of new muscle, it was possible to obtain in vivo a preparation of BL "ghosts" on which original synaptic sites were recognizable. When axons regenerated to these ghosts, they reinnervated original synaptic sites as specifically in the absence of muscle fibers as they would have in their presence (35), suggesting that the BL at synaptic sites retains some molecular trace recognized by regenerating axons. Furthermore, contact with BL induced axonal differentiation: Axonal stretches abutting synaptic BL lost features typical of growing axons and ac-

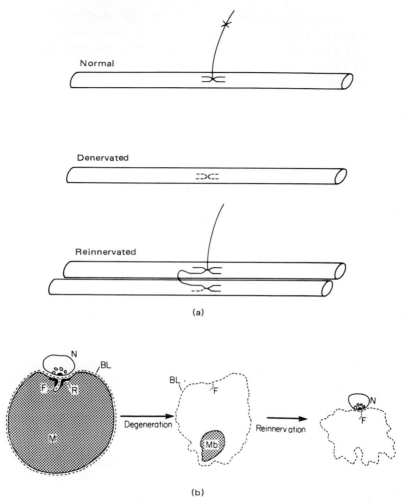

FIGURE 3. (*a*) Axons preferentially reinnervate original synaptic sites on denervated muscle fibers. (*b*) Axons preferentially reinnervate original sites on muscle BL sheaths from which muscle fibers have been removed, indicating that some important cues are associated with BL. N, nerve terminal; M, muscle fiber; F, junctional folds; R, acetylcholine receptors; Mb, myoblastic satellite cells.

quired morphological (Fig. 4*a* and *b*), immunohistochemical, and electrical properties of presynaptic nerve terminals (15). Thus, it appears that adhesive interactions of axons with Schwann cell and synaptic BL promote different behaviors—neurite extension and synapse formation, respectively.

The clear prediction of these experiments is that there is some molecule concentrated in synaptic BL that axons recognize. Immunohistochemical methods have been used to identify candidates: A variety of monoclonal and polyclonal antibodies have been prepared that stain synaptic BL far more intensely than adjacent extrasynaptic BL, thus defining "synaptic" antigens

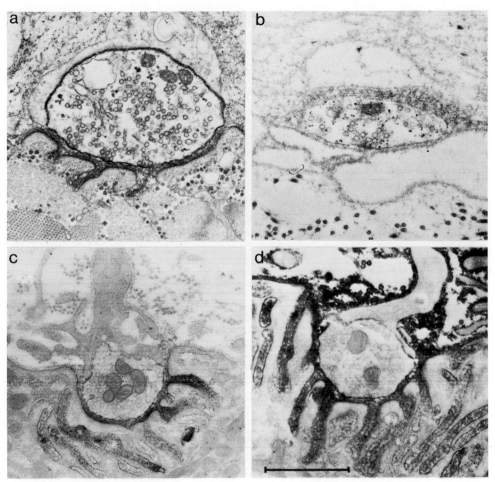

FIGURE 4. Electron micrographs of neuromuscular junctions. (*a*) A normal neuromuscular junction. (*b*) A nerve terminal that is reinnervating muscle fiber BL after the muscle was damaged and denervated. The axon has reconstituted its normal geometry by forming an active zone opposite the strut of BL that once lined a junctional fold. (From Ref. 15.) (*c*) This muscle had been stained with antibodies to s-laminin, a synapse-specific BL antigen. Reaction product is concentrated in the BL of the synaptic cleft. (*d*) This muscle had been stained with anti-laminin; reaction product is present both synaptically and extrasynaptically.

(Fig. 4*c* and *d*). One promising candidate (see Ref. 34 for an enumeration of others) is a 185×10^3 dalton protein called "junction-specific 1" or JS-1, which is recognized by a set of four monoclonal antibodies that all stain synaptic BL selectively (18). Neurons adhere to isolated JS-1 in short-term assays, indicating that it is an adhesive molecule, and molecular cloning has revealed that JS-1 is approximately 40 percent homologous to one of the subunits of laminin (17), already known to be a particularly potent inducer of neurite extension (see above). Because of its homology to laminin, we are renaming the JS-1 antigen s-laminin (17). It is intriguing to speculate that different members of a single gene family might induce axons to grow through nerve trunks (laminin itself) or to stop and differentiate at synaptic sites (s-laminin).

Axons Distinguish Denervated from Innervated Muscle

To induce formation of "ectopic" synapses on previously extrasynaptic regions of muscle fibers, the cut distal stump of a motor nerve can be sutured into a muscle a long distance (many millimeters) away from the original synaptic sites (Fig. 5a). The most interesting feature of this ectopic synaptogenesis is that no new synapses form until the original nerve is damaged (13, 21). This observation demonstrates that muscle fibers control their susceptibility to innervation in accordance with their current state of innervation. There must therefore be some signal that passes from nerve to muscle (so that the muscle "knows" whether or not it is innervated) and a second signal that passes from muscle to axons. There is strong evidence that the first signal is electrical in nature: implanted axons form synapses on innervated but pharmacologically paralyzed muscle fibers but fail to form synapses on denervated muscle fibers that are kept active by direct electric stimulation (20). The second signal, on the other hand, appears to be chemical: Innervated (or active) and denervated (or paralyzed) muscles presumably produce different amounts of some factors

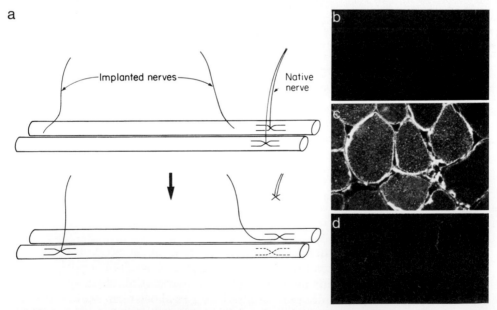

FIGURE 5. (a) Implanted axons form synapses on denervated but not innervated muscle fibers, indicating that muscles control their susceptibility to innervation in accordance with their current state of innervation. Cross-sections of (b) normal, (c) denervated, and (d) reinnervated muscle stained with anti-NCAM.

that axons sense. One such factor might be a secreted growth factor that promotes axonal growth (40), but several observations (3, 21, 39) suggest that the muscle fiber surface also changes upon denervation in ways that axons recognize.

A reasonable starting point for seeking muscle-associated molecules that signal susceptibility or refractoriness to innervation is to ask whether levels of any known adhesive molecules change on denervation. A number of relevant immunohistochemical observations are summarized in Table 2. In fact, levels of one adhesion molecule, NCAM, vary in the predicted manner (7, 8). Embryonic myotubes, which readily accept innervation, are rich in NCAM. As development proceeds and muscle fibers become refractory to innervation, NCAM is lost from extrasynaptic areas and becomes concentrated at and near synapses. Following denervation or paralysis, NCAM reappears at high levels in extrasynaptic areas; following reinnervation, extrasynaptic NCAM is lost (Fig. 5b to d). Denervation of adult muscles also leads to a reappearance of N-cadherin (J. Covault, L. Scott, and J. R. Sanes, unpublished observation). In contrast, several adhesive components of the extracellular matrix (e.g., laminin, collagen IV, fibronectin) are abundant in the BL of both innervated and denervated muscle fibers, while several cell adhesion molecules (e.g., uvomorulin and L1) are absent from extrasynaptic areas both before and after denervation. Levels of the laminin-proteoglycan epitope that neurons recognize (4) do increase following denervation, but as this increase occurs fairly late after denervation, it is unlikely to be of great importance. Thus, NCAM and possibly N-cadherin emerge as candidate mediators of the muscle's susceptibility to innervation. In vitro antibody-blocking experiments are consistent with this idea: Both anti-NCAM and anti-N-cadherin can disrupt interactions of embryonic neurons with myotubes (2, 31).

Axons Find the Original Synaptic Sites They Preferentially Reinnervate

If axons are capable of forming ectopic synapses, how do they find the small sites that they preferentially reinnervate? Since the original sites occupy only approximately 0.1 percent of the muscle fiber surface, axons might not readily find them by an unguided process of random search. Original nerve trunks presumably provide some guidance in this regard by delivering axons to synaptic areas (see above), but even axons growing outside of nerve trunks are able to reject extrasynaptic areas and choose synaptic areas. A second, intramuscular guidance mechanism is, therefore, likely to exist.

Immunohistochemical studies of denervated muscle suggest a basis for this second source (36). Following denervation, deposits of four adhesive molecules accumulate in interstitial areas surrounding the original synaptic sites: J1/tenascin, NCAM, fibronectin, and a heparan sulfate proteoglycan (Table 2). Three of these molecules are also present throughout the muscle fiber surface: NCAM on denervated fibers, and fibronectin and the proteoglycan on both innervated and denervated fibers. However, for all three molecules,

TABLE 2 Distribution of Potential Adhesion-Recognition Molecules in Adult Skeletal Muscle

Molecule	Normal muscle			Denervated muscle		
	Synaptic sites	Extrasynaptic surface	Perisynaptic interstitial	Synaptic sites	Extrasynaptic surface	Perisynaptic interstitial
NCAM	+	−	−	+	+	+
N-Cadherin	+	−	−	±	+	−
L1/NgCAM	+*	−	−	−	−	−
E-Cadherin/uvomorulin	−	−	−	−	−	−
J1/tenascin	−	−	−	−	−	+
Fibronectin	+	+	±	+	+	+
Heparan sulfate proteoglycan	+	+	−	+	+	+
Laminin	+	+	−	+	+	−
INO antigen	±	±	−	+	+	−
s-Laminin	+	−	−	+	−	−
C21 laminin antigen	−	+	−	−	+	−

*Terminal Schwann cells

442

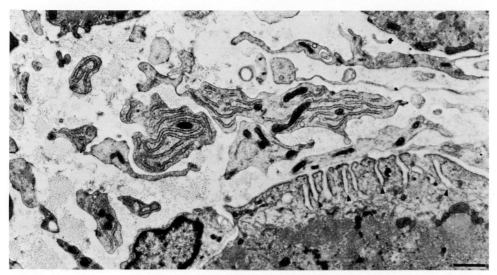

FIGURE 6. Electron micrograph of a denervated synaptic site (identifiable by junctional folds) showing the fibroblasts that accumulate perisynaptically. (From Ref. 14.)

the interstitial deposits induced by denervation are concentrated in perisynaptic areas. For J1/tenascin, the distribution is more striking: J1/tenascin is virtually undetectable in normal muscle, and in denervated muscles it is almost entirely confined to—and thereby serves as an excellent marker of—perisynaptic interstitial areas.

Recent studies implicate fibroblasts as the cellular source of the perisynaptic deposits of all four adhesive molecules. Fibroblasts are numerous in normal muscle, and it is known that these cells divide following denervation. It turns out that these divisions are more numerous near to than far from synaptic sites, leading to a perisynaptic accumulation of fibroblasts (Fig. 6) that is correlated spatially and temporally with the accumulation of these adhesion molecules. Furthermore, fibroblasts cultured from perisynaptic areas of denervated muscle synthesize all four molecular species in culture. Further studies in culture suggest that the perisynaptic fibroblasts are not unique in their ability to synthesize these molecules, but rather that they are ordinary fibroblasts responding to a localized signal or environment. Thus, it seems that denervation leads to a change in the perisynaptic milieu which stimulates fibroblasts to divide and which, in turn, leads to synthesis and secretion of several adhesion molecules (14). As regenerating axons traverse these spaces, they are likely to encounter the fibroblasts, and could be influenced by the adhesive (or other) molecules they provide.

Axons Distinguish among Muscles

Groups or "pools" of spinal motor neurons project to specific muscles. Yet, both "native" (or appropriate) and "foreign" (or inappropriate) axons can (1) form synapses on denervated but not innervated muscle fibers and (2) preferentially reinnervate original synaptic sites. Thus, the cues described so far are not "specific" in the classical sense, i.e., they do not promote distinctions between targets. In lower vertebrates, clear evidence has been obtained that given a choice, axons *prefer* to reinnervate appropriate muscles and appropriate types of muscle fibers (fast or slow) within muscles (12, 44). Until recently, attempts to demonstrate selective reinnervation of adult mammalian muscle have been unsuccessful: Native and foreign axons appeared to be comparably successful in reinnervating all muscles when matters of access were taken into account. Recently, however, a weak but consistent selectivity has been documented in electrophysiological experiments: It turns out that axons derived from particular rostrocaudal positions in the spinal cord preferentially reinnervate muscles from matched levels in the rostrocaudal axis (24, 45). Thus, there must be muscle-associated cues, possibly graded, that vary among muscles and that are recognized by axons. Are these cues adhesive molecules? No one knows, but it seems likely. Important challenges for the future will be to identify candidate mediators of these more selective interactions and to determine how they interact with the more widely distributed cues described above.

Summary

Several of the immunohistochemical results described above are sketched in Fig. 7. This summary suggests the idea that regenerating axons encounter increasing numbers of attractive molecules with decreasing distance from the original synaptic sites. Thus, denervated muscles may secrete soluble growth

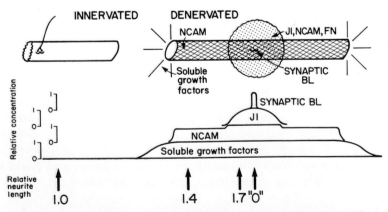

FIGURE 7. Schematic distribution of molecules that muscle might use to influence regenerating axons: (*Top and middle*) Summary of results described in the text and illustrated in Figs. 4 to 6. (*Bottom*) Relative neurite lengths summarize results from cryoculture experiments described in the text and in Fig. 8. (From Ref. 6.)

factors that axons sense from a distance; denervated fibers within a muscle are rich in NCAM; perisynaptic areas bear interstitial deposits of several adhesion molecules; and the BL at original synaptic sites contains molecules that provide the most highly localized—and perhaps the most attractive— cues. In addition, although not depicted, nerve trunks provide additional guidance that directs axons toward the most attractive area. Finally, as yet unidentified differences among muscles allow axons to choose appropriate synaptic partners if the axons are given a choice.

Thus, even as simple a phenomenon as the reinnervation of skeletal muscle may involve a bewildering array of target-derived cues and a comparable number of axon-associated sensors. It has been difficult enough to use immunohistochemical methods and in vitro cell adhesion assays to identify candidate mediators of these interactions. Will it be possible to learn how (or if) they fit together to account for the precise and orderly processes of reinnervation? The next sections consider a possible strategy aimed at this goal.

A Functional Assay for Adhesion Molecules in Adult Muscle

The binding of monovalent antibodies to their antigens can produce steric inhibition of adhesion molecule interactions, thus providing a means to assay their bioactivity. Indeed, immunological blockade in simple in vitro adhesion systems formed the basis for the initial identification of NCAM, NgCAM, N-cadherin, cytotactin, and integrin. Unfortunately, in vitro cocultures of dissociated muscle and nerves do not reproduce the topographical distribution of adhesion molecules with respect to end plates seen in denervated muscle in vivo (Fig. 7). Injection of antibodies into intact animals provides a potentially more direct test (see, for example, Ref. 23), but such experiments present great technical problems of execution as well as numerous uncontrolled variables that complicate the interpretation of results. To bridge this gap between studies of axon regeneration in culture versus in intact animals, we have recently developed a technique which exploits the advantages of in vitro methods to examine axon guidance by organized tissue elements (6).

Our method is based on the work of Stamper and Woodruff, who showed that lymphocytes bind selectively to venules in cryostat sections of lymph nodes (41). Using this binding as a short-term bioassay, they and others isolated cell surface molecules that mediate adhesion of lymphocytes to endothelial cells (47). We adapted the approach by using neurons instead of lymphocytes, monitoring outgrowth rather than attachment, and studying events occurring over hours and days instead of minutes. Dissociated neurons are plated on 5 to 10 μm thick cryostat sections of adult or embryonic tissues in appropriate media and cultured for 1 to 2 days to allow neurite regeneration. The "cryocultures" are then fixed and doubly labeled immunohistochemically to visualize neuronal processes and components of the cryostat section.

In most of our experiments we have used chick ciliary ganglion neurons, an easily isolated group of cholinergic motor neurons, although other types of neurons can also be used. Neurons extend processes for 2 to 3 days and readily survive for up to a week on sections of both innervated and denervated muscle as well as on several other tissues. In each case neurites grow preferentially along sectioned cell surfaces; several results of these studies (detailed in Ref. 6) argued that this outgrowth is guided by specific interactions of neurites with cell surface molecules rather than by mechanical inhomogeneities in the cryostat section.

Analysis of neurite outgrowth on sections of innervated and denervated muscle provided striking parallels with the biology of muscle reinnervation in vivo as well as with the distribution of adhesion-recognition molecules. First, neurites were, on average, longer, broader, and more highly branched on sections of denervated muscle than on sections of innervated muscle (Fig. 8a). Since large pools of soluble factors, which along with adhesion molecules may promote axon regeneration in vivo, are unlikely to be maintained in such thin tissue sections, this result provided direct evidence that differences between innervated and denervated muscle fiber *surfaces* are, in fact, recognized by growing axons. Second, neurite outgrowth was greater on sections cut from end plate–rich areas of denervated muscle than those cut from end plate–free areas. Thus, cell surface or matrix-bound factors enriched in perisynaptic areas can provide additional stimulation for axon growth in these regions and may, thereby, increase the likelihood that regenerating axons will contact original synaptic sites. Finally, neurites displayed a specific response—cessation of growth—upon contacting end plates on cryostat sections. End plates can be easily recognized in sections by their selective staining with rhodamine-α-bungarotoxin. Despite the relative scarcity of end plates (< 1 percent of the muscle fiber surface even in sections from end plate–rich areas), we observed 30 examples of neurite–end plate contact in a series of eight cryocultures. In 80 percent of these instances the regenerating neurite terminated precisely at the former synaptic site (Fig. 8b). Apparently, as occurs in vivo, regenerating neurites also specifically recognize components of synaptic sites in cryostat sections in vitro.

Neurite outgrowth on cryostat sections of muscle thus reveals differences in surface-bound neurite-outgrowth-regulating activities between innervated and denervated muscle, in synapse-rich and synapse-poor regions of denervated muscle, and in synaptic and extrasynaptic portions of muscle fiber surfaces. Figure 7 combines these data with the immunohistochemical distribution of adhesion-recognition molecules discussed above to show that regional variations in neurite outgrowth parallel specific features of the molecular topography of denervated muscle. Together, these results support the idea that denervated muscles use cell surface and extracellular matrix molecules to influence the behavior of regenerating axons. In addition, the cryoculture system provides an assay for immunologic tests of specific roles that adhesion molecules present in denervated muscle may play in guiding

Innervated

Denervated

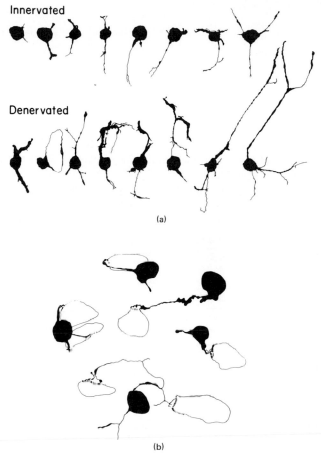

(a)

(b)

FIGURE 8. (a) Neurons plated on cryostat sections of muscle grow longer neurites on sections of denervated muscle than on sections of innervated muscle. (From Ref. 6.) (b) Neurites growing on cryostat sections stop when they encounter a synaptic site. (Synaptic sites are shown as stippled areas on muscle fiber cross-sections.)

reinnervation. A challenge for the future will be to use this assay, and others, to relate the adhesive properties of individual molecular species to the phenomena that occur as neurons differentiate, extend axons, and form synapses.

General References

RUTISHAUSER, V. and JESSELL, T. M. 1988. Cell adhesion molecules in vertebrate neural development. *Physiol. Rev.* **68**:819–857.

SANES, J. R. 1989. Extracellular matrix molecules that influence neural development. *Annu. Rev. Neurosci.* **12**:491–516.

References

1. Bixby, J. L., Lilien, J., and Reichardt, L. F. 1988. Identification of the major proteins that promote neuronal process outgrowth on Schwann cells *in vitro*. *J. Cell Biol.* **107**:353–361.

2. Bixby, J. L., Pratt, R. S., Lilien, J., and Reichardt, L. F. 1987. Neurite outgrowth on muscle cell surfaces involves extracellular matrix receptors as well as Ca-dependent and -independent cell adhesion molecules. *Proc. Natl. Acad. Sci. U.S.A.* **84**:2555–2559.

3. Brown, M. C., Holland, R. L., Hopkins, W. G., and Keynes, R. J. 1980. An assessment of the spread of the signal for terminal sprouting within and between muscles. *Brain Res.* **210**:145–151.

4. Chiu, A. Y., Matthew, W. D., and Patterson, P. H. 1986. A monoclonal antibody that blocks the activity of a neurite regeneration-promoting factor: studies on the binding site and its localization *in vivo*. *J. Cell Biol.* **102**:1383–1398.

5. Covault, J. 1989. Molecular biology of cell adhesion in neural development. In: *Molecular Neurobiology*, (D. M. Glover, and B. D. Hames, eds.), (in press).

6. Covault, J., Cunningham, J. M., and Sanes, J. R. 1987. Neurite outgrowth on cryostat sections of innervated and denervated skeletal muscle. *J. Cell Biol.* **105**:2479–2488.

7. Covault, J., and Sanes, J. R. 1986. Distribution of N-CAM in synaptic and extrasynaptic proteins of developing and adult skeletal muscle. *J. Cell Biol.* **102**:716–730.

8. Covault, J., and Sanes, J. R. 1985. Neural cell adhesion molecule (N-CAM) accumulates in denervated and paralyzed skeletal muscles. *Proc. Natl. Acad. Sci. U.S.A.* **82**:4544–4548.

9. Cunningham, B. A., Hemperly, J. J., Murray, B. A., Prediger, E. A., Brackenbury, R., and Edelman, G. M. 1987. Neural cell adhesion molecule: structure, immunoglobulin-like domains, cell surface modulation, and alternative RNA splicing. *Science* **236**:799–806.

10. Dickson, G., Gower, H. J., Barton, C. H., Prentice, H. M., Elson, V. L., Moore, S. E., Cox, R. D., Quinn, C., Putt, W., and Walsh, F. S. 1987. Human muscle neural cell adhesion molecule (N-CAM): identification of a muscle-specific sequence in the extracellular domain. *Cell* **50**:1119–1130.

11. Edelman, G. M. 1988. Morphoregulatory molecules. *Biochem.* **27**:3533–3539.

12. Elizalde, A., Huerta, M., and Stefani, E. 1983. Selective reinnervation of twitch and tonic muscle fibers of the frog. *J. Physiol. (Lond.)* **340**:513–524.

13. Elsberg, C. A. 1917. Experiments on motor nerve regeneration and the direct neurotization of paralyzed muscles by their own and by foreign nerves. *Science* **45**:318–320.

14. Gatchalian, C. L., Schachner, M., and Sanes, J. R. 1989. Fibroblasts that proliferate near denervated synaptic sites in skeletal muscle synthesize the adhesive molecules tenascin (Jl), N-CAM, fibronectin, and a heparan sulfate proteoglycan. *J. Cell Biol.* **108**:1873–1890.

15. Glicksman, M., and Sanes, J. R. 1983. Development of motor nerve terminals formed in the absence of muscle fibers. *J. Neurocytol.* **12**:661–671.

16. Gundersen, R. W. 1987. Response of sensory neurites and growth cones to patterned substrata of laminin and fibronectin *in vitro*. *Dev. Biol.* **121**:423–431.

17. Hunter, D. D., Shah, V., Merlie, J. P., and Sanes, J. R. 1989. A laminin-like adhesive protein concentrated in the synaptic cleft of the neuromuscular junction. *Nature* **338**:229–234.

18. Hunter, D. D., Sanes, J. R., and Chiu, A. Y. 1987. An antigen concentrated in the basal lamina of the neuromuscular junction. *Soc. Neurosci. Abstr.* **13**:375.

19. Ide, C., Tohyama, K., Yokota, R., Nitatori, T., and Onodera, S. 1983. Schwann cell basal lamina and nerve regeneration. *Brain Res.* **288**:61–75.

20. Jansen, J. K. S., Lomo, T., Nicolaysen, K., and Westgaard, R. H. 1973. Hyperinnervation of skeletal muscle fibers: dependence on muscle activity. *Science* **181**:559–561.

21. Korneliussen, H., and Sommerschild, H. 1976. Ultrastructure of the new neuromuscular junctions formed during reinnervation of rat soleus muscle by a "foreign" nerve. *Cell Tissue Res.* **167**:439–452.

22. Kunemund, V., Jungalwala, F. B., Fischer, G., Chou, D. K. H., Keilhauer, G., and Schachner, M. 1988. The L2/HNK-1 carbohydrate of neural cell adhesion molecules is involved in cell interactions. *J. Cell Biol.* **106**:213–223.

23. Landmesser, L., Dahm, L., Schultz, K., and Rutishauser, U. 1988. Distinct roles for adhesion molecules during innervation of embryonic chick muscles. *Dev. Biol.* **130**:645–670.

24. Laskowski, M. B., and Sanes, J. R. 1988. Topographically selective reinnervation of adult mammalian muscles. *J. Neurosci.* **8**:3094–3099.

25. Letourneau, P. C. 1975. Cell-to-substratum adhesion and guidance of axonal elongation. *Dev. Biol.* **44**:92–101.

26. Martini, R., and Schachner, M. 1986. Immunoelectron microscopic localization of neural cell adhesion molecules (L1, NCAM, and MAG) and their shared carbohydrate epitope and myelin basic protein in developing sciatic nerve. *J. Cell Biol.* **103**:2439–2448.

27. Mayne, R., and Burgeson, R. E. 1987. *Structure and Function of Collagen Types.* Academic Press, New York.

28. Ramón y Cajal, S. Reprinted 1968. *Degeneration and Regeneration of the Nervous System.* Hafner Publishing, London.

29. Rathjen, F. G., and Rutishauser, U. 1984. Comparison of two cell surface molecules involved in neural cell adhesion. *EMBO J.* **3**:1–465.

30. Rouslahti, E., and Pierschbacher, M. D. 1987. New perspectives in cell adhesion: RGD and integrins. *Science* **238**:491–497.

31. Rutishauser, R., Grumet, M., and Edelman, G. M. 1983. Neural cell adhesion molecule mediates initial interactions between spinal cord and muscle cells in culture. *J. Cell Biol.* **97**:145–152.

32. Rutishauser, U., Acheson, A., Hall, A. K., Mann, D. M., and Sunshine, J. 1988. The neural cell adhesion molecule (NCAM) as a regulator of cell-cell interactions. *Science* **240**:145–152.

33. Rutishauser, U., and Jessell, T. M. 1988. Cell adhesion molecules in vertebrate neural development. *Physiol. Rev.* **68**:819–857.

34. Sanes, J. R. 1989. Extracellular matrix molecules that influence neural development. *Annu. Rev. Neurosci.* **12**:491–516.

35. Sanes, J. R., Marshall, L. M., and McMahan, U. J. 1978. Reinnervation of muscle fiber basal lamina after removal of myofibers. Differentiation of regenerating axons at original synaptic sites. *J. Cell Biol.* **78**:176–198.

36. Sanes, J. R., Schachner, M., and Covault, J. 1986. Expression of several adhesive macromolecules (N-CAM, L1, J1, NILE, uvomorulin, laminin, fibronectin and a heparan sulfate proteoglycan) in embryonic, adult, and denervated adult skeletal muscle. *J. Cell Biol.* **102**:420–431.

37. Seilheimer, B., and Schachner, M. 1988. Studies of adhesion molecules mediating interactions between cells of peripheral nervous system indicate a major role of L1 in mediating sensory neuron growth on Schwann cells in culture. *J. Cell Biol.* **107**: 341–351.

38. Simmons, D., Makgoba, M. W., and Seed, B. 1988. ICAM, an adhesion ligand of LFA-1, is homologous to the neural cell adhesion molecule NCAM. *Nature (Lond.)* **331**:624–627.

39. Slack, J. R., and Pockett, S. 1981. Terminal sprouting of motoneurons is a local response to a local stimulus. *Brain Res.* **217**:368–374.

40. Slack, J. R., Hopkins, W. G., and Pockett, S. 1983. Evidence for a motor nerve growth factor. *Muscle & Nerve* **6**:243–252.

41. Stamper, H. B., and Woodruff, J. J. 1976. Lymphocyte homing into lymph nodes: *in vitro* demonstration of the selective affinity of recirculating lymphocytes for high-endothelial venules. *J. Exp. Med.* **144**:828–833.

42. Takeichi, M. 1988. The cadherins: cell-cell adhesion molecules controlling animal morphogenesis. *Development (Camb.)* **102**:639–655.

43. Wasserman, P. M. 1987. The biology and chemistry of fertilization. *Science* **235**: 553–560.

44. Wigston, D. J. 1986. Selective innervation of transplanted limb muscles by regenerating motor axons in the axolotl. *J. Neurosci.* **6**:2757–2763.

45. Wigston, D. J., and Sanes, J. R. 1985. Selective reinnervation of rat intercostal muscles transplanted from different segmental levels to a common site. *J. Neurosci.* **5**:1208–1221.

46. Williams, A. F., and Barclay, A. N. 1988. The immunoglobulin superfamily-domains for cell surface recognition. *Annu. Rev. Immunol.* **6**:381–405.

47. Woodruff, J. J., Clarke, L. M., and Chin, Y. H. 1987. Specific cell-adhesion mechanisms determining migration pathways of recirculating lymphocytes. *Annu. Rev. Immunol.* **5**:201–222.

Questions for Discussion with the Editor

1. *How varied and complex do you expect the cell adhesion molecule story to become over the next decade? Will the development of most tissues and organs (e.g., liver, spleen, ear, eye, etc.) employ their own unique cell adhesion molecule, or will combinatorial usage patterns prevail?*

Opinion on this critical issue is now divided. Some workers feel that combinations of a few dozen molecules could easily "code" many tissues as unique. Variations in the distribution, density, glycosylation, and so on of a few gene products can also clearly generate diversity. Others feel that studies to date have been biased in favor of finding "major" adhesion molecules, and that there is no telling how many low-abundance, cell type–specific markers remain to be discovered. We remain dogmatically "agnostic" on this point.

2. *Your first point, that the term "adhesion molecule" refers to an assay and not to a "biologic function", calls out for attention! Please sketch out—in prose—a spectrum which might illustrate the range of function you expect adhesion molecules to exhibit.*

The point here is that students of intercellular adhesion and students of intracellular mediators have, so far, worked separately. This is not unexpected—until recently, adhesion molecules were not sufficiently characterized to profitably consider the consequences of their occupancy. Now, it is becoming possible to ask what occurs inside a cell when it adheres to something and whether adhesion via different molecules has different effects. In terms of mediators, people are looking at cytoskeletal rearrangements, second messengers, gene activation, and so on. In terms of function, the question is how adhesion can lead to different effects—differentiation, shape changes, neurite outgrowth, myelination, and so on—in different circumstances.

CHAPTER 20

Gap Junctions: Formation and Role in Embryogenesis

Ida Chow

Introduction

CELL-CELL INTERACTIONS play an essential role during embryonic development since they provide the means for cell recognition, migration, and aggregation. Cellular interaction may occur via secreted substances or via direct contact between cell surface molecules. In addition, neighboring embryonic cells may establish direct communication between their cytoplasms via low electrical resistance pathways found in the cell membrane, the gap junctions. The presence of these intercellular channels can be detected by the passage of injected electric current, fluorescent dyes, or radioactive molecules from one cell into neighboring cells. Gap junctions are abundant during particular stages of development and disappear progressively as differentiation takes place within the embryo. A possible function for their timely presence is to provide a means of direct transfer between adjacent cells of diffusible cytoplasmic molecules that have molecular weights (M_r) of up to 1000. Such molecules might be morphogens, nucleotides, instructive or regulatory molecules, or other, as yet to be identified, modulatory signals critical in pattern formation and normal embryogenesis (5, 9, 24, 26, 46, 47, 59, 60, 66).

Although such an attractive hypothesis of diffusion of signals and morphogens through gap junctions during development was proposed many years ago (15, 65), and this view is generally accepted, only recently has it been possible to demonstrate more directly the role of these intercellular channels during embryogenesis. In addition, little is known about the process of formation of cell-cell coupling. Questions dealing with the number of junctional channels that open at the onset of coupling and the regulators of this process during the course of normal development are currently topics of intensive investigation.

Morphology of Gap Junctions

Gap junctions are regions of membrane in close apposition (2 to 4 nm) between adjacent cells that form pathways through which ions and low molecular weight molecules diffuse from one cell to another. They are abundant during embryogenesis and, in most tissues, disappear with development. They are found in a variety of mature vertebrate tissues, such as liver, lens, epithelial, neural and connective tissues, and smooth and cardiac muscles. They are also present in invertebrates where their morphology may differ from that found in the vertebrates. Detailed description of the gap junction structure can be found in the recent review by Zampighi (69).

Thin section transverse electron microscope (EM) views of tissue fixed in glutaraldehyde-osmium tetroxide solution, followed with uranyl acetate–block staining show that the junctions have a heptalamellar structure, with a central narrow gap that is 2 to 3 nm wide (arrows, Fig. 1A; from Ref. 69). Transverse views of gap junctions fixed in glutaraldehyde–tannic acid solutions show the frequently seen 15 to 16 nm wide pentalamellar structure with rows of 4 to 5 nm diameter densities spaced about 8 nm apart in the central band (arrows, Fig. 1B; from Ref. 70). EM views of isolated, negatively stained gap junctions show that they are formed by plaques of channels in hexagonal arrays with 8 to 9 nm center-to-center spacings. Each channel is a hollow hydrophilic cylinder about 6 nm in diameter with a 1 to 2 nm inner pore (Fig. 1C).

Freeze fracture studies show that the protoplasmic fracture face of the gap junction contains clusters of intramembrane particles, 6 to 7 nm in diameter and spaced 8 to 9 nm apart, and the external fracture face contains complementary pits. These particles presumably represent the junctional channels, and they form clusters with or without crystalline order. Although some investigators have suggested that the crystalline arrangement in hexagonal arrays represent uncoupled junctions and the disordered clusters are the replica of coupled junctions (see Ref. 45), other studies have found that the functional state of the junctions and the aggregation pattern of the particles are unrelated (27, 41).

Detection of Intercellular Communication

The presence of intercellular communication via electric signals was first demonstrated by Furshpan and Potter in the crayfish nervous system (23). The first observations of the passage of fluorescent dyes between salivary gland cells (32, 33) and between neurons (55) further confirmed this direct cytoplasmic communication between cells. The presumptive structure responsi-

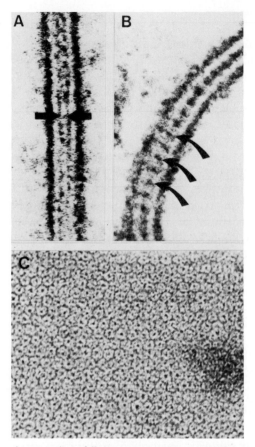

FIGURE 1. Electron micrographs of liver gap junctions. (*A*) Thin-section transverse view from liver stained in block with uranyl acetate showing the heptalamellar structure. Arrows point to the central 2 to 3 nm wide gap (×650,000). (*B*) Thin-section transverse view from liver fixed in glutaraldehyde–tannic acid solutions showing the pentalamellar structure with rows of densities 4 to 5 nm in diameter, about 8 nm apart, in the central band (arrows) (×650,000). (*C*) Isolated gap junction negatively stained with uranyl acetate to show the hexagonal lattice arrangement of the channels spaced 8 to 9 nm apart (×470,000).

ble for this coupling, the gap junction, was first described by Robertson; it was also reported to exist between giant axons in the crayfish (49) and in more detail in the goldfish (50).

In live preparations—usually cell cultures, freshly dissociated cells, or dissected tissues—detection of electrical or dye coupling is an indication of the existence of low-resistance pathways between the cells, presumably the gap junction (see Ref. 53). The absence of measured coupling, however, does not absolutely rule out the presence of these cell-to-cell passageways, since each of the techniques has certain limitations that are described below.

Passage of Fluorescent Dyes

A typical procedure is to inject a fluorescent dye into one cell through a micropipette by applying pressure or voltage and to observe simultaneously its direct passage to many other neighboring cells. The temporal correlation of dye transfer provides an estimate of the diffusion rate across the gap junctions. Lucifer Yellow (M_r of 457) and carboxyfluorescein (M_r of 376) are two popular dyes used today, and the latter has also been used for experiments using the photobleaching recovery method (58). However, one has to be cautious in interpreting the absence of detectable dye passage as lack of cell coupling because of an inherent limitation of this technique: Low sensitivity for detection of small amounts of the fluorescent dye that diffuse from one cell to the neighboring ones under conditions of low coupling. In order to increase the sensitivity of this procedure, an image-intensifying procedure combined with computer-aided analysis is widely used at present (see Ref. 51). Control experiments are always necessary to rule out the possibility of dye entry from bath solution rather than through intercellular channels.

Passage of Electric Signals

INTRACELLULAR RECORDING

Electrophysiological methods using intracellular recordings are more sensitive for detecting low levels of coupling than those using fluorescent dyes, but usually data are obtained from only two cells rather than the many cells examined with dye techniques. An equivalent circuit for the passage of electric currents from one cell to the next contacting one is illustrated in Fig. 2. The principle is based on the fact that the current injected (i_1) into one cell will flow into the bath through two paths: across the cell 1 membrane and across the gap junctions into cell 2. This gap junction current will then flow out across the membrane of cell 2, producing a change in membrane potential (V_2). As a first approximation, the ratio between the voltage changes in the receiving cell 2 and that of the injecting cell 1 (V_2/V_1) will provide an indica-

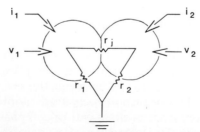

FIGURE 2. Measurement of electrical coupling with double intracellular microelectrodes in each cell. See details in Ref. 4.

tion of the extent of coupling between the two cells. One needs to keep in mind, however, that this "coupling ratio" does not give the exact quantitative measurement of the extent of coupling. It can only be used to obtain an estimate of the degree of coupling between two contacting cells or between two cells under different experimental conditions. The ratio of the voltage in the opposite cell to the applied current is called "transfer resistance" (r_{12} and r_{21}) and is equal to

$$r_{12} = V_2/i_1 = r_1 r_2/(r_1 + r_j + r_2)$$

$$r_{21} = V_1/i_2 = r_1 r_2/(r_1 + r_j + r_2)$$

where r_j = junctional resistance
    r_1 and r_2 = nonjunctional cell membrane resistances
    i_1 and i_2 = injected currents
    V_1 and V_2 = cell membrane potentials

Bennett (4) has developed a procedure to measure more accurately the extent of coupling. The procedure involves alternating the passage of current from cell 1 and cell 2, and it takes into consideration the other electrical parameters such as the membrane capacitance and transfer input impedance. Results are expressed as junctional conductance or its inverse-junctional resistance r_j.

Since the resting membrane potentials of the cells may not remain constant, the total membrane impedance may change, thus, also changing the amount of current that passes into the neighboring cell. In order to avoid this possibility the impaled cells can be held under voltage clamp conditions, i.e., the cell membrane potential is fixed at a constant value. This procedure is more amenable for large cells since good voltage clamp is usually achieved with the use of two microelectrodes in a cell. For smaller diameter cells the two-microelectrode technique may become a problem. Fortunately, with the development of the gigaohm seal patch-clamp technique (29), the study of electrical coupling between small cells (less than 30 μm diameter) can be more accurately done. In addition, because of the high resolution of the patch technique, events at the single gap junction channel level can be detected and analyzed.

Patch recording

Figure 3 shows the schematic representation of the equivalent circuit of both patch-clamp (right cell) and intracellular microelectrode (left cell) techniques used in our coupling formation studies (13). The patch pipette is in the whole-cell clamp mode, i.e., the interior of the pipette is open directly into the cell cytoplasm and holding the cell at its resting potential (-80 to -100 mV). This combination of recording techniques proved to be highly advantageous for studies of gap junction formation because of the amenability of the microelectrode, which allows the investigator to produce cell-cell contact at will. Although the ideal situation is to use patch pipettes on both cells, this is

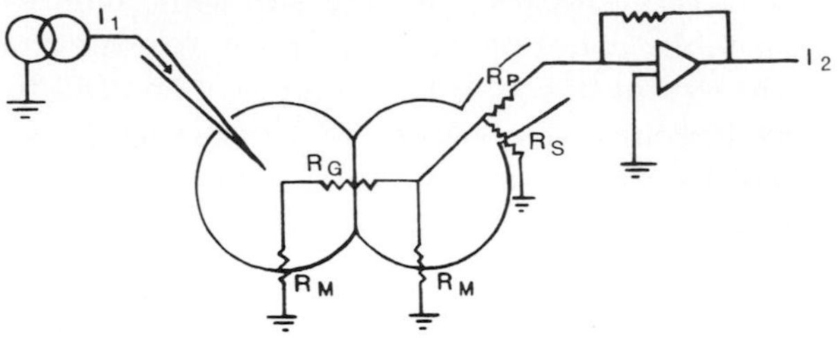

FIGURE 3. Diagram showing the recording system using both intracellular microelectrode (left cell) and patch clamp in whole-cell mode (right cell). Current pulses (I_1) are applied into cell 1 through the microelectrode; a fraction of this current passes through gap junction channels into cell 2 and is recorded by the patch pipette (I_2). Abbreviations used are R_g = gap junction resistance; R_m = nonjunction cell membrane resistance (75 MΩ); R_p = patch pipette access resistance (5 MΩ); R_s = seal resistance (10 GΩ); I_1 = 500 pA; I_2 = 3.75 pA. R_g = 500 pA/3.75 pA $\times$ 75 MΩ = 10,000 MΩ and the junctional conductance = $1/R_g$ = 100 pS.

practically very difficult because it is almost impossible to make a whole-cell patch on a cell and lift it off the substratum without excising the membrane off the cell. This translocation of the cell is significantly more manageable with an intracellular microelectrode impaled into the cell. Current pulses can then be injected into the cell through the microelectrode after the cells are manipulated into contact.

The high resolution of current measurement in the patched cell is provided by the very high resistance of the seal between the pipette and the cell membrane (R_s), which is two to three orders of magnitude higher than that of the pipette (R_p). R_p, in turn, is 10 times smaller than the cell membrane resistance (R_m). Therefore, the values of the current that flows through the gap junction and that actually gets recorded by the patch pipette can be considered nearly equivalent. This means that current fluctuations recorded with the patch pipette during injection of constant amplitude current pulses from the intracellular microelectrode will indicate the extent of the opening of gap junctional channels between the two cells.

It is important to note here that current pulses be used instead of a continuous dc current application in order to ascertain that the current pulses recorded are synchronous with those injected. This will happen only when the gap junction channels are the direct means of communication between the cells. If the recorded pulses are not simultaneous with the injected pulses, they may not have come through the intercellular channels but through some cell membrane sources.

Presence of Gap Junctions during Development

The presence of gap junctions as determined by electrical or dye coupling has been found between blastomeres during the very early stages of development.

For example, widespread electrical coupling between yolk cells and cells of the squid embryo was found at stage 10 (1 day) but had decreased significantly at stage 24, 4 days before hatching (47). The cell-to-cell passage of fluorescent dyes was detected in the 8-cell-stage mouse embryo but not before, and it may bear temporal relationship with compaction of the embryo (25, 35, 40).

Coupling persists within particular regions of the embryo throughout the process of differentiation and may disappear after morphogenesis is completed. For example, in the 32-cell-stage embryo of the South African clawed frog *Xenopus laevis*, the dorsal cells have a higher tendency to transfer Lucifer Yellow (M_r 457) than the cells in the ventral regions (28). At later stages, mesodermal cells are found to be electrically coupled. These cells uncouple during segmentation, and after this process is finished, coupling is detectable again between myotomal cells both within a segment as well as between adjacent segments (6). Segmentation of the embryo proceeds craniocaudally, followed by innervation of the myotomes in the same direction. During these stages the *Xenopus* larva becomes responsive to external stimuli and is capable of bending the body in a coordinated manner because the myotomal muscle cells are electrically coupled, although they are not all innervated. Finally, coupling is lost permanently after innervation of the myotomes is complete (2).

Junctional communication that is restricted within distinct developmental compartments has been described in insects (7, 61, 63) as well as in mouse embryos (35). This compartmentalization of junctional communication becomes more and more restricted as the embryos develop, until the communication ceases between different groups of cells. Whether final differentiation is a result of uncoupling or a cause of uncoupling remains to be elucidated. On the other hand, communication between different segments of the body seems to be necessary for the appropriate development of other organisms. For example, morphogenetic patterning during development and regeneration of the coelenterate *Hydra* requires the passage of a diffusible head-forming inhibitory substance from the head down the body column via gap junctions between the epithelial cells (21). This suggests that the interruption in intercellular communication is related to the different fates of the different tissues.

In addition, transient intercellular coupling can be found during synaptic-pathway formation as described in *Xenopus* eyes (17) and between optic axons from an ommatidium and the neuroblasts of the optic lamina in the *Daphnia* visual system (38). Grasshopper "pioneer" neurons also become coupled with "guidepost" cells when they first come into contact (56). This suggests that this brief intercellular communication may determine the direction of subsequent development.

Occasional electrical coupling between developing chick spinal neurons and myotubes in culture was first described by Fischbach (20). More recently, both electrical and dye coupling have been found in cocultures of chick myoblasts and neurons from the spinal cord, ciliary ganglion, and dorsal root ganglion (8). Bonner (8) found that dye-coupling was more frequent between a differentiating myoblast and a neuron than between an undifferentiated myoblast and a neuron, suggesting that the specificity of gap junction formation is primarily myogenic.

Transfer of Lucifer Yellow from muscle cells to neurons has also been demonstrated in *Xenopus* cocultures, and it was blocked when antibodies to gap junction was added (1). In this study the muscle cells were first "loaded" with the fluorescent dye by permeabilizing the muscle cell membrane with dimethyl sulfoxide and then the cells were dissociated and plated in culture with unlabeled spinal neurons. The fluorescent dye was later found in neurons that came into contact with the fluorescent muscle cells. This dye passage was significantly blocked when antibody to gap junction was coloaded with Lucifer Yellow into the muscle cells. Occasional transient electrical coupling was also found between cocultured *Xenopus* spinal neurons and myotomal muscle cells (68). The significance of the presence of a transient electrical synapse in addition to a long-lasting chemical synapse between nerve and muscle in culture remains to be determined. One possibility is that transient gap junction communication between neurons and muscle cells may be necessary for the establishment of a stable neuromuscular junction because it allows the passage of modulatory signals for the formation of synaptic structures. But the sequence of events is still unknown, i.e., whether the gap junction forms before, after, or simultaneously with the chemical synapse.

Effects of Junctional Blockade during Development

All these observations suggest that gap junctions play a role during development. Direct evidence for this function of the gap junctions has been obtained recently by investigators using specific antibodies to portions of the gap junction molecules. Warner et al. (62) injected antibodies to rat liver gap junction 27×10^3 dalton protein into identified *Xenopus* dorsal blastomeres of an 8-cell-stage embryo. These antibodies blocked both electrical coupling and dye passage between dorsal cells in the 32-cell-stage embryo. When the injected embryos were allowed to develop, absence or malformations of the structures presumably derived from the injected blastomeres was observed: (1) development of varying degrees of right-left asymmetry, (2) failure to form the right eye, (3) gross underdevelopment of the right side of the brain, and (4) occasional shift of the notochord to the uninjected side or (5) failure of brain development. Such abnormalities were not seen in embryos injected with preimmune sera, antibody to an extracellular matrix glycoprotein, or control buffer solutions which did not cause significant uncoupling between the blastomeres.

Fraser et al. (22) found that the gradient of the head-forming inhibitory factor in *Hydra* during regeneration was disrupted when antibodies to gap junction protein were introduced into the cytoplasm of epithelial cells. Their experimental paradigm was based on MacWilliams' original grafting experiments (39) that demonstrated the host head's ability to inhibit the formation of a second head after an excised apical piece of tissue had been grafted between the head and the bud of a normal animal. A second head and body axis will form if the host head is removed before grafting, indicating that the inhibitory effect is no longer present. Fraser and colleagues grafted untreated tissue onto antibody-treated host with its original head and found that the incidence of a second axis increased significantly. These results suggest that the disruption of gap junction communication by the antibodies decreased the head-forming inhibitory effect at the graft site, thus allowing the formation of a second body axis. The percentage of two-headed animals in control experiments using untreated normal hosts as well as grafts placed on preimmune sera–treated animals was expectedly low. These morphological abnormalities show that prevention of intercellular communication through gap junctions may disrupt normal pattern formation.

More recently, in their study of neural cell adhesion molecule (NCAM)–mediated gap junctional communication between chick neuroectodermal cells, Keane et al. (34) reported a decline in junctional transfer of fluorescent dyes coincidental with the differentiation of these neurons in culture. They found that tetanus toxin receptors, neurofilament protein, and neurite outgrowth failed to appear in neurons infected with the temperature-sensitive mutant Rous sarcoma virus which blocked junctional communication. In contrast, NCAM expression in the infected neurons was not prevented. They also reported that chronic treatment of the normal neurons with polyclonal anti-NCAM Fab fragments (monovalent) significantly suppressed the development of junctional transfer of fluorescent dyes to second-order neighbors within 5 min of dye injection. In contrast, no significant difference in the extent of dye transfer was found between first-order neighbors in antibody-treated cultures. These results suggest that NCAM-mediated adhesion may play a transient role in the establishment of extensive dye transfer between neuroectodermal neurons. More interestingly, they show that the expression of some neuronal characteristics can be prevented by blocking gap junctional communication between these cells.

These three studies constitute the initial, necessary experimental foundation for a more direct examination of the role of gap junctions in development. Interruption of junctional communication led to the absence or abnormal expression of tissue characteristics. What is still unknown is the exact nature(s) of the signals that pass through these intercellular channels and promote the proper morphogenesis and pattern formation. Also still undetermined are the mechanisms that initiate and control the formation and closure of the gap junctions at very precise times, events that are so crucial to the process of normal development.

Coupling Formation

Most of the presently available studies concern situations where junctional communication is already present; thus, little is known about the process of coupling formation. The rate of coupling formation has been shown to be very fast, on the order of a few minutes, in cell lines kept in culture (31, 43, 52) and between embryonic cells (12, 30). This suggests that the junctional channel precursors are readily available in the contacting cell membrane surfaces and that, at least initially, simple lateral diffusion allows for sufficient movement of these protochannels to the contact site so they can bind with their complementary halves in the apposing membrane and form the channel. This may also be the case for mouse blastomere junctions because inhibition of protein synthesis by cycloheximide at the 4-cell stage did not block the formation of gap junctions at the 8-cell stage (40). Also, neither inhibition of protein synthesis nor severe deprivation of ATP prevented junction formation in Novikoff cells (19). In addition, pretreatment of older cultured *Xenopus* muscle cells with colchicine, metabolic inhibitors, calcium- and magnesium-free saline, or trypsin significantly increased the rate of electrical coupling formation to a level that is close to that of younger cells, suggesting that the presence of extramembranous constraints rather than lack of junctional precursors slows coupling formation in the membrane of older cells (12). Apparently a simple, low-energy-requiring diffusion process is sufficient for the initial formation of the gap junctions after the cells come into contact, although the nature of the triggering signals remains unknown.

Another question is the number of protochannels which are needed to bind together and open up at the onset of coupling. Revel et al. (48) hypothesized that gap junction precursor hexamers (proconnexons) aggregate in formation plaques, align with the apposing ones when the cells come into contact, and after proteolytic cleavage, these protochannels undergo conformational changes leading to the formation of a channel. They did not, however, speculate whether only one or many channels open at the onset of coupling. As yet, there is no direct demonstration that this proposed mechanism actually occurs during the course of gap junction formation.

Loewenstein et al. (37), using a phase-sensitive recording method, reported discrete steplike increments of junctional conductance during coupling formation between *Xenopus* blastomeres, which they interpreted as the opening of channels. Still, they did not have a direct measurement at the single junctional channel level.

Formation of Electrical Coupling between Embryonic Muscle Cells

Intracellular Recording

Chow and Poo (12) demonstrated that in embryonic *Xenopus* muscle cell culture, electrical coupling formed rapidly after two isolated cells were manipulated into contact. Somites were isolated from young embryos, and myotomal muscle cells were dissociated and plated on clean, uncoated glass microscope slides. After 1, 2, or 5 days of culture, isolated, spherical (due to the lack of a strongly adhesive surface) muscle cells were impaled with a microelectrode and detached from the slide surface. They were then translocated and pushed into contact while the cell resting membrane potentials were monitored continuously with the intracellular microelectrode. The extent of electrical coupling between the cells after various contact periods was measured by monitoring the electrotonic spread of membrane potential changes induced either by intracellular current injections or by extracellular iontophoretic application of acetylcholine on the surface of one of the cells.

Electrical coupling was undetectable immediately after contact. In 1-day-old cultures, coupling was detected after a few minutes of contact, increased rapidly, and reached a plateau in 30 min (Fig. 4a). The rate of coupling formation was age-dependent. The percentage of cell pairs that established de-

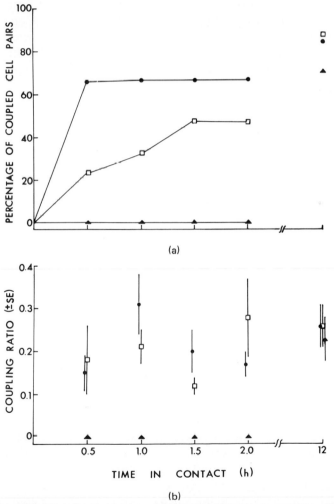

FIGURE 4. Coupling formation in 1-, 2-, and 5-day-old *Xenopus* muscle cultures. (a) Percentage of coupled cell pairs after 0.5, 1, 1.5, 2, and 12 h of manipulated contact. (b) Average coupling ratios (±S.E.M.) of the coupled cell pairs. The total number of cell pairs measured ranged from 15 to 38 (1-day-old ●), 17 to 34 (2-day-old □) and 8 to 16 (5-day-old ▲).

tectable coupling within 30 min of contact decreased from 66 percent in 1-day-old cultures to 24 percent in 2-day-old and 0 in 5-day-old cultures. The older cells, however, did develop the same extent of coupling (coupling ratio = 0.25) after prolonged (12 h) contact periods (Fig. 4b). These results suggest that the final total number of channels that formed or the number of channels that opened is the same once the maximum extent of coupling is reached. This also suggests that some sort of physical limiting factor(s) is restraining or delaying the onset of coupling in older cells.

Pretreatment of older (2- and 5-day) cells with colchicine, metabolic inhibitors, calcium- and magnesium-free saline, or trypsin significantly increased the rate of coupling formation to a level that was close to that of younger cells. These results suggest that the reduced rate of coupling formation in the older, untreated cell pairs was not due to a lack of channel precursors in the membrane but was probably due to the appearance of extramembranous constraints that do not permit close contact between the cells or that impair the channel assembly process. In fact, the development of a cell coat and cytoplasmic filamentous materials adjacent to the cell membrane with culture age has been described in these cells (64), and the drugs that were used are known to affect the integrity or disrupt the synthesis of these extramembranous structures.

Still, this study could not provide information about the channel properties during coupling formation at the single-channel level because the intracellular recording method that was used did not have the resolution necessary for such measurements. It did, nonetheless, tell us that coupling can be formed rapidly, that the channel precursors are readily available in the cell membrane, and that once close contact between the cells is established, the hemichannels pair up and form the gap junctions.

Single-Channel Recording

The first direct measurements of single gap junction channel currents were obtained recently from adult rat lacrimal cells (42) and chick embryonic heart cells (57). Both studies dealt with cells that were already coupled. The mean channel conductances were 120 and 165 pS, respectively, and channel opening from both cell types was voltage-dependent. The mean open times were long: 2 s for the lacrimal-cell channels and 332 ms for the heart-cell channels. Transition steps were described between the open and closed states of the rat lacrimal cells, whereas transition was very fast for the heart cells. Since these studies were carried out on cells that were already coupled, there was no information on the kinetics of coupling formation.

The study of coupling formation between two cells at the level of the single junctional channel was achieved by using cultured *Xenopus* myotomal muscle cells. These cells are amenable to whole-cell patch recording which has high-resolution single-channel recording properties, and these cells can be experimentally manipulated into contact, which allows precise timing of the onset of coupling (13).

Two isolated, spherical muscle cells were manipulated into contact at the start of the experiment. Hyperpolarizing current pulses of constant amplitude (500 pA) and duration (300 ms) were injected into one of the cells through an intracellular microelectrode, and any current passing into the second cell was recorded by the patch-clamp amplifier in the whole-cell configuration. The passive spread of the injected current across the junction between the two cells was detected within 20 min of contact in four out of seven cell pairs. The amplitude of the current that passed into the second cell remained constant and then, with increasing cell-cell contact times, increased in integral multiples of the smallest current level (average of 3.7 pA). This suggests that when the gap junction was formed only one channel opened at a time. This sequence from two experiments is shown in diagrammatic form in Fig. 5. The time of initial cell-cell contact is indicated by the vertical arrows, which also point to the baseline when no channels were open. After a certain delay the first channel opened, followed by the opening of a second channel many seconds later. In one of the two experiments a third channel was added later on. The time required for the first appearance of the first channel was variable, ranging from 30 to 395 s in four cell pairs. The times of appearance of subsequent channels also spanned a wide range—23 to 184 s.

Once formed, the junctional channels tended to remain open, closing only infrequently for brief periods. The transitions times between open and closed states were very short, i.e., less than 1 ms, which is the time resolution of the measuring system. This is also exemplified in Fig. 5: the second inset (above solid arrow) shows the temporary closure from three open channels to two open channels, followed 2.2 s later by an opening back to three open channels. The estimated long open times ranged from 10 to 40 s, and the brief closed times ranged between 0.5 and 1 s.

For an average cell membrane resistance of 75 MΩ and 3.7 pA current passing through the single channel out of a total of 500 pA injected, a channel conductance value of 100 pS was estimated. This figure is very close to the one measured with the double patch-clamp method from an already coupled cell pair, 100 to 200 pS range and an average value of 120 pS (67).

Estimate of the Total Number of Channels at the Newly Formed Gap Junctions

In previous intracellular studies, Chow and Poo (12) found that the degree of coupling between two *Xenopus* muscle cells reaches a plateau with an average coupling ratio of 0.25, which corresponds to about 3300-pS conductance across the entire gap junction. Using a value of 100 pS for the single-channel conductance, this total conductance translates to 33 open channels.

Although the absolute number of open channels is small, 33, the channels have high conductance, 100 pS, and an average of 90 percent of the channels are open at any time for long periods of time, 10 to 40 s (13). If one uses a diffusion coefficient of 1×10^{-7} to 4×10^{-7} cm^2/s for molecules of approximately 400 daltons (51) which cross the gap junction channels, a significant

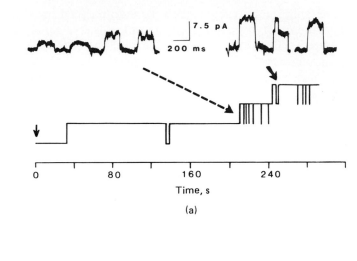

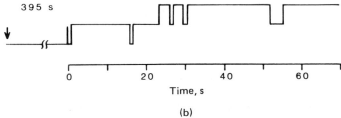

FIGURE 5. Diagrams showing times of new gap junctional channel formation in two experiments. The vertical arrows point to the time of initial cell-cell contact, at the baseline, without coupling (no junctional channels). Channel amplitudes are in arbitrary units. (*A*) Tracings of current pulses from two selected times after cell-cell contact are shown along with the entire kinetic history of one experiment. There was a delay of 30 s after cell contact before the first channel opened. The channel remained open for long periods of time [mean open time 30.9 ± 8.9 s (±S.E.M. of mean)] and closed for brief periods of time (0.8 ± 0.2 s). The first inset shows the transition between one open channel and two open channels (dotted line). The second inset shows the temporary closure from three open to two open channels and an opening back to three open channels 2.2 s later (solid line). (*B*) In this second experiment the first channel opened after 395 s of cell contact. The estimated mean open time is 15.0 ± 5.6 s, and the estimated mean closed time is 0.8 ± 0.3 s.

amount of morphogens or regulatory molecules can cross to neighboring cells under the conditions described above. As a result, gap junction communication may in fact contribute significantly to the process of tissue differentiation during development in the fast-growing embryos.

Lateral Mobility of Membrane Surface Molecules

Many glycoproteins, especially those in the embryonic plasma membrane, are capable of rapid lateral migration in the plane of the membrane (18). This suggests that lateral migration of membrane molecules may establish a specific cell-cell interaction; conversely, a specific intercellular interaction may be responsible for inducing lateral rearrangement of surface components. In the previously described experiments on *Xenopus* muscle cells, the direct cell-cell surface contact site between muscle cells could serve as a trap for the rapidly diffusing proconnexons in the plane of the cell membrane. Since the rate of coupling formation has been found to be fast, on the order of a few minutes, it is quite possible that the hemichannel precursors are already present in each of the cell membrane surfaces. Once the cells come into contact, the proconnexons diffuse lateraly and bind with their counterparts on the apposing cell membrane. Although no direct proof for such a sequence of events has yet been found, several other studies have shown that a number of cell membrane surface lectin receptors do aggregate at the cell-cell contact area and that the aggregation has a similar time course to that of coupling formation.

Substantial redistribution of cell surface lectin receptors occurred within 30 min after contact was made between embryonic *Xenopus* muscle cells in 1-day old cultures (11). Many soybean agglutinin (SBA) receptors, presumably glycoproteins containing D-galactose and/or *N*-acetyl-α-D-galactosamine residues, accumulated at the contact area, whereas concanavalin A receptors, glycoproteins containing α-D-glucose and/or α-D-mannose residues, moved away from it. Receptors to other lectins (peanut agglutinin, wheat germ agglutinin, and *Dolichos biflorus* agglutinin) accumulated to a lesser extent, and receptors to *Ricinus communis* agglutinin diffused away from the cell-contact site. The accumulation of SBA receptors at the contact site was greatly decreased by precontact binding with SBA, galactose, or cytochalasin B or by lowering the temperature of the incubation medium. It was unaffected by treatment with metabolic inhibitors, suggesting that the redistribution is a passive process of lateral diffusion of the receptors in the plane of the plasma membrane. In fact, the observed range of lateral diffusion coefficient of SBA receptors is 1.2×10^{-9} to 2.7×10^{-9} cm^2/s, fast enough to account for their rapid accumulation within 30 min at the cell-cell contact site by a simple "diffusion trap" mechanism previously shown to occur in 2-day old muscle cells (10).

As demonstrated in 2- and 5-day-old cultures, a mixture of metabolic inhibitors (dinitrophenol, sodium azide, and sodium fluoride), colchicine, and calcium- and magnesium-free saline increased the rate of electrical coupling between muscle cells, presumably by disruption of synthesis and integrity of extramembranous structures (12). This increased rate of electrical coupling could be a result of the breakdown of extracellular material which facilitated the lateral mobility of channel precursors in the plane of the cell membrane to quickly accumulate at the cell-cell contact site. This diffusion trap mechanism is similar to that proposed for SBA receptor accumulation. Such a trap model has also been proposed for the formation of gap junction by Loewenstein (36).

Modulation of Electrical Coupling Formation

Much work has been done on the regulation of the permeability of already formed gap junctions in different tissues, and many reviews can be consulted on this matter (e.g., Refs. 16, 36, 44, and 54). For example, depending on the tissue or the species of the animal, the conductance of the gap junctions can be blocked by a decrease in cytoplasmic pH, by an increase in cytoplasmic calcium ion concentration (maybe by some other cations also), or by changes in transjunctional voltage. The effects of second messengers such as cAMP, of components of transmembrane signal pathways such as diacylglycerol, and of calmodulin (45) or hormones on the gating of the junctional channels have been subjects of intensive investigation in several laboratories, and the results may vary greatly according to the preparations.

On the other hand, much less is known about the factors that modulate the formation of these intercellular junctions. Apparently in the early mouse embryo, gap junction formation is under strict temporal control beginning at the 8-cell stage and is possibly correlated with compaction of the embryo (25, 35, 40). Coupling formation takes place rapidly (12, 30, 31), and it still occurs in the absence of protein synthesis (40). The development of coupling between dissociated macroblastomeres of newt embryo brought into contact was not prevented by treatment with colchicine or with cytochalasin B, but it was blocked by treatment with dinitrophenol (30). In 2- and 5-day old *Xenopus* myotomal muscle cell cultures, pretreatment with colchicine, a mixture of metabolic inhibitors (dinitrophenol, sodium azide, and sodium fluoride), calcium- and magnesium-free saline, or trypsin all increased the rate of coupling formation after cell-cell contact to a value that was close to that found in 1-day-old cultures (12). These studies suggest that the development of cell-to-cell coupling seems to be dependent on the absence of physical constraints for the lateral mobility of the hemichannels. Whether second messengers known to regulate the gating (opening and closing) of "mature" gap junction channels also affect the formation of these junctions remains to be determined.

The measurement of single gap junction channel conductance of *Xenopus* muscle cells already in contact that was obtained from currents passing through the last gap junction channel during the process of uncoupling was possibly due to the "wash-out" effect of the whole-cell patch recording technique (Fig. 3). With time, the intensity of the junctional current dropped as the amplitude of voltage pulses in the current-injected cell rose. This uncoupling was complete after an average time of 16 min after the establishment of the whole-cell clamp. Since the patch pipette solution, which is directly in contact with the muscle cell cytoplasm, was strongly buffered for both pH (20 mM HEPES, pH 7.8) and Ca^{2+} ions (11 mM EGTA, 1 mM $CaCl_2$), two factors known to influence junctional conductance, this uncoupling effect suggests that other factors in the cytoplasm which modulate channel conductance are being washed out by the pipette solution.

The technique can, nevertheless, be used for the study of the possible intracellular factors regulating the maintenance and the formation of gap junctions. Taking advantage of this technique, cyclic AMP (1 mM) added to the intracellular solution was found to be effective in preventing uncoupling of already coupled cells, but cyclic AMP did not promote gap junction formation of cell pairs newly manipulated into contact (67). On the other hand, Azarnia et al. (3) reported promotion of gap junction formation by this nucleotide in a transformed cell line. In another study, addition of Leibovitz medium, consisting of various amino acids, salts, and vitamins, was essential for the formation of electrical coupling when at least one patch pipette was used (67). These results suggest that initiation of electrical coupling may be regulated by factors different from those necessary for the maintenance of junctional communication. Many of the presently known second messengers may affect the gap junction either directly or indirectly, and the details of this process remain to be elucidated. It is also possible that a cell surface signal triggers the initiation of gap junction formation, especially in view of a recent report on the partial block of dye transfer between cultured chick neuroectodermal cells by anti-NCAM (34). Incubation of *Xenopus* muscle cells with polyvalent rat brain anti-NCAM (gift of Dr. U. Rutishauser) did not, however, block the formation of electrical coupling formation between these muscle cells (I. Chow, unpublished data), although it greatly decreased the formation of neuromuscular synapse in these cultures (14).

Conclusions

Presently, it is no longer a matter of speculation whether gap junctions play a role during the course of normal development. In experiments in which intercellular passage of putative signals was prevented by blocking the opening or formation of gap junctions, abnormal development or the absence of derived structures resulted. The nature of the transferred signals is still unknown, however, as is the signal's mode of action. Whether the transferred molecules are the morphogens themselves or messengers for initiation of other cellular events necessary in the process of morphogenesis remains to be elucidated. With the availability of specific gap junction antibodies and the accessibility of the cell interior to the patch pipette, these questions may be answered in the near future.

Another important point to be clarified is the exact mechanism(s) underlying the formation of the gap junctions, especially at crucial periods of embryogenesis. Since the protochannels seem to be readily available on the cell membrane at various stages of development, the questions to be asked are (1) when and what triggers the cell to synthesize these proconnexons and (2) what triggers the pairing up of the two hemichannels and the ultimate opening of the complete gap junction channel? Once answers are found we will have a more complete understanding of the process of cell-cell interaction during embryogenesis. We now have the tools that can be used for the exploration of these questions.

Acknowledgments

I wish to thank Dr. G. Zampighi for providing the pictures used in Fig. 1 and for stimulating discussions on this topic; Dr. S. H. Young for helpful comments on the manuscript and collaboration on some of the described experiments; and Dr. A. D. Grinnell for his continuing interest and support. Supported by grants from the Muscular Dystrophy Association of America to the author and from NIH to Dr. A. D. Grinnell.

General References

GUTHRIE, S. C.: 1987. Intercellular communication in embryos. In: *Cell-to-Cell Communication.* (W. C. De Mello, ed.), pp. 223–244, Plenum, New York.

HERTZBERG, E. L., and JOHNSON, R. G.: 1988. *Gap Junctions,* Modern Cell Biology series vol. 7. Liss, New York. This book contains a section on "Gap junctions in development" with six chapters dealing with the various aspects of intercellular communication during development, as well as other sections on gap junction physiology, morphology, and biochemistry.

References

1. Allen, F. 1986. Gap junctional communication during neuromuscular junction formation in *Xenopus* cultures. *J. Physiol.* **377**:77P.

2. Armstrong, D. L., Turin, L., and Warner, A. E. 1983. Muscle activity and the loss of electrical coupling between striated muscle cells in *Xenopus* embryos. *J. Neurosci.* **3**:1414–1421.

3. Azarnia, R., Dahl, G., and Loewenstein, W. R. 1981. Cell junction and cyclic AMP: III. Promotion of junctional membrane permeability and junctional membrane particles in a junction-deficient cell type. *J. Membr. Biol.* **63**:133–146.

4. Bennett, M. V. L. 1966. Physiology of electrotonic junctions. *Annu. N. Y. Acad. Sci.* **137**:509–531.

5. Bennett, M. V. L., Spray, D. C., and Harris, A. J. 1981. Gap junctions and development. *Trends Neurosci.* **4**:159–163.

6. Blackshaw, S. E., and Warner, A. E. 1976. Low resistance junctions between mesoderm cells during development of trunk muscles. *J. Physiol.* **255**:209–230.

7. Blennerhassett, M. G., and Caveney, S. 1984. Separation of developmental compartments by a cell type with reduced permeability. *Nature (Lond.)* **309**:361–364.

8. Bonner, P. H. 1988. Gap junction form in culture between chick embryo neurons and skeletal muscle myoblasts. *Dev. Brain Res.* **38**:233–244.

9. Caveney, S. 1985. The role of gap junctions in development. *Annu. Rev. Physiol.* **47**:319–335.

10. Chao, N-m., Young, S. H., and Poo, M-m. 1981. Localization of cell membrane components by surface diffusion into a "trap." *Biophys. J.* **36**:139–153.

11. Chow, I., and Poo, M-m. 1982. Redistribution of cell surface receptors induced by cell-cell contact. *J. Cell Biol.* **95**:510–518.

12. Chow, I., and Poo, M-m. 1984. Formation of electrical coupling between embryonic *Xenopus* muscle cells in culture. *J. Physiol.* **346**:181–194.

13. Chow, I., and Young, S. H. 1987. Opening of single gap junction channels during formation of electrical coupling between embryonic muscle cells. *Dev. Biol.* **122**: 332–337.

14. Chow, I., Haubach, C., and Grinnell, A. D. 1988. Cell surface contact interaction during initial stage of synaptogenesis. *Soc. Neurosci. Abstr.* **14**:823.

15. Crick, F. 1970. Diffusion in embryogenesis. *Nature (Lond.)* **225**:420–422.

16. De Mello, W. C. 1987. Modulation of junctional permeability. In: *Cell-to-cell Communication* (W. C. De Mello, ed.), pp. 29–64, Plenum, New York.

17. Dixon, J. S., and Cronley-Dillon, J. R. 1972. The fine structure of the developing retina in *Xenopus laevis. J. Embryol. Exp. Morphol.* **28**:659–666.

18. Edidin, M., and Fambrough, D. 1973. Fluidity of the surface of cultured muscle fibers. Rapid lateral diffusion of marked surface antigens. *J. Cell Biol.* **57**:27–37.

19. Epstein, M. L., Sheridan, J. D., and Johnson, R. G. 1977. Formation of low resistance junctions in vitro in the absence of protein synthesis and ATP production. *Exp. Cell Res.* **104**:25–30.

20. Fischbach, G. D. 1972. Synapse formation between dissociated nerve and muscle cells in low density cultures. *Dev. Biol.* **28**:407–429.

21. Fraser, S. E., and Bode, H. R. 1981. Epithelial cells of *Hydra* are dye coupled. *Nature (Lond.)* **294**:356–358.

22. Fraser, S. E., Green, C. R., Bode, H. R., and Gilula, N. B. 1987. Selective disruption of gap junctional communication interferes with a patterning process in *Hydra. Science* **237**:49–55.

23. Furshpan, E. J., and Potter, D. D. 1959. Transmission of the giant synapse of the crayfish. *J. Physiol.* 145:289–325.

24. Gilula, N. B. 1980. Cell-to-cell communication and development. In: *The Cell Surface: Mediator of Developmental Processes.* (S. Subtelny and N. K. Wessells, eds.), pp. 23–41, Academic Press, New York.

25. Goodall, H., and Johnson, M. N. 1982. Use of carboxyfluorescein diacetate to study formation of permeable channels between mouse blastomeres. *Nature (Lond.)* **295**: 524–526.

26. Green, C. R. 1988. Evidence mounts for the role of gap junctions during development. *Bioessays* 8:3–6.

27. Green, C. R., and Severs, N. J. 1984. Gap junction connexon configuration in rapidly frozen myocardium and isolated intercalated disks. *J. Cell Biol.* **99**:453–463.

28. Guthrie, S. C. 1984. Patterns of junctional communication in the early amphibian embryo. *Nature (Lond.)* **311**:149–151.

29. Hamill, O. P., Marty, A., Neher, E., Sakman, B., and Sigworth, F. J. 1981. Improved patch-clamp techniques for high-resolution current recording from cells and cell-free patches. *Pflueger's Arch. Eur. J. Physiol.* **391**:85–100.

30. Ito, S., Sato, E., and Loewenstein, W. R. 1974. Studies on the formation of a permeable cell membrane junction. I. Coupling under various conditions of membrane contact. Colchicine, cytochalasin B, dinitrophenol. *J. Membr. Biol.* **19**:305–337.

31. Johnson, R., Hammer, M., Sheridan, J., and Revel, J.-P. 1974. Gap junction formation between reaggregated Novikoff hepatoma cells. *Proc. Natl. Acad. Sci. U.S.A.* **71**:4536–4543.

32. Kanno, Y., and Loewenstein, W. R. 1964. Intercellular diffusion. *Science* **143**:959–960.

33. Kanno, Y., and Loewenstein, W. R. 1966. Cell-to-cell passage of large molecules. *Nature (Lond.)* **212**:629–630.

34. Keane, R. W., Mehta, P. P., Rose, B., Honig, L. S., Loewenstein, W. R., and Rutishauser, U. 1988. Neural differentiation, NCAM-mediated adhesion, and gap junctional communication in neuroectocerm. A study in vitro. *J. Cell Biol.* **106**: 1307–1319.

35. Lo, C. W., and Gilula, N. B. 1979. Gap junctional communication in the preimplantation mouse embryo. *Cell* **18**:399–409.

36. Loewenstein, W. R. 1981. Junctional intercellular communication: the cell-to-cell membrane channel. *Physiol. Rev.* **61**:829–913.

37. Loewenstein, W. R., Kanno, Y., and Socolar, S. J. 1978. Quantum jumps of conductance during formation of membrane channels at cell-cell junction. *Nature (Lond.)* **274**:133–136.

38. Lopresti, V., Macagno, E. R. and Levinthal, C. 1974. Structure and development of neuronal connections in isogenic organisms: transient gap junctions between growing neurons and lamina neuroblasts. *Proc. Natl. Acad. Sci. U.S.A.* **71**:1098–1102.

39. MacWilliams, H. K. 1983. *Hydra* transplantation phenomena and the mechanism of *Hydra* head regeneration. I. Properties of the head inhibition. *Dev. Biol.* **96**:217–238.

40. McLachlin, J., Caveney, S., and Kidder, G. M. 1983. Control of gap junction formation in early mouse embryo. *Dev. Biol.* **98**:155–164.

41. Miller, T. M., and Goodenough, D. A. 1985. Gap junction structures after experimental alteration of junctional channel conductance. *J. Cell Biol.* **101**:1741–1748.

42. Neyton, J., and Trautmann, A. 1985. Single-channel currents of an intercellular junction. *Nature (Lond.)* **317**:331–335.

43. O'Lague, P., and Delan, H. 1974. Low resistance junctions between normal and between virus transformed fibroblasts in tissue culture. *Exp. Cell Res.* **86**:374–382.

44. Peracchia, C. 1987. Permeability and regulation of gap junction channels in cells and in artificial lipid bilayers. In: *Cell-to-Cell Communication.* (W. C. De Mello, ed.), pp. 65–102, Plenum, New York.

45. Peracchia, C., and Bernardini, G. 1984. Gap junction structure and cell-to-cell coupling regulation: Is there a calmodulin involvement? *Fed. Proc.* **43**:2681–2691.

46. Pitts, J. D., and Simms, J. W. 1977. Permeability of junctions between animal cells. Intercellular transfer of nucleotides, but not of macromolecules. *Exp. Cell Res.* **104**:153–163.

47. Potter, D. D., Furshpan, E. J., and Lennox, E. S. 1966. Connections between cells of the developing squid as revealed by electrophysiological methods. *Proc. Natl. Acad. Sci. U.S.A.* **55**:328–336.

48. Revel, J.-P., Griepp, E. A., Finbow, M., and Johnson, R. 1978. Possible steps in gap junction formation. *Zoon* **6**:139–144.

49. Robertson, J. D. 1953. Ultrastructure of two invertebrate synapses. *Soc. Exp. Biol. Med.* **82**:219–223.

50. Robertson, J. D. 1963. The occurrence of a subunit pattern in the unit membranes of club endings in Mauthner cell synapses in goldfish brain. *J. Cell Biol.* **19**:210–232.

51. Safranyos, R. G. A., and Caveney, S. 1985. Rates of diffusion of fluorescent molecules via cell-to-cell membrane channels in developing tissue. *J. Cell Biol.* **100**: 736–747.

52. Sheridan, J. D. 1971. Dye movement and low resistance junctions between reaggregated embryonic cells. *Dev. Biol.* **26**:627–636.

53. Socolar, S. J., and Loewenstein, W. R. 1979. Methods for studying transmission through permeable cell-to-cell junctions. In: *Methods in Membrane Biology*, vol. 10. (E. D. Korn, ed.), pp. 123–179, Plenum, New York.

54. Spray, D. C., and Bennett, M. V. L. 1985. Physiology and pharmacology of gap junctions. *Rev. Physiol.* **47**:281–303.

55. Stewart, W. W. 1978. Functional connections between cells as revealed by dye coupling with a highly fluorescent naphthalimide tracer. *Cell* **4**:741–759.

56. Taghert, P. H., Bastiani, M. J., Ho, R. K., and Goodman, C. S. 1982. Guidance of pioneer growth cones: filopidial contacts and coupling revealed with antibody to Lucifer Yellow. *Dev. Biol.* **94**:391–399.

57. Veenstra, R. D., and DeHaan, R. L. 1986. Measurement of single channel currents from cardiac gap junctions. *Science* **233**:972–974.

58. Wade, M. H., Trosko, J. E., and Schindler, M. 1986. A fluorescence photobleaching assay of gap junction-mediated communication between human cells. *Science* **232**:525–528.

59. Warner, A. E. 1973. The electrical properties of the ectoderm in the amphibian embryo during induction and early development of the nervous system. *J. Physiol.* **235**:267–286.

60. Warner, A. E. 1985. The role of gap junctions in amphibian development. In: *Early Amphibian Development.* (J. Slack, ed.) *J. Embryol. Exp. Morphol.* **89**(Suppl.):365–380.

61. Warner, A. E., and Lawrence, P. A. 1982. Permeability of gap junctions at the segmental border in insect epidermis. *Cell* **28**:243–252.

62. Warner, A. E., Guthrie, S. C., and Gilula, N. R. 1984. Antibodies to gap junctional protein selectively disrupt junctional communication in the early amphibian embryo. *Nature (Lond.)* **311**:127–131.

63. Weir, M. P., and Lo, C. W. 1984. Gap junction communication compartments in the *Drosophila* wing imaginal disc. *Dev. Biol.* **102**:130–146.

64. Weldon, P. R., and Cohen, M. W. 1979. Development of synaptic ultrastructure at neuromuscular contacts in an amphibian cell culture system. *J. Neurocytol.* **8**:239–259.

65. Wolpert, L. 1969. Positional information and the spatial pattern of cellular differentiation. *J. Theor. Biol.* **25**:1–47.

66. Wolpert, L. 1978. Gap junctions: Channels for communication in development. In: *Intercellular Junctions and Synapse.* (J. Feldman, N. B. Gilula, and J. D. Pitts, eds.), pp 83–98, Chapman and Hall, London.

67. Young, S. H., and Chow, I. 1987. Modulation of formation and maintenance of electrical coupling between muscle cells in culture. *Int. Conf. Gap Junction Abstr.* 37.

68. Young, S. H., Chow, I., and Grinnell, A. D. 1988. Electrical coupling between nerve and muscle cell of *Xenopus* in culture. *Soc. Neurosci. Abstr.* **14**:823.

69. Zampighi, G. 1987. Gap junction structure. In: *Cell-to-cell Communication.* (W. C. De Mello, ed.), pp. 1–28, Plenum, New York.

70. Zampighi, G., Corless, J. M., and Robertson, J. D. 1980. On gap junction structure. *J. Cell Biol.* **86**:190–198.

Questions for Discussion with the Editor

1. *Are all gap junctions—the world around—exactly the same? That is, is there a standard type of gap junction which is common to all cells?*

No. Physiologically speaking, the regulation of these channels may differ from tissue to tissue and from species to species. For example, some gap junctions are voltage sensitive and others are sensitive to H^+ ions; some are down-regulated by an increase in cytoplasmic cAMP, whereas this nucleotide may increase junctional permeability in other cell types. Gap junctions are also different morphological by and biochemically. The majority of gap junctions (liver, smooth and cardiac muscles, epithelia, and nervous tissue of vertebrates and some invertebrates) display a similar structural characteristic: hexagonal arrangement of the channels with 8- to 9-nm center-to-center spacing. Arthropod gap junctions also have a hexagonal display, but the channels are slightly larger and spaced farther apart. A third type, found only in vertebrate mature lens, has channels arranged in square arrays with a 6- to 7-nm spacing. Increasing number of studies have shown amino acid sequence homologies among gap junction proteins of liver and heart tissues of vertebrates, although their molecular weights may differ (47,000 for cardiac muscle, 27,000 and 21,000 for liver). Again, lens junction protein does not share such similarities, and some investigators question whether it is really an intercellular communicating junction. Antibodies against liver gap junction proteins have been shown to block intercellular dye passage between blastomers and between some epithelial and liver cells, suggesting the presence of similar immunogenic sites in the cytoplasmic side of the molecule. All this indicates that the gap junctions are highly preserved structures across the animal phyla, but they also have their own separate characteristics, rendering the appearance of different properties and regulatory mechanisms.

2. *Please speculate on what types of cues might serve to signal cells to form gap junctions. For example, would those cues frequently be cAMP-like molecules?*

We all would like to know the answer to this question, and many of us are working on it. Certainly several different events must be considered. I tend to imagine that the cells themselves are genetically regulated to express these hemichannels in their membrane at particular stages of their lives. If it happens that a cell comes into contact with another cell that also has such channels freely diffusing in the plane\of the membrane, the hemichannels will pair up and bind together, and the channel then opens up. The pairing up and binding of these molecules may simply be under physicochemical control; but what triggers the opening of the completed channel is a crucial matter and remains unknown.

As described in the text, it seems that at least in the *Xenopus* myotomal muscle cell system, addition of cAMP alone to the cells is not sufficient to promote coupling formation, but it can prevent uncoupling between already coupled cells held under whole-cell patch as a result of the "wash-out" phenomenon. Instead, some still unidentified component present in the L-15 medium was found to be essential for coupling formation when one of the cells was perfused with a patch pipette. Interestingly, neither additional cAMP nor L-15 medium was necessary for coupling formation and maintenance when both cells are impaled with intracellular microelectrodes. Putting these facts together, I suspect that whatever is needed inside the cell for coupling formation to proceed and which is washed out by the patch pipette solution can be replaced by the L-15 medium, and it acts by either diffuses across the cell membrane or as result of receptor binding or triggering of a second messenger other than cAMP. I am currently in the process of identifying this component(s) and any guess on its nature would only be a guess.

Index

Index

ABOUT THE EDITOR

GEORGE M. MALACINSKI, Ph.D., is a professor of
biology at Indiana University. He is also the editor of the
Axolotl Newsletter, has organized several international
symposia, and has published over 100 scientific papers.